Molecular Biology: Concepts and Applications

Molecular Biology: Concepts and Applications

Editor: Erik Pierre

www.callistoreference.com

Callisto Reference,
118-35 Queens Blvd., Suite 400,
Forest Hills, NY 11375, USA

Visit us on the World Wide Web at:
www.callistoreference.com

ISBN: 978-1-63239-915-1 (Hardback)

Cataloging-in-Publication Data

Molecular biology : concepts and applications / edited by Erik Pierre.
   p. cm.
Includes bibliographical references and index.
ISBN 978-1-63239-915-1
1. Molecular biology. 2. Biology. I. Pierre, Erik.
QH506 .M65 2018
572.8--dc23

Table of Contents

Permissions

List of Contributors

Index

Preface

Over the recent decade, advancements and applications have progressed exponentially. This has led to the increased interest in this field and projects are being conducted to enhance knowledge. The main objective of this book is to present some of the critical challenges and provide insights into possible solutions. This book will answer the varied questions that arise in the field and also provide an increased scope for furthering studies.

Molecular biology studies those activities of cellular molecules, such as proteins and nucleic acids, which are essential for cell maintenance and cell functioning. The mapping of transfer of biological sequential information is the central concern of this field. Molecular cloning, DNA copying through polymerase chain reaction (PCR), molecular blotting and probing, the preparation of microarrays are some of the modern applications of molecular biology. Some of the diverse topics covered in this book address the varied branches that fall under this category. Most of the topics introduced herein cover new techniques and applications of this field. For all those who are interested in molecular biology, this book can prove to be an essential guide.

I hope that this book, with its visionary approach, will be a valuable addition and will promote interest among readers. Each of the authors has provided their extraordinary competence in their specific fields by providing different perspectives as they come from diverse nations and regions. I thank them for their contributions.

Editor

NMR-Based Detection of Hydrogen/Deuterium Exchange in Liposome-Embedded Membrane Proteins

Xuejun Yao[1], Ulrich H. N. Dürr[1], Zrinka Gattin[1], Yvonne Laukat[1], Rhagavendran L. Narayanan[1], Ann-Kathrin Brückner[1], Chris Meisinger[2], Adam Lange[1], Stefan Becker[1], Markus Zweckstetter[1,3,4]*

1 Max Planck Institute for Biophysical Chemistry, Göttingen, Germany, **2** Institut für Biochemie und Molekularbiologie, ZBMZ and BIOSS Centre for Biological Signalling Studies, Universität Freiburg, Freiburg, Germany, **3** German Center for Neurodegenerative Diseases (DZNE), Göttingen, Germany, **4** Center for Nanoscale Microscopy and Molecular Physiology of the Brain (CNMPB), University Medical Center, Göttingen, Germany

Abstract

Membrane proteins play key roles in biology. Determination of their structure in a membrane environment, however, is highly challenging. To address this challenge, we developed an approach that couples hydrogen/deuterium exchange of membrane proteins to rapid unfolding and detection by solution-state NMR spectroscopy. We show that the method allows analysis of the solvent protection of single residues in liposome-embedded proteins such as the 349-residue Tom40, the major protein translocation pore in the outer mitochondrial membrane, which has resisted structural analysis for many years.

Editor: Michael Massiah, George Washington University, United States of America

Funding: ERC (grant agreement number 282008) to M Zweckstetter; Marie Curie fellowship within the 7th EU Framework Programme to Z Gattin; DFG collaborative research center 803 and Emmy Noether fellowship to A Lange. The funders had no role in study design, data collection and analysis, decision to publish, or preparation of the manuscript.

Competing Interests: The authors have declared that no competing interests exist.

* Email: Markus.Zweckstetter@dzne.de

Introduction

Membrane proteins have important biological functions and are the targets of over 50% of all modern medicinal drugs. However, only a small number of membrane protein structures have been solved up to now (http://blanco.biomol.uci.edu/mpstruc/#Latest). Structure determination of membrane proteins by either nuclear magnetic resonance (NMR) spectroscopy or X-ray crystallography is complicated by the need to solubilize membrane proteins in native-like environments [1]. In addition, solution-state NMR spectroscopy of membrane proteins relies on the ability to find detergents, bicelles or nanodiscs, in which the native structure of the protein is retained and relaxation losses are minimized [2]. Solid-state NMR spectroscopy on the other hand can investigate membrane proteins reconstituted into liposomes or uniformly oriented bilayers [3–9], but requires sufficient spectral quality to enable sequence-specific resonance assignment and structure determination.

Hydrogen/Deuterium (H/D) exchange has long been used to probe protein structures [10]. The success of H/D exchange is based on the strong influence of hydrogen bonds on amide proton exchange rates. Due to its power to provide single-residue information, NMR spectroscopy is optimally suited to monitor site-specific H/D exchange rates of proteins [10–13]. For membrane proteins, H/D exchange rates are particularly important, as it is often difficult to obtain a sufficient number of distance restraints [14–16]. In addition, H/D exchange coupled to solution-state NMR is useful for high-molecular weight systems

that would otherwise not be accessible to solution-state NMR. A particularly important application is the investigation of the structure of protein aggregates, in which the H/D exchange profile of the protein aggregate is detected with the help of the denatured monomer [17–19]. H/D exchange in membrane proteins can be also analyzed by mass spectrometry [20,21], although generally not at the residue resolution achievable by NMR spectroscopy. In addition, a solid-state NMR H/D exchange experiment performed on a helical membrane protein showed that the amide protons in an amphipathic helix are more slowly exchanging than those in the transmembrane helix in a four helix bundle with an aqueous pore [22].

Here we demonstrate that the solvent protection of single residues in liposome-embedded transmembrane proteins can be studied using solution-state NMR spectroscopy. Using a dedicated H/D exchange protocol, the information of the membrane-embedded state is transferred to the denatured state and analyzed using multidimensional NMR. The method is applied to the 349-residue protein Tom40 that forms the protein translocation pore in the outer mitochondrial membrane and has resisted structural analysis for many years [23–25].

Materials and Methods

Sample preparation

Tom40 from *neurospora crassa* (ncTom40) was expressed, refolded and purified as previously described [25]. ^{15}N- and ^{13}C-labelled protein was expressed in minimal medium with $^{15}NH_4Cl$

as nitrogen source and $^{13}C_6$-D-glucose as carbon source. Amino acid-selective labelled ncTom40 protein was expressed according to a recently published protocol [26]. For solid-state NMR measurements and H/D exchange the protein was reconstituted in DMPC at a protein/lipid ratio of 1:50 (mol/mol). H/D exchange for ncTom40 in liposomes was performed in 10 mM MOPS, 10 mM KCl, pD 7.0, 100% D_2O. Back-exchange was monitored in a dissolution buffer containing 4 M guanidinium thiocyanate (GdnSCN), 0.4% formic acid, pD 2.5.

NMR spectroscopy

Experiments were carried on Bruker 800 and 900 MHz spectrometer equipped with cryogenic probes. 6D HNCOCANH and 5D CBCACONH automated projection spectroscopy (APSY) experiments were recorded at 295K and 278K, while 7D HNCO(CA)CBCANH and 5D CBCACONH were measured at 310K [27–29]. APSY spectra were processed using PROSA [30]. Peaks on each projection spectrum were picked and the final peak list was calculated using GAPRO [28]. A 3D HNN experiment [31] was recorded at 278K. Assignment was performed in an iterative manner using MARS [32], manual inspection of the 3D HNN experiment [31] and HSQCs of amino acid selectively ^{15}N-labelled samples (Table S1). Back-exchange in dissolution buffer was monitored using two-dimensional [^1H, ^{15}N]-HSQC spectra recorded with a Bruker BEST-HSQC [33] pulse sequence at 278K with a recycle delay of 0.5s. HSQCs were processed and analyzed using NMRPipe [34] and Sparky 3 (University of California, San Francisco).

Solid-state NMR experiments were conducted using 4 mm triple-resonance magic-angle spinning (MAS) probeheads at a static magnetic field of 18.8 T and using 8.33 kHz MAS. Sample temperatures were +5°C for ^{13}C-^{13}C proton-driven spin diffusion (PDSD) correlation experiments and +15°C for INEPT-type experiments. Initial cross-polarization time for PDSD was set to 600 μs for ^1H-^{13}C transfer. ^{13}C-^{13}C mixing was accomplished by PDSD for 15 ms to obtain intra-residual correlations. The ^{13}C-^{13}C INEPT-TOBSY [35] correlation spectrum was recorded with decoupling field strength of 2.5 kHz.

Results

^{13}C/^{15}N-labeled ncTom40 was prepared recombinantly and subjected to refolding in detergent. Circular dichroism spectra of our ncTom40 preparation in decylmaltoside (Figure 1a) closely resembled previous CD spectra of recombinantly produced and refolded ncTom40, which had been demonstrated to be functional [25]. Using different algorithms the β-structure content had been estimated to be approximately 30–40%, consistent with a β-barrel structure (the expected percentage of β-structure in the 349-residue ncTom40 is 44% assuming the presence of 19 strands with a length of 8 residues on average) [25]. Despite screening many different conditions, however, it was not possible to obtain high-quality solution NMR spectra of ncTom40 solubilized in detergent (Figure S1). This might be due to the strong propensity of Tom40 to form homooligomers [23].

Next, ncTom40 was reconstituted into 1,2-Dimyristoyl-*sn*-Glycero-3-Phosphocholine (DMPC) liposomes. In a two-dimensional [^{13}C,^{13}C] solid-state NMR spectrum of liposome-embedded ncTom40 a large number of cross-peaks was observed (Figure 1b). The NMR signal distribution was similar to that previously observed for the 283-residue human voltage-dependent anion channel (isoform 1; hVDAC1), which shares a 30% sequence similarity with ncTom40. The structure of hVDAC1 is composed of 19 β-strands and an N-terminal α-helix [36–38]. The solid-state

NMR spectrum of ncTom40 is of lower quality than that of hVDAC1 (Figure 1b), but contained a larger number of defined cross peaks than isoform 2 of VDAC (hVDAC2), both of which were shown to be functional [4,39,40]. The similarity of the ^{13}C-^{13}C correlation spectra of ncTom40, hVDAC1 and hVDAC2 (Figure 1b and [39]) together with the CD profile of refolded ncTom40 (Figure 1a and [25]) indicates that ncTom40 folds into a β-barrel in both detergent and liposomes. Notably, neither for hVDAC1/2 or a β-barrel protein of similar size, the sequence-specific resonance assignment of its β-barrel when inserted into liposomes has been reported till date. This highlights the need to develop methods that allow characterization of the structure of large liposome-embedded proteins at single-residue resolution.

To address this need, we designed the H/D exchange approach outlined in Figure 2. The protein containing liposomes are centrifuged at 168,000×g for 1 hr at 4°C. The pellet is then transferred to 100% D_2O and incubated with agitation for 10 minutes at room temperature. During this time, amide protons will undergo exchange. At the end of the incubation time, the pellet is collected through centrifugation for 15 minutes and put into dissolution buffer that contains 4 M GdnSCN, 0.4% formic acid at pD 2.5 [41], with either 75% or 100% D_2O. In the dissolution buffer lipids immediately precipitate, while the protein stays in solution. Subsequently, [^1H, ^{15}N] heteronuclear single quantum coherence (HSQC) spectra are recorded (Figure S2) [41]. The dead time before starting the NMR experiment was approximately 20 minutes. To minimize the influence of back-exchange [42], the temperature was set to 278K and the experiments were recorded using the BEST scheme [12,33].

To obtain the sequence-specific resonance assignment of the 349 residues of ncTom40 in the dissolution buffer, we performed APSY experiments and assignment using the program MARS [28,32,43]. APSY enables the measurement of high-dimensional spectra, reducing NMR signal overlap [27,28]. High-dimensional APSY experiments [27,28] were recorded at 278 K, 295 K and 310 K on ^{13}C/^{15}N-labeled ncTom40 in 4 M GdnSCN, 0.4% formic acid (Figure 3). Most residue-specific assignments were obtained at 310 K (see Table 1), consistent with the increase in relaxation times at higher temperature. To further increase the assignment coverage, six-dimensional peak lists from high and low temperature spectra were automatically matched [43]. In combination with the sequential connectivity found in the APSY spectra recorded at 278 K and manual inspection of a 3D HNN experiment [31] 326 out of 339 non-proline residues of ncTom40 could be assigned. ncTom40 in GdnSCN is thus one of the largest unfolded proteins for which the backbone assignment was obtained [43–46].

The residue-specific assignment of ncTom40 in 4 M GdnSCN allows the analysis of residual structure in the denatured state. This is important as the presence of residual structure in the dissolution buffer might influence the back-exchange process. Comparison of carbon chemical shifts in ncTom40 in 4 M GdnSCN with random coil values showed that for most residues the secondary structure propensity is below 0.2 [47]. No region exists where more than three consecutive residues exceed the secondary structure propensity value of 0.3 (Figure 4), indicating that very little residual structure, which could potentially influence back-exchange, is present in ncTom40 in 4 M GdnSCN.

Although 326 residues could be assigned in high-dimensional spectra, signals were severely overlapped in two-dimensional [^1H, ^{15}N]-HSQC spectra (Figure S2). To allow analysis of the back-exchange curves of a large number of residues, amino acid specific labeling was used (Table S1). The combination of amino acid types in each sample was chosen to minimize NMR signal overlap

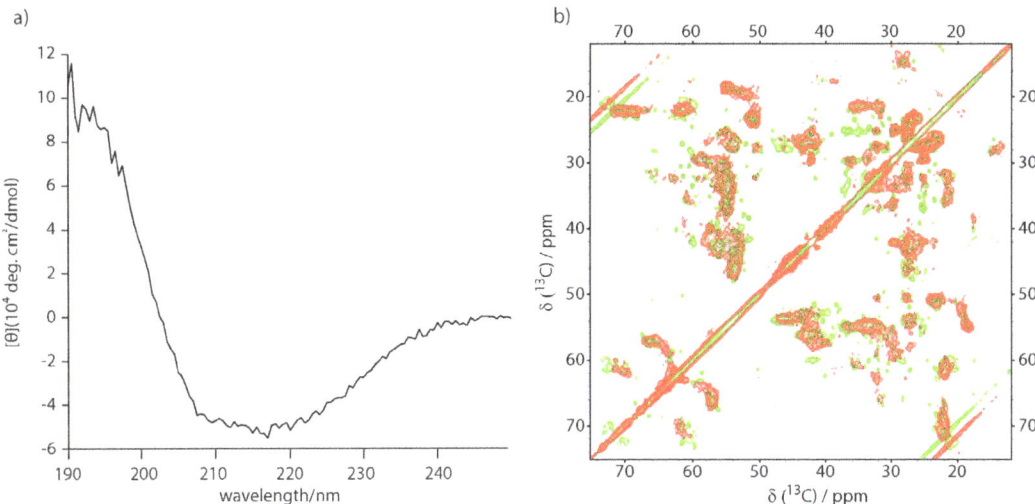

Figure 1. Recombinant ncTom40 has a β-barrel structure. (a) Far UV CD spectra of ncTom40 in decylmaltoside. (b) Superposition of ^{13}C-^{13}C proton driven spin diffusion spectra of ncTom40 (red) and hVDAC1 (green; reproduced from [4]), both in DMPC liposomes. The mixing time was 15 ms.

(Figure S3). Figure 5a shows the isoleucine region of the $[^1H, \, ^{15}N]$-HSQC spectra recorded on a selectively A/H/I/M/T ^{15}N-labeled sample of ncTom40 that had been exposed to the HD exchange protocol using 75% D_2O dissolution buffer. Six HSQC spectra were measured consecutively, with each spectrum recorded for three hours. In the first spectrum (labelled "0h" in Figure 5a) the signal intensity varied between different residues. For example, the signal intensity of I347 is lower than that of I137 and I328, suggesting that the amide proton of I347 has undergone more exchange during the H/D exchange period and is therefore less protected from solvent in the native state than I137 and I328. A distinct solvent protection along the sequence of ncTom40 was

also supported by normalized signal intensities in the first HSQC spectrum after dissolution in a 100% D_2O dissolution buffer (Figure 5b).

Identification of slow and fast solvent exchange in the native state on the basis of signal intensities in the first spectrum after dissolution relies on the assumption that (i) the dead time before starting the NMR experiment is zero and (ii) that the intrinsic exchange rates of different residues are identical. To overcome these limitations we further analysed NMR signal intensities during back-exchange in 75% D_2O dissolution buffer (Figure 5c). When amide protons slowly exchanged with deuterium in the liposome state and therefore retained a protonation level of more

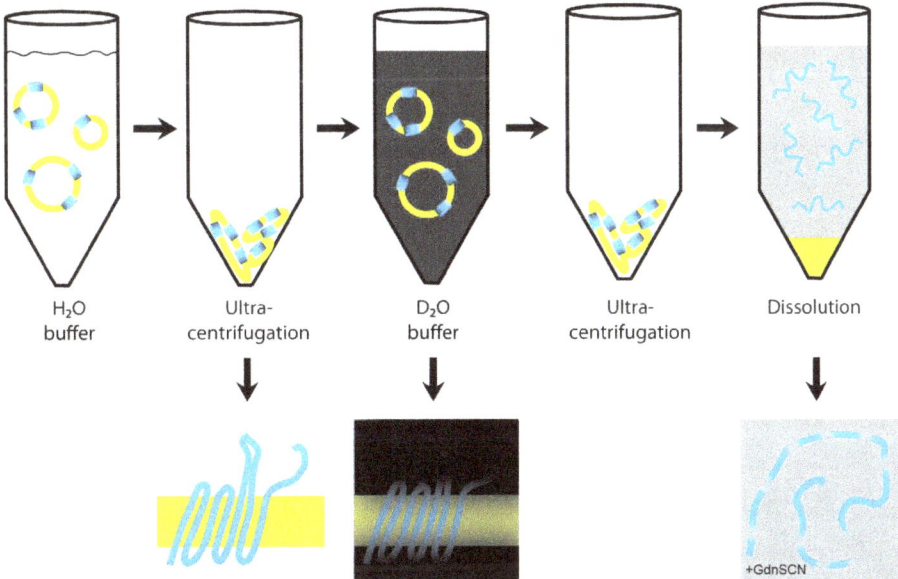

Figure 2. Scheme illustrating the H/D exchange strategy developed for membrane proteins (blue) reconstituted into liposomes (yellow). A white color indicates H_2O buffer, black color 100% D_2O buffer and grey color the dissolution buffer, which contains 4 M GdnSCN. During the incubation period in 100% D_2O solvent exposed residues will exchange amide protons against deuterium (lower row, middle panel). They will therefore not be visible in the denatured monomer (lower right panel).

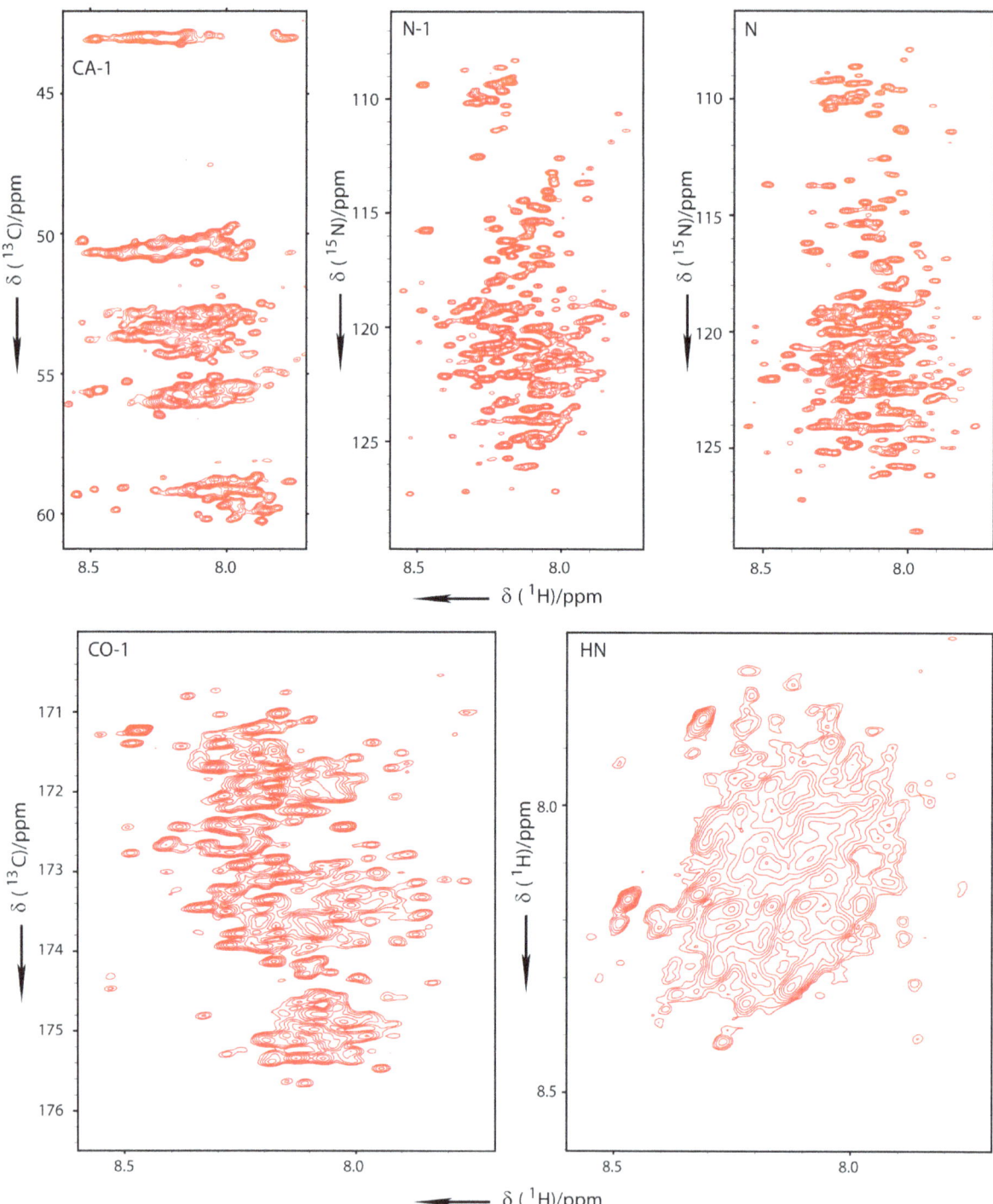

Figure 3. High-dimensional NMR experiments for assignment of ncTom40 in the denatured state. 2D projections (CA_{i-1}-HN_{i-1}, N_{i-1}- HN_{i-1}, N_i-HN_{i-1}, CO_{i-1}-HN_{i-1}, HN_i-HN_{i-1}) of the 6D APSY-HNCOCANH experiment, recorded on ncTom40 in 4 M GdnSCN, 0.4% formic acid. The measurement temperature was 278 K.

than 25%, the cross peak intensity decays as for example for residues I105, I137, I286, A311 and I328 (Figure 5c). In contrast, amide protons, which rapidly exchanged during the forward exchange period, will have low protonation levels when the dissolution is started. When their protonation levels are below 25%, they therefore will gain signal intensity due to back-exchange in the dissolution buffer. This is for example seen for I44, I47, A98 and I347 (Figure 5c). Notably, when the back-exchange rate in the dissolution buffer is very fast the signal intensity will remain constant in consecutive HSQCs. These residues will therefore be excluded from the analysis. Analysis of the back-exchange curves has the further advantage, that the protonation level at time 0 in the dissolving buffer can be estimated by an exponential decay curve that takes into account the dead time for the setup of the

Table 1. APSY experiments recorded at different temperatures and assignments obtained for denatured ncTom40 (339 non-proline residues) by MARS [31].

Temperature	Experiments	Numbers of assigned residues	
		Reliable	Low reliable
310 K	7D HNCO(CA)CBCANH	312	11
	5D CBCACONH		
295 K	6D HNCOCANH	249	26
	5D CBCACONH		
278 K	6D HNCOCANH	128	58
	5D CBCACONH		

Assignments classified by MARS as low are not reliable and were excluded from further analysis.

NMR experiment (approximately 20 minutes) [41]. In this way, estimates for the protonation level at the end of the solvent exchange time in the native state become accessible. The approach is distinct from H/D exchange measurements, in which increasing durations of forward exchange are used to determine protection factors with the following advantages: (i) at the end of the back-exchange the protonation level is 75% providing an internal reference that can be used to back-calculate the protonation level at the end of the forward exchange time; (ii) different samples can be compared on the basis of the internal reference; (iii) variations due to sample differences and dead time in NMR experiments are minimized.

Because a larger number of amino acid specific labeled samples was used for dissolution in 75% D_2O buffer, more residues could

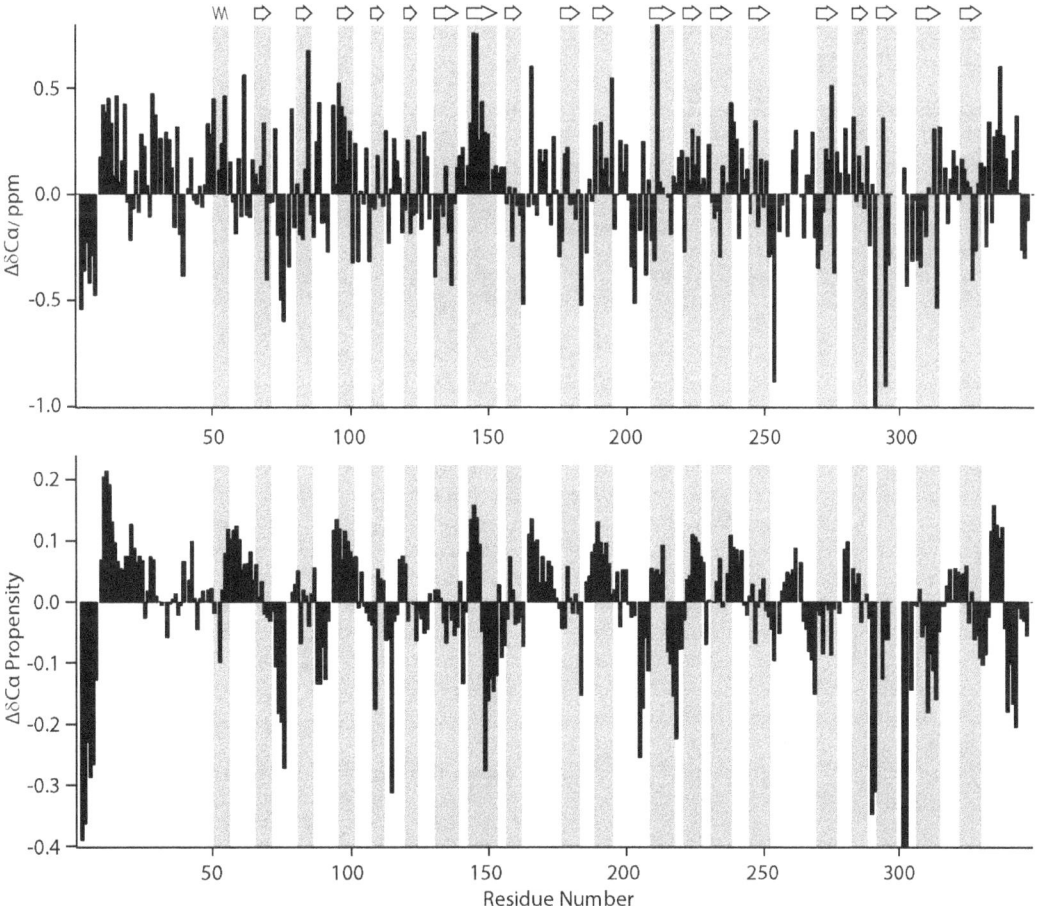

Figure 4. Cα secondary chemical shifts (upper chart) and Cα secondary structure propensities (lower chart) of ncTom40 obtained from APSY experiments recorded at 295K. Secondary structure propensities were calculated using SSP [42]. The predicted topology of ncTom40 is shown on top with secondary structure elements highlighted in grey. Only assignments classified by MARS [27] as reliable were used.

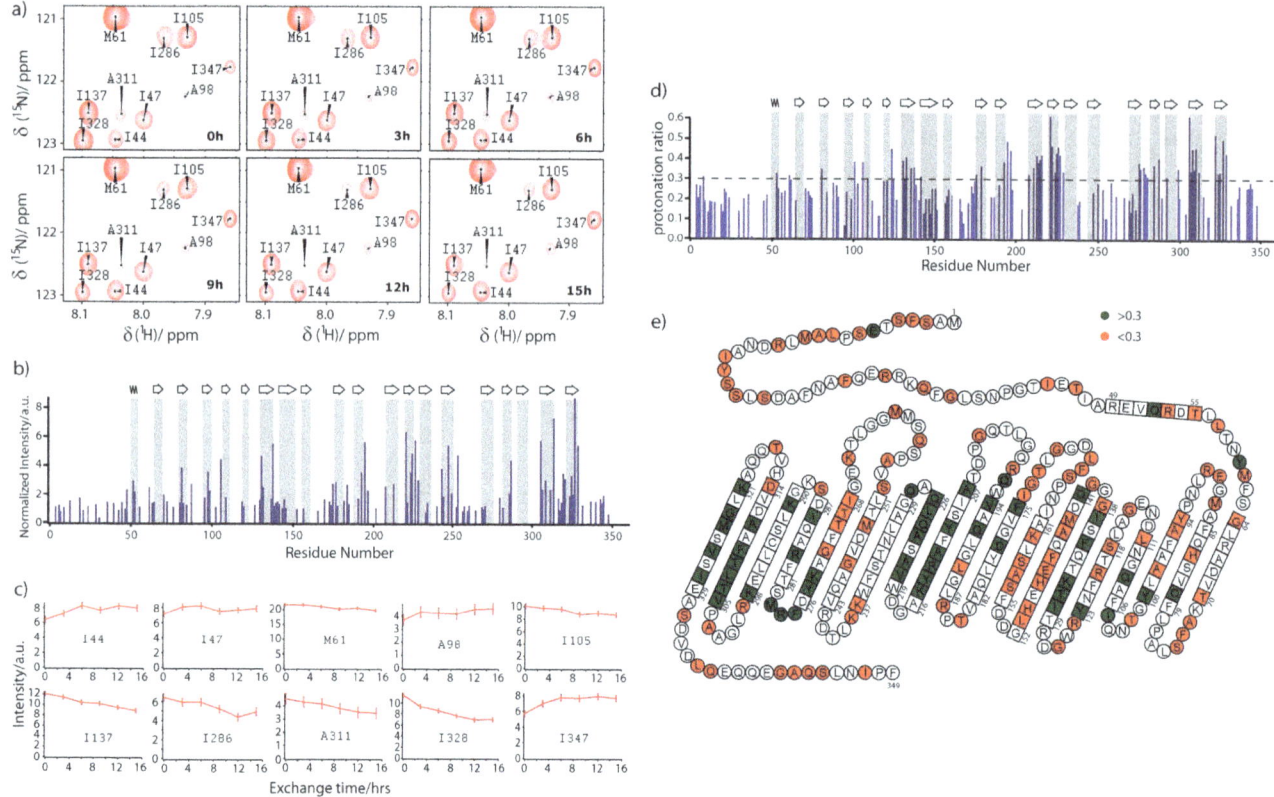

Figure 5. Structural characterization of liposome-embedded ncTom40 by H/D exchange coupled to solution-state NMR. (a) Enlarged spectral regions of [^{1}H, ^{15}N]-HSQC spectra at increasing back-exchange times. To reduce signal overlap, ncTom40 was selectively ^{15}N-labeled at ALA, HIS, ILE, MET, THR. Time points indicate the time after start of the first HSQC. The dissolution buffer contained 75% D$_2$O. Sequence-specific resonance assignments are indicated. (b) Sequence-specific signal intensities in the first HSQC after dissolution in 100% D$_2$O buffer. (c) NMR signal intensity change of residues in panel (a) during back-exchange in 75% D$_2$O. Intensity values were normalized on the basis of the noise level in the spectra. Error bars are based on signal-to-noise. (d) Protonation ratios for residues of Tom40 at the beginning of back exchange. (e) Protonation ratios shown in (d) were mapped onto the topology model of ncTom40, which was predicted on the basis of its homology to hVDAC1. Residues predicted to be in a β-strand or α-helix are boxed. Green-shaded (red-shaded) residues have protonation ratios larger (lower) than 0.3. Residues shown in white were not analyzed due to signal overlap, low signal-to-noise or missing resonance assignment.

be analysed when compared to dissolution in 100% D$_2$O (Table S1 and Figure 5b, d). At the same time, some residues, which are present in Figure 5b, were excluded from the back-exchange analysis as their back-exchange curves could not be described by a single exponential function. Based on the 145 residues with non-overlapping cross-peaks, we found that within the first ~50 residues the estimated protonation levels were close to or below 0.3 (Figure 5d). A high solvent exposure of the N- and C-terminal regions of liposome-embedded ncTom40 is in agreement with the signal intensities in the first HSQC spectrum after dissolution in 100% D$_2$O (Figure 5b) and with sequence-based analysis that predicts the first 50 and last 15 residues of ncTom40 to be disordered (Figure 5e). In between residues 51–330, exponential fits to the back-exchange curves resulted in protonation values exceeding 0.3 for residues 52, 60, 80, 100, 105, 108, 123, 130–132, 135, 137, 141,175, 178, 192, 194, 196, 207, 210, 212–215, 220–221,223–226, 228, 274–275, 277–279, 283, 286, 291, 298, 305–309, 311, 321, 323–324, 326, 328 (Figure 5d, e).

Both the signal intensities in the first HSQC spectrum after dissolution in 100% D$_2$O buffer (Figure 5b) and the protonation levels estimated from back-exchange curves in 75% D$_2$O buffer (Figure 5d, e) identified residues that exchange more slowly with solvent when ncTom40 is embedded into liposomes. As the slow solvent exchange is not restricted to a specific amino acid type, it

suggests that the observed variable degrees of protection are not purely caused by differences in intrinsic solvent exchange rates. In addition, certain amino acid types such as isoleucine are found protected when located in the central part of ncTom40, but solvent accessible when located near the N- or C-terminus. H/D exchange rates in liposome-embedded ncTom40 might be influenced by several other factors such as the type of secondary structure making a quantitative interpretation of the detected protonation levels challenging. However, a multitude of studies have shown that the strongest effect on solvent exchange is exerted by the presence and absence of hydrogen bonds. The higher protonation values – that is less efficient H/D exchange– observed for distinct residues in liposome-embedded ncTom40 therefore suggest that these residues are hydrogen-bonded. Notably, the fact that Tom40 forms a water-filled channel [48], the membrane insertion itself is unlikely to be responsible for the observed decrease in solvent exchange of membrane-embedded ncTom40. Moreover, a few residues, which are not predicted to be part of a β-strand or the N-terminal α-helix such as Y60 and ^{277}FRM^{279}, appear partially protected from H/D exchange (Fig. 5e), suggesting that hydrogen bonds stabilize the conformation in these region of the protein.

CD spectroscopy and solid-state NMR spectroscopy supports a β-barrel structure of ncTom40 (Figure 1). In addition, secondary structure analysis of ncTom40 on the basis of the known 3D structure of hVDAC1[36–38] using the software iTasser[49] predicts 19 β-strands and an N-terminal α-helix in ncTom40 (Figure 5e). The residues that were identified to be solvent protected support the location of several of the predicted β-strands (Figure 5d, e). These are particularly the predicted strands β6 (residues 129–138), β12 (residues 219–226), β18 (residues 305–314) and β19 (residues 321–329), which contains the membrane insertion signal of Tom40 [50]. For the other predicted strands as well as for the predicted α-helix at the N-terminus, the H/D exchange measurements do not provide clear support. This is partially due to the fact that only a subset of residues could be analyzed due to signal overlap and missing resonance assignments. At the same time, however, it is surprising that the amide protons of some of the predicted β-strands rapidly exchanged with solvent. For example, no amide proton with slow H/D exchange was observed for residues 142–161, the residues that are predicted to comprise β7 and β8. This could have several reasons such as the involvement in oligomer formation (in ^{145}QFEHEH^{150} every other residue faces the lipid environment such that at best two hydrophilic histidine residues point into the lipid environment) or the presence of chemical exchange [51,52]. For example, we previously showed that the N-terminal part of the β-barrel of hVDAC1 experiences dynamics on multiple time scales [53]. The amide protons in this region of the hVDAC1 barrel rapidly exchange with solvent, despite the fact that NOE contacts proof the presence of β-strands and defined β-strands were observed in the crystal structure of mouse VDAC1 [38]. H/D exchange measurements performed on detergent solubilized OmpA also had shown that its barrel does not behave like a solid block, but some strands are more mobile or accessible than others [54]. In addition, conformational exchange in the liposome-embedded state could cause a multi-exponential solvent exchange behavior. Finally, the number and/or exact location of the secondary structure elements predicted by iTasser may not be fully correct and therefore account for the differences between the H/D profile and the predicted location of β-strands.

Conclusions

We demonstrated that the solvent accessibility of single residues in large transmembrane proteins that are embedded in the near native environment of liposomes can be probed using a combination of H/D exchange and solution-state NMR spectroscopy. The method is applicable to highly challenging systems such as Tom40, which has resisted structural analysis by X-ray crystallography and NMR spectroscopy for many years.

Supporting Information

Figure S1 Two-dimensional [^{1}H,^{15}N]-HSQC spectrum of uniformly ^{15}N-labeled ncTom40 reconstituted into lauryldimethylamineoxide.

Figure S2 Two-dimensional [^{1}H,^{15}N]-HSQC spectrum of uniformly ^{15}N-labeled ncTom40 in dissolution buffer containing 4M GdnSCN, 0.4% formic acid.

Figure S3 [^{1}H,^{15}N]-HSQC spectra of ncTom40 with amino acid selective ^{15}N-labeling. Shown are the spectra at the end of back-exchange, i.e. time point 15 hr, in dissolution buffer containing 75% D$_2$O. Spectra were recorded at 278K, to slow down back-exchange. In all spectra, the contour level is set to five times the noise level as estimated by Sparky. Variations in overall signal intensity are due to differences in protein concentration.

Table S1 Amino acid selective ^{15}N-labeled samples of ncTom40 and their contribution to the H/D exchange experiments.

Author Contributions

Conceived and designed the experiments: MZ. Performed the experiments: XY UHND AKB YL. Analyzed the data: XY ZG RN CM AL SB. Wrote the paper: MZ XY.

References

1. Cross TA, Sharma M, Yi M, Zhou HX (2011) Influence of solubilizing environments on membrane protein structures. Trends Biochem Sci 36: 117–125.
2. Nietlispach D, Gautier A (2011) Solution NMR studies of polytopic alpha-helical membrane proteins. Curr Opin Struct Biol 21: 497–508.
3. Hong M, Zhang Y, Hu F (2012) Membrane protein structure and dynamics from NMR spectroscopy. Annu Rev Phys Chem 63: 1–24.
4. Schneider R, Etzkorn M, Giller K, Daebel V, Eisfeld J, et al. (2010) The native conformation of the human VDAC1 N terminus. Angew Chem Int Ed Engl 49: 1882–1885.
5. Knight MJ, Pell AJ, Bertini I, Felli IC, Gonnelli L, et al. (2012) Structure and backbone dynamics of a microcrystalline metalloprotein by solid-state NMR. Proc Natl Acad Sci U S A 109: 11095–11100.
6. Murray DT, Das N, Cross TA (2013) Solid state NMR strategy for characterizing native membrane protein structures. Acc Chem Res 46: 2172–2181.
7. Gopinath T, Mote KR, Veglia G (2013) Sensitivity and resolution enhancement of oriented solid-state NMR: application to membrane proteins. Prog Nucl Magn Reson Spectrosc 75: 50–68.
8. Marassi FM, Das BB, Lu GJ, Nothnagel HJ, Park SH, et al. (2011) Structure determination of membrane proteins in five easy pieces. Methods 55: 363–369.
9. Radoicic J, Lu GJ, Opella SJ (2014) NMR structures of membrane proteins in phospholipid bilayers. Q Rev Biophys 47: 249–283.
10. Krishna MM, Hoang L, Lin Y, Englander SW (2004) Hydrogen exchange methods to study protein folding. Methods 34: 51–64.
11. Bockmann A, Juy M, Bettler E, Emsley L, Galinier A, et al. (2005) Water-protein hydrogen exchange in the micro-crystalline protein crh as observed by solid state NMR spectroscopy. J Biomol NMR 32: 195–207.
12. Schanda P, Forge V, Brutscher B (2007) Protein folding and unfolding studied at atomic resolution by fast two-dimensional NMR spectroscopy. Proc Natl Acad Sci U S A 104: 11257–11262.
13. Van Melckebeke H, Schanda P, Gath J, Wasmer C, Verel R, et al. (2011) Probing water accessibility in HET-s(218–289) amyloid fibrils by solid-state NMR. J Mol Biol 405: 765–772.
14. Qureshi T, Goto NK (2012) Contemporary methods in structure determination of membrane proteins by solution NMR. Top Curr Chem 326: 123–185.
15. Fernandez C, Hilty C, Wider G, Guntert P, Wuthrich K (2004) NMR structure of the integral membrane protein OmpX. J Mol Biol 336: 1211–1221.
16. Dempsey CE (1994) Amide-resolved hydrogen-deuterium exchange measurements from membrane-reconstituted polypeptides using exchange trapping and semiselective two-dimensional NMR. J Biomol NMR 4: 879–884.
17. Hoshino M, Katou H, Hagihara Y, Hasegawa K, Naiki H, et al. (2002) Mapping the core of the beta(2)-microglobulin amyloid fibril by H/D exchange. Nat Struct Biol 9: 332–336.
18. Ippel JH, Olofsson A, Schleucher J, Lundgren E, Wijmenga SS (2002) Probing solvent accessibility of amyloid fibrils by solution NMR spectroscopy. Proc Natl Acad Sci U S A 99: 8648–8653.
19. Vilar M, Wang L, Riek R (2012) Structural studies of amyloids by quenched hydrogen-deuterium exchange by NMR. Methods Mol Biol 849: 185–198.
20. Hebling CM, Morgan CR, Stafford DW, Jorgenson JW, Rand KD, et al. (2010) Conformational analysis of membrane proteins in phospholipid bilayer nanodiscs by hydrogen exchange mass spectrometry. Anal Chem 82: 5415–5419.
21. Rey M, Man P, Clemencon B, Trezeguet V, Brandolin G, et al. (2010) Conformational dynamics of the bovine mitochondrial ADP/ATP carrier isoform 1 revealed by hydrogen/deuterium exchange coupled to mass spectrometry. J Biol Chem 285: 34981–34990.

22. Tian C, Gao PF, Pinto LH, Lamb RA, Cross TA (2003) Initial structural and dynamic characterization of the M2 protein transmembrane and amphipathic helices in lipid bilayers. Protein Sci 12: 2597–2605.

23. Neupert W, Herrmann JM (2007) Translocation of proteins into mitochondria. Annu Rev Biochem 76: 723–749.

24. Perry AJ, Rimmer KA, Mertens HD, Waller RF, Mulhern TD, et al. (2008) Structure, topology and function of the translocase of the outer membrane of mitochondria. Plant Physiol Biochem 46: 265–274.

25. Becker L, Bannwarth M, Meisinger C, Hill K, Model K, et al. (2005) Preprotein translocase of the outer mitochondrial membrane: reconstituted Tom40 forms a characteristic TOM pore. J Mol Biol 353: 1011–1020.

26. Tong KI, Yamamoto M, Tanaka T (2008) A simple method for amino acid selective isotope labeling of recombinant proteins in E. coli. J Biomol NMR 42: 59–67.

27. Hiller S, Wasmer C, Wider G, Wuthrich K (2007) Sequence-specific resonance assignment of soluble nonglobular proteins by 7D APSY-NMR spectroscopy. J Am Chem Soc 129: 10823–10828.

28. Hiller S, Fiorito F, Wuthrich K, Wider G (2005) Automated projection spectroscopy (APSY). Proc Natl Acad Sci U S A 102: 10876–10881.

29. Fiorito F, Hiller S, Wider G, Wuthrich K (2006) Automated resonance assignment of proteins: 6D APSY-NMR. J Biomol NMR 35: 27–37.

30. Güntert P, Dötsch V, Wider G, Wuthrich K (1992) Processing of multi dimensional NMR data with the new software PROSA. J Biomol NMR: 619–629.

31. Panchal SC, Bhavesh NS, Hosur RV (2001) Improved 3D triple resonance experiments, HNN and HN(C)N, for HN and 15N sequential correlations in (13C, 15N) labeled proteins: application to unfolded proteins. J Biomol NMR 20: 135-147.

32. Jung YS, Zweckstetter M (2004) Mars – robust automatic backbone assignment of proteins. J Biomol NMR 30: 11–23.

33. Schanda P, Van Melckebeke H, Brutscher B (2006) Speeding up three-dimensional protein NMR experiments to a few minutes. J Am Chem Soc 128: 9042–9043.

34. Delaglio F, Grzesiek S, Vuister GW, Zhu G, Pfeifer J, et al. (1995) NMRPipe: a multidimensional spectral processing system based on UNIX pipes. J Biomol NMR 6: 277–293.

35. Baldus M, Meier BH (1996) Total correlation spectroscopy in the solid state. The use of scalar couplings to determine the through-bond connectivity. J Magn Reson Ser A 121: 65–69.

36. Bayrhuber M, Meins T, Habeck M, Becker S, Giller K, et al. (2008) Structure of the human voltage-dependent anion channel. Proc Natl Acad Sci U S A 105: 15370–15375.

37. Hiller S, Garces RG, Malia TJ, Orekhov VY, Colombini M, et al. (2008) Solution structure of the integral human membrane protein VDAC-1 in detergent micelles. Science 321: 1206–1210.

38. Ujwal R, Cascio D, Colletier JP, Faham S, Zhang J, et al. (2008) The crystal structure of mouse VDAC1 at 2.3 A resolution reveals mechanistic insights into metabolite gating. Proc Natl Acad Sci U S A 105: 17742–17747.

39. Bauer AJ, Gieschler S, Lemberg KM, McDermott AE, Stockwell BR (2011) Functional model of metabolite gating by human voltage-dependent anion channel 2. Biochemistry 50: 3408–3410.

40. Eddy MT, Ong TC, Clark L, Teijido O, van der Wel PC, et al. (2012) Lipid dynamics and protein-lipid interactions in 2D crystals formed with the beta-barrel integral membrane protein VDAC1. J Am Chem Soc 134: 6375–6387.

41. Cho MK, Kim HY, Fernandez CO, Becker S, Zweckstetter M (2011) Conserved core of amyloid fibrils of wild type and A30P mutant alpha-synuclein. Protein Sci 20: 387–395.

42. Molday RS, Englander SW, Kallen RG (1972) Primary structure effects on peptide group hydrogen exchange. Biochemistry 11: 150–158.

43. Narayanan RL, Durr UH, Bibow S, Biernat J, Mandelkow E, et al. (2010) Automatic assignment of the intrinsically disordered protein Tau with 441-residues. J Am Chem Soc 132: 11906–11907.

44. Novacek J, Janda L, Dopitova R, Zidek L, Sklenar V (2013) Efficient protocol for backbone and side-chain assignments of large, intrinsically disordered proteins: transient secondary structure analysis of 49.2 kDa microtubule associated protein 2c. J Biomol NMR 56: 291–301.

45. Solyom Z, Schwarten M, Geist L, Konrat R, Willbold D, et al. (2013) BEST-TROSY experiments for time-efficient sequential resonance assignment of large disordered proteins. J Biomol NMR 55: 311–321.

46. Csizmok V, Felli IC, Tompa P, Banci L, Bertini I (2008) Structural and dynamic characterization of intrinsically disordered human securin by NMR spectroscopy. J Am Chem Soc 130: 16873–16879.

47. Marsh JA, Singh VK, Jia Z, Forman-Kay JD (2006) Sensitivity of secondary structure propensities to sequence differences between alpha- and gamma-synuclein: implications for fibrillation. Protein Sci 15: 2795–2804.

48. Endo T, Yamano K (2010) Transport of proteins across or into the mitochondrial outer membrane. Biochim Biophys Acta 1803: 706–714.

49. Zhang Y (2008) I-TASSER server for protein 3D structure prediction. BMC Bioinformatics 9: 40.

50. Kutik S, Stojanovski D, Becker L, Becker T, Meinecke M, et al. (2008) Dissecting membrane insertion of mitochondrial beta-barrel proteins. Cell 132: 1011–1024.

51. Chill JH, Naider F (2011) A solution NMR view of protein dynamics in the biological membrane. Curr Opin Struct Biol 21: 627–633.

52. Chill JH, Louis JM, Delaglio F, Bax A (2007) Local and global structure of the monomeric subunit of the potassium channel KcsA probed by NMR. Biochim Biophys Acta 1768: 3260–3270.

53. Villinger S, Briones R, Giller K, Zachariae U, Lange A, et al. (2010) Functional dynamics in the voltage-dependent anion channel. Proc Natl Acad Sci U S A 107: 22546–22551.

54. Catoire LJ, Zoonens M, van Heijenoort C, Giusti F, Guittet E, et al. (2010) Solution NMR mapping of water-accessible residues in the transmembrane beta-barrel of OmpX. Eur Biophys J 39: 623–630.

2

Identification of a Novel Drug Lead That Inhibits HCV Infection and Cell-to-Cell Transmission by Targeting the HCV E2 Glycoprotein

Reem R. Al Olaby[1], Laurence Cocquerel[2], Adam Zemla[3], Laure Saas[2], Jean Dubuisson[2], Jost Vielmetter[4], Joseph Marcotrigiano[5], Abdul Ghafoor Khan[5], Felipe Vences Catalan[6], Alexander L. Perryman[7], Joel S. Freundlich[7,8], Stefano Forli[9], Shoshana Levy[6], Rod Balhorn[10*¤], Hassan M. Azzazy[1]

1 Department of Chemistry, The American University in Cairo, New Cairo, Egypt, 2 Center for Infection and Immunity of Lille, CNRS-UMR8204/Inserm-U1019, Pasteur Institute of Lille, University of Lille North of France, Lille, France, 3 Pathogen Bioinformatics, Lawrence Livermore National Laboratory, Livermore, CA, United States of America, 4 Protein Expression Center, Beckman Institute, California Institute of Technology, Pasadena, CA, United States of America, 5 Department of Chemistry and Chemical Biology, Rutgers University, Piscataway, NJ, United States of America, 6 Department of Medicine, Stanford University Medical Center, Stanford, CA, United States of America, 7 Department of Medicine, Division of Infectious Diseases, Center for Emerging & Re-emerging Pathogens, Rutgers University-New Jersey Medical School, Newark, NJ, United States of America, 8 Department of Pharmacology and Physiology, Rutgers University-New Jersey Medical School, Newark, NJ, United States of America, 9 Department of Integrative Structural and Computational Biology, The Scripps Research Institute, La Jolla, CA, United States of America, 10 Department of Applied Science, University of California Davis, Davis, CA, United States of America

Abstract

Hepatitis C Virus (HCV) infects 200 million individuals worldwide. Although several FDA approved drugs targeting the HCV serine protease and polymerase have shown promising results, there is a need for better drugs that are effective in treating a broader range of HCV genotypes and subtypes without being used in combination with interferon and/or ribavirin. Recently, two crystal structures of the core of the HCV E2 protein (E2c) have been determined, providing structural information that can now be used to target the E2 protein and develop drugs that disrupt the early stages of HCV infection by blocking E2's interaction with different host factors. Using the E2c structure as a template, we have created a structural model of the E2 protein core (residues 421–645) that contains the three amino acid segments that are not present in either structure. Computational docking of a diverse library of 1,715 small molecules to this model led to the identification of a set of 34 ligands predicted to bind near conserved amino acid residues involved in the HCV E2: CD81 interaction. Surface plasmon resonance detection was used to screen the ligand set for binding to recombinant E2 protein, and the best binders were subsequently tested to identify compounds that inhibit the infection of Huh-7 cells by HCV. One compound, 281816, blocked E2 binding to CD81 and inhibited HCV infection in a genotype-independent manner with IC50's ranging from 2.2 µM to 4.6 µM. 281816 blocked the early and late steps of cell-free HCV entry and also abrogated the cell-to-cell transmission of HCV. Collectively the results obtained with this new structural model of E2c suggest the development of small molecule inhibitors such as 281816 that target E2 and disrupt its interaction with CD81 may provide a new paradigm for HCV treatment.

Editor: Vladimir N. Uversky, University of South Florida College of Medicine, United States of America

Funding: This work was conducted as part of the first authors PhD thesis work. This work was supported by a Yousif Jameel PhD Fellowship from The American University in Cairo awarded to Reem Al Olaby. The funders had no role in study design, data collection and analysis, decision to publish, or preparation of the manuscript.

Competing Interests: Reem Al Olaby, Dr. Hassan Azzazy and Dr. Rod Balhorn are co-inventors on a patent application related to the described work that has been submitted by The American University in Cairo, Cairo, Egypt. The title for the application is Ligands That Target Hepatitis C Virus E2 Protein. None of the other coauthors have competing interests.

* Email: rod@shaltech.com

¤ Current address: SHAL Technologies Inc., Livermore, CA, United States of America

Introduction

Hepatitis C virus (HCV) is a global public health problem [1] in which nearly 85% of affected individuals have acute HCV infections and exhibit no symptoms. In addition, more than three-quarters of these cases will advance to chronic disease, which include liver cirrhosis and liver cancer [2]. The current standard of care treatment for HCV (Peg-interferon/Ribavirin, PR) can cause

deleterious side effects, and a sustained virologic response (SVR) is achieved in less than 50% of genotype-1 patients [3]. The FDA approved protease inhibitors Telaprevir (TVR) and Boceprevir (BOC) have been shown to provide higher SVR rates in genotype 1 patients [3,4] when each is combined with PR. However the poor safety profile of TVR and BOC reported in the Week 16 analysis of the French Early Access Program suggest there is still a need for better HCV drugs [5]. The two most recent FDA approvals have been for the oral drugs Simeprevir and Sofosbuvir, inhibitors that target the HCV NS3/4A protease and polymerase, respectively [6]. Semiprevir, which needs to be administered with Ribavirin and Peg-interferon, has a number of undesirable side effects [7]. The efficacy of Semiprevir has also been shown to be diminished significantly, due to viral breakthrough (HCV RNA rebounds and becomes detectable in the patient before treatment is completed), in patients infected by HCV genotypes 4–6 containing the Q80K, R155K and D168E/V polymorphisms in the NS3 protease [7]. Recommendations for the use of Sofosbuvir indicate it should be administered with Ribavirin in HCV genotype 2 and 3 infections and that Peg-Interferon should be included in the treatment when infections involve genotypes 1 and 4. While Sofosbuvir is considered the Holy Grail in HCV treatment by some, it is recommended that treatments be limited to 12 weeks [6]. Its high cost ($1,000 USD/pill) also puts it out of reach of many HCV infected patients. This has led many of the larger pharmaceutical companies to continue developing new drugs that target one or more steps in the HCV life cycle and block virus invasion, processing of the pro-protein or replication of the viral genome.

Since its identification as the first putative receptor for HCV [8], the tetraspanin CD81 has been demonstrated to be a key player in HCV entry [9]. In particular, its large extracellular loop (CD81-LEL) is involved in the binding to the HCV envelope glycoprotein E2 [10,11]. Zhang et al. [12] elucidated a separate, additional function for CD81 in the HCV life cycle. These studies revealed that CD81-LEL is important for efficient HCV genome replication. In addition, the E2-CD81-LEL interaction has been determined to induce several immuno-modulatory effects such as the production and release of pro-inflammatory cytokine gamma interferon from T-cells. In addition, this interaction has also been shown to down regulate T-cell receptors and suppress the activity of natural killer (NK) cells [13]. Therefore, it is tempting to speculate that blocking the CD81-LEL:HCV E2 interaction might also contribute to arresting disease progression to liver cirrhosis.

Following the discovery of the E2 glycoprotein's role in HCV infection and disease progression, several approaches have been used to attempt to develop anti-HCV drugs and vaccines that target the HCV E2 glycoprotein [14–17] located on the surface of viral particles. These efforts have had to deal with challenges that relate to the genomic diversity and heterogeneity of HCV, limitations in animal models used to test vaccines and drugs, and the lack of a resolved crystal structure for the HCV E2 glycoprotein. Recently, two crystal structures have been reported for the core ectodomain of the HCV E2 protein [18,19]. Kong et al. [18] obtained the structure of amino acid residues 384–746 (E2c) by designing and expressing 41 soluble HCV E2 constructs and selecting 15 to screen against E2-specific Fab fragments in crystallization trials. Using a combination of x-ray crystallography and negative stain-electron microscopy, Kong et al. [18] discovered the structures they obtained for E2 were globular and very different from the predicted models of E2 that were created using class II fusion protein templates containing three β-sheet domains. Additionally, they were able to identify key CD81-binding residues through mutational studies. Important CD81 binding sites were determined to be in the epitope recognized by the neutralizing antibody AR3C, along one side of the β-sandwich (an isolated region of the CD81-binding loop) and a front layer consisting of loops, short helices and β-sheets [18–20]. AR3C was also found to cross-neutralize HCV genotypes by blocking CD81 binding to HCV E2 [21]. A second structure was reported for E2c (amino acid residues 492–649) by Khan et al. [19]. This new structure, which was obtained by crystallizing E2c in complex with a Fab

fragment of the mouse monoclonal antibody 2A12, is very similar to the previously reported structure. In addition to providing a second structure for the E2 core from a different HCV genotype (2a), new information was also reported on the accessibility of the E2 core amino acids within the structure using a combination of limited proteolytic degradation and deuterium exchange.

Despite the advances that have been made in the field of HCV drug development, our current drugs offer little protection against the emergence of genetic variants (escape variants) of HCV – a feature of HCV biology that complicates both drug and vaccine development. Drugs that target only one step in the HCV life cycle will be the least effective in treating patients that become infected with these emerging variants. The FDA approved drugs for HCV are good examples, as they are only effective against a subset of genotypes. In an effort to identify a suitable drug candidate that targets the majority of the existing HCV genotypes, we created an HCV E2 homology model based on the new HCV E2 core crystal structure reported by Kong et al. [18] that contains three peptide segments that were not present in the reported structure, and we have used this model to identify small molecule drug leads that target highly conserved sites on the HCV E2 glycoprotein located within the region bound by CD81. AutoDock was used to perform virtual screening runs against 1,715 small molecules and 34 of the best compounds were tested experimentally using surface plasmon resonance (Biacore T100) to identify a set of small molecules that bind to the recombinant E2 protein. The compounds showing binding activities were then tested for their ability to block HCV infection of Huh-7 cells. One compound, 281816, was found to block infection of the cells by each of the HCV genotypes and subtypes tested (1a, 1b, 2a, 2b, 4a and 6a) in a dose-dependent manner. Experiments with Huh-7 cells have shown that both mechanisms that lead to HCV infection, cell-free and cell-to-cell transmission, are abrogated by 281816. Inhibition of cell-free infection is limited to the viral attachment step, as well as interactions occurring during viral internalization and fusion; 281816 appears to have no effect on post-entry processes.

Materials and Methods

Creation of the homology model of E2 used for docking

A crystal structure of E2c deposited in the PDB under a code 4MWF was solved by Kong et al. [18] at a resolution of 2.65 Angstroms. However, upon examination of the structure file prior to docking, the set of reported atom coordinates of the protein was found to be incomplete. In addition to the coordinate file containing structural information for only 171 residues out of the 363 amino acids present in the full-length protein, structural information was missing for several peptide segments or loops within the structural core of the protein. In order to prepare a more complete version of the structure for docking, we have performed several homology modeling and structure analysis tasks using the coordinates of E2c as a template. The final structural model was created using the AS2TS system [21] based on atom coordinates from the PDB chains 4mwf_C and 4mwf_D and extensively manually edited. A structural search for similar fragments in proteins in the PDB that could be used to model missing loop regions was performed using the StralSV algorithm [22], which identifies protein structures that exhibit structural similarities despite low primary amino acid sequence similarity. The side-chain prediction was accomplished using SCWRL [23] when residue-residue correspondences did not match. Residues that were identical in the template and E2 protein were copied from the template onto the model. Potential steric clashes were identified in the unrefined model using a contact-dot algorithm in the MolProbity software package [24], and the constructed model was finished with relaxation using UCSF Chimera [25].

Virtual screen of the NCI Diversity Set III to the HCV E2 protein model

AutoDock VINA 1.1.2 (VINA) [26] was used to perform a virtual screen of the NCI Diversity Set III against the homology model that was created using the new crystal structure solved by Kong et al. [18] (PDB ID: 4MWF) as a template. The model of the protein was prepared for docking using the MolProbity Server (to add all of the hydrogen atoms and to flip the HIS/ASN/GLN residues if doing so significantly lowered the energy) and AutoDockTools 4.2 (which added the Gasteiger-Marsili charges and merged the non-polar hydrogens onto their respective heavy atoms) [27,28]. The NCI Diversity Set III library containing 1,715 models of compounds was obtained from the ZINC server (http://zinc.docking.org) [29]. The multi-molecule "mol2" files from ZINC were prepared for docking calculations using Raccoon [30], which added the Gasteiger-Marsili charges, merged the non-polar hydrogen atoms onto their respective heavy atoms, and determined which bonds should be allowed to freely rotate during the calculations, to generate the "pdbqt" docking input format.

Four different, overlapping grid boxes were used in this virtual screen to enable the docking calculations to explore almost the entire surface of this E2 model. Those amino acids missing from the E2c crystal structure whose modeled coordinates were known with the lowest degree of certainty, such as residues E454 and L456–E482 located in the large missing loop and residues G575–L580 and F586–K588 in the two other two missing segments, were not included in the boxes. By defining the boxes to exclude these residues, we were able to minimize the impact of these less accurate parts of the model on ligand docking. Since large grid boxes were used in these calculations, the "exhaustiveness" setting in VINA was increased to 20. Each calculation used 8 CPUs on the Linux cluster at Rutgers University-NJMS. The first box, which included P490, was centered at 38.829, 12.968, −40.958 (x, y, z) and had the following dimensions: 24.0×35.0×30.0 (x, y, z in Angstroms). The second grid box, which included G436, was centered at 48.401, 11.791, −14.449 and had a size of: 32.0×36.0×24.0. The third grid box, which included S528, was centered at 51.644, 25.877, −27.795 and encompassed 30.0×30.0×30.0 Angstroms^3. The fourth grid box, which was selected to include the side of E2 not covered by the previous three grid boxes, was centered at 57.777, 12.968, −34.067 and enclosed 24.0×35.0×32.0 Angstroms^3.

The docking outputs generated by VINA were processed and filtered using python scripts from Raccoon2 and Fox [30]. The top-ranked VINA mode from each docking calculation was harvested, and 17 different sets of energetic and interaction-based filters (Table 1) were investigated to harvest the most promising docking results for visual inspection. Different sites have different numbers and arrangements of hydrogen bond donors, hydrogen bond acceptors, and aromatic rings. They also have very different geometries (i.e., van der Waal surfaces and solvent accessibility patterns and percentages). Consequently, several different filters were tested for the docking results against each site in order to harvest a"reasonable" number of docked modes for visual inspection against each site. This is a subjective process, guided by extensive experience with virtual screening. If the same filters are used against each site, then for some sites (or for filters that are not restrictive enough), too many compounds are obtained for visual inspection (i.e., the process is less efficient and a larger number of false positive results are likely to occur). For other sites (or for filters that are too restrictive), an insufficient number of

compounds will be obtained for visual inspection. This would increase the chance of missing promising candidates (having too many false negatives) [31]. The following parameters were explored in the filtering process: -e indicates the minimum estimated Free Energy of Binding from the VINA score in kcal/mol, -l is the minimum ligand efficiency value in kcal/mol/heavy atom, -S is the minimum number of hydrogen bonds between the ligand and target, and -H indicates that the ligand had to form a hydrogen bond with either a backbone amino group (::N) or a backbone carbonyl oxygen (::O) of any residue in that grid box.

For the results with grid box 1, filters 12 and 13 each harvested 70 and 51 compounds, respectively. Those filtered sets were pooled together to form a set of 96 unique compounds for visual inspection. Filters 14 (which harvested 11 compounds), 15 (which harvested 21 compounds), and 1 (which harvested 34 compounds) were pooled together from the results with grid box 2, in order to identify 52 compounds for visual inspection. Similarly, for the results with grid box 3, filters 1 (which harvested 25 compounds), 14 (which identified 20 candidates), and 15 (which harvested 13 compounds) were pooled to obtain 34 compounds for visual inspection. To identify candidates in the results with grid box 4, filters 1 (which harvested 26 compounds), 14 (which harvested 19 compounds), and 15 (which harvested 14 compounds) were pooled to obtain 42 compounds.

These four different pools of potentially promising compounds were then visually inspected to select the ligands to be tested experimentally for binding to recombinant E2 protein. Both the structure of the compound and the nature of its predicted interaction with the protein were examined. Compounds were considered good hits and suitable for testing if they 1) were small (molecular weight ~200–600 Da), 2) contained a single free amine or carboxyl to facilitate their potential conjugation to other ligands, 3) were not highly charged or highly hydrophobic, 4) did not contain iodine, disulfide bonds or highly reactive functional groups, 5) did not contain multiple conjugated aromatic ring systems, 6) exhibited multiple contacts to the protein surface, and 7) had conformers that bound to the protein surface near one or more E2 amino acid residues that have been shown to participate in or be required for binding to CD81. Detergent-like molecules were avoided and only commercially available compounds were considered for screening.

Expression and purification of the HCV E2 protein Con1eE2

A construct containing a sequence encoding amino acids 384–656 of the Con1 envelope protein 2 ectodomain (eE2) [19], a genotype 1 E2 sequence, was cloned into a lentiviral expression vector containing a carboxy-terminal Protein A tag separated by a PreScission Protease cleavage consensus sequence. eE2-ProtA was stably expressed in HEK293T cells using lentiviral infection. The protein was secreted into the media and supernatants were purified using IgG Sepharose (GE Healthcare, Piscataway, NJ). eE2-ProtA was eluted with 100 mM sodium citrate and 20 mM KCl at pH 3 directly into tubes containing 1M Tris pH 9 for immediate neutralization. PreScission Protease (GE Healthcare, Piscataway, NJ) was added to the eluted sample at a ratio of 1:50 (enzyme:eE2), and the digest was then dialyzed into 20 mM HEPES pH 7.5, 250 mM NaCl, 5% glycerol. eE2 was separated from the cleaved tag and the PreScission Protease by ion exchange chromatography [19].

Table 1. Energetic and interaction-based filters used to harvest the most promising results from the ligand docking runs.

Filter	Parameter Set
1	-e −6.5 -l −0.29 -S 3
2	-e −7.0 -l −0.29 -S 3
3	-e −7.5 -l −0.29 -S 3
4	-e −8.0 -l −0.29 -S 3
5	-e −7.0 -l −0.29 -S 4
6	-e −7.5 -l −0.29 -S 4
7	-e −8.0 -l −0.29 -S 4
8	-e −6.5 -l −0.29 -S 3 -H ::N
9	-e −6.5 -l −0.29 -S 3 -H ::O
10	-e −7.0 -l −0.29 -S 3 -H ::N
11	-e −7.0 -l −0.29 -S 3 -H ::O
12	-e −7.0 -l −0.29 -S 4 -H ::N
13	-e −7.0 -l −0.29 -S 4 -H ::O
14	-e −7.0 -S 3 -H ::N
15	-e −7.0 -S 3 -H ::O
16	-e −7.0 -S 4 -H ::N
17	-e −7.0 -S 4 -H ::O

Experimental analysis of ligand binding to recombinant E2 and CD81-LEL by surface plasmon resonance (SPR) detection

A set of 34 of the ligands predicted by AutoDock to bind to E2 were tested experimentally to determine if they bound to recombinant E2 protein immobilized on a chip using surface plasmon resonance detection. The SPR analyses were performed using a Biacore T100 workstation (GE Healthcare, NJ, USA) and recombinant HCV E2 protein. 1 μM HCV E2 was diluted into 10 mM sodium acetate buffer pH 5 and immobilized for 15 min at a flow speed of 5 μl/min onto a CM5 sensor chip using amine coupling (EDC-NHS). Approximately 10,000 response units (RU) of protein were immobilized on the chip. His-CD81-LEL (Bioclone Inc., San Diego, CA) binding to HCV E2 was tested as a positive control prior to injecting the ligands to confirm the E2 protein was functional and would bind CD81-LEL. In a typical experiment with CD81, 1 μl of his-CD81 (50 nM) in 114 μl PBS was injected into channel 2 and 106.4 RUs of CD81 bound to the E2 on the chip. This was followed by testing the binding of the 34 virtual screening hits where the ligands were prepared as 200 μM solutions in PBS and they were introduced to the protein using a pre-programmed 3 min association and 1 min dissociation interval. The response was measured at two time points during dissociation, 10 and 50 seconds, to obtain information on the rate of ligand dissociation from E2.

Two single cycle kinetic studies were also performed to compare the binding of 281816 to the recombinant E2 and his-tagged CD81-LEL proteins. In both studies, the proteins were diluted to a concentration of 1 μM in 10 mM sodium acetate buffer pH 4.5 and immobilized for 15 min on a CM5 sensor chip using amine coupling (EDC-NHS). Data on the kinetics and affinity of 281816 binding was obtained by flowing five concentrations of the 281816 ligand (2.5 μM, 7.4 μM, 22.2 μM, 66.7 μM and 200 μM) over the chip sequentially at a flow rate of 30 μl/min. Equilibrium binding curves were generated for each protein and the data were fitted using a monovalent binding model to determine the Kd for 281816 binding to E2 and CD81-LEL.

HCV infection assays

Pseudotyped retroviral particles harboring HCV envelope proteins (HCVpp) from different genotypes were produced as described previously [32,33] with plasmids kindly provided by F.L. Cosset, J. Ball, and R. Bartenschlager. A plasmid encoding the feline endogenous virus RD114 glycoprotein [34] was used for the production of RD114pp. Both HCVpp and RD114pp expressed *Firefly* luciferase.

The cell culture-produced HCV particles (HCVcc) used in this study were based on the JFH1 strain [35] and were prepared as described previously [36,37]. They were engineered to express the A4 epitope, titer-enhancing mutations and *Gaussia* luciferase [36,37].

To identify ligands that inhibit HCV infection, Huh-7 cells were seeded in 96-well plates and treated the day after with six different concentrations of each ligand diluted in DMSO in duplicate using a Zephyr automated liquid handling workstation (Caliper BioSciences, Hopkinton, MA). The final concentration of DMSO (1%) was adjusted to be the same for all ligand concentrations. Cells treated with DMSO were used as negative controls. Cells treated with different concentrations of anti-CD81 antibody (JS-81 from BD Pharmingen, San Jose, CA) 1 hr before infection, were also used as positive controls. The third day, RD114pp, HCVpp or HCVcc were inoculated and incubated for 30 hr at 37°C. *Firefly* and *Gaussia* luciferase assays were performed as indicated by the manufacturer (Promega, San Luis Obispo, CA).

The analysis of the effect of the 281816 ligand on Huh-7 infection by HCVpp bearing envelope proteins from different genotypes was performed in 24-well plates using the method described above. This ligand was also screened for toxicity to the cells using the MTS (3-(4,5-dimethylthiazol-2-yl)-5-(3-carboxy-methoxyphenyl)-2-(4-sulfophenyl)-2Htetrazolium) assay [38] and was found to not be toxic under the conditions used in the infection assays.

Inhibition of recombinant E2 binding to native CD81

Two different assays were performed to test for the inhibition of E2 binding to CD81 by 281816. In a cell binding assay, the human B cell line Raji (ATCC, Manassas, VA), which expresses high levels of CD81 on its surface [39,40], was used to determine if ligand 281816 inhibits the binding of HCV-E2 protein to native human CD81. Purified HCV-E2 protein (4 µg) was pre-incubated with 1,5,15, 50, 100 or 400 µM of the ligand 281816 for 25 min at RT. After pre-incubation the E2-ligand complex was added to the cells and incubated for 25 min. The complexes were washed from the cells and 0.5 µg of mouse anti E2 antibody (clone H53) was added followed by a FITC-conjugated anti-mouse antibody (Southern Biotechnology, Birmingham, AL). The cells were washed, fixed with 3% paraformaldehyde, and analyzed by flow cytometry (BD FACSCalibur, software: Cell Quest Pro). The mean fluorescence intensity (MFI) was calculated using Flowjo software (TreeStar, www.flowjo.com).

The second test used an ELISA assay to determine if E2 binding to a recombinant CD81 protein is inhibited by the presence of ligand 281816. In this assay, a 96 well plate was coated with GST-tagged human CD81-LEL (5 µg/ml) overnight as previously described [10], then washed with PBS, 0.5% Triton X-100 and blocked with 2% milk in PBS for 1 hr. HCV E2 protein (5 µg/ml) was pre-incubated with different concentrations of 281816 for 30 min before adding to the plate, then HCV-E2 protein (with or without the ligand) was added to the GST-tagged human CD81-LEL coated plate and incubated for 1 hr at room temperature to allow the protein to bind. To detect HCV-E2 binding, a primary mouse anti-E2 antibody (H53 clone, 5 µg/ml) was added and incubated for 1 hr followed by a secondary goat anti-mouse-horse radish peroxidase (HRP) antibody (Southern Biotechnology Associates, Birmingham, AL) diluted 1:5000. Substrate was added (citrate buffer pH 4.0, 3.5 µl hydrogen peroxide and 100 µl 2,2'-azino-bis(3-ethylbenzothiazoline-6-sulphonic acid)) and the absorbance was measured at 405 nm.

Inhibition of anti-CD81 5A6 antibody binding to CD81-LEL

To determine if ligand 281816 binds to the E2 binding site on CD81, a competition binding experiment was run using the ligand and an anti-CD81 antibody (5A6 clone) that has been shown previously to block E2 binding [57,58]. In this assay, a 96 well plate was coated with GST-tagged human CD81-LEL overnight and then blocked with 2% milk in PBS for 1 hr at RT. The plate was then incubated for 40 min with 281816 (1 µM) or PBS as a control. The indicated concentrations of mouse anti-human CD81 antibody (5A6) were added to the plate and incubated for an additional 1 hr, followed by anti-mouse IgG-HRP. The absorbance was then measured at 405 nm.

HCVcc cell-to-cell transmission assay

Cell-to-cell transmission was measured as described previously [41,42]. Briefly, Huh-7 cells were seeded on coverslips and infected at low multiplicity of infection with HCVcc for 2 hr at 37°C. After washing, cells were cultured in medium containing neutralizing anti-E2 antibody (3/11; 50 µg/ml) to block cell-free transmission and 281816 at the indicated concentrations. Cells cultured in the presence of DMSO or Epigallocatechin-3-gallate (EGCG, 50 µM) [42] were used as negative and positive controls of inhibition, respectively. Three days post-infection, cells were

fixed with formalin solution (formaldehyde 4%, Sigma, St Louis, MO) and stained by indirect immunofluorescence using the anti-E1 monoclonal antibody A4 and Alexa555-conjugated anti-mouse immunoglobulins. Cell-to-cell transmission was quantified by counting the number of infected cells per focus. As a control, cells were infected with HCVcc pre-incubated with 50 µg/ml 3/11 antibody to confirm cell-free transmission is blocked by the antibody as reported previously [41].

Kinetics of entry

Cells treated with 281816 at 10 µM or with DMSO were infected with HCVcc for 1 hr at 4°C (attachment/binding period). Virus was removed, cells were washed with medium and incubated again for 1 hr at 4°C (post-attachment/binding period). Cells were then washed and incubated for 1 hr at 37°C (endocytosis/fusion period). Lastly, cells were washed and incubated in complete culture medium for 21 hr. Infection levels were monitored by measuring luciferase activities. To confirm 281816 was not toxic to the cells, an MTS assay [38] was also performed after incubating the cells with 10 µM 281816 for the same lengths of time (1 hr, 2 hr or 3 hr) the ligand was exposed to the cells in the entry experiments.

Antibodies

Mouse anti-E1 A4 [43], anti-E2 H53 [44] and rat anti-E2 3/11 [45] were produced in vitro using a MiniPerm apparatus (Heraeus). FITC-conjugated and Alexa555-conjugated anti-mouse immunoglobulins were obtained from Southern Biotechnology (Birmingham, AL) and Jackson Immunoresearch (West Grove, PA), respectively.

Results

Structural model of E2

In order to maximize the likelihood that these experiments would lead to the discovery of small molecules that bind to E2 and block E2's binding to CD81, we created a homology model of the core of the E2 protein to use as our docking target. This model was created using the HCV genotype 1a protein sequence NP_751921.1, which corresponds to isolate H77, and the crystal structure of E2c as the primary template (PDB entry: 4MWF). Using a model, rather than the E2c crystal structure, was important because the reported crystal structure of E2c has three large gaps in which atom coordinates for 57 amino acids, or one quarter of the E2c structure, is missing. The coordinates listed in PDB chains 4mwf_C and 4mwf_D provide structural information for only 169 and 171 residues respectively out of the 363 amino acids present in the full-length E2 protein. Within each of the deposited PDB chains, three stretches of amino acid sequence (large loop P453-P491 containing 39 amino acids, T542-G547 or V574-N577, and F586-R596) are missing from the structure (Figure 1A). Docking to structures lacking such a large proportion of their amino acids can be problematic because the missing peptide segments are usually located on the protein's surface, and the underlying amino acid residues packed in the interior of the protein are exposed and incorrectly presented as the surface during the docking. Unfortunately, similar regions are also not present in the crystal structure of the genotype 2a HCV E2c protein (PDB chain: 4nx3_D) reported by Kahn et al. [19] which provides atom coordinates for only 119 amino acids. Structural superposition of 4mwf_C and 4nx3_D (Figure 1B) shows strong conformational similarities between the experimentally solved structures of the E2 proteins with a root mean square deviation of 1.07 Angstroms measured on 98 residues for which distances

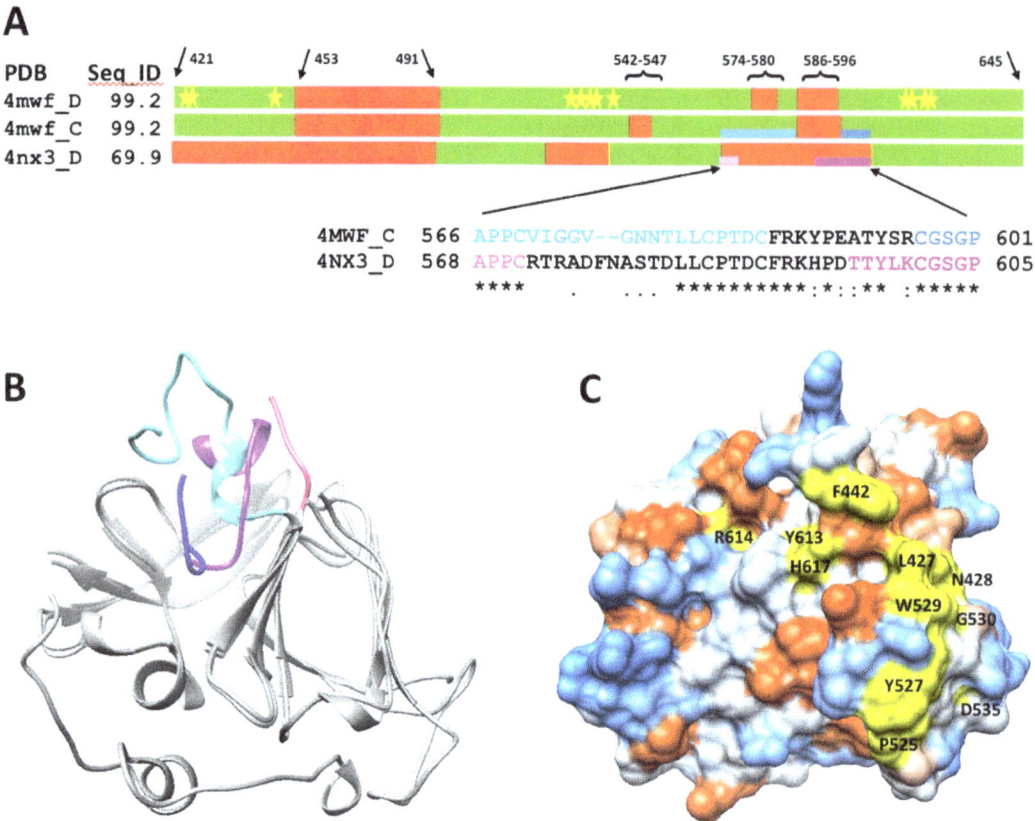

Figure 1. Comparison of structural templates used for modeling the HCV E2c protein. (A) Bar representation of E2 sequence showing the structural similarities between crystal structures 4MWF chains C and D (E2c structure, genotype 1a), and 4NX3 chain D (genotype 2a). Regions reported in the coordinates span amino acid residues from H421 to N645. The percent sequence identities between amino acid sequences taken from coordinates and corresponding sequence fragments from HCV E2 protein of genotype 1a are shown in the column Seq_ID. In green are colored regions where structural deviations are below 3 Ångstroms measured as Cα-Cα distances between corresponding residues from the superimposed structures. In red are regions where structural data is missing or deviations are greater than 3 Angstroms. The locations of amino acid residues that have been reported to be important for E2 binding to CD81 are marked with yellow stars. (B) Structural superposition of 4mwf_C and 4nx3_D shows strong conformational similarities between experimentally solved structures of E2 proteins for which the level of sequence identity is 69%. In blue and purple are colored structural fragments where two structures 4mwf_C (566–601; Blue: light-dark) and 4nx3_D (568–605; Purple: light-dark) significantly differ. (C) Surface presentation of the 4mwf_D structure showing the amino acid residues identified to be important for E2 binding to CD81 (yellow). The other amino acid residues are color coded with the most hydrophilic residues being colored blue, the most hydrophobic residues colored red orange, and intermediate residues colored white.

between corresponding Cα atoms are under 3 Angstroms. The most significant structural deviations are observed in the region 566–601 (numbering from 4mwf_C) which corresponds to the region that also exhibits the greatest variation in sequence (see sequence alignment in Figure 1A).

Exhaustive structure similarity searches of 90 residue structural fragments of E2 conducted using the entire PDB database (255,302 PDB chains) revealed that no additional structural homologs could be found at the level of calculated structure similarities by LGA score [46] higher than LGA_S = 45%, suggesting that the HCV E2 protein represents a novel fold in the current PDB. Thus, the modeling of the structure of the insertions needed to fill in missing regions in the experimentally solved structures and to complete the model was a difficult task, and it was completed with a very low degree of confidence. By applying a combination of structural modeling and analysis methods to the E2 crystal structure (see Materials and Methods section), we were able to construct a model (Figure 2) that contains the 57 amino acids that are missing in the E2c structure, including an amino acid known to be critical for E1 binding (W487), key

amino acids known to participate in CD81 binding (Figure 1C), as well as the exact sequence for the HCV genotype 1a E2 protein. Three regions of the protein that have been identified by others to be critical for E2 binding to CD81 [47–49] are contained in the model in their entirety. Currently, however, only three of the twenty-one Region 1 amino acids (H421–N423) are present in the model. A comparison of our model to the two E2c structures (see bar plots in Figure 1A and superposition of the E2c structure and the model in Figure 2) shows the main core regions are, as one would expect, very similar. The differences that are observed in the core region are small and appear to reflect only minor local deviations between experimentally solved structural templates. The large region that does differ corresponds to the missing peptide segments.

Ligands predicted to bind to CD81 binding sites on E2

Five ligand-binding sites on the HCV E2 homology model (Figure 3) were identified by docking the National Cancer Institute's Diversity Set III library of ligands to the E2 model. Each of these sites is associated with or positioned next to one or

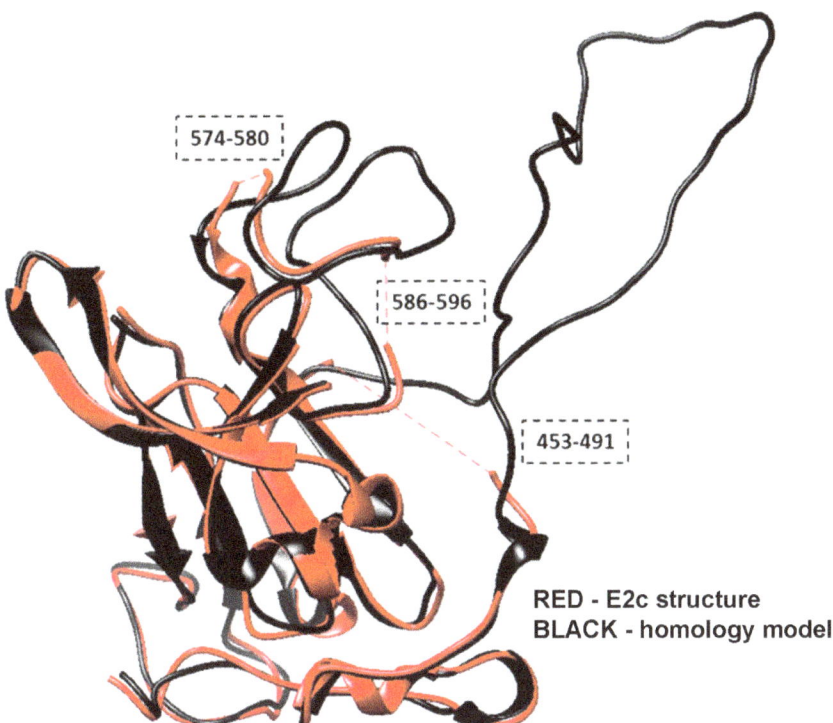

Figure 2. Comparison of the crystal structure of E2c with the homology model. Structural superposition between E2c crystal structure from the PDB chain 4mwf_D (red) and the homology model (black) is illustrated using a ribbon representation. The crystal structure and homology model overlap in most of the regions, except the fragments where coordinates in the experimental structure are missing (red dashed lines).

more of the amino acid or peptide sequences that have been identified by others to either participate in E2 binding to CD81, E2 binding to E1, or to be important for HCV infectivity. While the accuracy of the structure of the modeled segments missing from E2c may be low, the docking and visual inspection processes focused on the regions of the target that were based on the crystal structure. The majority of the amino acids that make up or surround each of the cavities used for ligand docking and the neighboring amino acids that play a role in E2 binding to CD81 are all located in the core region of E2. The structure of this region of the model is known with high confidence, as it is essentially identical to the two recent crystal structures of the E2 protein core determined by two different groups [18,19]. The locations of the grid boxes were also defined in such a manner that the amino acids in the modeled segments missing from the crystal structure of E2c would only be marginally considered during the docking. Only those residues in close contact with the core E2c structure were included in the boxes. In this way, the regions of the homology model with the least well-defined structures had a minimal impact on the docking results.

The first sequence of importance is the peptide segment Q412–N423 that was identified to bind to the broadly neutralizing antibody AP33 [20,50]. Alanine mutagenesis studies have shown all of the amino acids in this region appear to be important for HCV infectivity [48]. The model used in this study currently contains only three of the amino acids that correspond to this segment, H421, I422 and N423. Sequence 2 spans the second hyper-variable domain of E2, extending from amino acid Y474 to R492 [13,47–49]. The majority of amino acids in this sequence have been shown to have no effect on E2 binding to CD81 when mutated [51], but antibodies binding to this region of the protein do inhibit HCV infectivity [49] and CD81 binding [50]. One

amino acid located within sequence 2, W487, does however appear to be critical for E2 binding to E1. This amino acid is the first residue in one of the WHY motifs that have been reported to play a role in E1:E2 dimerization [47]. The third sequence spans amino acids S522–G551 [20,47–49] and the fourth sequence of importance is comprised of amino acids P612–P619 [49]. Mutations of residues Y527, W529, D535, Y613, R614, W616, H617 and Y618 in these two regions have all been shown to eliminate E2 binding to CD81 [47,49]. Mutating all but three of these amino acids (D535, R614 and W616) appears to eliminate specific interactions with CD81. W616 is the first amino acid in another WHY motif that is located in a region (G600–C620) that has been shown to be involved in fusion [52]. Alanine mutagenesis of D535, R614 and W616 was found to disrupt the structure of the AR3A epitope and indirectly impact CD81 binding [49].

These five binding sites were used to guide to our selection of the top virtual screening hits to be tested experimentally for binding to recombinant E2 protein. All five sites are cavities in the protein surface that would be expected to be accessible to ligands because they contain or are surrounded by amino acid residues known to participate in E2 binding to CD81 or they are located within the epitopes of antibodies that inhibit HCV infectivity or block CD81 binding. While there is still some debate regarding the importance of the entire regions bound by neutralizing antibodies, amino acid mutagenesis studies have provided a great deal of insight into those amino acids located within the epitopes that participate in E2 binding to CD81. Based on this information, we have used the set of amino acids W420–I422, S424, G523, Y527, W529, G530, D535, P612–R614 and W616–P619, whose mutation has been shown to eliminate E2 binding to CD81, to identify locations within these sites (Figure 3) where ligand binding would be expected to disrupt E2's ability to bind to CD81.

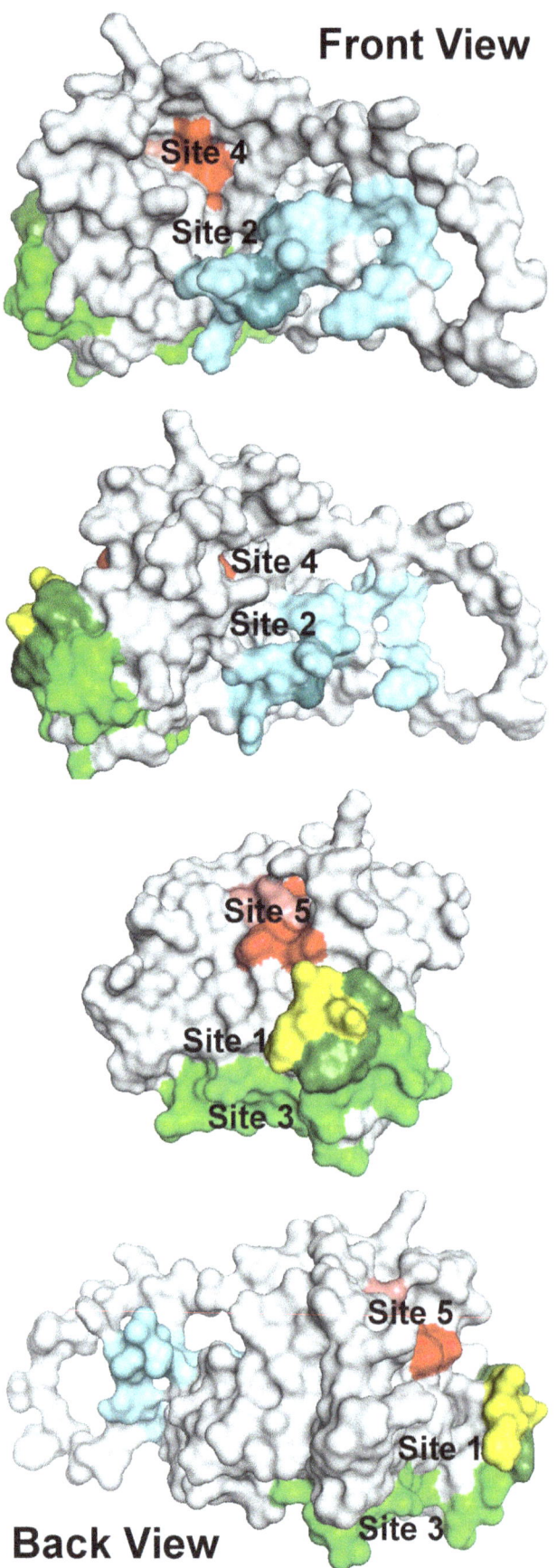

Front View

Site 4
Site 2

Site 4
Site 2

Site 5
Site 1
Site 3

Site 5
Site 1
Site 3

Back View

Figure 3. Location of ligand-binding sites on the E2 homology model used to select ligands for testing. Each of these sites either covers or is located immediately adjacent to amino acids or peptide segments of the E2 protein known to be important for HCV infectivity. H421–N423 (yellow): each amino acid in this region is important for infectivity. Antibodies binding to amino acids Y474–R492 (light cyan) have been shown to prevent infectivity, but this region of the protein has no effect on E2 binding CD81. W487 (dark cyan) is a key amino acid that is involved in E2 binding to E1. S522–G551 (light green) and Y527 and W529 (dark green) are critical for E2 binding to CD81. Site 4: P612, Y613, and H617–P619 (red) are critical for E2 binding to CD81; mutations to R614–W616 (pink) disrupt the structure of the region. The four views show the structure as it is rotated counterclockwise from left to right. Movie S1 shows the rotating structure.

Thirty-four of the highest scoring ligands were selected from the docking run for experimental analysis. Docked conformations of each of the ligands were predicted to bind to one or more of these five binding sites. The best ligands were considered to be those that exhibited the lowest free energy of binding and were predicted to interact with or bind nearby one or more of the E2 amino acids within the sites that were reported to be critical for E2 binding to CD81. The free energy of binding predicted for the best bound ligand conformations, shown in Table 2, ranged from −6.2 to −8.7 kcal/mol. Additional criteria used to select among the group of ligands predicted to bind include the number of contact points/ interactions (such as hydrogen bonds, salt bridges, van der Waals interactions) with amino acids in the model (the larger number of contacts or interactions the better) and the chemical structure of the ligands (preference is given to those that contain a free amino or carboxyl group that is exposed to solvent). Ligands with free amino or carboxyl groups can easily be linked to other ligands to create higher affinity or more selective second-generation inhibitors. Compounds that have been reported previously to be highly toxic were excluded.

Experimental confirmation of ligand binding to HCV E2

Each of the 34 ligands was tested experimentally using surface plasmon resonance (SPR) detection (on a Biacore T100 instrument) to determine if it would bind to recombinant HCV E2 protein and to obtain an assessment (relative to the other ligands) of how well it binds. Twenty-three of the molecules provided a positive change in response units (RUs) indicating they bound to the E2 protein immobilized on the chip (Table 3). The measured responses for the ligands that bound varied from 54 to 276 RUs. Data was also obtained on the rate of ligand dissociation by measuring the amount of ligand remaining bound at two time points, dissociation 1 (10 seconds) and dissociation 2 (50 seconds), during the rinsing of the chip with buffer (Figure 4). The majority of the ligands dissociated quickly, as one might expect for small molecules that bind to the surface of a protein. A few, such as ligands 121861, 4429, 158413, 81462, and 57103, exhibited slower off rates when compared to others.

Inhibition of HCV entry

The 23 compounds that were observed to bind to recombinant E2 protein were then tested to determine if they would block HCV infection of Huh-7 cells. Pseudotyped retroviral particles harboring the envelope protein of an endogenous feline retrovirus (RD114pp) were first used to determine the specificity and the safety of molecules. We excluded from a further characterization the molecules for which the half maximal inhibitory concentration (IC50) against RD114pp was lower than 10 μM or the molecules that significantly increased RD114pp infection (Table 4). The remaining ligands were next tested against pseudotyped retroviral

Table 2. Ligands predicted to bind to the HCV E2 protein by blind docking of the NCI Diversity Set III small molecule library to the HCV E2 structural model and their predicted free energies of binding.

Ligand ID (NSC#)	Free Energy of Binding (kcal/mol)	Ligand ID	Free Energy of Binding (kcal/mol)
670283	−7.69	211490	−8.7
86467	−7.47	113486	−6.26
639174	−7.81	144694	−7.27
81462	−6.81	4429	−7.3
403379	−7.58	133071	−7.5
213700	−7.89	163910	−7.4
359472	−7.91	54709	−7.3
146554	−7.67	135618	−8.7
204232	−8.54	281254	−6.5
281816	−8.64	319990	−7.4
308835	−8.4	369070	−6.3
60785	−7.48	59620	−7.3
84100	−6.99	38968	−3.9
158413	−7.9	171303	−5.8
57103	−6.36	228155	−8.7
121861	−8.16	13316	−6.8
3076	−7.71	117268	−7.6

particles harboring genotype-2a HCV envelope proteins (HCVpp 2a), cell culture produced HCV particles (HCVcc) or RD114pp. As a positive control, an anti-CD81 antibody (JS-81) was included in the assays. One compound, 281816, showed an inhibitory effect on both HCVpp and HCVcc infection with IC50's of 1.02 μM and 3.95 μM, respectively (Table 4 and Figure 5A), indicating that this molecule inhibits the entry step of the HCV lifecycle, probably through a specific effect on the virus's interaction with CD81. Huh-7 cell toxicity was not observed over the range of ligand 281816 concentrations tested in these assays.

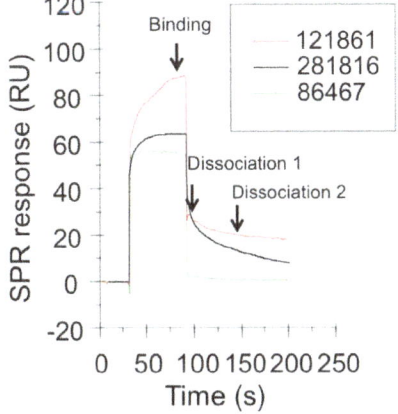

Figure 4. Surface plasmon resonance sensorgrams of ligands binding to recombinant E2 protein (Biacore T100). This figure shows sensorgrams (binding and dissociation plots) for three of the ligands that bound to the recombinant E2 protein immobilized on a CM5 chip, 281816 (black), 86467 (green) and 121861 (red), and the three reference points that are used to measure the binding and dissociation (dissociation 1 and dissociation 2) of the compound expressed in response units (RU).

To determine if 281816 would inhibit HCV genotypes other than 2a, a series of infection assays was performed with HCVpp bearing envelope proteins from a number of different HCV genotypes. Interestingly, 281816 was found to be equally effective in inhibiting Huh-7 infection by all the HCV genotypes tested (1a, 1b, 2a, 2b, 4a and 6a, Figure 5B). The IC50 values ranged from 2.2 μM to 4.6 μM (Table 5).

To confirm that 281816 inhibits HCV entry with no further effect on post-entry steps, 281816 (10 μM) was added at different time points (Figure 6A) before (−2 to 0 hr, b), during (0 to 2 hr, c), or after (2 to 24 hr, d) inoculation of Huh-7 cells with HCVcc, as previously described [53]. Cells treated with dimethylsulfoxide (DMSO) and cells treated continuously (−2 to 24 hr, a) with 281816 were used as controls. The results clearly showed that 281816 significantly inhibits HCVcc infection when present during virus infection (Figure 6A, c). The decrease in HCVcc infection that was observed in condition b is likely to be due either to some 281816 remaining bound to the cell after the washing step or its entering into the cells and acting on the entry step (Figure 6A, b). Similarly, a slight decrease was also observed in condition d (Figure 6A), which is likely related to 281816 acting on the entry of the remaining particles (those entering after 2 hr). Together, these results confirm that 281816 inhibits the entry step of HCV lifecycle.

After attachment to the cell surface and binding to entry factors, HCV virions are internalized by clathrin-mediated endocytosis [54,55]. Following internalization, HCV is transported to early endosomes along actin stress fibers, where fusion seems to take place [55,56]. To determine which step in HCV entry is impaired by 281816, we administered the ligand at different intervals during the early phase of infection. Virus attachment and binding were performed at 4°C (Figure 6B, Steps 1 and 2), Then, cells were shifted to 37°C to allow endocytosis and fusion (Figure 6B, Step 3). Cells treated with JS-81 were used as controls. The addition of 281816 during step 2 and step 3 led to the strongest inhibition of

Table 3. Magnitude of surface plasmon resonance binding response obtained for the 23 ligands that were identified to bind to recombinant E2 protein immobilized on a CM5 sensor chip.

Ligand ID (NSC#)	Binding (RU)	Dissociation 1 (RU)	Dissociation 2 (RU)
670283	54.3	4	1.4
86467	54.9	1.9	0.8
639174	55.4	2.3	0.6
81462	57.2	9.2	6.5
403379	58	2.8	1.1
213700	62	3.1	0.8
359472	62	2.5	0.8
146554	63.4	3.1	0.8
204232	63.4	2.5	0.4
281816	64.5	3.7	0.9
308835	64.8	7.1	5.2
60785	70.4	2.8	0.6
84100	71.2	4.2	2.2
158413	71.2	10.3	8.5
57103	81.6	11.4	2.5
121861	88.4	26.1	20.4
117268	88.5	4.1	1.2
3076	92.2	3.2	1.6
211490	102.9	6.1	2.1
113486	104.7	7	2.6
144694	118.8	6	2.3
4429	155.3	28.9	14.2
133071	276.3	1.8	−2

The rate of ligand dissociation is assessed by measuring the response units at two time points (10 sec and 50 sec) after the chip with bound ligand is rinsed with buffer.

HCV infection, as strong as the one observed when 281816 was present during all three steps. We also observed a significant inhibition of HCV infection when 281816 was added during the early attachment/binding steps (Figure 6B, Step 1). An MTS assay performed with 10 μM 281816 for each length of time the cells were treated with 281816 (1 hr, 2 hr, and 3 hr) showed the compound was not cytotoxic to the cells under the conditions used in the assay (Figure 7). Together, these results indicate that 281816 inhibits HCV infection by acting on more than the first (attachment/binding) step of viral entry. These data suggest the ligand also affects interactions during HCV internalization and fusion.

Blocking of E2 binding to CD81

Ligand 281816 was originally selected for testing based on the prediction by docking that it would bind to a site on the HCV E2 protein where CD81 binds. The infection assay conducted with Huh-7 cells demonstrated 281816 is effective in inhibiting the entry step in the HCV life cycle. To confirm that the binding of 281816 to E2 inhibits the HCV E2-CD81 interaction, flow cytometry was used to monitor the binding of a recombinant form of the E2 protein to native CD81 overexpressed on Raji cells as a function of 281816 concentration. The results in Figure 8 show binding of the E2 protein to Raji cells is inhibited by 281816 in a dose dependent manner. Using a second technique (an ELISA-based assay), we observed a similar dose-dependent effect of 281816 on the inhibition of the E2 protein binding to recombinant

CD81-LEL immobilized on micro titer plates (Figure 9). While an IC50 for 281816 blocking the binding of E2 to CD81 could not be determined from the flow cytometry data, the ELISA results indicate the IC50 is in the range of 0.2–0.5 μM.

A blind docking experiment with 281816 has also suggested the ligand may bind to several sites on CD81, including one that is located within the region bound by E2. 281816 binding to CD81 or other cellular proteins could explain the 281816 retention observed in washed cells in the HCV entry experiments. Such binding would likely be of little consequence, unless the ligand were to be bound within the E2 binding site on CD81 and were to block the E2:CD81 interaction by targeting both proteins.

To determine if ligand 281816 also binds to CD81, surface plasmon resonance was used to test for 281816 binding to recombinant CD81-LEL protein immobilized on a chip. As shown in Figure 10, 281816 does bind to CD81-LEL. A kinetic analysis of this binding has shown that the ligand binds to CD81-LEL (Kd = 57 μM) almost as well as it binds to E2 (Kd = 41 μM). However, a competition ELISA experiment that used the anti-CD81 antibody, 5A6, that blocks E2 binding to CD81 (Figure 11) revealed that 281816 does not bind to the E2 binding site on CD81. In this experiment, CD81-LEL was immobilized on a micro titer plate and the binding of the 5A6 antibody was monitored in the presence of 1 μM 281816 as a function of antibody dilution. The antibody 5A6 has been shown previously to bind to the same site on CD81 recognized by E2 [57,58] with an affinity (75 nM) [59] about 1/10th that of E2 (4–10 nM) [60,61].

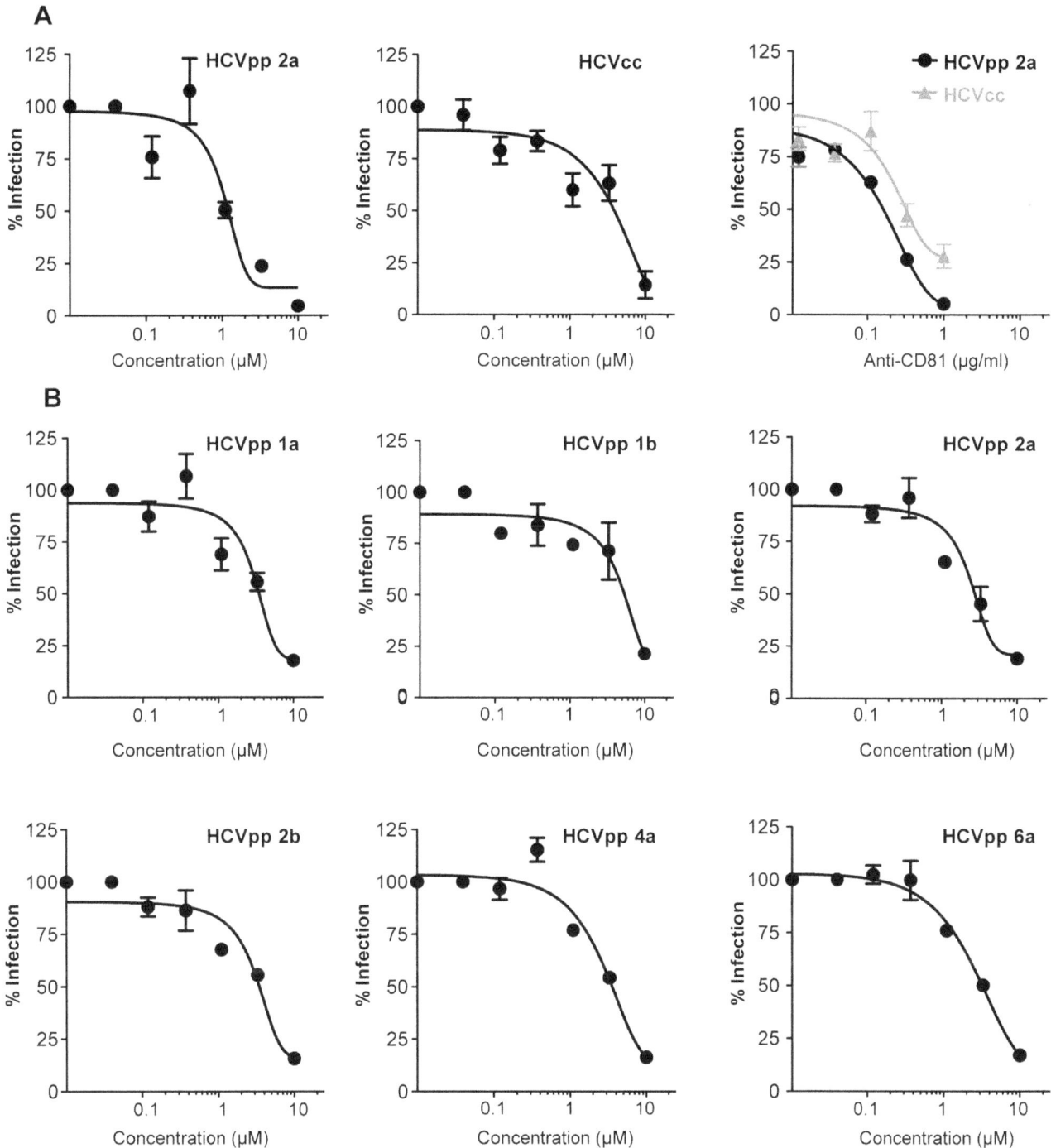

Figure 5. 281816 inhibits HCV entry in a genotype-independent manner. (A) Huh-7 cells in 96-well plates were pre-treated with 281816 (left and middle panels) or anti-CD81 antibody (right panel) at the indicated concentrations and then infected with HCVpp 2a or HCVcc. (B) Huh-7 cells in 24-well plates were pre-treated with 281816 at the indicated concentrations and infected with HCVpp expressing envelope proteins from the indicated genotype. After 30 hr of infection, cells were lysed and luciferase activities quantified. HCVpp infections were normalized to RD114pp infections.

Working at the same concentration of 281816 (1 μM) that provided the best inhibition of E2 binding to CD81-LEL, 281816 did not inhibit the 5A6 antibody binding to CD81 even when the antibody concentration was reduced to 1/2000th the concentration of the ligand (Figure 11). The amount of 5A6 antibody bound to CD81 remained the same in the presence and absence of the ligand, demonstrating that 281816 does not bind sufficiently well to the E2 binding site on CD81 to block 5A6 binding.

Table 4. The IC50 values obtained for the 23 ligands screened for their ability to inhibit HCVcc, HCVpp and RD114pp infection of Huh-7 cells.

Ligand ID (NSC#)	IC50 (μM)		
	RD114pp	HCVpp	HCVcc
670283	3	ND	ND
86467	>10	>10	>10
639174	0.03	ND	ND
81462	>10	>10	>10
403379	>10	>10	>10
213700	>10	>10	>10
359472	>10	>10	>10
146554	>10	ND	ND
204232	>10	>10	>10
281816	**>10**	**1.02**	**3.95**
308835	>10	>10	>10
60785	3.5	ND	ND
84100	>10	>10	>10
158413	>10	>10	>10
57103	0.3	ND	ND
121861	>10	>10	>10
117268	0.1	>10	>10
3076	0.25	ND	ND
211490	0.5	ND	ND
113486	>10	>10	>10
144694	>10	>10	>10
4429	>10	>10	>10
133071	0.10	ND	ND
Anti-CD81	>10*	0.17*	0.36*

ND refers to molecules that were not assayed because the molecule was not specific for HCV (it inhibited RD114pp infection). *IC50 values for anti-CD81 antibody are in μg/ml.

281816 abrogates HCV cell-to-cell transmission

In addition to cell-free infection, HCV can also be transmitted to neighboring cells via cell-to-cell contact by a mechanism that is not completely understood [41,42,62]. Indeed, HCV is transmitted in the presence of monoclonal antibodies, such as the anti-E2 antibody 3/11, or patient-derived antibodies that are able to neutralize virus-free infectivity [42,62]. Since cell-to-cell transmission has been suggested to be a major route of transmission for HCV [41], we next analyzed the effect of 281816 on this process.

For this purpose, Huh-7 cells were infected at low multiplicity of infection with HCVcc for 2 hr and then cultured with neutralizing anti-E2 antibody (3/11), which blocks infection by free particles as shown in Figure 12 [41], and in the presence of 281816 (1 μM and 10 μM). Cells cultured in the presence of 3/11 and solvent (DMSO) or Epigallocatechin-3-gallate (EGCG, 50 μM) [42] were used as negative and positive controls of inhibition, respectively. Three days post-infection, cells were fixed and foci of infected cells were visualized by immunofluorescence. Cell-to-cell transmission

Table 5. Genotype independent inhibition of HCVpp infection of Huh-7 cells by ligand 281816.

Subtypes	IC50 (μM)
HCVpp 1a	2.95
HCVpp 1b	4.66
HCVpp 2a	2.22
HCVpp 2b	2.93
HCVpp 4a	3.44
HCVpp 6a	3.30

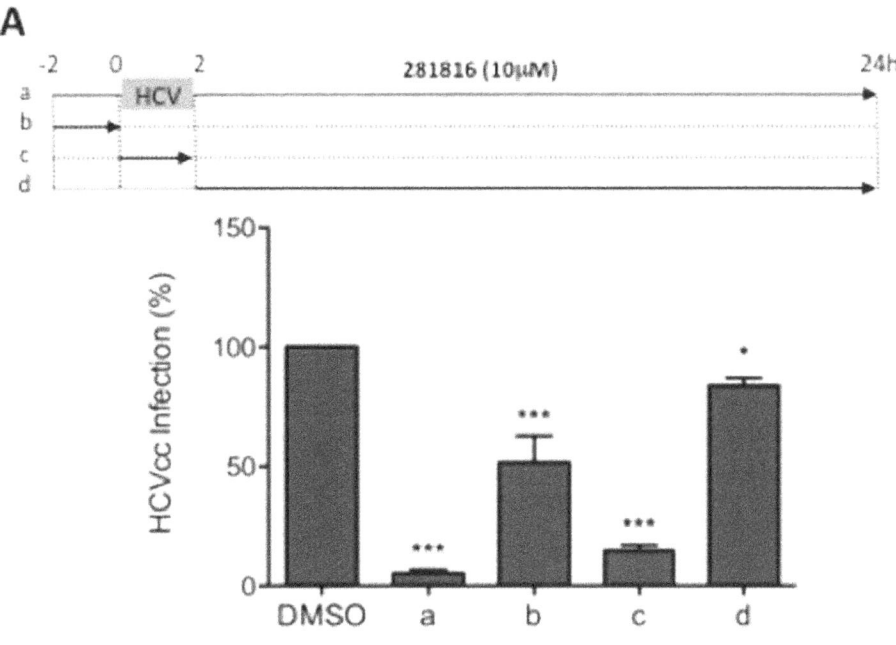

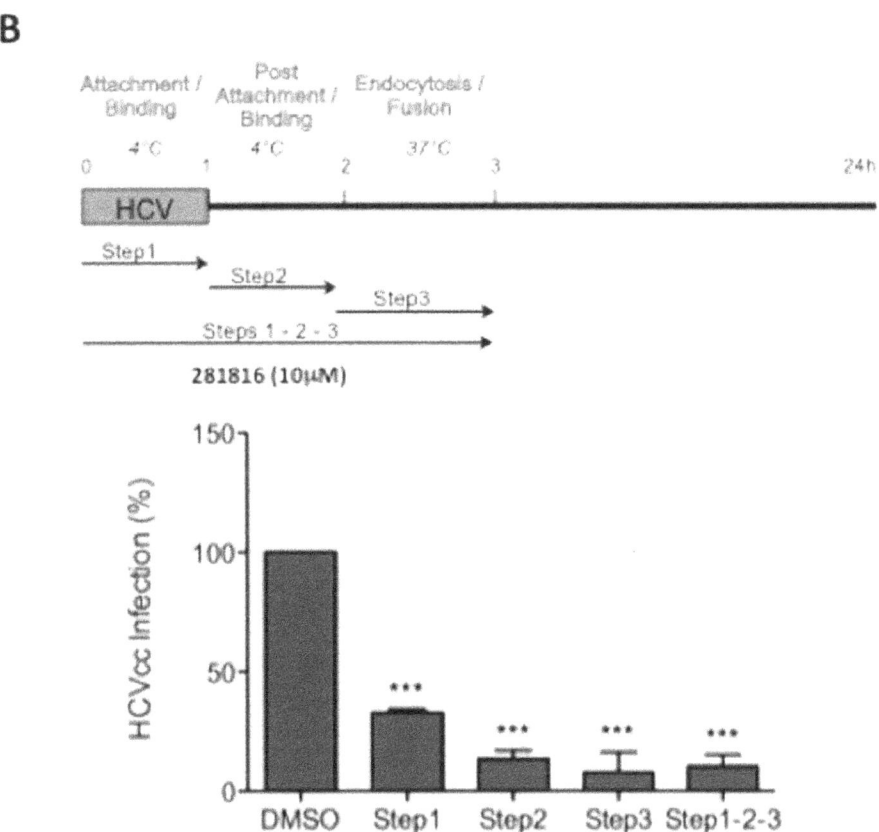

Figure 6. 281816 inhibits HCV entry. (A) Huh-7 cells in 24-well plates were treated at different time points with 281816 at 10 µM and infected with HCVcc for 2 hr at 37°C. 281816 was added full-time during the experiment (a), 2 hr before virus inoculation (b), 2 hr during virus inoculation (c), or full-time after virus inoculation (d). (B) Huh-7 cells were infected with HCVcc for 1 hr at 4°C (Step 1: attachment/binding), then virus was removed and cells incubated again at 4°C for 1 hr (Step 2: post-attachment/binding). Finally, cells were shifted at 37°C for 1 hr (Step 3: endocytosis/fusion) and left at 37°C for 21 hr. 281816 was added at 10 µM either during the Step 1, Step 2, Step 3 or Steps 1-2-3. * and *** indicate p values below 0.05 and 0.0001, respectively.

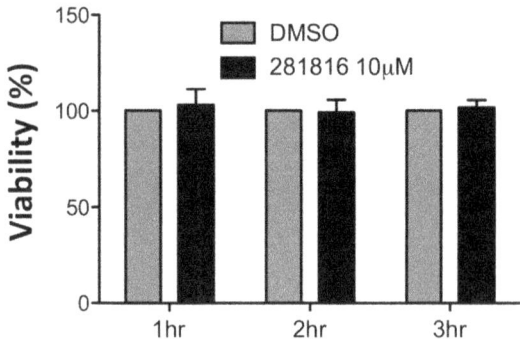

Figure 7. Viability of Huh-7 cells treated with 281816 in the HCV entry experiments. An MTS assay [38] was used to determine the viability of cells treated with 10 μM 281816 in DMSO (and DMSO alone, as a control) for 1 hr, 2 hr, or 3 hr under the same conditions used in the HCV entry experiments. 281816 is not toxic under any of the conditions used in this assay. There were no significant differences between the 281816 treated and control samples (p values <0.05).

was measured by counting the number of infected cells per focus. The results showed that 281816 led to a significant reduction of the number of infected cells per focus in a dose-dependent manner (Figure 13). Together, these results indicate that 281816 also inhibits cell-to-cell transmission of HCV.

Discussion

While it has been known for some time that the E2 envelope glycoprotein plays an important role in the life cycle of HCV, we are only now beginning to learn details about the structure of the E2 and how it functions. This has been attributed to the challenging intrinsic properties of the protein, such as the presence of multiple flexible loops, its tendency to form disulfide aggregates in solution and the high level of N-linked glycosylation, all of which make it difficult to determine it's structure. Neutralizing antibody epitope analyses and mutation studies, in contrast, have provided a great deal of information about the regions of the E2 protein and specific amino acids that participate in CD81 binding and are important for HCV infectivity [9].

The recent determination of two HCV E2 protein core crystal structures [18,19] and our use of the deposited coordinates to create a new homology model of the protein's structure containing the majority of conserved amino acids and peptide segments known to be important for viral invasion of hepatocytes has made it possible to use computational docking and structure-based drug design methods to begin developing anti-HCV drugs that target the conserved regions of E2 and block its interaction with host receptors. Our docking of a library of diverse small molecules to this homology model led to the identification of a set of ligands that were predicted to bind to sites near key amino acids known to participate in CD81 or E1 binding or to block HCV infection, and 23 of the 34 compounds were confirmed by experiment to bind to recombinant E2 protein. When these 23 ligands were tested for activity in blocking HCV infection of Huh-7 cells, only ligand 281816 was found to inhibit HCV infection using both HCVcc and HCVpp based assays. Upon analyzing the activity spectrum of HCV using HCVpp bearing envelope proteins from different HCV genotypes (1a, 1b, 2a, 2b, 4a and 6a), 281816 was found to inhibit the infection of all tested genotypes with IC50's ranging from 2.2 μM to 4.6 μM (Table 5), indicating that this small molecule inhibits HCV infection in a genotype-independent manner. Ligand 281816 was also observed to block the binding

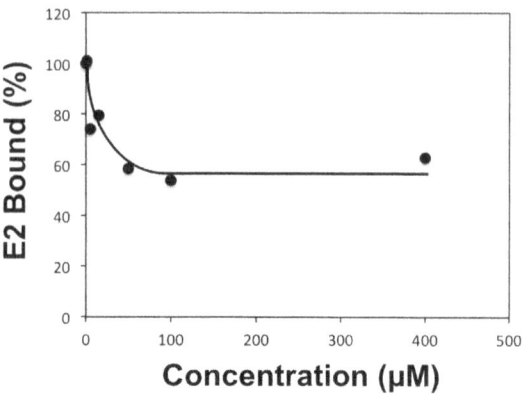

Figure 8. 281816 inhibition of HCV E2 protein binding to native CD81 on Raji cells. Flow cytometry was used to quantify recombinant HCV E2 protein binding to native CD81 over-expressed on Raji cells. Binding of the recombinant E2 protein to native CD81 on the surface of Raji cells was detected using the mouse monoclonal E2 antibody clone H53 followed by staining with a secondary FITC anti-mouse antibody. E2 binding is inhibited by 281816 in a dose-dependent manner up to 100 μM.

of HCV E2 protein to CD81-LEL protein and to Raji cells expressing CD81.

The docking experiments conducted with 281816 identified the two binding sites on E2 shown in Figure 14. One cluster of 281816 conformers bound deep inside a cavity positioned directly above Y618 and P619, two amino acids in site 4 (Figure 3) that are

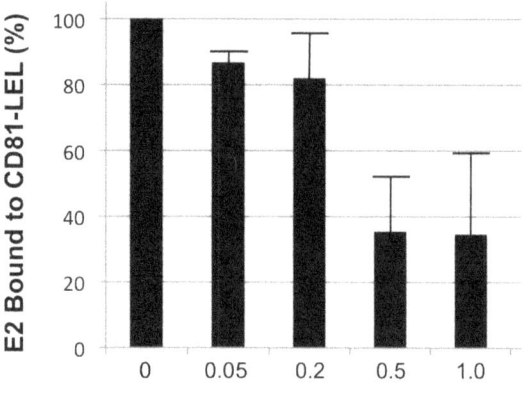

Figure 9. Ligand 281816 inhibits HCV-E2 binding to recombinant CD81-LEL. Binding of recombinant E2 protein to GST-tagged human CD81-LEL immobilized on a 96 well plate was determined using an ELISA assay. The plate was coated with GST-tagged human CD81-LEL (5 μg/ml) overnight as previously described [10], HCV E2 protein (5 μg/ml) was pre-incubated with different concentrations of 281816 for 30 min before adding to the plate, and the HCV-E2 protein (with or without the ligand) was then added to the GST-tagged human CD81-LEL coated plate and incubated for 1 hr. HCV-E2 binding was detected using a primary mouse anti-E2 antibody (H53 clone) and a secondary goat anti-mouse-HRP antibody by measuring the absorbance at 405 nm. The results, which are plotted as percent of E2 protein bound to CD81-LEL relative to E2 binding observed in the absence of the ligand (buffer control), show a dose-dependent effect of 281816 on the inhibition of the E2 protein binding to immobilized recombinant CD81-LEL. P values for the 0.05, 0.2, 0.5 and 1.0 μM 281816 samples are 0.0069, 0.0195, 0.0006 and 0.0009 respectively.

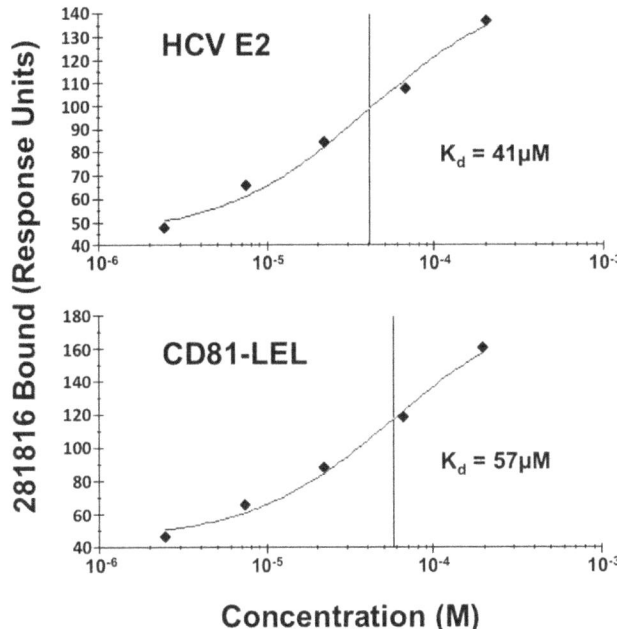

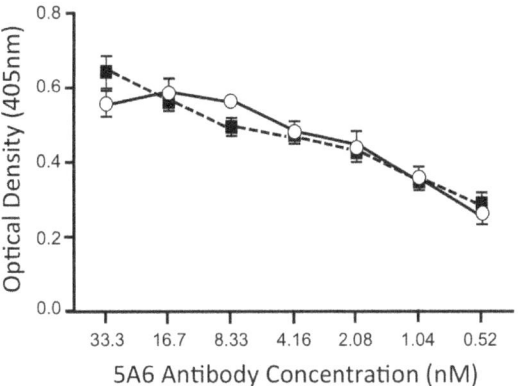

Figure 11. Ligand 281816 does not compete with binding of an anti-CD81 antibody to human CD81-LEL. Binding of anti-CD81 monoclonal antibody 5A6 to recombinant GST-tagged human CD81-LEL protein was determined by ELISA. Serial dilutions of 5A6 antibody were incubated with the CD81-LEL protein in the presence of 1 µM ligand 281816 (black squares) or PBS as a control (open circles). The amount of 5A6 antibody bound to CD81 remained the same in the presence and absence of the ligand, demonstrating that 281816 does not bind sufficiently well to the E2 binding site on CD81 to block 5A6 binding, even when the antibody concentration was reduced to 1/2000th the concentration of the ligand.

Figure 10. Single cycle kinetics of 281816 binding to recombinant HCV E2 and his-tagged CD81-LEL. Using surface plasmon resonance detection, ligand 281816 is observed to bind to both HCV E2 and CD81-LEL proteins immobilized on a CM5 chip. Analyses of the binding kinetics were used to obtain dissociation constants for 281816 binding to recombinant E2 ($K_d = 41$ µM) and recombinant his-tagged CD81-LEL ($K_d = 57$ µM) protein.

known to contribute to E2's binding to CD81 [47]. The two strongest 281816 ligand binding modes are shown bound to this site. 281816 was also predicted to bind to a shallow cavity on the opposite side of the protein. These conformers were predicted to bind to site 1 near residues V515, G517, P515 and H421–N423. H421–N423 is part of a larger segment of E2 that has been shown to bind to CD81 [19]. As expected, the ligand positioned above Y618 and P619 in the deeper cavity was predicted to bind more strongly to this region of the protein (free energy of binding of the best bound ligand = −8.64 kcal/mol) than when it was bound to the shallow cavity on the other side of the protein (free energy of binding = −6.39 kcal/mol).

A subset of the 281816 conformers in the cluster observed to bind near site 4 overlapped into site 2 and bound immediately adjacent to D481–P490, part of the epitope targeted by antibodies that block HCV infectivity and E2 binding to CD81 [50]. W487, a residue within this peptide segment whose mutation has been shown to disrupt E2:E1 dimerization [63], is also located near site 2. Other conformers in the cluster binding to site 1 also overlapped into site 5 and bound near amino acid residues P612 and Y613. These docking results illustrate one interesting and unique feature of the 281816 ligand; a number of its conformers are predicted to bind immediately above or next to both exposed faces of the P612–P619 amino acid residues that are known to participate in E2 binding to CD81 [47].

One factor that can have a significant impact on the accessibility of these sites to the binding of 281816 and other ligands is the oligomeric state of the native E2 protein. Analyses of the E2 protein and its complexes have shown that the protein exists in several different oligomeric states [43,64–73]. These include non-covalent heterodimers of E1 and E2, large disulfide cross-linked E2 complexes and aggregates, as well as monomers and disulfide

linked E2 homodimers. E1E2 non-covalent heterodimers are formed in infected cells [70–72], and it has been proposed that the two proteins remain as a complex on the virus surface, perhaps covalently linked through an intermolecular disulfide bond located in their transmembrane regions [69]. Large covalent complexes containing E2 stabilized by multiple disulfide bonds have also been observed to be associated with the surface of infective virions [71]. It has been hypothesized that the disulfide crosslinks in these large complexes may contribute to the structural stability of the virion. It has also been proposed that these large complexes may play a role in budding. Other large disulfide cross-linked aggregates of E2 have been found in recombinant E2 expression systems [74–78] and in the endoplasmic reticulum, but these aggregates do not

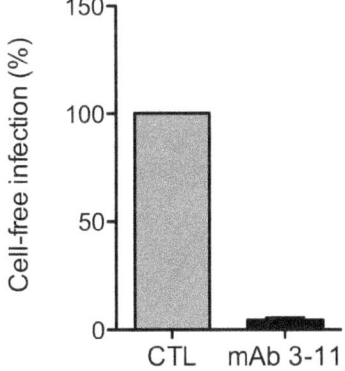

Figure 12. Cell-free infection of Huh-7 cells by HCVcc is blocked by the anti-E2 antibody 3/11. HCVcc were pre-incubated for 1 hr with neutralizing anti-E2 monoclonal antibody 3/11 at a concentration of 50 µg/ml and next inoculated to Huh-7 cells for 2 hr. Three days post-infection, cells were fixed, stained with the mouse anti-E1 antibody A4 and Alexa555 conjugated anti-mouse IgG, and the number of infected cells was counted. The results show the anti-E2 antibody 3/11 effectively blocks cell-free infection of Huh-7 cells by HCVcc.

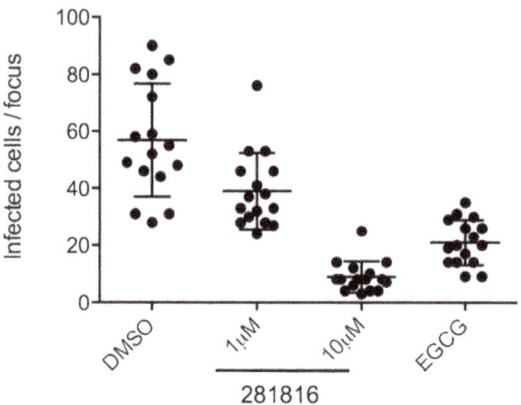

Figure 13. 281816 blocks HCV cell-to-cell transmission. Huh-7 cells were seeded on coverslips and infected with HCVcc for 2 hr at 37°C. Cells were then washed and cultured for 72 hr at 37°C in culture medium containing the 3/11 neutralizing antibody (50 μg/ml) in presence or in absence of 281816 at indicated concentrations. Cells cultured in presence of DMSO or EGCG at 50 μM were used as controls. The number of infected cells per focus was determined by A4 indirect immunofluorescence. The results show treatment with 281816 significantly reduces the number of infected cells per focus in a dose-dependent manner. Mean p values were below 0.001 for 1 μM 281816 and below 0.0001 for the 10 μM 281816 and EGCG treatment groups.

appear to represent a functional or biologically relevant oligomeric state.

E2 is known to be heavily glycosylated [69,79,80]. In the HCVcc system E2 has been reported to have 11 glycosylated sites that collectively account for nearly half the mass of the protein [69,79,80]. Glycosylation of a number of these sites has been shown to prevent the binding of neutralizing antibodies to E2 and to block E2 binding to CD81 [81], Although the ligands identified in this study are much smaller than antibodies or CD81, the presence of the glycans could also prevent small molecules from binding to the protein's surface if the glycosylated amino acid residues are located close to the ligand binding site.

While the oligomeric structure of E2 on the surface of the HCV virus is not known, the E2 present in the virus and pseudoparticles does bind to both recombinant CD81-LEL and the native CD81 receptor present on hepatocytes. Since each of the five ligand binding sites on E2 we used in our docking experiments contain or are located immediately adjacent to amino acid residues that are known to participate in E2 binding to CD81, ligands targeting these sites should have access to bind in the cavities. In addition, a number of these binding sites are located within epitopes recognized by antibodies that inhibit HCV infectivity, providing an additional confirmation of the accessibility of the sites. Only one of the 11 known glycosylated amino acid residues, E2N7, is located near a ligand docking site (Site 3). Our docking studies have identified 281816 ligand conformers that are predicted to bind to sites 1 and 4 and a few that overlap into sites 2 and 5, but none are expected to bind to Site 3. This suggests that the binding

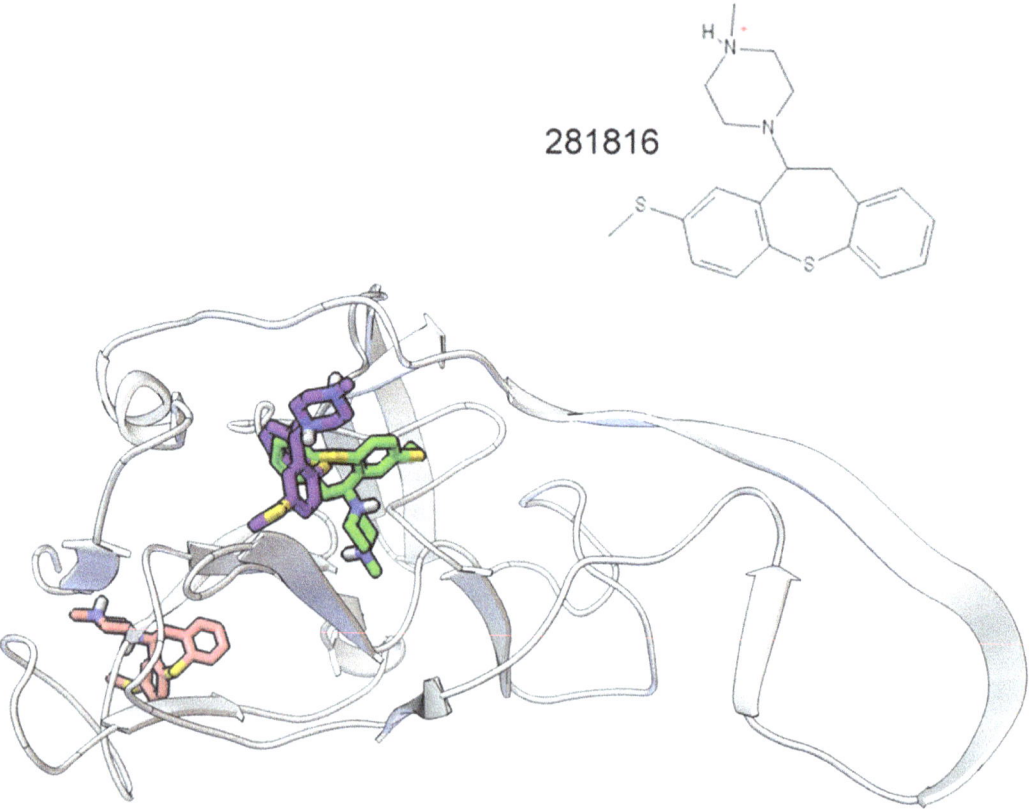

Figure 14. Relative location of 281816 binding sites 1 and 4 on HCV E2. 281816 (structure, top) is predicted to bind to two sites on the E2 protein. Two conformers of 281816 with the lowest free energy of binding are shown bound to site 4. The conformer with the lowest free energy of binding to site 1 is also shown. A video showing the surface structure of the E2 homology model with the three 281816 conformers bound that rotates 360° can be found in Movie S1.

of 281816 and its analogs should not be affected by glycosylation. The limited proteolysis and deuterium exchange experiments conducted with the E2 protein core and reported by Khan et al. [19] also indicate that each of the five ligand docking/binding sites is accessible and exposed to solvent – an important prerequisite for ligand binding.

To probe more deeply into the inhibition of the infection process by 281816, experiments were conducted to determine if the inhibition of cell-free infection by 281816 might be limited to viral entry, which step in the entry process might be affected by the compound, and what impact, if any, 281816 might have on cell-to-cell transmission of HCV. Analyses of Huh-7 cells inoculated with HCVcc before, during or after treatment with 281816 revealed the compound only blocks HCV entry and does not inhibit post-entry processes in the HCV life cycle. A kinetic analysis of the effect of 281816, coupled with a temperature block to endocytosis and fusion, was used to examine the cell-free entry steps in more detail and showed 281816 inhibits not only the initial attachment/binding step, but it also has an effect on interactions that occur later during viral internalization and fusion. Ligand 281816 was also observed to abrogate the cell-to-cell transmission of HCV. 281816 treatment of Huh-7 cells cultured in the presence of the anti-E2 neutralizing antibody 3/11 not only led to a dose dependent reduction in the number of cells forming foci, but it was found to be more effective in blocking cell-to-cell transmission that the Epigallocatechin-3-gallate [42] used as a positive control.

The observation that some 281816 remained bound to cells after washing in the HCV entry experiments suggests that 281816 may bind to other cellular proteins. This should not be surprising, since this compound has been reported to have other activities [82–87]. A blind docking experiment performed with 281816 also predicted the ligand could bind to CD81. Three potential binding sites were identified, one located within the E2 binding site on CD81 and two others in regions that would not be expected to impact E2 binding. Collectively these observations suggested the exciting possibility that 281816 might play a dual role in blocking E2 binding to CD81 by binding not only to the CD81 binding site on E2, but also by binding to the E2 binding site on CD81. In support of this idea, results obtained in an SPR binding study showed 281816 bound to CD81 almost as well as it bound to E2. However, a subsequent competition experiment conducted with 281816 and a monoclonal antibody (5A6) known to bind to the E2 binding site on CD81 revealed that the ligand did not compete with the antibody. One micromolar 281816, which effectively blocks E2 binding to CD81-LEL and inhibits viral invasion of Huh-7 cells, had no effect on antibody 5A6 binding to CD81-LEL.

In addition to identifying a promising new small molecule drug lead for treating HCV that targets the E2 glycoprotein, this study also demonstrates the utility of our new E2 homology model in the discovery of small molecules that bind to important sites on E2. By targeting sites containing amino acid residues identified by others to participate in CD81 binding and CD81-dependent processes that impact HCV infectivity, a small molecule was identified that not only blocks E2 binding to CD81 and the cell-free entry process, but it is also effective in blocking the cell-to-cell transmission of HCV – the predominant mechanism of transmission that contributes to the persistence of infections [41] but for which the precise mechanism needs to be defined. Recently, it has been shown that exosomes produced by HCV infected hepatic cells can transfer viral RNA to plasmacytoid dendritic cells [88] and might transmit infection to naïve hepatic cells [89]. Although several entry factors have been implicated in this process, the viral determinants, entry factor requirements and molecular mechanisms involved in cell-to-cell transmission route still need to be

further characterized. In particular, the role played by CD81 has remained controversial with studies reporting HCV cell-to-cell transmission as a CD81-dependent pathway [41,90,91], whereas others demonstrated a CD81-independent transmission [56,62,92]. However, a recent study has highlighted the coexistence of CD81-dependent and CD81-independent cell-to-cell transmission [93]. The inhibition of cell-to-cell transmission of HCV by 281816, which blocks E2 binding to CD81, is consistent with other reports of a CD81-dependent cell-to-cell transmission process [93–94] that can be blocked by anti-CD81 antibodies [41,94] and soluble CD81 [56], both of which also block E2 binding to CD81. While it is possible that E2 binding to CD81 may play a role in the cell-to-cell transmission of HCV, it is also possible the 281816 that binds to CD81, which does not inhibit E2 binding, may have a totally unrelated effect that impacts the interaction of CD81 with other proteins or molecular structures in the tetraspanin web [95,96] and blocks fusion related events involving CD81 that occur during the cell-to-cell transmission process.

281816, known as methiothepin or 1-methyl-4-(3-methylsulfanyl-5,6-dihydrobenzo[b] [1]benzothiepin-5-yl)piperazine, is also interesting because it has been determined previously to block dopamine [82] and serotonin [83] receptors and has been reported to inhibit a number of other biological activities, which include the binding or entry of two other unrelated viruses into cells (Lassa [84], Marburg [85]), *Plasmodium falciparum* proliferation [86], and *Mycoplasmodium tuberculosis* infections [87]. Numerous structural analogs of 281816 have been tested and shown to be effective in treating a wide variety of neurological diseases (schizophrenia [97,98], Parkinson and dementia-related psychoses [99,100]), bipolar disorders [101,102], and depression [103,104]. While we have not found experimental studies that report the membrane permeability of 281816, the logP (log of octanol/water partition coefficient) has been calculated to be 4.14, which indicates the compound is likely to exhibit good membrane permeability. Since the logP is less than 5, according to Chris Lipinski/Pfizer's Rule of 5 the compound could also be orally active, as are a number of 281816 structural analogs (octoclothepine, loxapine, amoxapine, clozapine, quetiapine, olanzapine, and amitriptyline) that have been used to treat a variety of neurological disorders.

Acknowledgments

We thank Lucie Feneant for providing the HCVpp used in this work. Biacore support and instrument use was provided by the Protein Expression Center at the California Institute of Technology, Pasadena, CA. We thank Leslie P. Michelson and Ryan Novosielski of the Rutgers University Office of Information Technology, High Performance and Research Computing, for developing and maintaining the Linux cluster at Rutgers University-NJMS and for assisting us with its use. The ligands tested in this study were provided by the National Cancer Institute through its NCI/Developmental Therapeutics Program (DTP) Open Chemical Repository (http://dtp.cancer.gov).

Author Contributions

Conceived and designed the experiments: RRA LC AZ JD JV JM JSF SF SL RB HMA. Performed the experiments: RRA AZ LS AGK FVC ALP SF. Analyzed the data: RRA LC AZ JV AGK FVC ALP SF RB. Contributed reagents/materials/analysis tools: RRA JD JV JM JSF SL HMA. Contributed to the writing of the manuscript: RRA LC AZ RB.

References

1. Anwar MI, Rahman M, Hassan MU, Iqbal M (2013) Prevalence of active hepatitis C virus infections among general public of lahore, Pakistan. Virol J 10(1): 351-422X-10-351. 10.1186/1743-422X-10-351; 10.1186/1743-422X-10-351.

2. Blackard JT, Shata MT, Shire NJ, Sherman KE (2008) Acute hepatitis C virus infection: A chronic problem. Hepatology 47(1): 321–331. 10.1002/hep.21902.

3. Zeuzem S, Berg T, Moeller B, Hinrichsen H, Mauss S, et al. (2009) Expert opinion on the treatment of patients with chronic hepatitis C. J Viral Hepat 16(2): 75–90. 10.1111/j.1365-2893.2008.01012.x.

4. Marks KM, Jacobson IM (2012) The first wave: HCV NS3 protease inhibitors telaprevir and boceprevir. Antivir Ther 17(6 Pt B): 1119–1131. 10.3851/IMP2424. 10.3851/IMP2424.

5. Colombo M, Fernandez I, Abdurakhmanov D, Ferreira PA, Strasser SI, et al. (2013) Safety and on-treatment efficacy of telaprevir: The early access programme for patients with advanced hepatitis C. Gut. 10.1136/gutjnl-2013-305667; 10.1136/gutjnl-2013-305667.

6. Asselah T (2014) Sofosbuvir for the treatment of hepatitis C virus. Expert Opin Pharmacother 15(1): 121–130. 10.1517/14656566.2014.857656; 10.1517/14656566.2014.857656.

7. Lenz O, Vijgen L, Berke JM, Cummings MD, Fevery B, et al. (2013) Virologic response and characterization of HCV genotype 2–6 in patients receiving TMC435 monotherapy (study TMC435-C202). J Hepatol 58(3): 445–451. 10.1016/j.jhep.2012.10.028; 10.1016/j.jhep.2012.10.028.

8. Pileri P, Uematsu Y, Campagnoli S, Galli G, Falugi F, et al. (1998) Binding of hepatitis C virus to CD81. Science 282: 938.

9. Fenecant L, Levy S, Cocquerel L (2014) CD81 and hepatitis C virus (HCV) infection. Viruses 6: 535–572. Doi:10.3390/v6020535.

10. Higginbottom A, Quinn ER, Kuo CC, Flint M, Wilson LH, et al. (2000) Identification of amino acid residues in CD81 critical for interaction with hepatitis C virus envelope glycoprotein E2. J Virol 74(8): 3642–3649.

11. Drummer HE, Wilson KA, Poumbourios P (2002) Identification of the hepatitis C virus E2 glycoprotein binding site on the large extracellular loop of CD81. J Virol 76(21): 11143–11147.

12. Zhang YY, Zhang BH, Ishii K, Liang TJ (2010) Novel function of CD81 in controlling hepatitis C virus replication. J Virol 84(7): 3396–3407. 10.1128/JVI.02391-09.

13. Ahlenstiel G (2013) The natural killer cell response to HCV infection. Immune Netw 13(5): 168–176. 10.4110/in.2013.13.5.168.

14. El-Awady MK, Tabll AA, El-Abd YS, Yousif H, Hegab M, et al. (2009) Conserved peptides within the E2 region of hepatitis C virus induce humoral and cellular responses in goats. Virol J 6: 66-422X-6-66. 10.1186/1743-422X-6-66; 10.1186/1743-422X-6-66.

15. Carlsen TH, Scheel TK, Ramirez S, Foung SK, Bukh J (2013) Characterization of hepatitis C virus recombinants with chimeric E1/E2 envelope proteins and identification of single amino acids in the E2 stem region important for entry. J Virol 87(3): 1385–1399. 10.1128/JVI.00684-12; 10.1128/JVI.00684-12.

16. Li YP, Kang HN, Babiuk LA, Liu Q (2006) Elicitation of strong immune responses by a DNA vaccine expressing a secreted form of hepatitis C virus envelope protein E2 in murine and porcine animal models. World J Gastroenterol 12(44): 7126–7135.

17. Ray R, Meyer K, Banerjee A, Basu A, Coates S, et al. (2010) Characterization of antibodies induced by vaccination with hepatitis C virus envelope glycoproteins. J Infect Dis 202(6): 862–866. 10.1086/655902; 10.1086/655902.

18. Kong L, Giang E, Nieusma T, Kadam RU, Cogburn KE, et al. (2013) Hepatitis C virus E2 envelope glycoprotein core structure. Science 342(6162): 1090–1094. 10.1126/science.1243876.

19. Khan AG, Whidby J, Miller MT, Scarborough H, Zatorski AV, et al. (2014) Structure of the core ectodomain of the hepatitis C virus envelope glycoprotein 2. Nature. 10.1038/nature13117.

20. Law M, Maruyama T, Lewis J, Giang E, Tarr AW, et al. (2008) Broadly neutralizing antibodies protect against hepatitis C virus quasispecies challenge. Nat Med 14(1): 25–27. 10.1038/nm1698.

21. Zemla A, Zhou CE, Slezak T, Kuczmarski T, Rama D, et al. (2005) AS2TS system for protein structure modeling and analysis. Nucleic Acids Res 33(Web Server issue): W111-5. 10.1093/nar/gki457.

22. Zemla AT, Lang DM, Kostova T, Andino R, Ecale Zhou CL (2011) StralSV: assessment of sequence variability within similar 3D structures and application to polio RNA-dependent RNA polymerase. BMC Bioinformatics 12: 226.

23. Krivov GG, Shapovalov MV, Dunbrack Jr RL (2009) Improved prediction of protein side-chain conformations with scwrl4. Proteins 77: 778.

24. Chen VB, Arendall 3rd WB, Headd JJ, Keedy DA, Immormino RM, et al. (2010) MolProbity: all-atom structure validation for macromolecular crystallography, Acta Crystallogr D Biol Crystallogr 66: 12.

25. Pettersen EF, Goddard TD, Huang CC, Couch GS, Greenblatt DM, et al. (2004) UCSF chimera-a visualization system for exploratory research and analysis, J Comput Chem 25: 1605.

26. Trott O, Olson AJ (2010) AutoDock VINA: Improving the speed and accuracy of docking with a new scoring function, efficient optimization, and multi-threading. J Comput Chem 31(2): 455–461. 10.1002/jcc.21334.

27. Chen VB, Arendall WB, Headd JJ, Keedy DA, Immormino RM et al. (2010) MolProbity: All-atom structure validation for macromolecular crystallography, Acta Cryst D66: 12.

28. Morris GM, Huey R, Lindstrom W, Sanner MF, Belew RK, et al. (2009) AutoDock4 and AutoDockTools4: Automated docking with selective receptor flexibility. J Comput Chem 30(16): 2785–2791. 10.1002/jcc.21256.

29. Irwin JJ, Sterling T, Mysinger MM, Bolstad ES, Coleman RG (2012) ZINC: A free tool to discover chemistry for biology. J Chem Inf Model 52: 1757.

30. Forli S Raccoon. Available: http://autodock.scripps.edu/resources/raccoon. Accessed 2013. Molecular Graphics Laboratory, The Scripps Research Institute, La Jolla, CA, 2010.

31. Perryman AL, Santiago DN, Forli S, Santos-Martins D, Olson AJ (2014) Virtual screening with AutoDock VINA and the common pharmacophore engine of a low diversity library of fragments and hits against the three allosteric sites of HIV integrase: participation in the SAMPL4 protein-ligand binding challenge. J Comput Aided Mol Des 28: 429-41. doi: 10.1007/s10822-014-9709-3.

32. Bartosch B, Bukh J, Meunier JC, Granier C, Engle RE, et al. (2003) In vitro assay for neutralizing antibody to hepatitis C virus: Evidence for broadly conserved neutralization epitopes. Proc Natl Acad Sci U S A 100(24): 14199–14204. 10.1073/pnas.2335981100.

33. Op De Beeck A, Voisset C, Bartosch B, Ciczora Y, Cocquerel L, et al. (2004) Characterization of functional hepatitis C virus envelope glycoproteins. J Virol 78(6): 2994–3002.

34. Sandrin V, Boson B, Salmon P, Gay W, Negre D, et al. (2002) Lentiviral vectors pseudotyped with a modified RD114 envelope glycoprotein show increased stability in sera and augmented transduction of primary lymphocytes and CD34+ cells derived from human and nonhuman primates. Blood 100(3): 823–832. 10.1182/blood-2001-11-0042.

35. Wakita T, Pietschmann T, Kato T, Date T, Miyamoto M, et al. (2005) Production of infectious hepatitis C virus in tissue culture from a cloned viral genome. Nat Med 11(7): 791–796. 10.1038/nm1268.

36. Rocha-Perugini V, Montpellier C, Delgrange D, Wychowski C, Helle F, et al. (2008) The CD81 partner EWI-2wint inhibits hepatitis C virus entry. PLOS One 3(4): e1866. 10.1371/journal.pone.0001866; 10.1371/journal.pone.0001866.

37. Delgrange D, Pillez A, Castelain S, Cocquerel L, Rouille Y, et al. (2007) Robust production of infectious viral particles in huh-7 cells by introducing mutations in hepatitis C virus structural proteins. J Gen Virol 88(Pt 9): 2495–2503. 10.1099/vir.0.82872-0.

38. Malich G, Markovic B, Winder C (1997) The sensitivity and specificity of the MTS tetrazolium assay for detecting the in vitro cytotoxicity of 20 chemicals using human cell lines, Toxicology 124: 179.

39. Ferrer M, Yunta M, Lazo PA (1998) Pattern of expression of tetraspanin antigen genes in Burkitt lymphoma cell lines. Clin Exp Immunol 113: 346–352.

40. Luo RF, Zhao S, Tibshirani R, Myklebust JH, Sanyal M, et al. (2010) CD81 protein is expressed at high levels in normal germinal center B cells and in subtypes of human lymphomas. Human Pathol 41(2): 271–280.

41. Vausselin T, Calland N, Belouzard S, Descamps V, Douam F, et al. (2013) The antimalarial ferroquine is an inhibitor of hepatitis C virus. Hepatology 58: 86–97. 10.1002/hep.26273.

42. Calland N, Albecka A, Belouzard S, Wychowski C, Duverlie G, et al. (2012) (−)-Epigallocatechin-3-gallate is a new inhibitor of hepatitis C virus entry. Hepatology 55(3): 720–729. 10.1002/hep.24803; 10.1002/hep.24803.

43. Dubuisson J, Hsu HH, Cheung RC, Greenberg HB, Russell DG, et al. (1994) Formation and intracellular localization of hepatitis C virus envelope glycoprotein complexes expressed by recombinant vaccinia and sindbis viruses. J Virol 68(10): 6147–6160.

44. Cocquerel L, Meunier, Pillez A, Wychowski C, Dubuisson J (1998) A retention signal necessary and sufficient for endoplasmic reticulum localization maps to the transmembrane domain of hepatitis C virus glycoprotein E2. J Virol 72: 2183–2191.

45. Flint M, von Hahn T, Zhang J, Farquhar M, Jones CT, et al. (2006) Diverse CD81 proteins support hepatitis C virus infection. J Virol 80(22): 11331–11342. 10.1128/JVI.00104-06.

46. Zemla A (2003) LGA: A method for finding 3D similarities in protein structures. Nucleic Acids Res 31(13): 3370–3374.

47. Owsianka AM, Timms JM, Tarr AW, Brown RJ, Hickling TP, et al. (2006) Identification of conserved residues in the E2 envelope glycoprotein of the

hepatitis C virus that are critical for CD81 binding. J Virol 80(17): 8695–8704. 10.1128/JVI.00271-06.

48. Roccasecca R, Ansuini H, Vitelli A, Meola A, Scarselli E, et al. (2003) Binding of the hepatitis C virus E2 glycoprotein to CD81 is strain specific and is modulated by a complex interplay between hypervariable regions 1 and 2, J Virology 77: 1856.

49. Rothwangl KB, Manicassamy B, Uprichard SL, Rong L (2008) Dissecting the role of putative CD81 binding regions of E2 in mediating HCV entry: Putative CD81 binding region 1 is not involved in CD81 binding. Virol J 5: 46-422X-5-46. 10.1186/1743-422X-5-46; 10.1186/1743-422X-5-46.

50. Lavillette D, Pecheur EI, Donot P, Fresquet J, Molle J, et al (2007) Characterization of fusion determinants points to the involvement of three discrete regions of both E1 and E2 glycoproteins in the membrane fusion process of hepatitis C virus, J Virology 81: 8752.

51. Tarr AW, Owsianka AM, Timms JM, McClure CP, Brown RJ, et al. (2006) Characterization of the hepatitis C virus E2 epitope defined by the broadly neutralizing monoclonal antibody AP33, Hepatology 43: 592.

52. Goueslain L, Alsaleh K, Horellou P, Roingeard P, Descamps V, et al (2010) Identification of GBF1 as a cellular factor required for hepatitis C virus RNA replication, J Virology 84: 773.

53. Blanchard E, Belouzard S, Goueslain L, Wakita T, Dubuisson J, et al. (2006) Hepatitis C virus entry depends on clathrin-mediated endocytosis. J Virol 80(14): 6964–6972. 10.1128/JVI.00024-06.

54. Meertens L, Bertaux C, Dragic T (2006) Hepatitis C virus entry requires a critical postinternalization step and delivery to early endosomes via clathrin-coated vesicles. J Virol 80(23): 11571–11578. 10.1128/JVI.01717-06.

55. Coller KE, Berger KL, Heaton NS, Cooper JD, Yoon R, et al. (2009) RNA interference and single particle tracking analysis of hepatitis C virus endocytosis. PLOS Pathog 5(12): e1000702. 10.1371/journal.ppat.1000702; 10.1371/journal.ppat.1000702.

56. Timpe JM, Stamataki Z, Jennings A, Hu K, Farquhar MJ, et al. (2008) Hepatitis C virus cell-cell transmission in hepatoma cells in the presence of neutralizing antibodies. Hepatology 47(1): 17–24. 10.1002/hep.21959.

57. Flint M, Maidens C, Loomis-Price LD, Shotton C, Dubuisson J, et al. (1999) Characterization of hepatitis C virus E2 glycoprotein interaction with a putative cellular receptor, CD81. J Virol 73 (8): 6235–6244.

58. VanCompernolle SE, Wiznycia AV, Rush JR, Dhanasekaran M, Baures PW, et al. (2003) Small molecule inhibition of hepatitis C virus E2 binding to CD81. Virol 314: 371–380.

59. Delandre C, Penabaz TR, Passarelli AL, Chapes SK, Clem RJ (2009) Mutation of juxtamembrane cysteines in the tetraspanin CD81 affects palmitoylation and alters interaction with other proteins at the cell surface. Exp Cell Res 315(11): 1953–1963. Doi:10.1016/j.yexcr.2009.03.013.

60. Takayama H, Chelikani P, Reeves PJ, Zhang S, Khorana HG (2008) High-level expression, single-step immunoaffinity purification and characterization of human tetraspanin membrane protein CD81. Plos One 3(6) e2314.

61. Rosa D, Campagnoli S, Moretto C, Guenzi E, Cousens L, et al. (1996) A quantitative test to estimate neutralizing antibodies to the hepatitis C virus: Cytofluorimetric assessment of envelope glycoprotein 2 binding to target cells. Proc Natl Acad Sci USA 93: 1759–1763.

62. Witteveldt J, Evans MJ, Bitzegeio J, Koutsoudakis G, Owsianka AM, et al. (2009) CD81 is dispensable for hepatitis C virus cell-to-cell transmission in hepatoma cells. J Gen Virol 90(Pt 1): 48–58. 10.1099/vir.0.006700-0; 10.1099/vir.0.006700-0.

63. Yi M, Nakamoto Y, Kaneko S, Yamashita T, Murakami S (1997) Delineation of regions important for heteromeric association of hepatitis C virus E1 and E2. Virology 231: 119.

64. Op de Beeck A, Cocquerel L, Dubuisson J (2001) Biogenesis of hepatitis C virus envelope glycoproteins. J Gen Virol 82: 2589–2595.

65. Dubuisson J, Rice CM (1996) Hepatitis C virus glycoprotein folding: disulfide bond formation and association with calnexin. J Virol 20(2): 778–86.

66. Grakoui A, Wychowski C, Lin C, Feinstone SM, Rice CM (1993) Expression and identification of hepatitis C virus polyprotein cleavage products. J Virol 67(3): 1385–1395.

67. Lanford RE, Notvall L, Chavez D, White R, Frenzel G, et al. (1993) Analysis of hepatitis C virus capsid, E1, and E2/NS1 proteins expressed in insect cells. Virol 197(1): 225–35.

68. Ralston R, Thudium K, Berger K, Kuo C, Gervase B, et al. (1993) Characterization of hepatitis C virus envelope glycoprotein complexes expressed by recombinant vaccinia viruses. J Virol 67(11): 6753–6761.

69. Whidby J, Mateu G, Scarborough H, Demeler B, Grakoui A, et al. (2009) Blocking hepatitis C virus infection with recombinant form of envelope protein 2 ectodomain. J Virol 83(21): 11078–11089.

70. Drummer HE, Poumbourios P (2004) Hepatitis C virus glycoprotein E2 contains a membrane-proximal heptad repeat sequence that is essential for E1E2 glycoprotein heterodimerization and viral entry. J Biol Chem 279: 30066–30072.

71. Vieyres G, Thomas X, Descamps V, Duverlie G, Patel AH, et al. (2010) Characterization of the envelope glycoproteins associated with infectious hepatitis C virus. J Virol 84(19): 10159–68. doi: 10.1128/JVI.01180-10.

72. Op De Beeck A, Montserret R, Duvet S, Cocquerel L, Cacan R, et al. (2000) The transmembrane domains of hepatitis C virus envelope glycoproteins E1 and E2 play a major role in heterodimerization. J Biol Chem 275: 31428–31437. doi: 10.1074/jbc.M003003200.

73. Deleersnyder V, Pillez A, Wychowski C, Blight K, Xu J, et al. (1997) Formation of native hepatitis C virus glycoprotein complexes. J Virol 71(1): 697–704.

74. Brazzoli M, Helenius A, Foung SK, Houghton M, Abrignani S, et al. (2005) Folding and dimerization of hepatitis C virus E1 and E2 glycoproteins in stably transfected CHO cells. Virol 332: 438–453.

75. Liu J, Zhu L, Zhang X, Lu M, Kong Y, et al. (2001) Expression, purification, immunological characterization and application of Escherichia coli-derived hepatitis C virus E2 proteins. Biotechnol Appl Biochem 34: 109–119.

76. Martinez-Donato G, Capdesuner Y, Acosta-Rivero N, Rodriguez A, Morales-Grillo J, et al. (2007) Multimeric HCV E2 protein obtained from Pichia pastoris cells induces a strong immune response in mice. Mol Biotechnol 35: 225–235.

77. Rodriguez-Rodriguez M, Tello D, Yelamos B, Gomez-Gutierrez J, Pacheco B, et al. (2009) Structural properties of the ectodomain of hepatitis C virus E2 envelope protein. Virus Res 139: 91–99.

78. Yurkova MS, Patel AH, Fedorov AN (2004) Characterization of bacterially expressed structural protein E2 of hepatitis C virus. Protein Expr Purif 37: 119–125.

79. Goffard A, Callens N, Bartosch B, Wychowski C, Cosset FL, et al. (2005) Role of N-linked glycans in the functions of hepatitis C virus envelope glycoproteins. J Virol 79: 8400–8409.

80. Helle F, Goffard A, Morel V, Duverlie G, McKeating J, et al. (2007) The neutralizing activity of anti-hepatitis C virus antibodies is modulated by specific glycans on the E2 envelope protein. J Virol 81(15): 8101–8111.

81. Helle F, Vieyres G, Elkrief L, Popescu, Wychowski C, et al. (2010) Role of N-linked glycans in the functions of hepatitis C virus envelope proteins incorporated into infectious virions. J Virol 84(22): 11905–11915.

82. HTS assay for allosteric agonists of the human D1 dopamine receptor: Primary screen for antagonists. NIH Molecular Libraries Probe Production Network. BioAssay AID 488983. Accessed on March 21, 2014. Available: https://pubchem.ncbi.nlm.nih.gov/assay/assay.cgi?aid=488983.

83. Antagonists at human 5-hydroxytryptamine receptor 5-ht1e. Extracted from literature and IUPHAR database. BioAssay 624232. Accessed on March 21, 2014. Available: https://pubchem.ncbi.nlm.nih.gov/assay/assay.cgi?aid=624232.

84. qHTS for inhibitors of binding or entry into cells for Lassa virus. NIH Molecular Libraries Probe Production Network. BioAssay 540256. Accessed on March 21, 2014. Available: https://pubchem.ncbi.nlm.nih.gov/assay/assay.cgi?aid=540256.

85. qHTS for inhibitors of binding or entry into cells for Marburg virus. NIH Molecular Libraries Probe Production Network. BioAssay 540256. Accessed on March 21, 2014. Available: https://pubchem.ncbi.nlm.nih.gov/assay/assay.cgi?aid=720532.

86. qHTS for inhibitors of Plasmodium falciparum proliferation. NIH National Institute of Allergy and Infectious Diseases, Xinzhuan Su. BioAssay 504749. Accessed on March 21, 2014. Available: https://pubchem.ncbi.nlm.nih.gov/assay/assay.cgi?mid=504749_53.

87. High throughput screen to identify inhibitors of Mycobacterium tuberculosis H37Rv. Southern Research Institute, Birmingham, AL. E. Lucile White. BioAssay AID 1332. Accessed on March 21, 2014. Available: https://pubchem.ncbi.nlm.nih.gov/assay/assay.cgi?aid=1332.

88. Dreux M, Garaigorta U, Boyd B, Decembre E, Chung J, et al. (2012) Short range exosomal transfer of viral RNA from infected cells to plasmacytoid dendritic cells triggers innate immunity. Cell Host Microbe 12(4): 558–570. doi:10.1016/j.chom.2012.08.010.

89. Ramakrishnaiah V, Thumann C, Fofana I, Habersetzer F, Pan Q, et al. (2013) Exosome-mediated transmission of hepatitis C virus between human hepatoma Huh7.5 cells. Proc Natl Acad Sci USA 110(32): 13109–13113.

90. Russell R, Meunier, Takikawa S, Faulk K, Engle RE, et al. (2008) Advantages of a single-cycle production assay to study cell culture-adaptive mutations of hepatitis C virus. Proc Natl Acad Sci USA 105(11): 4370–4375.

91. Potel J, Rassam P, Montpellier C, Kaestner L, Werkmeister E, et al. (2013) EWI-2wint promotes CD81 clustering that abrogates Hepatitis C virus entry. Cell Microbiol 15(7): 1234-52. doi: 10.1111/cmi.12112. Epub 2013 Feb 16.

92. Jones CT, Catanese MT, Law LMJ, Khetani SR, Syder AJ, et al. (2010) Real-time imaging of hepatitis C virus infection using a fluorescent cell-based reporter system. Nat Biotechnol 28(2): 167–171. doi:10.1038/nbt.1604.

93. Catanese MT, Loureiro J, Jones CT, Dorner M, von Hahn T, et al. (2013) Different requirements for scavenger receptor class B Type 1 in hepatitis C virus cell-free versus cell-to-cell transmission. J Virol 87 (15): 8282–8293.

94. Fofana I, Xiao F, Thumann C, Turek M, Zona L, et al. (2013) A novel monoclonal anti-CD81 antibody produced by genetic immunization efficiently inhibits hepatitis C virus cell-to-cell transmission. PLOS One 8(5): e64221.

95. Rubinstein E, Le Naour F, Lagaudriere-Gesbert C, Billard M, Conjeaud H, et al. (1996) CD9, CD63, CD81, and CD82 are components of a surface tetraspan network connected to HLA-DR and VLA integrins. Eur J Immunol 26(11): 2657–2665.

96. Shoham T, Rajapaksa R, Kuo, Haimovich J, Levy S (2006) Binding of the tetraspanin web: Distinct structural domains of CD81 function in different cellular compartments. Mol Cell Biol 26(4): 1373–1385. doi:10.1128/MCB.26.4.1373–1385.2006.

97. Uhlir F, Stevkova K, Kanczuka V (1975) A comparison of oxyperpine Winthrop and clorothepin (Clothepin Spofa) in schizophrenic psychoses. Activitas Nervosa Superior 17(4): 215.

98. Allen MH, Feifel D, Lesem MD, Zimbroff DL, Ross R, et al. (2011) Efficacy and safety of loxapine for inhalation in the treatment of agitation in patients with schizophrenia: a randomized, double-blind, placebo-controlled trial. J Clin Psychiatry 72(10): 1313–1321.

99. Maher AR, Theodore G (2012) Summary of the comparative effectiveness review on off-label use of atypical antipsychotics. J Manag Care Pharm 18(5 Suppl B): S1–20.

100. Weintraub D, Chen P, Ignacio RV, Mamikonyan E, Kales HC (2011) Patterns and trends in antipsychotic prescribing for Parkinson disease psychosis. Arch Neurol 68(7): 899–904. doi: 10.1001/archneurol.2011.139.

101. Kwentus J, Riesenberg RA, Marandi M, Manning RA, Allen MH, et al. (2012) Rapid acute treatment of agitation in patients with bipolar I disorder: a multi-center, randomized, placebo-controlled clinical trial with inhaled loxapine. Bipolar Disord 14(1): 31–40.

102. Thase ME (2008) Quetiapine monotherapy for bipolar depression. Neuropsychiatric Disease and Treatment 4(1): 21–31.

103. Ban TA, Fujimori M, Petrie WM, Ragheb M, Wilson WH (1982) Systematic studies with amoxapine, a new antidepressant. Int Pharmacopsychiatry 17(1): 18–27.

104. Hormazabal L, Omer LM, Ismail S (1985) Cianopramine and amitriptyline in the treatment of depressed patients – a placebo-controlled study. Psychopharmacology (Berl) 86(1–2): 205–8.

3

Structural Basis of Substrate Selectivity of *E. coli* Prolidase

Jeremy Weaver, Tylan Watts, Pingwei Li, Hays S. Rye*

Department of Biochemistry and Biophysics, Texas A&M University, College Station, Texas, United States of America

Abstract

Prolidases, metalloproteases that catalyze the cleavage of Xaa-Pro dipeptides, are conserved enzymes found in prokaryotes and eukaryotes. In humans, prolidase is crucial for the recycling of collagen. To further characterize the essential elements of this enzyme, we utilized the *Escherichia coli* prolidase, PepQ, which shares striking similarity with eukaryotic prolidases. Through structural and bioinformatic insights, we have extended previous characterizations of the prolidase active site, uncovering a key component for substrate specificity. Here we report the structure of *E. coli* PepQ, solved at 2.0 Å resolution. The structure shows an antiparallel, dimeric protein, with each subunit containing N-terminal and C-terminal domains. The C-terminal domain is formed by the pita-bread fold typical for this family of metalloproteases, with two Mg(II) ions coordinated by five amino-acid ligands. Comparison of the *E. coli* PepQ structure and sequence with homologous structures and sequences from a diversity of organisms reveals distinctions between prolidases from Gram-positive eubacteria and archaea, and those from Gram-negative eubacteria, including the presence of loop regions in the *E. coli* protein that are conserved in eukaryotes. One such loop contains a completely conserved arginine near the catalytic site. This conserved arginine is predicted by docking simulations to interact with the C-terminus of the substrate dipeptide. Kinetic analysis using both a charge-neutralized substrate and a charge-reversed variant of PepQ support this conclusion, and allow for the designation of a new role for this key region of the enzyme active site.

Editor: Annalisa Pastore, National Institute for Medical Research, Medical Research Council, London, United Kingdom

Funding: This work was supported by a grant from the National Institutes of Health (GM065421) to H.S.R. The funder had no role in the study design, data collection and analysis, decision to publish, or preparation of the manuscript.

Competing Interests: The authors have declared that no competing interests exist.

* Email: haysrye@tamu.edu

Introduction

Prolidases, also known as Xaa-Pro dipeptidases, are metalloproteases that catalyze the hydrolysis of dipeptides containing a C-terminal proline residue. These enzymes are conserved in prokaryotes and eukaryotes, including not only single-celled organisms, such as yeast, but also humans and higher plants [1–8]. In higher organisms, prolidase serves a critical role in the recycling of collagen, as the penultimate products of collagen catabolism include the dipeptides Ala-Pro and Gly-Pro [9–11]. In humans, specific mutant alleles of prolidase have been linked to a wide array of physiological problems, which are known collectively as prolidase deficiency [10–13]. Despite the importance of human prolidase and the disease states associated with various mutations of the gene, knockout and knockdown studies in several eukaryotic model organisms have yet to reveal an essential role for prolidase [14–17]. Therefore, further studies are required for insight into the role of prolidase in collagen metabolism and human health.

In contrast to the human enzyme, there are no observable phenotypes for *Escherichia coli* prolidase mutants [18]. While a physiological role for prolidase in bacteria remains to be established, the enzyme is known to possess protective activity against toxic organophosphates [6,19–21]. The *E. coli* enzyme may also play a role similar to that of human prolidase – the

breakdown of dipeptides stemming from protein catabolism – or an additional, regulatory role [22]. In support of this theory, *Mycoplasma* species possess Xaa-Pro peptidases [23–25]. These bacteria, which evolved to retain only those cellular functions essential to their parasitic lifestyle, import most amino acids and lipids from the host cell [26]. The fact that *Mycoplasma* retain an enzyme for cleaving Xaa-Pro bonds suggests that prolyl peptide catabolism plays a broad and generally important physiological role.

Prolidases share a number of conserved sequence and structural features. These enzymes possess an N-terminal domain and a C-terminal catalytic domain, and form dimers through contact between both domains in a head-to-tail arrangement [5]. The catalytic site features a binuclear metal cluster in the center of a pita-bread fold that is a canonical feature of this family of enzymes [22,27]. While the identity and configuration of the coordinating ligands are conserved, the types of metals found in the active site vary widely, though manganese, cobalt and zinc appear to be the most common metals used [28]. Such metal variability has been observed in other pita-bread fold proteins [29,30]. Interestingly, the human prolidase can utilize magnesium, though to significantly lower extent than manganese – a feature not commonly seen in other prolidases [2,4,28,31]. Crystal structures of various prolidases, particularly those with bound substrates or inhibitors, have provided important structural insights into how these enzymes bind substrate peptides and metals, though few members of this enzyme family have been thoroughly examined biochemically.

Members of the pita-bread fold family of proteins, which also includes other metalloproteases, share a number of sequence-specific features that permit robust structure/function prediction, despite the varying substrate specificities of different enzymes [27]. The first prolidase structure solved was from the archaea *Pyrococcus furiosus* [5,32], which confirmed that prolidases possess many of the structural features common to the pita-bread fold superfamily. However, four large regions of primary structure, ranging from 13–25 amino acids in length, are found in the human prolidase that do not appear in the *P. furiosus* sequence [5]. Some of these regions are also absent from related pita-bread fold members, including methionine aminopeptidases, which cleave N-terminal methionine residues, as well as proline aminopeptidases, which cleave N-terminal residues that are followed by proline, from both bacterial and human sources [27,33].

Interestingly, the peptide regions absent in *P. furiosus* are present in prolidases from Gram negative bacteria, including *E. coli* and *Alteromonas sp.* [5], and include eleven residues highly conserved between humans and these two bacteria. *E. coli* PepQ, the only prolidase found in this organism [18], was previously characterized for activity against dipeptides, organophosphates and other small molecules [6], though the lack of an atomic structure for PepQ has prevented a detailed comparison to other prolidases. Examination of the *Alteromonas* prolidase structure, however, reveals an arginine residue reaching into the active site from one of the additional peptide segments. This residue appears to be involved in positioning a structured water molecule and other active site residues and metals, and has been postulated to interact with the C-terminus of the substrate dipeptide [20], an interaction similar to that seen in a shifted location for proline aminopeptidase [34]. Because proline aminopeptidases cleave tripeptides, the positioning of this residue may have evolved to specify substrate length in pita-bread fold proteins.

Here we report the structure of *E. coli* PepQ, showing it to have the predicted pita-bread fold. We examine its ability to utilize various active site metals, including magnesium. Furthermore, we compare its sequence and structural similarity to proline aminopeptidase and other prolidases, showing that the position of the conserved arginine has, in fact, moved throughout evolution, likely to accommodate substrate peptide length. We further characterize the role of this arginine, demonstrating that it plays a critical role in substrate dipeptide binding.

Materials and Methods

Cloning, Expression and Purification of PepQ and PepQ Mutants

The PepQ gene was PCR amplified from purified, chromosomal *E. coli* DNA, using primers adding a 5′-NdeI restriction site and a 3′-XhoI restriction site. The PCR product was sub-cloned into the pET21a vector (Novagen) and the sequence of this construct was verified by DNA sequencing. The R370E mutation of the PepQ gene was created via site-directed mutagenesis of the wild-type construct and was verified by DNA sequencing. Either 6 or 12 L of LB-Amp (100 mg/L) were inoculated 1:500 with overnight cultures of BL21[DE3] cells transformed with either the wild-type or R370E PepQ plasmid. Upon reaching an $A_{600} = 0.6$–0.8, expression was induced with the addition of IPTG to a concentration of 400 μM. After four hours, the cells were

centrifuged and the pellets were resuspended in cell disruption buffer (20 mM Tris, pH 8, 1 μM $MnCl_2$, 20% (w/w) sucrose, 4 mM DTT). Cells were lysed using a gas-driven cell-disruptor (Microfluidics Corporation; Newton, MA) and clarified by ultracentrifugation. The supernatant was loaded onto a fast-flow Q (GE Healthcare) anion exchange column. The column was washed with Buffer A (50 mM Tris, pH 7.4, 1 μM $MnCl_2$, 2 mM DTT) and washed with Buffer A containing 100 mM NaCl. A linear gradient was then developed from 100 mM to 500 mM NaCl. The fractions of the greatest PepQ purity were concentrated by precipitation with 70% (w/v) ammonium sulfate. The pellet was resuspended in a small volume of Buffer A containing 500 mM ammonium sulfate and loaded on a phenyl-sepharose hydrophobic interaction column (GE Healthcare). After washing with Buffer A containing 1 M ammonium sulfate, a linear gradient was developed from 1 M to 300 mM ammonium sulfate. Fractions of the greatest PepQ purity were again concentrated by precipitation with 70% (w/v) ammonium sulfate. The pellet was resuspended with a small volume of Buffer A and dialyzed against Buffer B (25 mM Tris, pH 7.4, 25 mM KCl, 1 μM $MnCl_2$, 2 mM DTT). Following addition of glycerol to 15% (v/v), the sample was aliquoted, snap frozen with liquid nitrogen and stored at $-80°C$. Thawed samples showed no detectable loss of enzymatic activity.

Cloning, Expression and Purification of Alanine Dehydrogenase (AlaDH)

The AlaDH gene was PCR amplified from purified, chromosomal *Bacillus subtilis str. 168* DNA, using primers adding a 5′-NcoI restriction site (which required a mutation in the second codon, which was later reverted with site-directed mutagenesis) and a 3′-XhoI restriction site. The PCR product was sub-cloned into the pETDuet vector (Novagen) and the sequence of this construct was verified by DNA sequencing. Protein expression was conducted in 6 L of LB-Amp (100 mg/L) inoculated 1:500 with overnight cultures of BL21[DE3] cells transformed with the AlaDH plasmid. Upon reaching an $A_{600} = 0.6$–0.8, expression was induced with the addition of IPTG to a concentration of 400 μM. After four hours, the cells were centrifuged and the pellets were resuspended in cell disruption buffer (20 mM Tris, pH 8, 0.5 mM EDTA, 20% (w/w) sucrose, 4 mM DTT). Cells were lysed, clarified and loaded onto a fast-flow ion exchange column, as described above. The column was washed with Buffer C (50 mM Tris, pH 7.4, 0.5 mM EDTA, 2 mM DTT) containing 150 mM NaCl. A linear gradient was then developed from 150 mM to 500 mM NaCl. Fractions of the greatest AlaDH purity were concentrated by precipitation with 70% (w/v) ammonium sulfate. The pellet was then resuspended in a small volume of Buffer C containing 1 M ammonium sulfate. The sample was then loaded on a phenyl-sepharose hydrophobic interaction column (GE Healthcare). After washing with Buffer C containing 900 mM ammonium sulfate, a linear gradient was developed from 900 to 650 mM ammonium sulfate. Fractions of the greatest AlaDH purity were concentrated by precipitation with 70% (w/v) ammonium sulfate. The pellet was resuspended with a small volume of Buffer C and dialyzed against Buffer D (25 mM Tris, pH 7.4, 25 mM KCl, 0.5 mM EDTA, 2 mM DTT). Following addition of glycerol to 15% (v/v), the sample was aliquoted, snap frozen with liquid nitrogen and stored at $-80°C$. Thawed samples showed no detectable loss of enzymatic activity.

Crystallization and Refinement of PepQ

The PepQ sample was buffer-exchanged into 50 mM Tris, pH 7.4, 5 mM $MgCl_2$ and 5 mM DTT at a final concentration of

Table 1. Statistics of crystallographic analysis for pepQ.

PDB Entry	4QR8
Data collection	
Space group	$P2_12_12_1$
Cell dimensions	
a, b, c (Å)	72.57, 97.44, 126.94
α, β, γ (°)	90.0, 90.0, 90.0
Resolution (Å)	2.00 (2.03 to 2.00)[1,2]
[3]R_{sym} or R_{merge}	11.6% (0.695)
$I/\sigma I$	18.0 (2.0)
Completeness (%)	97.0 (92.3)
Redundancy	4.5 (3.5)
Refinement	
Resolution (Å)	2.0
No. reflections	59597
[4]R_{work}/[5]R_{free}	17.39%/21.1%
No. atoms	
Protein	7052
Ligand/ion	4
Water	1314
B-factors	
Protein	20.3
Ligand/ion	19.6
Water	31.2
R.m.s. deviations	
Bond lengths (Å)	0.004
Bond angles (°)	0.80

[1]One crystal was used to collect each of the dataset.
[2]Values in parentheses are for highest-resolution shell.
[3]$R_{sym} = \sum_h \sum_i |I_{i,hkl} - \langle I_{hkl} \rangle| / \sum_{hkl} \sum_i |I_{i,hkl}|$, where $I_{hkl,i}$ is the intensity measured for a given reflection with Miller indices h, k, and l, and $\langle I_{hkl} \rangle$ is the mean intensity of that reflection.
[4]$R_{work} = \sum ||F_o| - |F_c|| / \sum |F_o|$, where F_o and F_c are the observed and calculated structure-factor amplitudes, respectively.
[5]R_{free} was calculated as R_{work} using a randomly selected subset (10%) of unique reflections not used for structure refinement.

12 mg/ml. The protein was crystallized by the hanging drop vapor diffusion method at 4°C using 20% PEG MME 5000 in 0.1 M Bis-Tris buffer at pH 6.5. The crystals were transferred stepwise to a cryobuffer containing 30% PEG 400, 20% PEG MME 5000, 0.1 M Bis-Tris at pH 6.5 and flash frozen in liquid nitrogen. The diffraction data were collected at beamline 7.1 at the Stanford Synchrotron Radiation Lightsource (SSRL) using a Quantum 315R CCD detector. The diffraction data were processed with the HKL2000 package [35]. The structure was determined by molecular replacement using Phaser in the Phenix package [36]. A homology model of PepQ generated using Swiss-Model based on the crystal structure of *Alteromonas macleodii* OpaA structure (PDB 3RVA) [20] was used as search model. The model was fine-tuned with Coot [37] and refined using the Phenix package [36]. Statistics of data collection and refinement are shown in Table 1.

Metal Usage

Metal usage of PepQ was directly monitored by the decrease in absorbance at 222 nm upon cleavage of the substrate peptide bond [6]. *E. coli* PepQ was diluted to 12.5 μM into 50 mM Tris, pH 7.4 and 10 mM EDTA. Following incubation at 25°C for 30 min, this solution was then diluted 25-fold into 25 mM Tris, pH 7.4 containing either a divalent metal (1 mM), EDTA (5 mM) or no additional component. Samples were incubated at 25°C for an additional 10 min. This sample was then diluted 10-fold with 10 mM Tris, pH 8 and the substrate dipeptide AlaPro (TCI-America). The reaction was immediately assayed at 25°C. The final concentration of PepQ was 50 nM and AlaPro was 0.25 mM in a final volume of 1 mL. All assays were conducted using a Perkin Elmer Lambda 35 spectrophotometer with a PCB 1500 water Peltier temperature control system.

Docking Simulations

Preparation of structure files and docking was done as described [38]. In brief, substrate and protein structure files were prepared using MGL Tools, in which polar hydrogens were added and flexible bonds were designated. Autodock Vina was then used to simulate the interaction of the small molecules with the active site of PepQ.

Enzyme Quaternary Structure

The stability of the dimeric structure of wild-type and R370E PepQ was determined using analytical gel filtration. PepQ

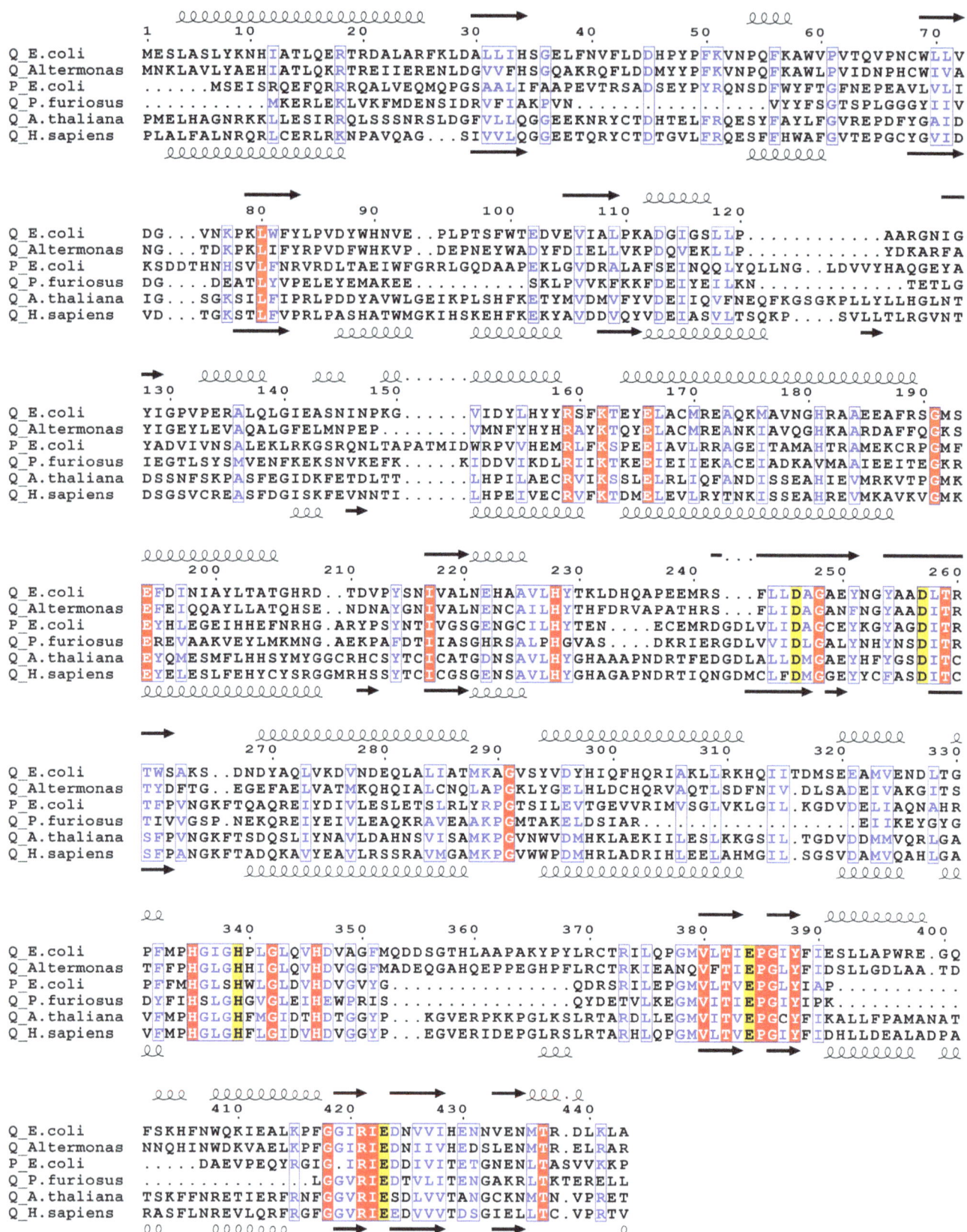

Figure 1. Sequence alignment of prolidases. Sequence alignment of *E. coli* PepQ (accession number P21165) with eukaryotic and prokaryotic pita-bread fold enzymes was performed using CLUSTALW [60] and graphically organized with ESPript [61]. Completely conserved residues are highlighted in red and highly conserved residues or regions are boxed and shown in blue. Metal-chelating residues are highlighted with yellow. Numbering shown is for *E. coli* PepQ. Secondary structure assignments shown above the alignment are those from *E. coli* PepQ, while those shown

below the alignment are from human PepD. The aligned proteins (with percent identity/similarity to *E. coli* PepQ, along with the number of aligned positions shown in parentheses; followed by the accession number of the sequence) are: *Alteromonas sp.* PepQ (50/67, 441), Q44238; *E. coli* PepP (31/46, 330), P15034; *Pyrococcus furiosus* PepQ (24/40, 337), P81535; *Arabidopsis thaliana* Xaa-Pro Dipeptidsae (34/51, 292), Q8L780; *Homo sapiens* PepD (29/45, 466), P12955. The degree of identity and similarity was determined by two-sequence alignment with BLAST [62].

(10 nM) in 50 mM Tris, pH 7.4, 50 mM KOAc, 10 mM $Mg(OAc)_2$ and 2 mM DTT was injected on a Superose 6 gel filtration column (GE), equilibrated in the same buffer, with a constant flow rate of 0.4 mL/min driven by an HPLC unit with a binary pump (Waters). The tryptophan fluorescence (excitation at 280 nm, emission at 340 nm) of the sample was measured using an in-line, post-column fluorescence detector (Waters).

Enzyme Stability

The thermodynamic stability of wild-type and R370E PepQ was determined by the red-shift in the tryptophan fluorescence peak as the protein unfolds with increasing concentrations of the chemical denaturant guanidinium-HCl. PepQ (50 nM) was incubated at room temperature for 60 minutes in solutions of 50 mM Tris, pH 7.4, 10 mM $Mg(OAc)_2$, 2 mM DTT and varying concentrations of guanidinium-HCl. The tryptophan fluorescence was measured using a PTI fluorometer with excitation at 295 nm and emission from 315–375 nm. Solutions of buffer and guanidinium at each concentration, without protein, were also measured to account for changes in scattered light. The peak maximum and corresponding wavelength was determined using Microsoft Excel (MAX and VLOOKUP functions).

Enzyme Kinetics

The PepQ reaction rate was monitored by coupling the hydrolysis of the dipeptide AlaPro to the NAD-dependent oxidation of alanine [39]. These reactions were conducted in a 1 mL volume in 50 mM Tris, pH 8 and 20 mM $Mg(OAc)_2$ at 25°C with varying concentrations of AlaPro-COOH (TCI America) or AlaPro-CONH$_2$ (Chem-Impex), supplemented with 1 μM AlaDH and 2 mM NAD$^+$ (Chem-Impex). The increase in absorbance at 340 nm was monitored as NADH was produced. All assays were conducted using a Perkin Elmer Lambda 35 spectrophotometer with a PCB 1500 water Peltier temperature control system.

Results

E. coli Prolidase Possesses an Expanded Sequence

To examine the extent of sequence conservation in the *E. coli* prolidase, PepQ, we collected primary structure information from organisms with sequenced genomes, including both higher plants and animals. Upon alignment (Figure 1), many regions of *E. coli* PepQ show sequence similarity (boxed) and identity (shaded) with the sequences of human and plant prolidase, illustrating the conservation of various elements of this protein family. Overall, *E. coli* PepQ shows high sequence identity (~30%) and similarity (~50%) with the eukaryotic prolidases. Furthermore, the *E. coli* sequence shows good coverage of the human gene, with only one region of 10–15 residues missing (Figure 1, between *E. coli* residues 120–125). Although these additional regions may be shifted in our alignment, in a previous alignment [5], four regions of at least ten residues appeared in *E. coli* and human prolidase, but did not appear in *P. furiosus* prolidase (*E. coli* residues 35–53, 303–321, 360–372 and 391–415). In these regions, eleven residues (*E. coli* residues Gly36, Asp45, Phe50, Leu309, Ser319, Glu321, Leu369, Arg370, Glu391, Leu393 and Leu394) are conserved. Of these residues, all but two (Ser319 and Glu391) are also conserved

among *E. coli*, humans and *Arabidopsis* (Figure 1). While shorter than ten residues, another additional region appears in all of the sequences, but not in *P. furiosus* PepQ – an N-terminal loop extension (94–101), though this region does not include any conserved residues.

To better understand the potential significance of sequence conservation between the *E. coli* and human prolidases, we solved the structure of the bacterial enzyme at 2.0 Å resolution (Figure 2A, Table 1). The protein is comprised of two sections – an N-terminal domain and a C-terminal catalytic domain. The catalytic domain features the predicted, canonical pita-bread fold common to this family of enzymes. At the center of the pita-bread fold is the active site, containing two metal ions chelated by five residues (metals shown in green). The asymmetric unit contains a single PepQ dimer, which is the native oligomeric structure of this protein [6], arranged head-to-tail with inter-dimer contacts made between both domains. With tertiary and quaternary features appearing as expected, we next focused our analysis on the regions of sequence not found in *P. furiosus* and the residues in those regions that are conserved in other sequences.

The conserved regions in PepQ consist of two helicies and two loop structures, and three of these structural features are in the catalytic domain (Figure 2B). The N-terminal loop (highlighted in red) makes significant contact with the same loop from the other subunit. The loop in the catalytic domain (highlighted in pink) extends into the active site. The two helices in the catalytic domain (highlighted in blue and yellow) are on the outside edge of the domain, with both helices in contact with each other and one also in contact with the loop in the catalytic domain (pink). Given the location of these regions of sequence, it is not surprising that only two of the nine residues found in regions absent from *P. furiosus* (but conserved from *E. coli* into the eukaryia), are located near the active site of the enzyme (Asp45 and Arg370). The Arg370 equivalent residue in *Alteromonas* (also Arg370) has been predicted to play a role in organizing water in the active site and, possibly, interacting with the C-terminus of the substrate dipeptide [20]. Asp45, which reaches into the active site of one monomer from a loop region in the N-terminal domain of the other monomer, is seen in *E. coli* to be within interaction distance of Arg370, with the charged ends of the side chains approximately 3.5Å apart. The conservation of this interaction suggests co-evolution of these residues in support of additional known interactions in the active site.

The active site of *E. coli* PepQ also features canonical metal binding residues, Asp246, Asp257, His339, Glu384 and Glu423, chelating two metal ions. Because PepQ was crystallized in buffer containing magnesium, the density found in this region is most likely derived from magnesium ions (Figure 2C). Additionally, the mF$_0$-DFc difference map shows greater density for one of the two metal ions (chelated by His339), consistent with reports from other pita-bread fold peptidases that this binding site has a higher affinity for metal ions [34,40]. The decreased occupancy at the second metal site is surprising, given that the magnesium concentration during crystallization was in the millimolar range. This observation suggests that the affinity for magnesium of either PepQ in general, or this site in particular, is not as high as seen for the preferred manganese ion in related proteins, reported to be in the low- or sub-micromolar range [22,34,41]. However, metal

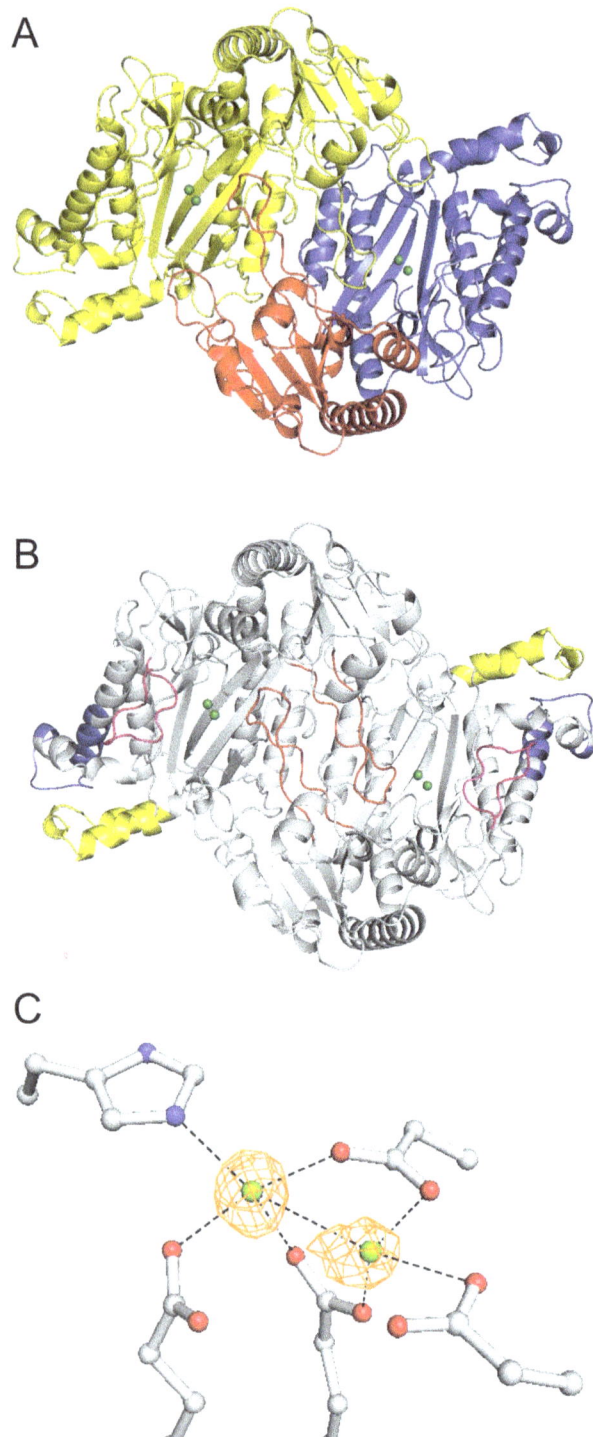

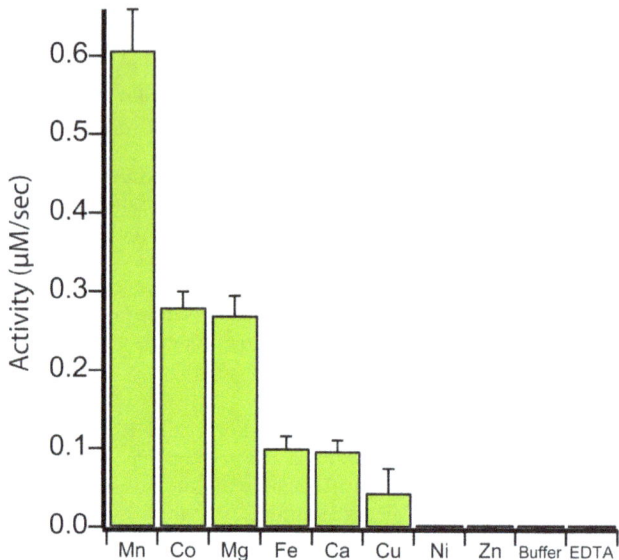

Figure 3. PepQ can utilize different divalent metals with varying efficiency. *E. coli* PepQ (50 nM) was assayed in the presence of various metals, in the absence of added metal (Buffer) or in the presence of EDTA. All metals used were in the form of metal-dichlorides. Error bars show the standard deviation of three independent samples.

binding by prolidase does not necessarily convey enzymatic activity, leaving the functionality of magnesium-bound PepQ unresolved.

E. coli PepQ Can Utilize Mutliple Metals for Catalysis

Despite the shared pita-bread fold, prolidases, methionine aminopeptidases and proline amino-peptidases from a range of taxa, display widely varying abilities to bind and utilize different metals for catalysis. The presence of magnesium ions in both metal binding sites of *E. coli* PepQ (Figure 2) suggests that this prolidase might be enzymatically active with this, though magnesium is not known to be the preferred metal of any pita-bread fold enzyme. We therefore examined the ability of PepQ to utilize various divalent cations – testing the dominant ions found in pita-bread fold proteases: manganese, zinc, cobalt, iron, nickel, copper, magnesium and calcium (Figure 3). As with many proteins in this family, manganese appears to be the optimal metal for PepQ activity, with cobalt a distant second. Nickel and copper are not generally employed by this family of proteases. Other metals, such as zinc and calcium, are known to require specific coordination and spacing regimes that are not easily accessed in the active site of many pita-bread fold proteins [34], leading to little or no activity, consistent with our observations with PepQ (Figure 3). Magnesium, a metal that only rarely conveys activity in other pita-bread proteases, displayed significant levels of activity with PepQ, similar to cobalt. As expected for a metalloprotease, the addition of EDTA abolished the activity of PepQ.

To control for contaminating metal in the buffer, as well as for metal that was not removed from the active site prior to the experiment, PepQ was also tested in buffer in the absence of any residual metal (Figure 3). An absence of enzymatic activity indicates that the pre-incubation of PepQ with EDTA effectively stripped any remaining bound metal. Whether the metals that convey little or no activity do not bind, or bind, but are incapable of supporting catalysis, is unknown. Zinc, for example, has been shown to bind to the active site of some pita-bread fold peptidases and still not convey activity [2,5,31,34]. It is possible that the

Figure 2. PepQ forms a canonical pita-bread fold with binuclear active site. (A) The PepQ dimer (PDB entry 4QR8) is shown with one monomer shown in yellow and one monomer colored by domain: N-terminal (residues 1–159, red) and catalytic (160–443, blue). The magnesium ions are colored green. The image was rendered in PyMOL [63]. (B) The PepQ dimer with new regions of sequence (those not in *P. furiosus*) highlighted (residues 35–53, red; 303–321, blue; 360–372, pink; 391–415, yellow). (C) Electron density shows conserved active site residues coordinating two magnesium ions.

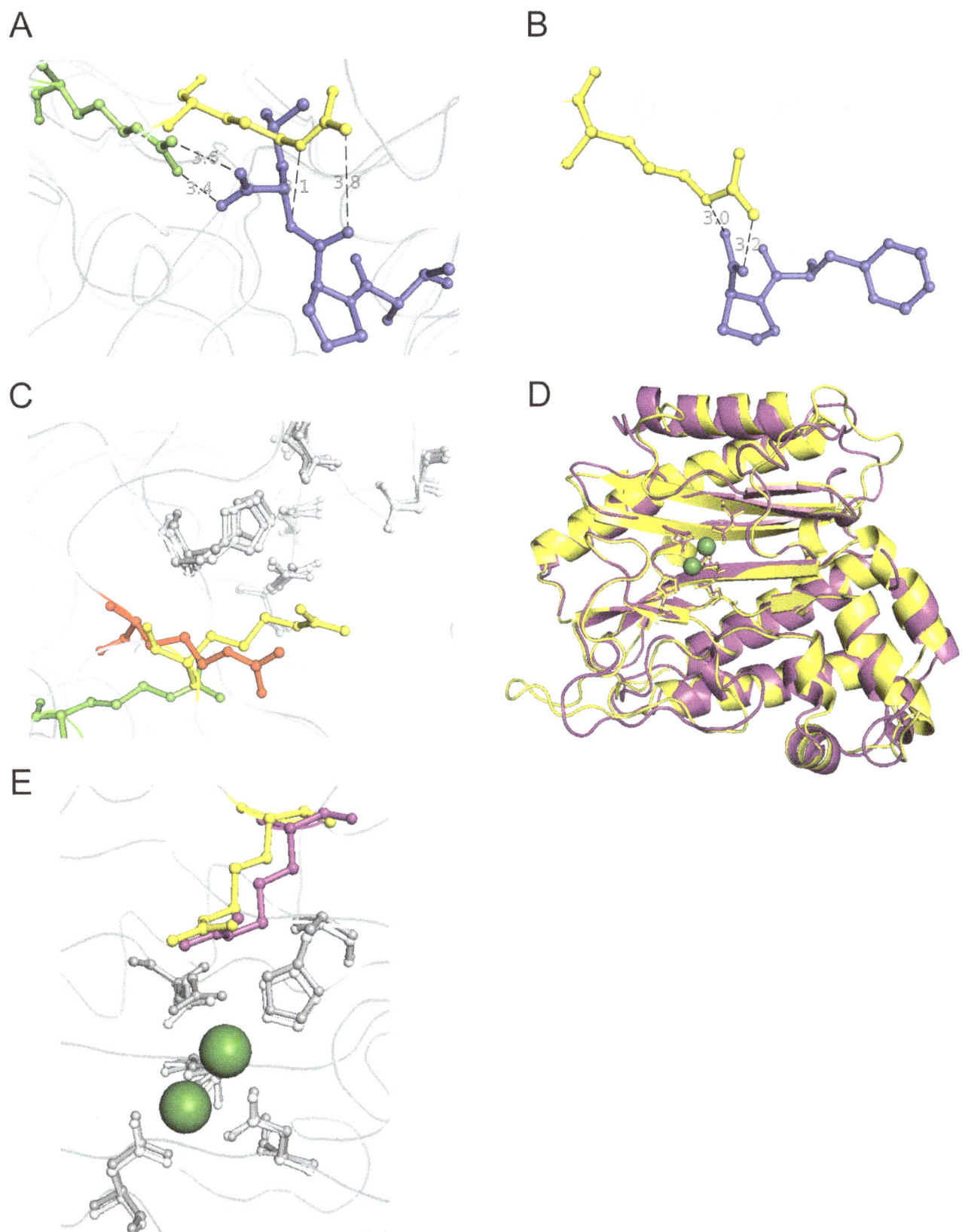

Figure 4. Structural alignment of prolidases reveals a conserved active site arginine. (A) The PepQ catalytic domain (residues 160–443) was aligned with *E. coli* PepP, the proline aminopeptidase, with the bound substrate tripeptide ValProLeu (PDB entry 2BHA, residues 175–425; RMSD = 1.05 Å, 1020 atoms aligned). PepQ R370 is shown in yellow, PepP R371 is shown in green and the tripeptide is colored blue. The distances between PepP R371 and the C-terminal oxygens of the tripeptide measured at 3.4 and 3.6 Å. The distances between PepQ R370 and the prolyl-leucyl

amide nitrogen and oxygen measured at 3.1 and 3.8 Å, respectively. (B) Docking simulations were performed between PepQ (yellow) and substrate dipeptides using AutoDock Vina [38]. Shown is the substrate PhePro (blue). The distances between R370 and the dipeptide C-terminal oxygens measured at 3.0 and 3.2 Å. (C) *E. coli* PepQ (yellow) and PepP (green) were aligned with *P. furiosus* prolidase (PDB entry 1PV9, residues 124–345, red). PepQ R370, PepP R371 and *P. furiosus* R295 are highlighted. ($RMSD_{EcoliQ-PfuriosusQ}$ = 0.92 Å, 816 atoms aligned; $RMSD_{EcoliP-PfuriosusQ}$ = 0.82 Å, 908 atoms aligned) (D) Structure alignment of catalytic domains of *E. coli* PepQ (yellow) and human PepD (PDB entry 2IW2, residues 187–470, purple; RMSD = 0.97 Å, 1179 atoms aligned). (E) R370 in PepQ (yellow) is sequentially and structurally conserved in humans (R398, purple). All structural alignments and distance measurements were performed with PyMOL [63].

inactivity of PepQ in the presence of zinc and nickel is the result of weak metal binding, which could, in principle, be examined by increasing in the concentrations of these metals in the PepQ assay. However, the concentrations of these metals are not thought to be higher *in vivo* than used here *in vitro* [42], implying that these metals are not likely used to support catalysis in the cell.

Ionic Interactions Favor the Substrate Peptide C-Terminus

While the metal-chelating residues of prolidase are well-described, the identity of these residues does not reliably predict metal usage. Likewise, *de novo* prediction of substrate specificity is limited to dipeptides, as well as certain small molecules such as organophosphates that are hydrolyzed with far lower efficiency than peptide substrates. Different prolidases have varying affinities for dipeptides, though cleavage of collagen-catabolism products, such as GlyPro or AlaPro, seems to be conserved [3,6,31,43]. How dipeptide specificity is enforced by prolidase, as well as why these enzymes display a total lack of activity toward longer peptides, is not obvious, particularly given the high structural similarity between prolidase, which cannot cleave peptides longer than two amino acids, and proline aminopeptidase, which can cleave tripeptides (Xaa-Pro-Xaa) at the N-terminal side of proline. To further examine these differences, the structures of PepQ and PepP, the *E. coli* proline aminopeptidase, were aligned for comparison (Figure 4A). In the structure of PepP, which includes a bound tripeptide, Arg371 (of PepP) interacts with the C-terminus of the tripeptide. PepQ Arg370, which is projected further into the active site on one of the loop regions conserved in prolidases from higher organisms, is placed far enough into the active site that it would physically impede the binding of longer peptides, as seen in the overlap between this arginine and the PepP-bound substrate tripeptide. PepQ R370 is, however, in an appropriate position for the guanidinium group of the arginine to interact with the C-terminus of the proline residue.

To further examine the potential role of R370 in dipeptide selection by PepQ, the structures of model dipeptides were docked into the active site of the PepQ structure [38]. Docking of dipeptides resulted in a configuration similar to that seen in PepP – the terminus of the dipeptide is in position to interact with Arg370 (Figure 4B). In order to experimentally test this interaction, a two-pronged approach was pursued. First, a charge-reversed mutant of PepQ (R370E) was made in which the predicted favorable interaction between the peptide carboxylate and R370 was replaced with an unfavorable interaction. The R370E mutant was expressed and purified following the protocols used for the wild-type enzyme, and eluted on gel filtration chromatography identically to wild-type PepQ (Figure 5A), suggesting that both the structure and dimer stability of the enzyme was not significantly compromised by the R370E mutation. To test this conclusion further, we examined the thermodynamic stability of the R370E mutant relative to wild type PepQ using guanidinium-induced unfolding at 25 °C. As shown in Figure 5B, the two proteins show essentially identical unfolding transitions, indicating that the thermodynamic stability of PepQ is not affected by the R370E

mutation. Consequently, any changes in the activity of R370E relative to the wild-type enzyme are not likely due to secondary effects of the mutation on protein structure or stability.

As a second approach to examining the role of R370, we examined the activity of PepQ toward a substrate dipeptide featuring a terminal amide, rather than a carboxylic acid. With this modified substrate, the predicted interaction with Arg370 should remain favorable, as hydrogen bonding could still occur, though the favorable ionic interaction would be lost. Due to the partial positive charge of the amide nitrogen, a potentially favorable ionic interaction between the modified amide terminus of this substrate dipeptide and the glutamate of the mutant R370E remained a possibility. Kinetic analysis of both wild-type and R370E PepQ with both AlaPro-COOH and AlaPro-CONH$_2$ strongly supports the proposed model for the role of R370 (Table 2). R370E displayed a considerably higher K_m for the substrate AlaPro-COOH than the wild-type protein, while actually having a lower K_m for AlaPro-CONH$_2$, when compared to wild-type prolidase. The reduction in k_{cat} seen in R370E is likely due the role of this residue in the organization of water and other residues in the active site [20]. The changes in K_m, with information from both a charge-reversed protein and a charge-neutralized substrate, strongly suggest an interaction between the substrate carboxylate group and R370.

The placement of the loop arginine evolved for substrate selectivity

The location of the key R370 residue in PepQ, and similar arginine residues in other pita-bread fold enzymes, may have been an important factor in the evolution of prolidase. To examine this idea, the structures of *E. coli* PepP, *E. coli* PepQ and *P. furiosus* PepQ were aligned (Figure 4C). The loop region containing this arginine is absent from *P. furiosus* prolidase. While the archaeal prolidase retains an arginine in the same spatial location of the active site (Arg295), it appears to be in an intermediate position, relative to PepP and PepQ from *E. coli*. The active site residues of the three enzymes are nearly super-imposable, indicating that this change in position is not an artifact of the structure alignment. It thus seems reasonable that the addition of the expanded peptide regions containing the arginine in Gram-negative bacteria and eukaryotes could have resulted from evolutionary fine tuning of the enzyme for high specificity dipeptidase activity. While the loop residues are conserved in the sequence of human prolidase, we sought to verify the placement of this residue as a potential means of selecting for dipeptides. The structures of the catalytic domains of human and *E. coli* prolidases were aligned, showing high conservation in the secondary structure elements and an RMSD of less than 1Å (Figure 4D, secondary structures shown in Figure 1). The critical arginine is observed in the active site of the human prolidase in a position nearly identical to the bacterial residue (Figure 4E). This suggests that after the initial evolution of the loop regions for the placement of this residue, no further optimization was necessary for the selection of dipeptides as the enzyme evolved further over the course of several billion years.

A

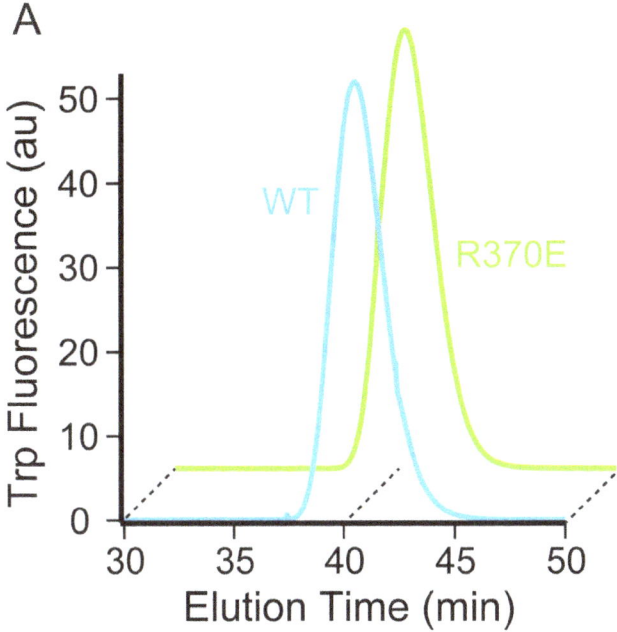

B

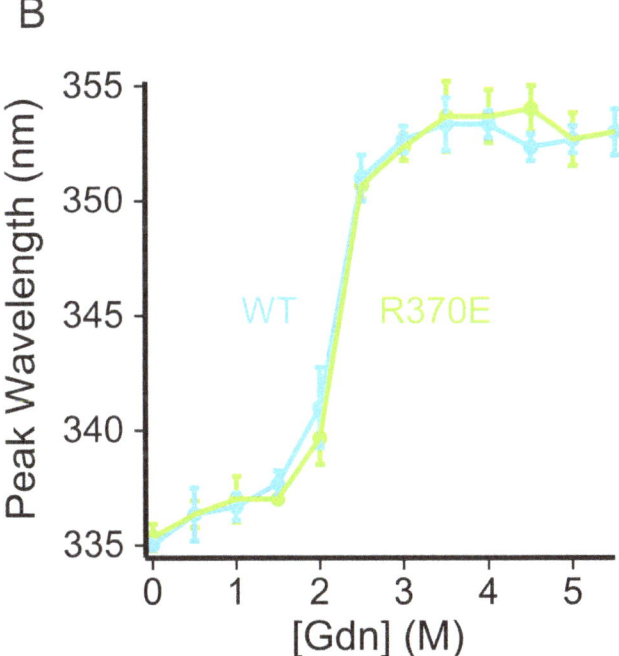

Figure 5. R370E mutation does not perturb PepQ structure or stability. (A) Wild-type (WT, blue) and R370E (green) PepQ (10 nM) were analyzed by analytical size exclusion chromatography. (B) Wild-type (WT, blue) and R370E (green) PepQ (50 nM) were incubated with varying concentrations of guanidinium-HCl and the peak position of the tryptophan fluorescence emission spectrum of each was determined. Error bars indicate the standard deviation from three independent samples.

Discussion

The results presented here support a role for substrate length specificity in pita-bread fold enzymes through the positioning of an active site arginine. With a high-resolution structure of *E. coli* PepQ in hand, we were able to compare it to related enzymes, both bioinformatically and structurally. We found that the position of the active site arginine has changed during the evolution in this family of proteins, with a shift further into the active site leading to selection against peptide substrates greater than two residues in length. Not only does the placement of this arginine physically occlude longer peptides, as seen structurally, but also, kinetic analysis demonstrates the important role of the ionic interaction between this positively charged residue and the negatively charged C-terminus of the substrate dipeptide. We have also found that while this protein is maximally active with manganese, it can utilize other metals, including magnesium, an uncommon property for this family of metalloproteins.

Although they are very similar proteins, the members of this family of enzymes vary in a number of significant ways. Of particular note is the presence of several large regions of additional residues in the prolidase sequences of Gram-negative bacteria, single-celled eukaryotes and higher plants and animals that are absent in the sequences of other bacterial prolidases, as well as proline aminopeptidase. When comparing the additional regions found in *E. coli* that are absent in *P. furiosus*, perhaps the most striking insert is the N-terminal loop. This loop not only makes significant contact with its counterpart on the adjacent subunit, but also contributes to the opening of the active site, relative to the loop-less *P. furiosus* structure. The role of the two helices inserted into the catalytic domain of *E. coli* is more difficult to surmise, given the distance from the active site and the other subunit. The structural rearrangements created through the insertion of the N-terminal loop may be stabilized by the presence of these helices, but this remains to be examined. However, both of these helices, as well as the N-terminal loop are found in the sequence and structure of *E. coli* PepP (Figure 1) [22,30,40]. This suggests that these changes may have occurred in an ancestor of this family before the divergence that led to separate substrate specificities of PepQ and PepP. The component found in neither PepP nor *P. furiosus* PepQ is the loop in the active site of the protein, which contains the conserved arginine.

Enzymes generally dictate specificity by utilizing binding pockets with specific interactions that favor some substrates and disfavor or occlude other substrates [44-47]. Pita-bread fold proteins are no exception – the occlusion of branched amino acids and selection against small amino acids in substrates has been observed previously in PepP [34] and charge interactions have been observed to dictate specificity in some prolidases [48]. Despite the high level of conservation among proteins with the pita-bread fold, these enzymes are very specific for their substrates, at least in terms of peptide length. While the evolutionary benefit of selecting for dipeptides stems from the availability of byproducts of protein catabolism, like those derived from collagen, the movement of this residue also levies an advantage against certain small molecules. *E. coli* PepQ can not only hydrolyze at least thirteen different dipeptides, but also an assortment of organophosphates and other small molecules [6]. While these substrates vary considerably on the N-terminal side of the scissile bond, the C-terminal end of all previously tested substrates shared a negatively charged group, either a carboxylate or a nitro group [6]. Reactivity toward these substrates is likely dictated by the positioning of R370 in the active site of the enzyme. We have shown that the addition of a loop in the catalytic domain, near the active site, allowed for the substrate peptide length-determining residue to be repositioned, altering the specificity of the enzyme. Utilization of an arginine at the designated position in either PepP or PepQ for this selection likely stems from the ability of arginine

Table 2. Kinetic parameters for the hydrolysis of the dipeptides AlaPro and AlaPro-NH$_2$ by wild-type and R370E PepQ.

	AlaPro-COOH			AlaPro-CONH$_2$		
	k_{cat}/Km (M^{-1}s^{-1})	k_{cat} (s^{-1})	K$_m$ (mM)	k_{cat}/Km (M^{-1}s^{-1})	k_{cat} (s^{-1})	K$_m$ (mM)
WT	1.2×10^5	139.1 ± 2.5	1.2 ± 0.1	1.2×10^4	80.7 ± 19.7	6.6 ± 0.5
R370E	5.1×10^1	6.5 ± 1.1	127 ± 2	2.3×10^2	1.1 ± 0.4	4.7 ± 0.2

to interact ionically with both oxygens in the C-terminus of the substrate peptide, as well as through hydrogen bonding. A lysine at this position is unlikely to interact with both oxygens due to spatial and angular limitations. Although the enzyme is still functional without this interaction, the activity is severely compromised, which is consistent with reports that *E. coli* PepP has minor activity against dipeptides [18]. Interestingly, the genomes of sequenced *Pyrococcus* species include PepQ sequences, but lack PepP annotations [49–55]. Given the intermediate positioning of the conserved arginine in *P. furiosus* PepQ, a dual functionality for cleaving di- and tripeptides may be predicted for that enzyme.

While many prolidases share various similarities, structural and biochemical data reveal that *E. coli* prolidase is more similar to the human enzyme than other enzymes. The catalytic domains of the *E. coli* and human prolidases align with an RMSD of less than 1.0 Å, and this bacterial enzyme utilizes magnesium to a similar extent as human prolidase, suggesting that the specific placement or conformational flexibility that influences metal coordination is shared between the enzymes of these two distantly related organisms. Variable metal usage has been postulated to serve a regulatory role in aminopeptidases [29]. Other similarities between these two proteins may allow for the *E. coli* protein to provide insights into the functionality of the human protein. Not only is the placement of the critical arginine residue unchanged in the human prolidase, but many other residues are conserved between the two proteins, including some that are associated with disease alleles, for example, E412K and G448R in human prolidase deficiency [41].

Interestingly, despite its role in substrate selectivity, no mutation of the equivalent R370 residue has yet been associated with the onset of prolidase deficiency in humans [13,56–58]. It is possible that mutation of the same residue in the human enzyme results in a reduction of enzymatic activity too small to yield an observable phenotype. However, this seems unlikely, given that losing R370 in PepQ results in a decrease in enzymatic activity that is orders of magnitude more severe than caused by single residue mutants in the human enzyme with known phenotypes [41]. Notably, many disease associated mutations also (i) decrease the stability of the enzyme, (ii) have a reduced abundance *in vivo*, and (iii) perturb the dimer binding constant so that formation of active enzyme requires protein concentrations that are much higher than needed for the wild-type protein [41]. It is possible that the impact of these mutations on folding and stability is, overall, more serious than the loss of activity seen with the arginine mutation alone, which has no effect on enzyme stability or folding. Alternately, loss of the active

site arginine might have such severe developmental consequences that homozygous and many heterozygous genotypes are simply not viable. It also remains possible that the number of studied cases of prolidase deficiency is yet too small to have sampled every disease-associated allele. Although the critical active site arginine residue has yet to be associated with the physiological outcomes of reduced activity, observed defects in conserved regions in one enzyme generally predict similar defects in other, highly homologous enzymes. Utilizing *E. coli* PepQ may, therefore, be an effective strategy for studying prolidases in general and deficiencies in human prolidase specifically.

Many proteins evolve through the addition of loops or domains to gain solubility, new interactions or new activity [59]. Although it is not necessary for a prolidase to have the additional catalytic domain loop in order to place specificity-defining residues in the active site, as seen in the *P. furiosus* PepQ, the positioning of this residue in Gram-negative bacteria and higher organisms has remained constant during billions of years of evolution, indicating a preferred or optimal placement for activity. While *E. coli* PepQ may serve as a tool for studying prolidases in general, other prolidases may provide further insight into the role of these enzymes beyond collagen recycling in humans. One such protein of interest is the Xaa-Pro peptidase from the nearly exclusively catabolic organism *Mycoplasma mobile*. Examination of this protein structurally and biochemically would reveal how this minimalist organism utilizes this enzyme, demonstrating the extent of its role in metabolism. Despite the continual advance of knowledge about prolidases – structurally, biochemically and genetically – much is left unknown about the role of these enzymes in metabolism and their connection to disease.

Acknowledgments

We thank the Raushel Lab (TAMU) and Dr. Margaret Glasner (TAMU) for helpful discussions, the Straight Lab (TAMU) for the gift of *B. subtilis* genomic DNA and many labs of the Department of Biochemistry and Biophysics (TAMU) for gifts of reagents. We also thank Dr. Chavela Carr (TAMU) for helpful discussions and editorial assistance.

Author Contributions

Conceived and designed the experiments: JW TW PL HSR. Performed the experiments: JW TW. Analyzed the data: JW TW PL HSR. Contributed reagents/materials/analysis tools: JW TW PL. Contributed to the writing of the manuscript: JW PL HSR.

References

1. Browne P, O'Cuinn G (1983) The purification and characterization of a proline dipeptidase from guinea pig brain. Journal of Biological Chemistry 258: 6147–6154.
2. Fernandez-Espla MD, Martin-Hernandez MC, Fox PF (1997) Purification and characterization of a prolidase from Lactobacillus casei subsp. casei IFPL 731. Appl Environ Microbiol 63: 314–316.
3. Jalving R, Bron P, Kester HC, Visser J, Schaap PJ (2002) Cloning of a prolidase gene from Aspergillus nidulans and characterisation of its product. Mol Genet Genomics 267: 218–222.
4. Lupi A, Della Torre S, Campari E, Tenni R, Cetta G, et al. (2006) Human recombinant prolidase from eukaryotic and prokaryotic sources. Expression,

purification, characterization and long-term stability studies. FEBS J 273: 5466–5478.

5. Maher MJ, Ghosh M, Grunden AM, Menon AL, Adams MW, et al. (2004) Structure of the prolidase from Pyrococcus furiosus. Biochemistry 43: 2771–2783.

6. Park MS, Hill CM, Li Y, Hardy RK, Khanna H, et al. (2004) Catalytic properties of the PepQ prolidase from Escherichia coli. Arch Biochem Biophys 429: 224–230.

7. Sjostrom H, Noren O, Josefsson L (1974) Purification and specificity of pig intestinal prolidase. Biochimica et Biophysica Acta 327: 457–470.

8. Arabidopsis Genome I (2000) Analysis of the genome sequence of the flowering plant Arabidopsis thaliana. Nature 408: 796–815.

9. Surazynski A, Miltyk W, Palka J, Phang JM (2008) Prolidase-dependent regulation of collagen biosynthesis. Amino Acids 35: 731–738.

10. Kitchener RL, Grunden AM (2012) Prolidase function in proline metabolism and its medical and biotechnological applications. J Appl Microbiol 113: 233–247.

11. Myara I, Charpentier C, Lemonnier A (1984) Prolidase and prolidase deficiency. Life Sci 34: 1985–1998.

12. Endo F, Tanoue A, Hata A, Kitano A, Matsuda I (1989) Deduced amino acid sequence of human prolidase and molecular analyses of prolidase deficiency. J Inherit Metab Dis 12: 351–354.

13. Lupi A, Tenni R, Rossi A, Cetta G, Forlino A (2008) Human prolidase and prolidase deficiency: an overview on the characterization of the enzyme involved in proline recycling and on the effects of its mutations. Amino Acids 35: 739–752.

14. Amsterdam A, Nissen RM, Sun Z, Swindell EC, Farrington S, et al. (2004) Identification of 315 genes essential for early zebrafish development. Proc Natl Acad Sci U S A 101: 12792–12797.

15. Kamath RS, Fraser AG, Dong Y, Poulin G, Durbin R, et al. (2003) Systematic functional analysis of the Caenorhabditis elegans genome using RNAi. Nature 421: 231–237.

16. Silva JM, Marran K, Parker JS, Silva J, Golding M, et al. (2008) Profiling essential genes in human mammary cells by multiplex RNAi screening. Science 319: 617–620.

17. White JK, Gerdin AK, Karp NA, Ryder E, Buljan M, et al. (2013) Genome-wide generation and systematic phenotyping of knockout mice reveals new roles for many genes. Cell 154: 452–464.

18. Miller CG, Schwartz G (1978) Peptidase-deficient mutants of Escherichia coli. J Bacteriol 135: 603–611.

19. Vyas NK, Nickitenko A, Rastogi VK, Shah SS, Quiocho FA (2010) Structural insights into the dual activities of the nerve agent degrading organophosphate anhydrolase/prolidase. Biochemistry 49: 547–559.

20. Stepankova A, Duskova J, Skalova T, Hasek J, Koval T, et al. (2013) Organophosphorus acid anhydrolase from Alteromonas macleodii: structural study and functional relationship to prolidases. Acta Crystallogr Sect F Struct Biol Cryst Commun 69: 346–354.

21. Cheng TC, Rastogi VK, Defrank JJ, Sawiris GP (1998) G-type nerve agent decontamination by Alteromonas prolidase. Annals of the New York Academy of Sciences 864: 253–258.

22. Lowther WT, Matthews BW (2002) Metalloaminopeptidases: common functional themes in disparate structural surroundings. Chem Rev 102: 4581–4608.

23. Fraser CM, Gocayne JD, White O, Adams MD, Clayton RA, et al. (1995) The minimal gene complement of Mycoplasma genitalium. Science 270: 397–403.

24. Himmelreich R, Hilbert H, Plagens H, Pirkl E, Li BC, et al. (1996) Complete Sequence Analysis of the Genome of the Bacterium Mycoplasma Pneumoniae. Nucleic Acids Research 24: 4420–4449.

25. Jaffe JD, Stange-Thomann N, Smith C, DeCaprio D, Fisher S, et al. (2004) The complete genome and proteome of Mycoplasma mobile. Genome Res 14: 1447–1461.

26. Pollack JD, Williams MV, McElhaney RN (1997) The comparative metabolism of the mollicutes (Mycoplasmas): the utility for taxonomic classification and the relationship of putative gene annotation and phylogeny to enzymatic function in the smallest free-living cells. Crit Rev Microbiol 23: 269–354.

27. Bazan JF, Weaver LH, Roderick SL, Huber R, Matthews BW (1994) Sequence and Structure Comparison Suggest That Methionine Aminopeptidase, Prolidase, Aminopeptidase-P, and Creatinase Share a Common Fold. Proceedings of the National Academy of Sciences of the United States of America 91: 2473–2477.

28. Alberto ME, Leopoldini M, Russo N (2011) Can human prolidase enzyme use different metals for full catalytic activity? Inorg Chem 50: 3394–3403.

29. Wilcox DE (1996) Binuclear Metallohydrolases. Chem Rev 96: 2435–2458.

30. Wilce MC, Bond CS, Dixon NE, Freeman HC, Guss JM (1998) Structure and mechanism of a proline-specific aminopeptidase from Escherichia coli. Proc Natl Acad Sci U S A 95: 3472–3477.

31. Wang SH, Zhi QW, Sun MJ (2005) Purification and characterization of recombinant human liver prolidase expressed in Saccharomyces cerevisiae. Archives of Toxicology 79: 253–259.

32. Willingham K, Maher MJ, Grunden AM, Ghosh M, Adams MWW, et al. (2001) Crystallization and characterization of the prolidase from Pyrococcus furiosus. Acta Crystallographica Section D: Biological Crystallography 57: 428–430.

33. Lowther WT, Matthews BW (2000) Structure and function of the methionine aminopeptidases. Biochim Biophys Acta 1477: 157–167.

34. Graham SC, Bond CS, Freeman HC, Guss JM (2005) Structural and functional implications of metal ion selection in aminopeptidase P, a metalloprotease with a dinuclear metal center. Biochemistry 44: 13820–13836.

35. Otwinowski Z, Minor W (1997) Processing of X-ray diffraction data collected in oscillation mode. 276: 307–326.

36. Adams PD, Afonine PV, Bunkoczi G, Chen VB, Davis IW, et al. (2010) PHENIX: a comprehensive Python-based system for macromolecular structure solution. Acta Crystallogr D Biol Crystallogr 66: 213–221.

37. Emsley P, Lohkamp B, Scott WG, Cowtan K (2010) Features and development of Coot. Acta Crystallogr D Biol Crystallogr 66: 486–501.

38. Trott O, Olson AJ (2010) AutoDock Vina: improving the speed and accuracy of docking with a new scoring function, efficient optimization, and multithreading. J Comput Chem 31: 455–461.

39. Ito Y, Watanabe Y, Hirano K, Sugiura M, Sawaki S, et al. (1984) A fluorometric method for dipeptidase activity measurement in urine, using L-alanyl-L-alanine as substrate. J Biochem 96: 1–8.

40. Zhang LB, Crossley MJ, Dixon NE, Ellis PJ, Fisher ML, et al. (1998) Spectroscopic identification of a dinuclear metal centre in manganese(II)-activated aminopeptidase P from Escherichia coli: implications for human prolidase. Journal of Biological Inorganic Chemistry 3: 470–483.

41. Besio R, Gioia R, Cossu F, Monzani E, Nicolis S, et al. (2013) Kinetic and structural evidences on human prolidase pathological mutants suggest strategies for enzyme functional rescue. PLoS One 8: e58792.

42. Outten CE, O'Halloran TV (2001) Femtomolar sensitivity of metalloregulatory proteins controlling zinc homeostasis. Science 292: 2488–2492.

43. Ghosh M, Grunden AM, Dunn DM, Weiss R, Adams MW (1998) Characterization of native and recombinant forms of an unusual cobalt-dependent proline dipeptidase (prolidase) from the hyperthermophilic archaeon Pyrococcus furiosus. J Bacteriol 180: 4781–4789.

44. Fersht A (1999) Structure and Mechanism in Protein Science: A Guide to Enzyme Catalysis and Protein Folding. New York: W. H. Freeman. 631 p.

45. Gerlt JA, Babbitt PC (2001) Divergent evolution of enzymatic function: mechanistically diverse superfamilies and functionally distinct suprafamilies. Annu Rev Biochem 70: 209–246.

46. Shao Z, Arnold FH (1996) Engineering new functions and altering existing functions. Curr Opin Struct Biol 6: 513–518.

47. Perona JJ, Craik CS (1995) Structural basis of substrate specificity in the serine proteases. Protein Sci 4: 337–360.

48. Hu K, Tanaka T (2009) S1 site residues of Lactococcus lactis prolidase affect substrate specificity and allosteric behaviour. Biochim Biophys Acta 1794: 1715–1724.

49. Cohen GN, Barbe V, Flament D, Galperin M, Heilig R, et al. (2003) An integrated analysis of the genome of the hyperthermophilic archaeon Pyrococcus abyssi. Mol Microbiol 47: 1495–1512.

50. Jun X, Lupeng L, Minjuan X, Oger P, Fengping W, et al. (2011) Complete genome sequence of the obligate piezophilic hyperthermophilic archaeon Pyrococcus yayanosii CH1. J Bacteriol 193: 4297–4298.

51. Jung JH, Lee JH, Holden JF, Seo DH, Shin H, et al. (2012) Complete genome sequence of the hyperthermophilic archaeon Pyrococcus sp. strain ST04, isolated from a deep-sea hydrothermal sulfide chimney on the Juan de Fuca Ridge. J Bacteriol 194: 4434–4435.

52. Kawarabayasi Y, Sawada M, Horikawa H, Haikawa Y, Hino Y, et al. (1998) Complete sequence and gene organization of the genome of a hyper-thermophilic archaebacterium, Pyrococcus horikoshii OT3. DNA Res 5: 55–76.

53. Lecompte O, Ripp R, Puzos-Barbe V, Duprat S, Heilig R, et al. (2001) Genome evolution at the genus level: comparison of three complete genomes of hyperthermophilic archaea. Genome Res 11: 981–993.

54. Lee HS, Bae SS, Kim MS, Kwon KK, Kang SG, et al. (2011) Complete genome sequence of hyperthermophilic Pyrococcus sp. strain NA2, isolated from a deep-sea hydrothermal vent area. J Bacteriol 193: 3666–3667.

55. Maeder DL, Weiss RB, Dunn DM, Cherry JL, Gonzalez JM, et al. (1999) Divergence of the hyperthermophilic archaea Pyrococcus furiosus and P. horikoshii inferred from complete genomic sequences. Genetics 152: 1299–1305.

56. Forlino A, Lupi A, Vaghi P, Cornaglia AI, Calligaro A, et al. (2002) Mutation analysis of five new patients affected by prolidase deficiency: The lack of enzyme activity causes necrosis-like cell death in cultured fibroblasts. Human Genetics 111: 314–322.

57. Ledoux P, Scriver CR, Hechtman P (1996) Expression and molecular analysis of mutations in prolidase deficiency. Am J Hum Genet 59: 1035–1039.

58. Ledoux P, Scriver C, Hechtman P (1994) Four novel PEPD alleles causing prolidase deficiency. Am J Hum Genet 54: 1014–1021.

59. Tawfik DS (2006) Biochemistry. Loop grafting and the origins of enzyme species. Science 311: 475–476.

60. Larkin MA, Blackshields G, Brown NP, Chenna R, McGettigan PA, et al. (2007) Clustal W and Clustal X version 2.0. Bioinformatics 23: 2947–2948.

61. Gouet P (2003) ESPript/ENDscript: extracting and rendering sequence and 3D information from atomic structures of proteins. Nucleic Acids Research 31: 3320–3323.

62. Altschul SF, Gish W, Miller W, Myers EW, Lipman DJ (1990) Basic local alignment search tool. J Mol Biol 215: 403–410.

63. Schrodinger LLC (2010) The PyMOL Molecular Graphics System, Version 1.3r1.

4

Operating Mechanism and Molecular Dynamics of Pheromone-Binding Protein ASP1 as Influenced by pH

Lei Han[1,2], Yong-Jun Zhang[3], Long Zhang[4], Xu Cui[5], Jinpu Yu[1], Ziding Zhang[2]*, Ming S. Liu[6]*

1 Centre for Cancer Molecular Diagnosis, Tianjin Medical University Cancer Institute and Hospital, National Clinical Research Center for Cancer, Tianjin, China, **2** State Key Laboratory of Agrobiotechnology, College of Biological Sciences, China Agricultural University, Beijing, China, **3** State Key Laboratory for Biology of Plant Diseases and Insect Pests, Institute of Plant Protection, Chinese Academy of Agricultural Sciences, Beijing, China, **4** Key Lab for Biological Control of the Ministry of Agriculture, China Agricultural University, Beijing, China, **5** Beijing Computing Center, Beijing, China, **6** CSIRO - Computational Informatics & Digital Productivity Flagship, Private Bag 10, Clayton South, Australia

Abstract

Odorant binding protein (OBP) is a vital component of the olfactory sensation system. It performs the specific role of ferrying odorant molecules to odorant receptors. OBP helps insects and types of animal to sense and transport stimuli molecules. However, the molecular details about how OBPs bind or release its odorant ligands are unclear. For some OBPs, the systems' pH level is reported to impact on the ligands' binding or unbinding capability. In this work we investigated the operating mechanism and molecular dynamics in bee antennal pheromone-binding protein ASP1 under varying pH conditions. We found that conformational flexibility is the key factor for regulating the interaction of ASP1 and its ligands, and the odorant binds to ASP1 at low pH conditions. Dynamics, once triggered by pH changes, play the key roles in coupling the global conformational changes with the odorant release. In ASP1, the C-terminus, the N-terminus, helix $\alpha2$ and the region ranging from helices $\alpha4$ to $\alpha5$ form a cavity with a novel 'entrance' of binding. These are the major regions that respond to pH change and regulate the ligand release. Clearly there are processes of dynamics and hydrogen bond network propagation in ASP1 in response to pH stimuli. These findings lead to an understanding of the mechanism and dynamics of odorant-OBP interaction in OBP, and will benefit chemsensory-related biotech and agriculture research and development.

Editor: Chandra Verma, Bioinformatics Institute, Singapore

Funding: This work was partially supported by the National Key Basic Research Program of China (2009CB918802 and 2012CB114104) and the State Education Ministry of China (20120008110014). The funders had no role in study design, data collection and analysis, decision to publish, or preparation of the manuscript.

Competing Interests: The authors have declared that no competing interests exist.

* Email: zidingzhang@cau.edu.cn (ZZ); ming.liu@csiro.au (MSL)

Introduction

Olfactory sensation is an essential capability for insects and mammals, enabling them to detect stimuli in the surroundings for prey, survival and reproduction [1,2]. In the chemical-to-sensation process, odorant binding proteins (OBPs) ferry small, primarily hydrophobic odorant and/or pheromone molecules through sensillar lymph to olfactory receptors (ORs), triggering a cascade of chemosensory events which lead to activate sensory neurons [3,4]. Signaling chemical molecule perception is particularly vital for many insects, such as social insect like honey bees, where the queen groups and controls the individual behaviors using sophisticated pheromone communication. Many studies have attempted to determine the key residues for OBP ligand recognition [5], binding and releasing [6–10]; however, the OBP's roles in delivering odorants has caused extensive debate [4,11]. Therefore, the mechanism and dynamic pathways on how OBP binds and releases pheromones *in vivo* need to be explored at molecular level. An atom-level dynamics understanding of OBPs' binding and unbinding of odorant ligands, especially how pH affects the interactions between OBPs and their ligands, will help us understand the operating mechanism and functions of OBPs, ORs and the chemosensory system. This will assist with disease prevention, pest-control [12], food processes and agricultural technologies [1,12].

The initial steps of chemo-sensing in honey bees involve the pheromone binding proteins (PBPs, one sub-type of OBP) binding to pheromone molecules, and carrying these ligands to ORs so as to activate ORs [3]. To date the accepted mechanism, as revealed by the crystal structures [7,13–15] (see Fig.1), is that the process of OBP binding and releasing pheromone is to some degree pH-dependent. The same PBP and their ligands can be crystallized either in *apo* (ligand-free) or *holo* (ligand-bound) states subject to varying pH. Meanwhile, in an aqueous environment, there are different conformational states of the same protein at different pH [14]. In addition to honey bees, BmorPBP1 (PBP from Bombyx mori) structures show that the *C*-terminal loop is an important region in the presence of changing pH conditions [13,14]. When BmorPBP1 is exposed to low pH condition (e.g. pH = 4.5), the *C*-terminal loop forms a new helix towards the binding cavity and pushes the odorant ligand out of the cavity. Conversely, at neutral pH condition (e.g. pH = 6.5), the unstructured *C*-terminus (*C*-ter) extends into the solvent and opens the binding cavity to host the ligand. Similar to BmorPBP1, ApolPBP (PBP from the giant silk moth *Antheraea polyphemus*) and AtraPBP1 (PBP from the navel orange worm *Amyelois transitella*) have the same long and unstructured *C*-ter as BmorPBP1, sharing a similar same mechanism in response to pH changes [16–18]. Unlike BmorPBP1, ASP1 (PBP from honeybee *Apis mellifera*) [19], AgamOBP1 (OBP from the malaria mosquito *Anopheles gambia*)

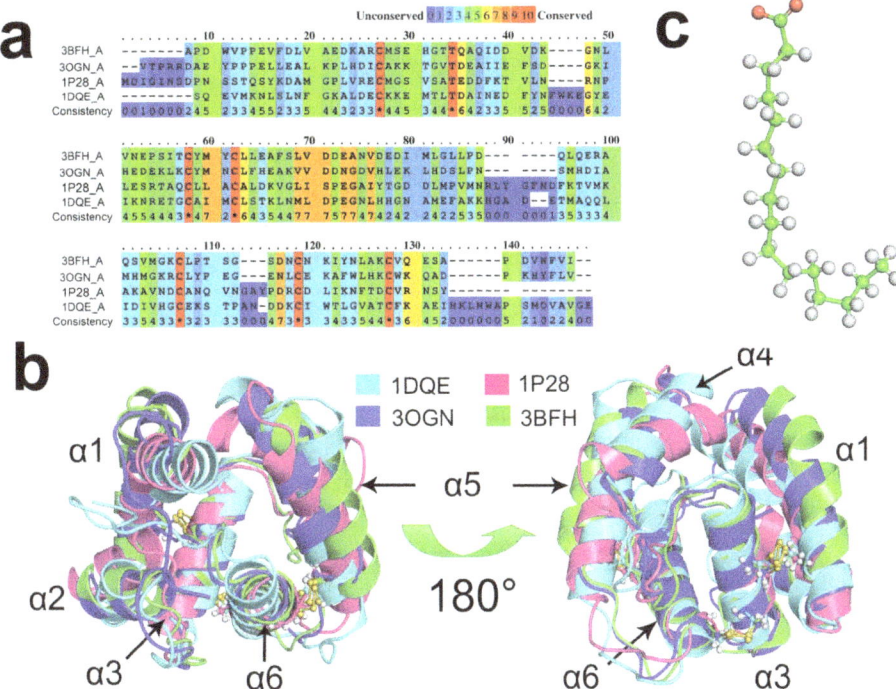

Figure 1. OBP sequences and structures. (A) The multi-sequence alignment for OBPs, 3BFH (PBP from honeybee *Apis mellifera*), 3OGN (OBP from the malaria mosquito *Anopheles gambia*), 1P28 (PBP from the cockroach *Leucophaea maderae*) and 1DQE (PBP from the silkmoth *Bombyx mori*). They represented the OBPs with different length of chain. The main difference of these OBPs is the length of C-terminal loop: 1DQE has a long C-ter, 1P28 has no C-ter while 3BFH and 3OGN have a mediate long C-ter. (B) The structure alignment for the above OBPs. Though they have low sequence similarity, they share almost identical tertiary structures, which imply that they share the same operating mechanism and dynamics. Three conserved disulfide bonds are shown as ball-and-stick models in yellow color. (C) The structure of palmitic acid, a typical odorant ligand.

[20], AaegOBP1 (OBP from the yellow fever mosquito *Aedes aegypti*) [21] and CquiOBP1 (OBP from southern house mosquito *Culex quinquefasciatus*) [9] do not contain a long loop at their C-termini, but their short loop could also fold into the binding cavity, occupy the binding site and disrupt the ligand's entry.

Earlier structural biology investigations [7,13–17] have mainly focused on the functionality of the C-ter loop and key residues such as Asp35 in response to mutation and pH changes. However, they neglected the detailed and complete dynamics pathways on how OBP protein binds and unbinds its odorants, particularly at different conditions of pH. Recently, a few molecular modellings attempted to tackle the OBP- odorant dynamics and interactions [22–27]. These works encourage further investigation due to the still missing mechanisms, the lack of dynamics details and considerable uncertainty about the structure-function rationale. In this work, we chose honeybee *Apis mellifera* ASP1 as a model system and undertook long-time all-atom molecular simulations in order to elucidate the molecular mechanism and dynamics interaction of OBPs and its odorant ligands. ASP1, in significant contrast to many other OBPs, was reported to bind its ligands at lower pH condition while releasing them at neutral or high pH [7]. At the same time, ASP1 is structurally and genetically aligned well with many other OBPs (Fig. 1a & 1b). In our work, we focus on how pH affects the interation of OBP with its ligands and the mechanisms of OBP releasing ligand at a favorable pH condition. Through quantitative analysis of the global, local conformational changes and other dynamic properties of apo- and holo-ASP1s at pH 4.5 or pH 7.0, we try to illustrate a complete dynamics picture of molecular process how pH affects odorant release. We examined the dynamics contribution, not only of C-ter but also

of N-ter, helix α2 and the region ranging from helices α4 to α5 (which form the entrance and core of the binding cavity), as well as its intrinsic disulfide bonds and hydrogen bond network.

Results and Discussion

Flexibility and fluctuation of the ASP1 structure

The global conformational changes in ASP1 with ligand (the *holo* state) and without ligand (the *apo* state) are depicted in Figure S2.1 in **Supporting Information S2**. The *holo* states are shown to have lower RMSD and RMSF than the *Apo* states (see Figure S2.1a & S2.1c in **Supporting Information S2**). Compared with the *apo* forms, the weaker conformation flexibility for the *holo* forms of ASP1 indicates that the ligand located in the hydrophobic cavity helps stabilize the overall structure of ASP1 and keep it dynamically tighter. The less flexible *holo* structures can help ASP1 carry the hydrophobic odorant molecules to the odorant receptor. The *apo* states of ASP1 at pH 4.5 condition appear to be more dynamic than the *holo* states (see Figure S2.1c in **Supporting Information S2**). This implies that the low pH condition provides the needed environment and dynamic condition for ligands to bind onto ASP1.

pH does have an effect on the flexibility of ASP1. As presented in Figure S2.1b, S2.1d & S2.1e in **Supporting Information S2**, in the *holo* state, RMSD and RMSF of ASP1 at pH 4.5 condition are definitely lower than the systems at pH 7.0 condition. The average values of RMSD (of 200 ns MD triplicates) for 'holo-3BFH' at low pH 4.5 and neutral pH 7.0 conditions are 1.56 and 2.02 Å, respectively, with standard deviation 0.15 and 0.29 Å. We found the same behavior in the 'holo-2H8V' state (see Figure S2.1a in **Supporting Information S2**): When set at pH 4.5 condition, its RMSD is about 1.37 Å with standard deviation 0.14 Å. At pH 7.0, RMSD fluctuates around 1.78 Å with a standard deviation of 0.33 Å. These observations indicate that the ASP1 structure at pH 7.0 undergoes greater conformational change and higher fluctuation during dynamics runs, and it is unfavorable for ligand docking to the cavity under pH 7.0.

The C-terminal loop was believed to be important for odorant binding and releasing [13,14]. For example in BmorPBP1, a pheromone-binding protein from the silk moth *Bombyx mori*, the transition from *holo*- to *apo*- BmorPBP1 comes when the C-terminal loop occupies the binding cavity in *apo* state [13]. A key question is whether the C-terminal loop folds into the binding cavity when the odorant ligand is released from the *holo*-ASP1, or it falls out of the cavity when the ligand is docked into the *apo*-ASP1? To explore these scenarios, we carefully looked at the trajectories simulated at pH 4.5 for the *apo* and *holo* states. Putting aside N-ter (a.a. 3 to 14) and C-ter (a.a. 111 to 119), as shown in Figure S2.1c in **Supporting Information S2**, the main fluctuating regions are the down-stream loop of $\alpha 2$ (a.a. 28 to 35), the loop between the helices of $\alpha 3$ and $\alpha 4$, and the region from $\alpha 4$ to $\alpha 5$ (a.a. 67 to 89). These fluctuating regions form the binding cavity, with the dynamic parts acting as a kind of 'entrance' of the cavity (see Fig. 2a, namely the 'entrance' formed by N-ter, C-ter and helix $\alpha 2$ and helices $\alpha 4$ and $\alpha 5$), which works to attract or eject the odorant ligands into/out of the cavity.

Given that RMSF only represents the flexibility residue by residue, we need to trace the forces of motion and other dynamic movement of the cavity during ASP1 actions. For this purpose, we monitored how the distances between the center of the cavity and four key cavity 'entrance' components change over the full trajectories. In this process, the centers of the total cavity (C_{total}) and each component of the cavity 'entrance' (C_{each}) were calculated using Eq. (2), then distances between C_{total} and C_{each} were measured as demonstrated in Fig. 2a. Fig. 2b shows that, except for helix $\alpha 2$, the other three components of the cavity entrance moved inwards the cavity in a coupling way and the binding cavity can contract without a ligand. This demonstrates again that C-ter is not the only region affected by the ligand, but the other parts of the whole cavity will respond to the ligand binding/releasing in a cooperative manner.

ASP1's fluctuations response to the pH stimuli while the protonated residues do not directly interact with cavity/ligands

Previous investigation argued that the pH condition might trigger the release of the odorant ligand [11,14,17]. For ASP1, the ligand could settle in the cavity at the low pH condition (e.g. pH = 4.5) and unbind from the cavity at the neutral pH condition [7]. The different pH conditions will of course induce a different protonation state of the ionizable residues (as described in the Methods section). The structural studies implied that the effect of pH is employed indirectly by Asp35 to bend or unbend C-ter against the cavity. Nonetheless, it is unclear how different pH conditions (with responsible residues protonated) trigger the OBP/PBP cavity to bind or unbind the ligand, or even further ferry odorant molecules through lymph to activate the odorant receptors. It would be straightforward to explain the pH effect if the protonated residues of ASP1 are located in the binding cavity and directly involved in the cavity-ligand interaction. However, as shown in Fig. 3a, due to the location and distribution of protonated residues in ASP1, the protonated residues are not able to directly mediate the interaction of ASP1 and its ligands. The residues which are around 5 Å of the ligand are mainly

hydrophobic and aromatic residues and the ionizable residues are distanced from the binding cavity (Fig. 3b). Then how did the ionizable residues mediate the interaction between ASP1 and its ligands at different pH levels? The only answer lies in that the change of conformational flexibility and fluctuation of ASP1, induced by pH changes, will propagate and induce the dynamic changes to the cavity for binding or releasing ligands.

ASP1 favors the low pH condition for binding the ligand

Radius of gyration (R_g) is a dynamic feature representing the structural compactness of protein [28]. As depicted in Fig. 4a, ASP1 at pH 4.5 has lower value of R_g than at pH 7.0 condition. The higher R_g value at pH 7.0 implies that the ASP1 structure becomes more dynamic at this pH condition, which is not favorable for keeping the ligand bound to the cavity. As also illustrated in Figure S2.1d in **Supporting Information S2**, the RMSF of ASP1 at pH 7.0 is larger than the case of at pH 4.5. The significant fluctuating regions, in addition to N-ter and C-ter, are $\alpha 2$ and the segment from $\alpha 4$ to $\alpha 5$. As discussed previously, these regions form an entrance and channel path for ASP1's cavity. The higher flexibility of these regions will make the cavity tend to unbind the ligand, instead of binding it.

From the above evidence, we conclude that the changes of global flexibility and fluctuation propagation as induced by pH changes define the operation mode of ASP1 and its cavity-ligand interactions. To test this hypothesis, we measured the time evolution of the cavity entrance's orientation and shape at different pH conditions. The details are shown in Fig. 2c. At pH 7.0, N-ter, helix $\alpha 2$ and helices $\alpha 4$ and $\alpha 5$ are far away from the center of cavity while C-ter is close to the center of the entrance, in contrast to their position at pH 4.5. These give us a quantitative picture of how ASP1 releases its ligand out of the cavity. During the ligand releasing, N-ter, helix $\alpha 2$ and helices $\alpha 4$ to $\alpha 5$ will open up, facilitating the ligand's unbinding. At the same time, C-ter folds into the center and occupies the binding cavity, while helping to unbind the ligand from the cavity. This mechanism enforces the previous finding that C-ter plays a main role in responding to the change of pH condition [13,14]. Generally, the pH changes should trigger a collaborative motion of not only C-ter, but also N-ter, helices $\alpha 2$, $\alpha 4$ and $\alpha 5$, and the coordination between C-ter and other cavity parts is vital condition.

Fig. 2c also shows that there is more than one peak for $\alpha 2$ and C-ter at pH 7.0 condition. This suggests that the ligand releasing is a dynamics propagating process. In other words, at pH 7.0 condition the cavity is more flexible, so as to transit in a series of conformation population (e.g. from opening to closing, and vice versa). The ligand is released from the binding pocket with certain higher probability. This assumption is confirmed by the conformation evolution as in Fig. 2d. In Fig. 2d, the total simulation time was divided into ten segments with the intervals of 20 ns. At pH 4.5, the cavity is very stable and dynamically constrained. But for ASP1 at pH 7.0, the four key components of the cavity, especially C-ter and helix $\alpha 2$, become very dynamic and undergo conformational transition. The higher flexibility of these components opens up the cavity (both volume and entrance) and shifts the dynamics probability towards the ligand releasing. At the same time, the opening of the cavity will reduce the interaction between ASP1 and its hydrophobic ligands. This means that the ligand will move more freely in favour of releasing.

ASP1, like most of OBPs, has three disulfide bonds in the core of the tertiary structure (Fig. 1b). Figure S2.1 in **Supporting Information S2** shows the regions near the disulfide bonds have low flexibility in comparison with other regions, either at pH 4.5 or at pH 7.0 condition. In contrast to the flexible cavity entrance,

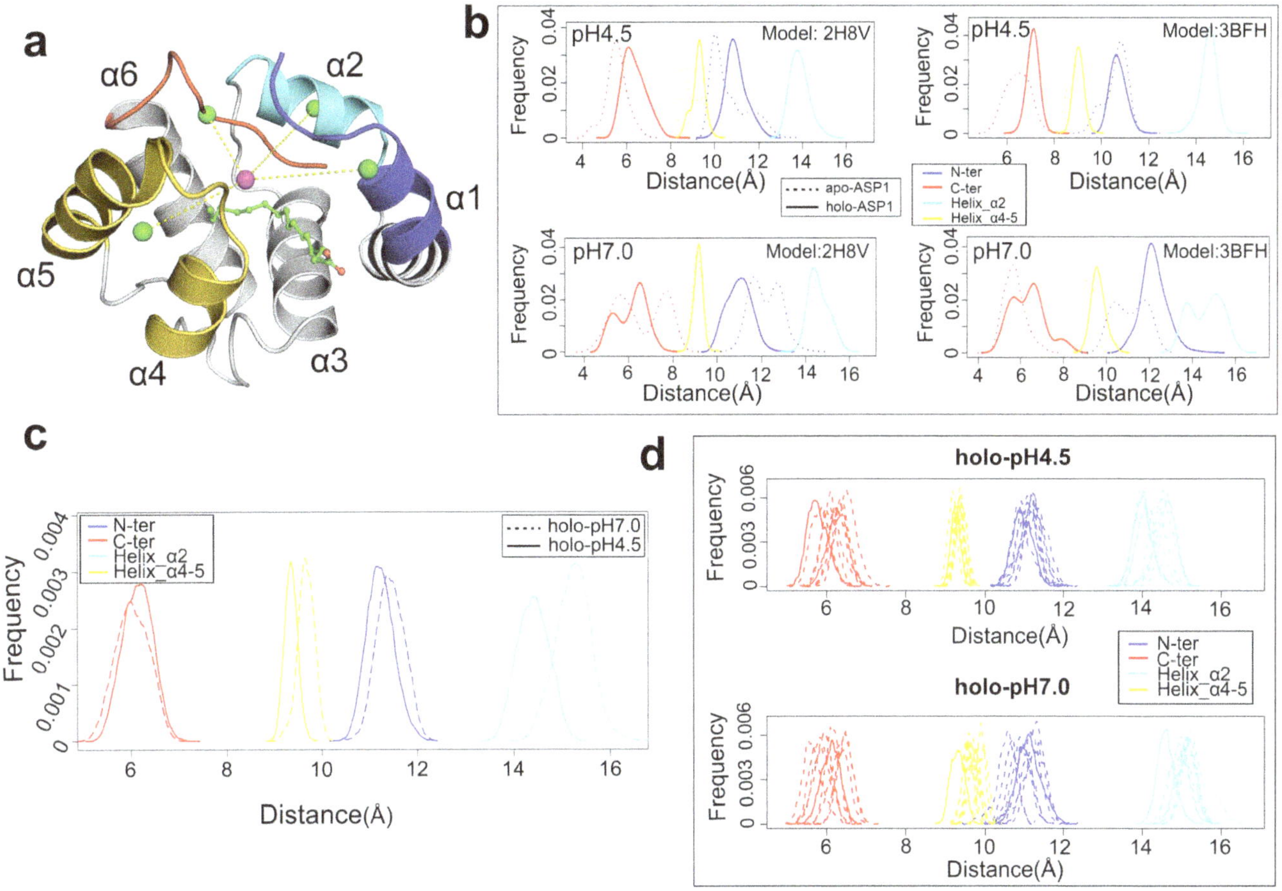

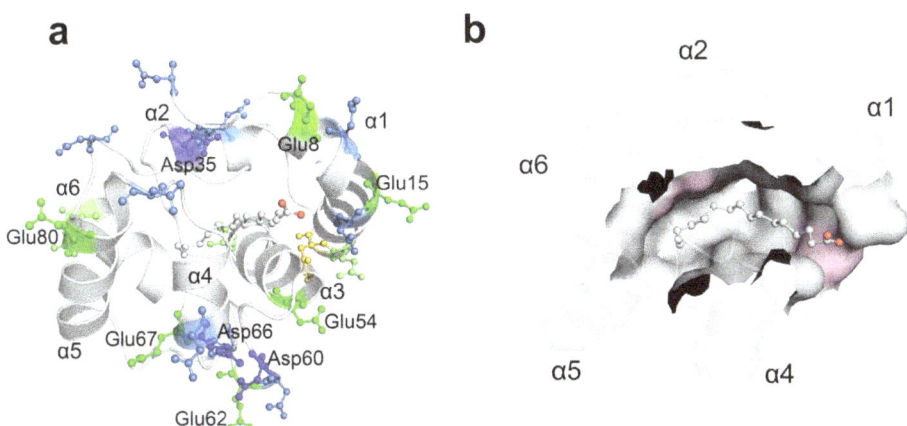

Figure 2. OBP binding cavity and its dynamics. (A) The 'entrance': four key components of the 'entrance' are drawn in different colors, with the center of the cavity represented in pink ball and the center of each key component depicted as green ball. The protein is presented by solid ribbons and the ligand molecule is in a ball-and-stick model. (B) The relative distribution of distance from the centers of each 'entrance' component to the cavity center for *apo-* and *holo-* states at same pH condition. (C) The relative distribution of distance between each entrance component and the entrance center for *holo*-state simulated at low pH 4.5 (solid lines) and neutral pH 7.0 (dash lines) conditions. (D) Time evolution of the relative distribution (with 20 ns interval) of each 'entrance' component to the center of cavity for *holo*-state at pH 4.5 (top) and pH 7.0 (below) conditions.

Figure 3. Distribution of ionizable residues versus the cavity. (A) Ionizable residues are highlighted by ball-and-stick model with different colors, aspartic acid (blue), glutamic acid (green) and histidine (gold); (B) The cavity residues, 5 Å from ligand, are shown as surface model. The hydrophobic residues are shown with white and the pink represented with aromatic residues. The ligand bound into the binding cavity is drawn in a ball-and-stick model.

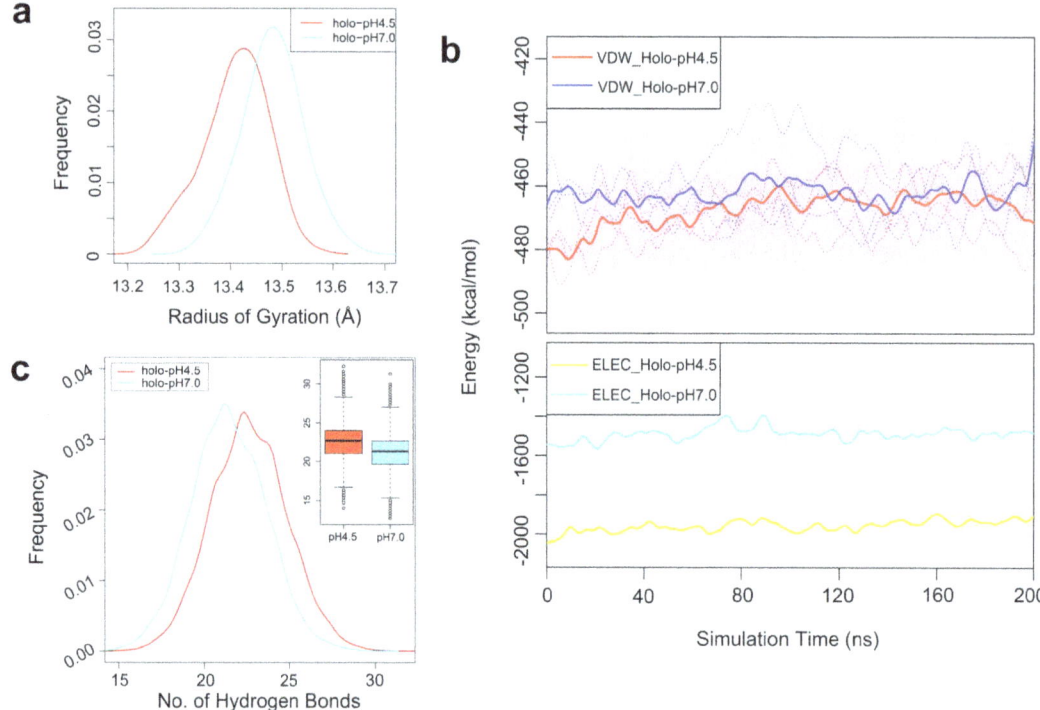

Figure 4. The radius of gyration R_g (A) and hydrogen bonds (C) distribution of *holo-* state simulated at pH 4.5 (solid red lines) and pH 7.0 (solid cyan lines) conditions. (B) The time evolution of van der Waals (top) and electrostatic (below) energy of *holo-* state simulated at pH 4.5 and pH 7.0 conditions, respectively. The dotted lines indicate individual trajectories of the three replicates.

the disulfide bonds form a unique geometry plane (Figure S2.2a in **Supporting Information S2**), consisting of supra scaffolds, plays the role to stabilize the whole ASP1 structure. Due to the special role of these disulfide bonds, the fluctuation of geometry plane of three disulfide bonds reflect the exact motion of different regions in ASP1. As depicted in Figure S2.2b in **Supporting Information S2**, at pH 7.0 condition, the geometry plane of disulfide bonds is more dynamic (with larger fluctuation of planar angle) than at pH 4.5. This indicates that pH 7.0 is a condition for larger flexibility and will have the cavity opening for ligand releasing.

To quantitatively assess the binding energy of ligand to ASP1 as impacted by different pH conditions, we calculated by free energy perturbation (FEP) the binding energy difference at pH 4.5 and pH 7.0 condition (see Eq. (1)). The measured energy difference$\Delta\Delta G^{bind}$ (for binding at pH 4.5 versus at pH 7.0) is -3.47 kcal/mol, see Figure S2.3 in **Supporting Information S2**. This binding energy difference clearly indicates that the ligand prefers binding ASP1 at low pH condition pH 4.5, rather than pH 7.0. Meanwhile, the relative small binding energy difference indicates that the interaction between ASP1 and its ligand is likely non-specific.

Hydrogen bond network does matter: How the pH changes apply its influence

Through the above analysis we learn that the global flexibility and fluctuation play key roles in regulating ligand binding/releasing in ASP1. To further explore this mechanism, we considered three key non-bonded inter- and intra- molecular forces: the electrostatic, van der Waals and hydrogen bonds interactions at different pH conditions. In term of van der Waals energy, the ASP1-ligand complex at pH 4.5 is slightly lower than

the counterpart at pH 7.0. However, the electrostatic energy of the ASP1-ligand (*holo* state) complex at pH 4.5 show much lower values than at pH 7.0 (Fig. 4b), indicating that the complex prefers the pH 4.5 condition. Hydrogen bond (H-bond) is another key factor reflecting the protein dynamics [29]. Fig. 4c shows that ASP1 possesses more H-bonds at pH 4.5 than at pH 7.0.

According to secondary structure analysis (with DSSP algorithm [30]), a beta sheet is formed between residues Leu58 and Asp66 at pH 4.5 condition, but this region is maintained as a unstructured loop at pH 7.0 (Fig. 5a). The H-bond between Leu58 and Asp66 is essential for the formation of the beta sheet. For example, at pH 4.5, Asp66 donates an H-bond to Leu58 over half of the trajectory time (Table 1 & Fig. 5c). However, there is no H-bond formation at pH 7.0. Structurally, the segment ranging from Leu58 to Asp66 connects α3 and α4 (Fig. 5b) and Leu58 to Asp66 are located at the two ends of this region. This specific H-bond can lock the beta sheet motif and thus maintain the rigidity of α4 and its neighboring region. As we know, α4 and its neighbor region are key components of the entrance of the binding cavity, so stabilizing this region can keep the cavity in the closed state for holding the ligand. At pH 7.0, the unprotonated Asp66 moves away from Leu58 and its chain stretches to the opposite direction to the configuration at pH 4.5 (Fig. 5b). With the loss of this H-bond 'lock', α4 and its neighbor part (as well as the cavity 'entrance'), are destabilized and ligand unbinding will be induced.

In addition to the Leu58-Asp66 H-bond, other H-bonds are formed by residues Trp4, Lys17, Val34, Asp35, Try48, Trp116, Val118 and Ile119, which connect *N*-ter, helix α2 and *C*-ter (Fig. 6a), the key components of ASP1's cavity 'entrance'. Among these H-bonds, the H-bond pairs contributed by Val34-Trp116 and Asp35-Ile119 are vital for stabilizing the relative position of *C*-ter and α2. At pH 4.5, the carboxylic side-chain of protonated

Asp35 can form an H-bond with oxygen atoms at carboxyl of Ile119. These two H-bonds appear to be complementary to each other (Fig. 6b) and present about 50% of the trajectory time. However, at pH 7.0, the distances between Asp35 and Ile119 are about 6 Å (Fig. 6b), thus these H-bonds can not form when Asp35 is in an unprotonated state at pH 7.0, until the moment that ligand starts leaving the cavity (e.g. about 40 ns in Fig. 6b). At the same time, the H-bond between Trp116 (N) and Val34 (O) occupies about 17% over the trajectory time at pH 4.5, in contrast to less than 1% at pH 7.0 (Table 1). The H-bond formed by Lys17 and Ile119 is another dynamic factor in stabilizing the cavity entrance of ASP1 (Fig. 6a). At pH 4.5 condition, the established H-bond occupies about 40% of simulation time while only 30% at pH 7.0 condition (Table 1).

It is important to note that the H-bonds contributed by Ile119 interplay with Lys17 and Asp35 to stabilize *C*-ter and *N*-ter (Figure S2.4 in **Supporting Information S2**). This interplay may help the domain swap of *N*-ter at pH neutral condition [31] or potential dimerization [9]. This hypothesis could be proved, since an H-bond between Trp4 and Val118 can be formed (At pH 4.5, this H-bond takes about 8% of simulation time while only near zero under the pH 7.0 condition), where a mutated ASP1

could only remain monomeric form [31]. Meanwhile, the H-bond between Tyr48 and Ile119, which is important in the crystal structure [7], is indeed found to be playing an active role during our dynamics simulation (12.17% occupation time at pH 4.5 vs. 2.92% at pH 7.0).

Through H-bond networks, for ASP1 at pH 4.5 condition, its *C*-ter is largely locked by H-bond formed by residues of Val118, Ile119 and Lys17, Asp35, Tyr48. H-bonds apparently constrain *C*-ter from folding into the binding pocket so as to keep the ligand bound in cavity. Nonetheless, once this H-bond network is broken at pH 7.0, particularly around the *C*-ter and α2 region, then the highly dynamic and flexible *C*-ter will open up the cavity and allow the release of the odorant ligands. This clearly solves the experimentally observed puzzle [7] on how pH affects the structures of ASP1. It demonstrates the pH-sensing 'lock'/'unlock' mechanism proposed by Leal and coworkers [9].

Conclusions

Comprehensive molecular simulations of OBP ASP1 in *apo* and *holo* states at different pH conditions were carried out. In *apo* state, *C*-ter will strike into the cavity and the global conformation undergoes large fluctuations. When the ASP1 and ligand complex

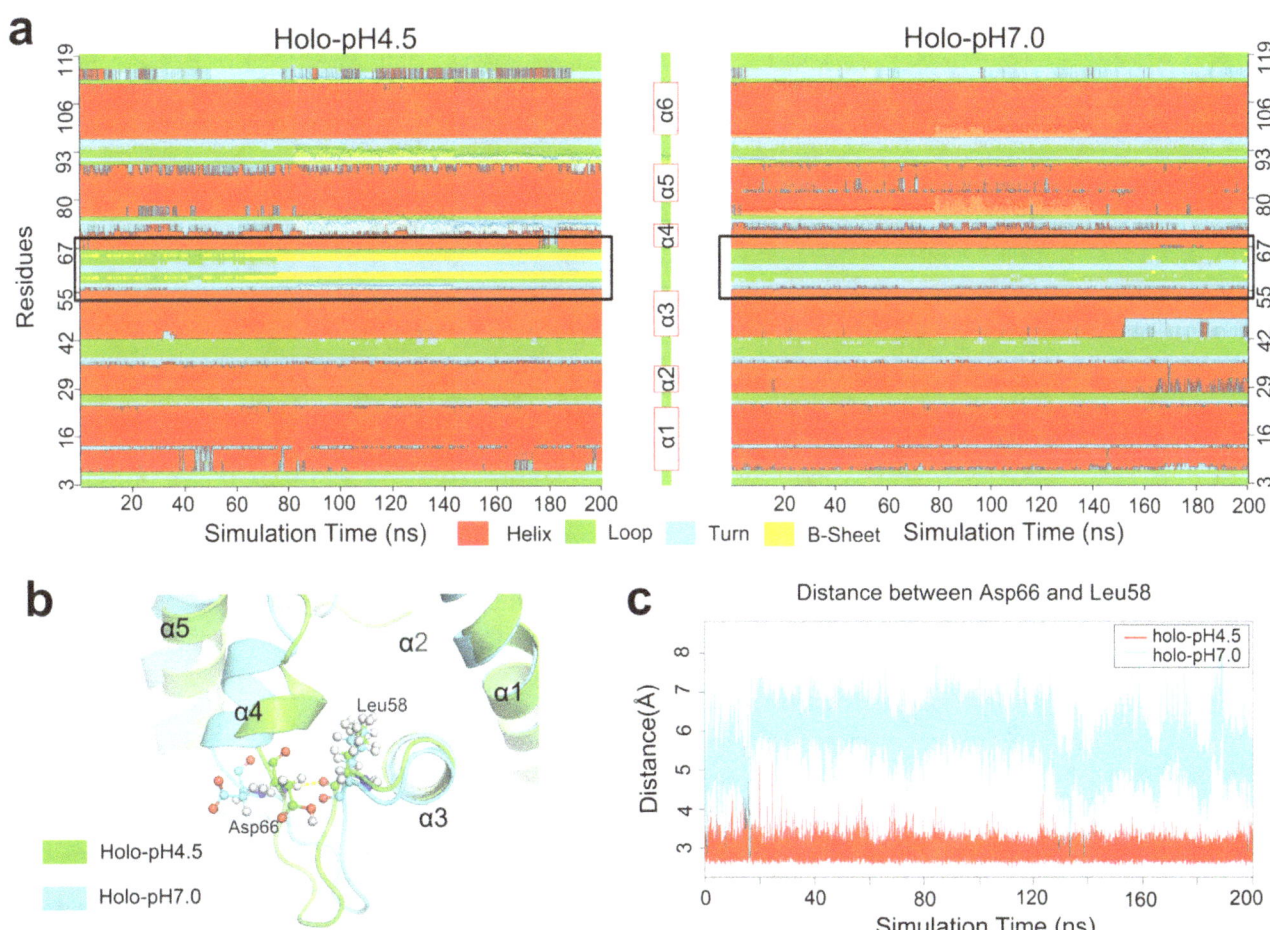

Figure 5. The time evolution of secondary structure and *H*-bond of *holo*-state simulated at pH 4.5 and pH 7.0 conditions. (A) The segment from Leu58 to Asp66 forms a beta sheet at pH 4.5 but always in a loop state at pH 7.0 condition. (B) A snapshot of the region from Leu58 to Asp66 of *holo* ASP1 at pH 4.5 (green) and pH 7.0 (cyan) conditions, respectively. Leu58 and Asp66 are highlighted in ball-and-stick, with its H-bond drawn as dotted line. (C) The hydrogen bond distances between Leu58 and Asp66 are depicted for pH 4.5 (red) and pH 7.0 (cyan) simulation conditions.

Table 1. Hydrogen bonds occupancy in ASP1 at different pH conditions.

Hydrogen bond type	Occupancy (%)	
	Holo-pH 4.5	Holo-pH 7.0
N@Asp66: O@Leu58	50.58	0.31
OD2@Asp35: OT1@Ile119	19.31	~0.00
OD2@Asp35: OT2@Ile119	18.23	~0.00
NZ@Lys17: OT1@Ile119	22.38	17.07
NZ@Lys17:OT2@Ile119	17.29	13.82
NE1@Trp4: O@Val118	8.43	~0.00
OH@Tyr48: OT1@Ile119	12.17	2.92
NE1@Trp116: O@Val34	16.65	0.96

were set at low or neutral pH conditions, we found that the ligand binding cavity formed dynamic 'entrance' geometry (by C-ter, N-ter, helix $\alpha2$ and helices $\alpha4$ and $\alpha5$) and it responds to the changing pH condition. Interestingly the ionizable residues, which answer to pH change with protonation or unprotonation, do not directly interact with the ASP1 ligands; rather, the protonation affects the overall flexibility and fluctuation. There is clearly a process of dynamics propagation in ASP1 (and other OBPs) in response to pH stimuli. ASP1 is found to bind in favor of low pH conditions. H-bonds formed at the cavity entrance play an important role in regulating the ligand release, as indicated by

ASP1 exposed from low pH to neutral pH conditions. The H-bond network carries on the dynamics change induced by varying the pH condition and passes on to the global change of flexibility and fluctuations of the OBP-ligand complex.

In summary, in OBPs/PBPs the ligand binding or releasing is very sensitive to pH conditions, and C-ter must cooperate with other key dynamic components for effective operation. Given there are different OBP families with characteristic C-ter, our finding can provide insightful molecular understanding of the mechanism and dynamics of OBPs, and further harness this understanding to biotechnology and agricultural applications.

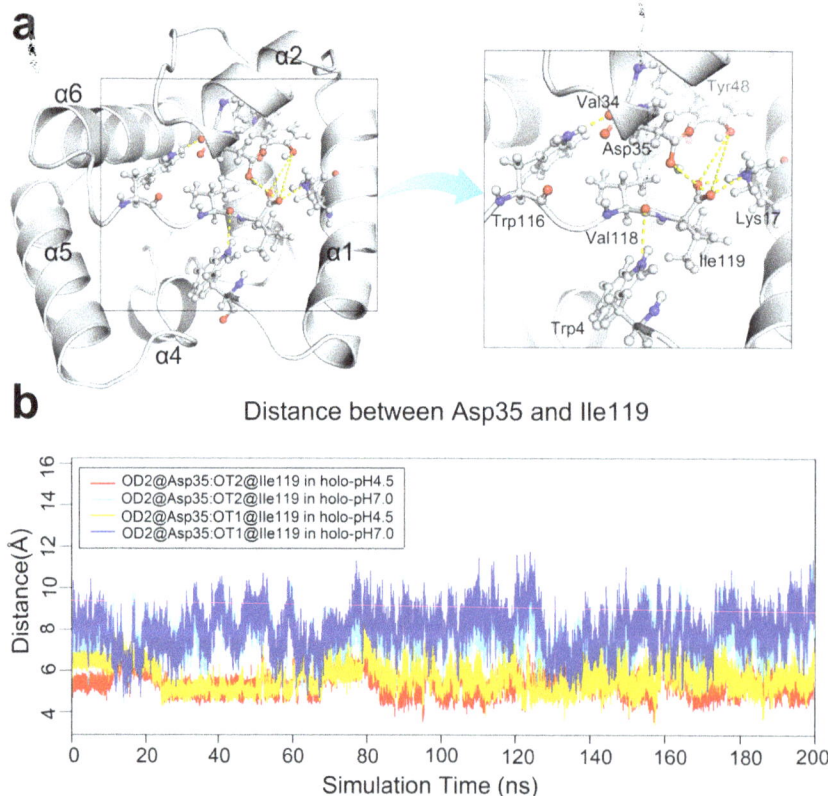

Figure 6. Dynamics of *H*-bond network. (A) Hydrogen bonds network formed by residues Trp4, Lys17, Val34, Asp35, Try48, Trp116, Val118 and Ile119. The residues participating in this hydrogen bond network are showed in the ball-and-stick models. (B) A time evolution of the bond distance between Asp35 and Ile119.

Materials and Methods

The ASP1 structures and starting states

The molecular OBP structures were set up based on the crystal structures of bee antennal pheromone-binding protein ASP1 [32]. ASP1 typically has six tightly packed helices linked by short unstructured loops (Fig. 1a & 1b, with the odorant molecule of palmitic acid as shown in Fig. 1c). There are about 20 crystal structures of ASP1 in the states of native *apo* or complex with ligands. A *holo* form of crystal structure with the palmitic acid ligand (PDB code: 3BFH [7]) and an *apo* one (PDB code: 2H8V [7]) were used as our initial structures. All the crystal water molecules and other small molecules in the protein structures were removed before modelling. The missing residues (Asp3) at *N*-terminus of the *holo* structure were rebuilt in order to match the full length sequence of *apo* ASP1 structures. To assess reasonable ASP1 states/structures with or without odorant ligand against the benchmark, two other model ASP1 structures were created for molecular dynamics simulation. Based on PDB:3BFH and PDB:2H8V, one mimic structure is the complex with the ligand docking into the *apo* ASP1 structure (i.e. PDB:2H8V), and the other is an *apo* state with the ligand deleted from the *holo* state (i.e. PDB:3BFH). Consequently, total of four *apo* or *holo* starting systems were set up for simulations either at low or neutral pH conditions: These are 'apo-2H8V' and 'holo-3BFH', two states starting from ASP1 crystal structures plus 'holo-2H8V' and 'apo-3BFH' the mimic states, with all residues set at desirable protonated states (see **Supporting Information S1** and **Supporting Information S3**).

Molecular docking and dynamics simulations

To create a ligand-bound state of ASP1, AutoDock 4.2 program [33] was used to dock palmitic acid into *apo* ASP1 structures (for more details please see **Supporting Information S1**). The lowest energy conformation of the complex structure, in which ligand had the similar conformation as the ligand in *holo* ASP1 structure, was selected as the initial structure for molecular dynamics (MD) simulation.

We performed all atom MD simulation on *apo* and *holo* states at different pH conditions to examine how pH affects the dynamics of ASP1 and its interactions with ligands. In order to capture the odorant releasing mechanism as per being influenced by pH, we focused our simulations on the *holo* states with varying pH conditions. All MD simulation systems were prepared and visualized with VMD [34].

MD simulations were performed on NAMD (version 2.8) [35] with the CHARMM27 force field [36,37]. For each system, 200 ns MD production was performed at NPT ensemble, keeping the temperature at 300k, and the conformations were conserved every 0.1ps for subsequent analysis. For the *holo* states of ASP1, three replicates of each pH condition with random initial velocities were executed in order to explore more conformational space of ASP1 More molecular simulation details can be found in **Supporting Information S1**.

Relative binding energy calculation

To quantitatively evaluate the perturbation of palmitic acid and ASP1, free energy perturbation (FEP) method can be used to calculate the binding energy difference of palmitic acid and ASP1 under different pH conditions. According the thermo-dynamics cycle (see Figure S2.3a in **Supporting Information S2**), the relative difference of binding energy could be measured as Eq. (1).

$$\Delta\Delta G^{bind} = \Delta G^{bind}_{pH4.5} - \Delta G^{bind}_{pH7.0} = \Delta G^{apo}_{pH7.0 \to pH4.5} - \Delta G^{holo}_{pH7.0 \to pH4.5} \ (1)$$

Thus a relative negative/positive value of $\Delta\Delta G^{bind}$ shall indicate whether binding at pH 4.5 is more or less preferable than at pH 7.0. For details on the FEP binding energy calculation, please see **Supporting Information S1**.

Dynamics analysis

The conformational changes with 0.1ps intervals were extracted from MD trajectories and analyzed using VMD and Wordom programs [38]. Backbone atoms root mean-square deviation (RMSD, versus the time) and root mean-square fluctuation (RMSF, versus the residues) were measured using Wordom. For hydrogen bond analysis, the distance and angle cutoff were set as 3.0 Å and 20 degree, respectively.

In this work, the center of selected residues was represented with the geometrical center of C^α atom of each residue, as in our previous work [39]. Each residue in the selected region was represented by its C^α atom and the atomic coordinates of the geometrical center of this region were calculated as follows:

$$(R_x, R_y, R_z) = (\frac{\sum_{i=1}^{N} x_i}{N}, \frac{\sum_{i=1}^{N} y_i}{N}, \frac{\sum_{i=1}^{N} z_i}{N}) \quad (2)$$

Where R_x, R_y and R_z are the coordinates of the center of the selected residues; x_i, y_i and z_i are the trajectory of the C^α atom in residue i; while N is the total number of selected residues.

To measure the compactness of the protein structure, the radius of gyration, R_g, was calculated for each protein conformation by,

$$R_g = \sqrt{\sum_{i=1}^{N} \frac{(r_i - R_C)^2}{N}} \quad (3)$$

where N represents the number of selected atoms. In our work, the backbone atoms of protein structure were chosen. R_C is the center of the protein structure as calculated using Eq. (2), and r_i is the position of each backbone atom.

For MD simulations of the *holo* and *apo* states at different pH condition, the analysis were carried out by the average of three replicates. To clear the tendency of data, the cubic smoothing spline was fitted to the time evolution data using the smooth.spline function in R program. The secondary structure of each conformation was analyzed with DSSP program [30]. The sequences were aligned using PRALINE server [40] and multiple structural alignments of OBP structures were done with MultiProt program [41]. All the figures were prepared using PyMol and R program.

Supporting Information

Movie S1 The full trajectories of 200 ns MD simulation of *holo* ASP1 under pH 4.5 and pH 7.0 conditions, respectively.

Movie S2 The full trajectories of 200 ns MD simulation of *holo* ASP1 under pH 4.5 and pH 7.0 conditions, respectively.

Supporting Information S1 Detailing the protonation, molecular dock, simulation procedures and calculation of binding energy by free energy perturbation (FEP) method.

Supporting Information S2 Showing the fluctuations, disulfide bonds, H-bond, and thermodynamic cycles utilized in FEP calculations as per influenced by pH conditions, respectively.

Supporting Information S3 The calculated pK_a values for ASP1 residues before protonation treatment.

References

1. Pelosi P, Maida R (1995) Odorant-binding proteins in insects. Comp Biochem Physiol B Biochem Mol Biol 111: 503–514.
2. Krieger J, Breer H (1999) Olfactory Reception in Invertebrates. Science 286: 720–723.
3. Pelosi P, Zhou J, Ban L, Calvello M (2006) Soluble proteins in insect chemical communication. Cell Mol Life Sci 63: 1658–1676.
4. Leal WS (2013) Odorant Reception in Insects: Roles of Receptors, Binding Proteins, and Degrading Enzymes. Annu Rev Entomol 58: 373–391.
5. Jiang Q-Y, Wang W-X, Zhang Z, Zhang L (2009) Binding specificity of locust odorant binding protein and its key binding site for initial recognition of alcohols. Insect Biochem Mol Biol 39: 440–447.
6. Wang SY, Gu SH, Han L, Guo YY, Zhou JJ, et al. (2013) Specific involvement of two amino acid residues in cis-nerolidol binding to odorant-binding protein 5 AlinOBP5 in the alfalfa plant bug, Adelphocoris lineolatus (Goeze). Insect Mol Biol 22: 172–182.
7. Pesenti ME, Spinelli S, Bezirard V, Briand L, Pernollet J-C, et al. (2008) Structural Basis of the Honey Bee PBP Pheromone and pH-induced Conformational Change. J Mol Biol 380: 158–169.
8. Kruse SW, Zhao R, Smith DP, Jones DNM (2003) Structure of a specific alcohol-binding site defined by the odorant binding protein LUSH from Drosophila melanogaster. Nat Struct Mol Biol 10: 694–700.
9. Mao Y, Xu X, Xu W, Ishida Y, Leal WS, et al. (2010) Crystal and solution structures of an odorant-binding protein from the southern house mosquito complexed with an oviposition pheromone. Proc Natl Acad Sci USA 107: 19102–19107.
10. Wang SY, Gu SH, Han L, Guo YY, Zhou JJ, et al. (2013) Specific involvement of two amino acid residues in cis-nerolidol binding to odorant-binding protein 5 AlinOBP5 in the alfalfa plant bug, Adelphocoris lineolatus (Goeze). Insect Molecular Biology 22: 172–182.
11. Leal WS (2005) Kinetics and molecular properties of pheromone binding and release. Proc Natl Acad Sci USA 102: 5386–5391.
12. Carey AF, Carlson JR (2011) Insect olfaction from model systems to disease control. Proc Natl Acad Sci USA 108: 12987–12995.
13. Sandler BH, Nikonova L, Leal WS, Clardy J (2000) Sexual attraction in the silkworm moth: structure of the pheromone-binding-protein bombykol complex. Chemistry & biology 7: 143–151.
14. Horst R, Damberger F, Luginbühl P, Güntert P, Peng G, et al. (2001) NMR structure reveals intramolecular regulation mechanism for pheromone binding and release. Proc Natl Acad Sci USA 98: 14374–14379.
15. Lartigue A, Gruez A, Spinelli S, Rivière S, Brossut R, et al. (2003) The Crystal Structure of a Cockroach Pheromone-binding Protein Suggests a New Ligand Binding and Release Mechanism. J Biol Chem 278: 30213–30218.
16. Mohanty S, Zubkov S, Gronenborn AM (2004) The Solution NMR Structure of Antheraea polyphemus PBP Provides New Insight into Pheromone Recognition by Pheromone-binding Proteins. J Mol Biol 337: 443–451.
17. Zubkov S, Gronenborn AM, Byeon I-JL, Mohanty S (2005) Structural Consequences of the pH-induced Conformational Switch in A.polyphemus Pheromone-binding Protein: Mechanisms of Ligand Release. J Mol Biol 354: 1081–1090.
18. Xu X, Xu W, Rayo J, Ishida Y, Leal WS, et al. (2010) NMR Structure of Navel Orangeworm Moth Pheromone-Binding Protein (AtraPBP1): Implications for pH-Sensitive Pheromone Detection. Biochemistry 49: 1469–1476.
19. Lartigue A, Gruez A, Briand L, Blon F, Bézirard V, et al. (2004) Sulfur Single-wavelength Anomalous Diffraction Crystal Structure of a Pheromone-Binding Protein from the Honeybee Apis mellifera L. J Biol Chem 279: 4459–4464.
20. Wogulis M, Morgan T, Ishida Y, Leal WS, Wilson DK (2006) The crystal structure of an odorant binding protein from Anopheles gambiae: Evidence for a common ligand release mechanism. Biochem Biophys Res Commun 339: 157–164.
21. Leite NR, Krogh R, Xu W, Ishida Y, Iulek J, et al. (2009) Structure of an Odorant-Binding Protein from the Mosquito Aedes aegypti Suggests a Binding Pocket Covered by a pH-Sensitive "Lid". PLoS ONE 4: e8006.
22. Gräter F, de Groot BL, Jiang H, Grubmüller H (2006) Ligand-Release Pathways in the Pheromone-Binding Protein of Bombyx mori. Structure 14: 1567–1576.
23. Chu W-T, Zhang J-L, Zheng Q-C, Chen L, Wu Y-J, et al. (2013) Constant pH molecular dynamics (CpHMD) and molecular docking studies of CquiOBP1 pH-induced ligand releasing mechanism. J Mol Model 19: 1301–1309.
24. Klusák V, Havlas Z, Rulíšek Lr, Vondrášek J, Svatoš A (2003) Sexual Attraction in the Silkworm Moth: Nature of Binding of Bombykol in Pheromone Binding Protein—An Ab Initio Study. Chemistry & Biology 10: 331–340.
25. Golebiowski J, Antonczak S, Fiorucci S, Cabrol-Bass D (2007) Mechanistic events underlying odorant binding protein chemoreception. Proteins 67: 448–458.
26. Hajjar E, Perahia D, Débat H, Nespoulous C, Robert CH (2006) Odorant Binding and Conformational Dynamics in the Odorant-binding Protein. J Biol Chem 281: 29929–29937.
27. Manoharan M, Fuchs PFJ, Sowdhamini R, Offmann B (2014) Insights on pH-dependent conformational changes of mosquito odorant binding proteins by molecular dynamics simulations. Journal of Biomolecular Structure and Dynamics 32: 1742–1751.
28. Lobanov MY, Bogatyreva NS, Galzitskaya OV (2008) Radius of gyration as an indicator of protein structure compactness. Molecular Biology 42: 623–628.
29. Myers JK, Pace CN (1996) Hydrogen bonding stabilizes globular proteins. Biophys J 71: 2033–2039.
30. Kabsch W, Sander C (1983) Dictionary of protein secondary structure: Pattern recognition of hydrogen-bonded and geometrical features. Biopolymers 22: 2577–2637.
31. Pesenti ME, Spinelli S, Bezirard V, Briand L, Pernollet J-C, et al. (2009) Queen Bee Pheromone Binding Protein pH-Induced Domain Swapping Favors Pheromone Release. J Mol Biol 390: 981–990.
32. Berman HM, Westbrook J, Feng Z, Gilliland G, Bhat TN, et al. (2000) The Protein Data Bank. Nucleic Acids Res 28: 235–242.
33. Morris GM, Huey R, Lindstrom W, Sanner MF, Belew RK, et al. (2009) AutoDock4 and AutoDockTools4: Automated docking with selective receptor flexibility. J Comput Chem 30: 2785–2791.
34. Humphrey W, Dalke A, Schulten K (1996) VMD: Visual molecular dynamics. J Mol Graph 14: 33–38.
35. Phillips JC, Braun R, Wang W, Gumbart J, Tajkhorshid E, et al. (2005) Scalable molecular dynamics with NAMD. J Comput Chem 26: 1781–1802.
36. MacKerell AD, Bashford D, Bellott, Dunbrack RL, Evanseck JD, et al. (1998) All-Atom Empirical Potential for Molecular Modeling and Dynamics Studies of Proteins†. J Phys Chem B 102: 3586–3616.
37. MacKerell AD, Banavali N, Foloppe N (2000) Development and current status of the CHARMM force field for nucleic acids. Biopolymers 56: 257–265.
38. Seeber M, Cecchini M, Rao F, Settanni G, Caflisch A (2007) Wordom: a program for efficient analysis of molecular dynamics simulations. Bioinformatics 23: 2625–2627.
39. Han L, Zhang Y-J, Song J, Liu MS, Zhang Z (2012) Identification of Catalytic Residues Using a Novel Feature that Integrates the Microenvironment and Geometrical Location Properties of Residues. PLoS ONE 7: e41370.
40. Simossis VA, Heringa J (2005) PRALINE: a multiple sequence alignment toolbox that integrates homology-extended and secondary structure information. Nucleic Acids Res 33: W289–W294.
41. Shatsky M, Nussinov R, Wolfson HJ (2004) A method for simultaneous alignment of multiple protein structures. Proteins 56: 143–156.

Acknowledgments

LH thanks the industry traineeship provided by CSIRO - Computational Modeling Program; MSL thanks China Agricultural University for hosting his visits and acknowledges the support from CSIRO - Computational and Simulation Sciences-TCP and CSIRO - CAFHS Capability Development Fund. The authors thank the National Computational Infrastructure (NCI) and Victorian Life Sciences Computation Initiative (VLSCI), Australia, for allocation of supercomputer time.

Author Contributions

Conceived and designed the experiments: LH MSL ZZ. Performed the experiments: LH MSL. Analyzed the data: LH MSL ZZ YJZ LZ XC JPY. Wrote the paper: LH ZZ MSL.

5

Single-Molecule Study on Histone-Like Nucleoid-Structuring Protein (H-NS) Paralogue in *Pseudomonas aeruginosa*: MvaU Bears DNA Organization Mode Similarities to MvaT

Ricksen S. Winardhi[1,2,3], **Sandra Castang**[4], **Simon L. Dove**[4]*, **Jie Yan**[2,3,5]*

1 NUS Graduate school for Integrative Sciences and Engineering, Singapore, Singapore, **2** Mechanobiology Institute, National University of Singapore, Singapore, Singapore, **3** Centre for Bioimaging Sciences, National University of Singapore, Singapore, Singapore, **4** Division of Infectious Diseases, Boston Children's Hospital, Harvard Medical School, Boston, Massachusetts, United States of America, **5** Department of Physics, National University of Singapore, Singapore, Singapore

Abstract

Pseudomonas aeruginosa contains two distinct members of H-NS family of nucleoid-structuring proteins: MvaT and MvaU. Together, these proteins bind to the same regions of the chromosome and function coordinately in the regulation of hundreds of genes. Due to their structural similarity, they can associate to form heteromeric complexes. These findings left us wondering whether they bear similar DNA binding properties that underlie their gene-silencing functions. Using single-molecule stretching and imaging experiments, we found striking similarities in the DNA organization modes of MvaU compared to the previously studied MvaT. MvaU can form protective nucleoprotein filaments that are insensitive to environmental factors, consistent with its role as a repressor of gene expression. Similar to MvaT, MvaU filament can mediate DNA bridging while excessive MvaU can cause DNA aggregation. The almost identical DNA organization modes of MvaU and MvaT explain their functional redundancy, and raise an interesting question regarding the evolutionary benefits of having multiple H-NS paralogues in the *Pseudomonas* genus.

Editor: Yang Zhang, University of Michigan, United States of America

Funding: This work was supported by grants MOE2008-T2-1-096 from the Ministry of Education of Singapore and internal funding from the Mechanobiology Institute Singapore (to JY), and by National Institutes of Health Grant AI069007 (to SLD). The funders had no role in study design, data collection and analysis, decision to publish, or preparation of the manuscript.

Competing Interests: The authors have declared that no competing interests exist.

* Email: phyyj@nus.edu.sg (JY); simon.dove@childrens.harvard.edu (SLD)

Introduction

The H-NS protein is an abundant nucleoid-associated protein in enteric bacteria, and plays an important role mainly in regulating gene transcription, silencing laterally acquired genes, and packaging the bacterial nucleoid [1–5]. The N-terminal region of H-NS forms a coiled coil structure that functions as an oligomerization domain, whereas the C-terminal region is responsible for DNA binding [6–9]. In *E. coli*, H-NS can associate with another H-NS-related protein called StpA, which shares 58% amino acid identity with H-NS, to form heteromeric complexes through their N-termini [1]. StpA is an H-NS paralogue that can functionally substitute for H-NS, has its own unique features, and may have distinct functions from that of H-NS [10–13].

H-NS paralogues also exist in other species of bacteria, some of which can functionally substitute for H-NS and complement an *E. coli* H-NS deficient mutant *in vivo* despite their lack of sequence similarity and identity to *E. coli* H-NS [14–16]. In *Pseudomonas*, MvaT and MvaU have been identified as the paralogues of H-NS that share structural and functional similarity to H-NS despite sharing <20% sequence identity to H-NS [16] (see Figure 1 for sequence alignment and predicted structural elements of MvaT

and MvaU). Transcriptional profiling using DNA microarrays revealed that MvaT from *Pseudomonas aeruginosa* regulates the expression of hundreds of genes [17–19]. Moreover, MvaU has been shown to function coordinately with MvaT, occupying the same regions of the chromosome and co-regulating the expression of ~350 genes [20]. The amount of genes they regulate is similar to that regulated by H-NS and StpA in E. coli, which have intracellular concentrations of few micromolar [21]. Therefore, we reason that MvaT and MvaU have similar intracellular concentrations. The deletion of either MvaT or MvaU leads to the increase in production of the other, indicating cross-regulation of the two proteins and that they can functionally compensate each other [22]. In addition, MvaT and MvaU show a binding preference for AT-rich regions of the chromosome, suggesting that these proteins are involved in xenogeneic DNA silencing [20]. This preferential binding to AT-rich regions seems to be shared among H-NS family proteins [23–26]. Other than the AT-rich preferential binding, in general H-NS family proteins bind DNA non-specifically, which is consistent with their role as abundant nucleoid associated proteins that play a crucial role in organizing chromosomes.

A

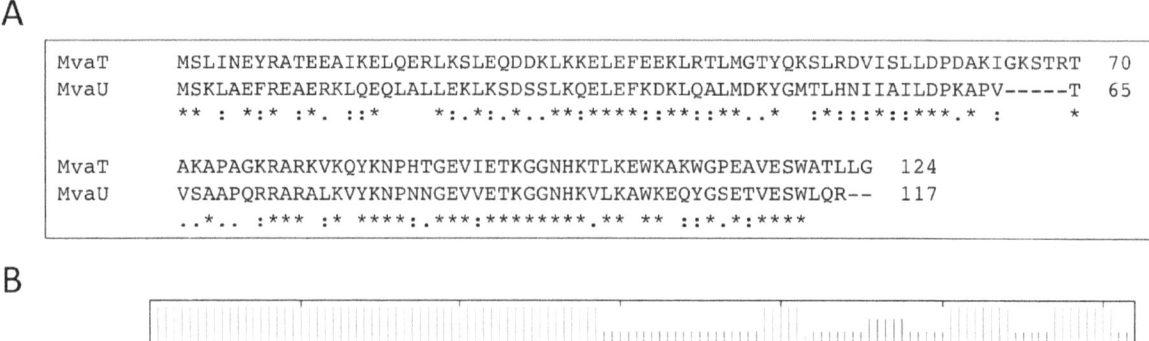

B

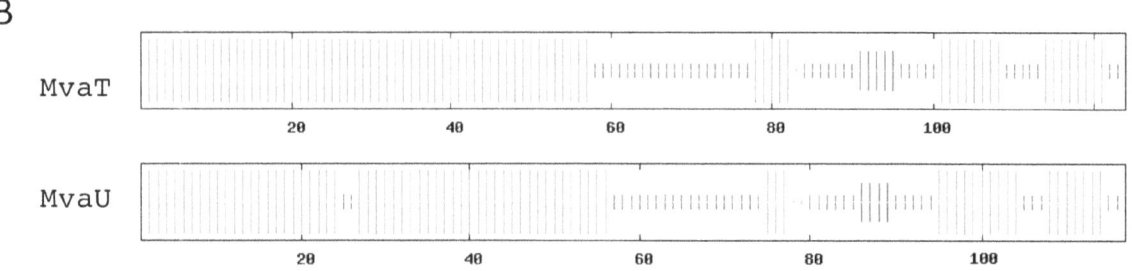

Figure 1. Sequence alignment and predicted secondary structure of MvaT and MvaU. (A) The sequence alignment was done with ClustalW2 [46,47], with 46% pairwise identity. An * (asterisk) indicates positions which have a single, fully conserved residue, a : (colon) indicates conservation between groups of strongly similar properties, and a . (period) indicates conservation between groups of weakly similar properties. (B) The secondary structures were predicted using a consensus prediction method [48]. Blue vertical bars represent a-helix, red vertical bars represent extended strand, purple vertical bars represent random coil, and gray vertical bars represent ambiguous states.

MvaT and MvaU are thought to be structurally similar to H-NS: the DNA binding domain located in the C-terminus of MvaT and MvaU is more conserved compared to H-NS, while their multimerization domain located in the N-terminus is more divergent [14]. It has been suggested that MvaT and MvaU are the functional counterparts of H-NS and StpA, as they have reciprocal regulatory interaction and form heteromeric complexes through their oligomerization domain in the N-terminus [22]. The majority of MvaU in the cell may be associated with MvaT as heteromeric complexes due to its low copy number compared to MvaT [22].

The molecular mechanism of how nucleoid-associated proteins organize DNA (i.e. DNA organization mode) may be related to how they perform their *in vivo* function. For example, the mechanism of gene silencing by H-NS and its anti-silencing activity are largely explained based on its DNA organization modes [27–30]. Previous AFM studies reported that MvaT can induce DNA bridging [31]. However, our recent study has revealed that MvaT can form rigid nucleoprotein filaments on double-stranded DNA, which can mediate DNA-bridging and form compact DNA structures at high protein concentrations [32]. In our single-molecule DNase1 cleavage assay, we found that formation of MvaT nucleoprotein filaments can protect DNA from DNase1 cleavage, while negligible protection was found when we used higher-order oligomerization defective mutants that failed to form nucleoprotein filaments (unpublished data). Since the MvaT mutants failed in both repressing *cupA* gene expression *in vivo* and forming nucleoprotein filaments [32,33], these data further support the relevance of filaments formation to the role of MvaT as a gene silencer. Filament formation may be a conserved property of H-NS-like gene-silencing proteins across prokaryotes, as it is widely found in StpA from *E. coli*, MvaT from *Pseudomonas aeruginosa*, Lsr2 from *Mycobacterium tuberculosis*, and Alba1 from *Sulfolobus solfataricus* [32,34–36]. Mutations that cause loss of gene-silencing function have been shown to disrupt filament formation by MvaT and H-NS [28,32], suggesting that

the formation of nucleoprotein filaments is essential for H-NS-like proteins to perform their gene silencing functions.

In spite of its importance as the paralogue of MvaT in the *Pseudomonas* genus, the binding properties of MvaU to DNA are much less well understood. Although MvaT and MvaU show some degree of functional redundancy, it does not necessarily mean that they have identical DNA binding properties. For instance, StpA, the paralogue of H-NS in *E. coli*, can cause DNA bridging while the nucleoprotein filament formed by StpA is insensitive to KCl concentration, temperature, and pH, unlike the nucleoprotein filament formed by H-NS [34]. This raises the question on the organization modes of MvaU to DNA, which may be relevant to how MvaU performs its *in vivo* function as the paralogue and binding partner of MvaT. In this work, we elucidate the binding properties of MvaU to DNA, and the conformations of MvaU-DNA complexes, which has implication on how gene silencing may be achieved.

Materials and Methods

Proteins

Plasmid pET24-MvaU-His6 was transformed into *E. coli* strain BL21 (DE3). Transformed cells were grown at 37°C in LB medium with 50 μg/mL of kanamycin. At OD_{600} of 0.6, the cells were induced with 1 mM IPTG for 4 hours at 37°C. The cells were harvested and resuspended in cell lytic solution for 2 hours on ice and centrifuged for 30 minutes at 10,000 r.p.m. to collect the supernatant. 6xHis-tagged MvaU was purified by gravity-flow chromatography. Nickel-charged resin (Ni-NTA Agarose, Qiagen, Singapore) was added to the supernatant and incubated for 2 hours on ice, and the sample was loaded on the purification column. The pelleted resin was washed several times with 50 mM potassium phosphate buffer (pH 8.0), 1 M NaCl, 25 mM Imidazole, before eluted with 50 mM potassium phosphate buffer (pH 8.0), 500 mM NaCl, 250 mM Imidazole. Protein purity and identity were verified by SDS-PAGE and mass spectrometry, and the concentration was measured using Nanodrop ND1000

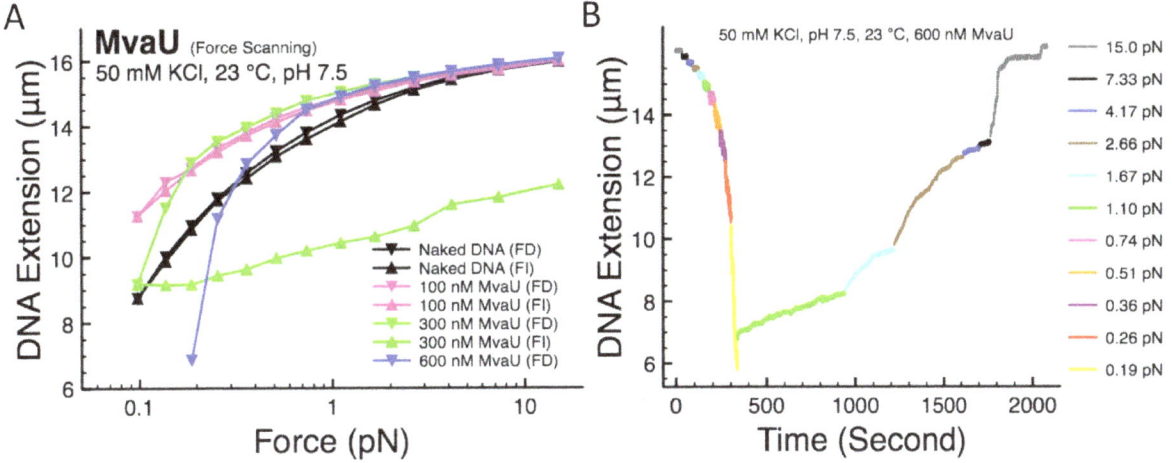

Figure 2. Single-molecule stretching experiments on MvaU-DNA. (A) In the presence of MvaU, the DNA was stiffened as indicated by higher extension compared to naked DNA at the same force. Along with DNA stiffening, increase in DNA folding at higher concentration was also observed, indicated by either hysteresis or DNA folding. The DNA folding at 600 nM MvaU occurred at higher force, and the force-increase scan was not recorded to prevent the DNA extension from dropping below the minimal measurable extension. (B) The time-course data of the stretching experiment in panel A, showing the dynamics of force-decrease that leads to progressive DNA folding, followed by force-increase that gradually unfold the MvaU-DNA.

(Wilmington, USA), with an extinction coefficient of $15,470 \ M^{-1}cm^{-1}$. Glycerol was added to the stock protein (40–50%) and the protein was stored at $-20°C$.

Transverse Magnetic Tweezers

λ-DNA labelled with biotin on both ends were stretched between streptavidin-coated cover slip and streptavidin-coated paramagnetic bead as described previously [32]. The position of the magnet was adjusted to control the amount of force applied on the DNA and the extension of DNA at each force value was recorded in real time using LabVIEW software. We performed force-scanning procedure as follows. The DNA was initially held at high force (~10 pN), and the force was reduced successively to several lower forces down to ~0.1 pN. The force-extension curves measured during decreasing force are referred to as force-decrease curves. Following the force-decrease scan, the force was increased successively at the same set of forces to obtain force-increase curves. Overlapping force-decrease and force-increase curves indicates that MvaU-mediated DNA organization has reached a steady state. If protein-DNA interaction is not at equilibrium, hysteresis between the two curves is expected. In the measurements, the DNA was held for ~30 seconds at each force and the extension of the DNA at a certain force was obtained from the average values of the last 15 seconds. Solid lines were drawn connecting the data points to guide the eye. The stretching experiments were done in 50 mM KCl, 10 mM Tris pH 7.5 at room temperature, unless otherwise stated, while varying the protein concentration. To determine the effect of protein binding on single DNA, for each condition the data presented on the manuscript is a representative experiment performed on the same DNA tether. The general trend is reproducible and consistent in multiple independent experiments (see Figure S1 for additional data obtained from independent experiments).

Atomic Force Microscopy

We used glutaraldehyde-functionalized mica to immobilize protein-DNA complexes for imaging experiments, as described previously [32]. 0.2 ng/μL of linearized ΦX174 were incubated with various amounts of MvaU in 50 mM KCl, 10 mM Tris pH 7.5 at room temperature for 1 minute before the sample was deposited on the glutaraldehyde-functionalized mica for 15 minutes. The mica was gently rinsed with deionized water and dried with nitrogen gas. Images were acquired on a Molecular Imaging 5500 AFM (Molecular Imaging, Agilent Technologies) using tapping mode with silicon cantilever (Photonitech, Singapore).

Results

MvaU binding causes DNA stiffening and DNA folding

To elucidate the molecular mechanism of how MvaU organize DNA (i.e. the DNA organization mode of MvaU) we performed stretching experiments on single λ-DNA molecules using force-scanning procedure as detailed in the Methods section. The force-extension curves of DNA incubated with various MvaU concentrations in 50 mM KCl, pH 7.5, at 23°C were investigated (Figure 2A). In the absence of MvaU, the force-decrease and force-increase curves of the naked DNA overlap each other, as expected for stretched molecules in equilibrium (Figure 2A, black data). In the presence of 100 nM MvaU (Figure 2A, magenta data), the DNA is stiffened as indicated by the higher extension compared to naked DNA at the lower force regime. At the high force regime, the curves nearly overlap with the naked DNA curves, showing that the nucleoprotein contour length remains unaffected. We observe no hysteresis between the force-decrease and force-increase curves, indicating that the protein binding and the resulting DNA stiffening have reached a steady state.

As the MvaU concentration in the buffer was increased to 300 nM (Figure 2A, green data), the force-decrease curve shows a similar level of DNA stiffening compared to 100 nM MvaU, indicating a saturation of the stiffening effect. The force-increase curve, however, does not overlap the force-decrease curve and falls below the naked DNA curve. The non-overlapping curves indicate a non-equilibrium state of the protein-DNA complexes while the lower extension of the force-increase curve indicates protein-induced DNA folding that results in the apparent hysteresis. However, the conformation of nucleoprotein complexes that corresponds to the observed protein-induced DNA folding cannot be distinguished in our stretching experiments. Further increase of the MvaU concentration to 600 nM results in a similar level of DNA stiffening at forces >0.8 pN (Figure 2A, blue data). At forces <0.2 pN, the DNA extension dropped below the extension of naked DNA due to DNA folding. The extension data at forces < 0.2 pN and the force-increase curve were not recorded to prevent the DNA from being folded below the minimum extension that can be recorded by our instrument (~2 μm). The occurrence of DNA stiffening at higher force prior to DNA folding at lower force in the presence of 600 nM MvaU suggests that the DNA organization mode of MvaU can be regulated by force. In addition, the DNA folding is enhanced when the MvaU concentration is increased.

In another independent experiment, the time-course data of MvaU-induced DNA folding and unfolding process in the presence of 600 nM MvaU is plotted in Figure 2B. In this experiment, the folding occurred at forces <0.51 pN and the extension dropped to ~6 μm, before the force was increased to 1.1 pN to stop the folding. Forces up to 15 pN are required to completely unfold MvaU-DNA to its initial extension. Since DNA folding and unfolding are non-equilibrium in nature, the magnitude of folding and unfolding force may vary across experiments (see Figure S1B for another independent time-course experiment performed at identical conditions). Due to this, we do not use the level of hysteresis as an absolute measure of DNA folding.

MvaU organizes DNA into bridges and compact structures in a concentration-dependent manner

The results of single DNA stretching experiments give us information on the impact of protein binding on DNA mechanical response. To complement these results, we performed AFM imaging experiments to study the conformations of MvaU-DNA formed at various protein concentrations. ΦX174 DNA was used for the imaging studies because of its convenient size (5,386 bp) for AFM imaging. In the absence of MvaU, the DNA assumed a random-coiled structure in solution (Figure 3A). At 30 nM MvaU (corresponding to 1 MvaU monomer for every 10 bp), the DNA was predominantly organized into DNA bridges that assume extended conformations (Figure 3B). We define DNA bridges as two DNA segments bridged by a track of proteins. The two DNA segments can be bridged from within the same DNA molecule through DNA looping or from different DNA molecule. The blue arrows in Figure 3B indicate small loops formed at the end of the DNA bridges. These end-loops are also observed previously for H-NS-DNA bridging, which we believe is caused by naked DNA segment that loops back and associate with MvaU-coated DNA segments of the same molecule [37]. The resulting sharp DNA curvature inside the loop is often unassociated with MvaU, which can be explained by a higher bending energy cost to form a curved MvaU nucleoprotein filament, which has an increased bending rigidity compared to naked DNA. This can hinder further MvaU binding, causing the naked DNA loops that we observe. Representative line profiles of naked DNA as well as loops and stems of the DNA bridges are shown in Figure 3E–F. The measured height of the loops is 0.37±0.08 nm, which is similar in comparison to 0.55±0.11 nm obtained for naked DNA, indicating that the loops of the DNA bridges are uncoated DNA molecule. The height obtained for naked DNA is lower than the expected dsDNA thickness of ~2 nm, which is a well-known anomaly in tapping mode AFM caused by adhesion at the sample surface [38,39]. In comparison, the stem of the DNA bridges has a measured height of 1.20±0.18 nm.

At a higher MvaU concentration of 300 nM (corresponding to 1 MvaU monomer for every bp), higher-order rod-like structures were observed along with compact nucleoprotein structures. The orange arrows indicate large loops formed at the end of the higher-order rod-like structures. Representative line profiles of the loops and stems of the rod-like structures are shown in Figure 3G. The height of the large loops is 2.03±0.44 nm, suggesting that they are coated by proteins. These loops are considerably larger than the stem of DNA bridges formed in 30 nM MvaU, suggesting that DNA inside the loop is organized into DNA bridges or higher order structures. The stem of these rod-like structures have a measured height of 4.30±1.32 nm. The nucleoprotein complexes vary considerably in size, suggesting that they are likely formed by adsorption of additional DNA or DNA-protein complexes to simple DNA bridges along the stem direction. Note that higher-order structures are also observed previously for H-NS-DNA complexes [37].

At 3 μM MvaU (corresponding to 10 MvaU monomers for every bp), we only found large and compact rod-like nucleoprotein complexes (Figure 3D). These structures are considerably larger with height generally >4 nm, which can still be explained by adsorption of additional DNA or DNA-protein complexes along the stem direction. Representative line profiles of the large rod-like structures formed at 3 μM MvaU are shown in Figure 3H. The increasing occurrence and size of the large rod-like structures as MvaU concentration was increased can be explained by free MvaU in solution that associates with nearby nucleoprotein filaments to form larger nucleoprotein structures (see Discussion for more details).

In general, our AFM data agree with our observation on the magnetic tweezers experiments. The apparent DNA stiffening suggests that the first step of MvaU binding is an assembly of a rigid nucleoprotein filament, which can mediate DNA bridging when it meets a naked DNA molecule to form extended DNA bridges as seen in AFM (Figure 3B). A higher MvaU concentration leads to DNA folding in single-DNA stretching experiments and formation of large rod-like structures in AFM imaging experiments. This higher-order DNA organization is similar to the nucleoprotein structures formed by MvaT and Lsr2 [32,35]. We propose that these observations are related to a concentration-dependent distribution of MvaU oligomerization states [22,33,40], which is elaborated upon in the Discussion.

Dependence of MvaU-DNA binding properties on environmental factors

Environmental factors, such as KCl concentration, pH, temperature, and MgCl₂ concentration, have been found to significantly affect the DNA organization modes of nucleoid-associated proteins such as H-NS, StpA, and IHF, and moderately affect DNA binding of MvaT and Lsr2 in physiological conditions [32,34,35,41,42]. The responses of protein-DNA complexes to environmental cues may be important to their roles *in vivo*. Thus, we would like to determine the impact of KCl concentration, pH, temperature, and MgCl₂ concentration on MvaU-DNA. The elasticity of naked DNA is insensitive to variation in these factors, thus we can attribute any observed changes in force-extension curve to changes in protein-DNA interaction (see Figure S2 for naked DNA control).

We fixed the concentration of MvaU to 300 nM while KCl concentration, pH, temperature, or MgCl₂ concentration was varied. Using the force-scanning procedure, we are able to gain rich insights on the impact of these environmental factors to MvaU-induced DNA stiffening and DNA folding. The level of DNA stiffening can be assessed based on the force-decrease curves

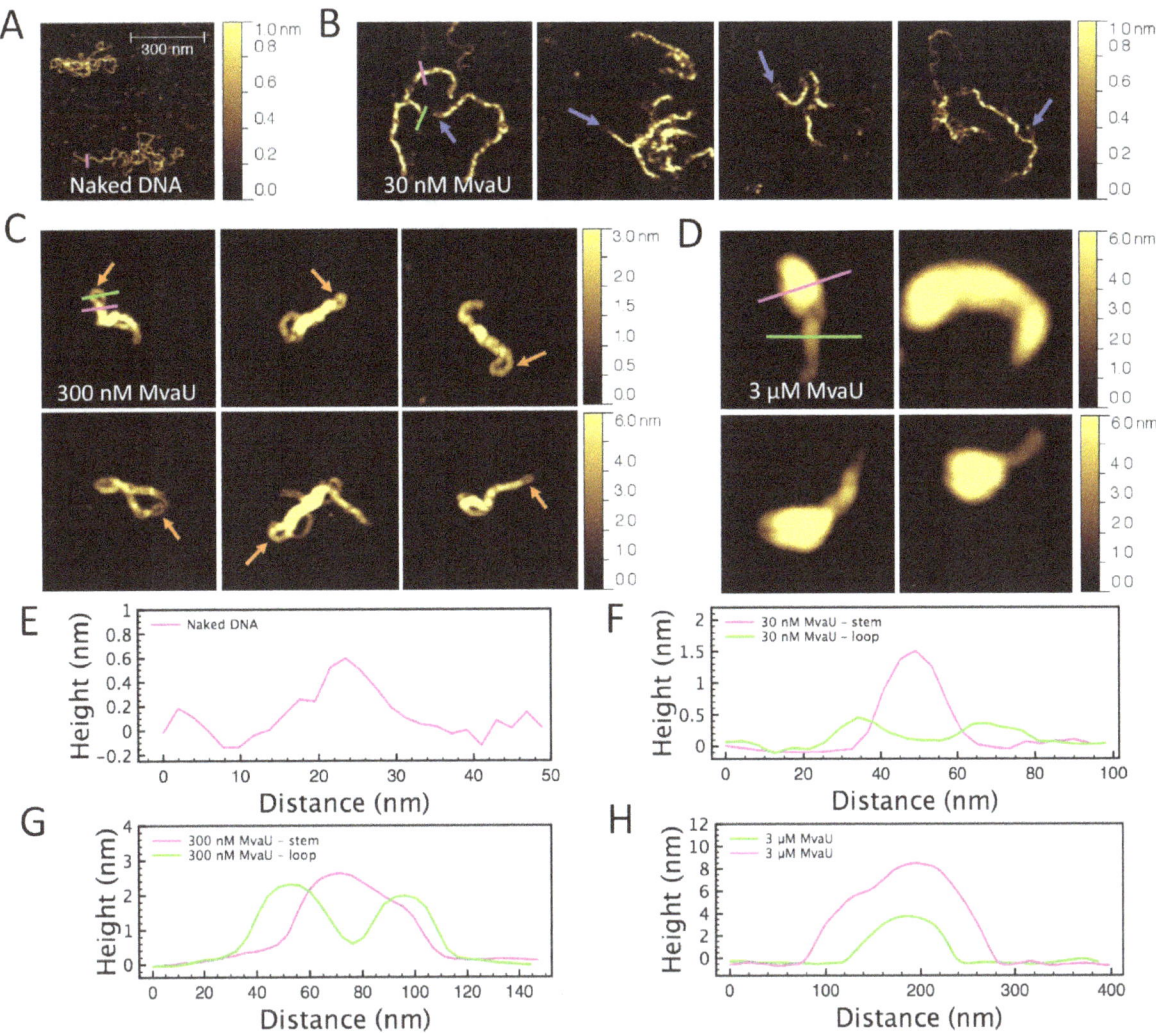

Figure 3. The conformations of MvaU-DNA at various protein concentrations in AFM imaging experiments. (A) Naked ΦX174 DNA adopting random coiled structure. (B–D) ΦX174 DNA complexed with 30 nM (B), 300 nM (C), and 3 μM (D) MvaU. Blue arrows in panel B indicate small uncoated DNA loops at the end of the DNA bridges. The orange arrows in panel D show large bridges formed through further association or compaction of thinner MvaU nucleoprotein filaments. (E–H) Line profiles of the loop end structures as indicated by the magenta and green lines in panel A–D. The surface area for all the images are 0.7 μm×0.7 μm.

at higher force values (>1 pN), as MvaU folding may interfere with DNA extension measurement at lower force values (<1 pN). In addition, the level of DNA folding is based on the observation of hysteresis between the force-decrease and force-increase curves, i.e. protein binding have not reached a steady state due to the non-equilibrium nature of DNA folding. As the level of hysteresis may vary from one experiment to another due to the non-equilibrium nature of DNA folding, we only assess the general trend of increased or decreased level of DNA folding in response to variation in solution conditions.

In 200 mM KCl, MvaU stiffens the DNA with slight hysteresis between the force-decrease and force-increase curves (Figure 4A, magenta data). Maintaining the same MvaU concentration of 300 nM, we reduced the KCl concentration to 100 mM, resulting in a similar level of DNA stiffening and hysteresis (Figure 4A, green data). In 50 mM KCl, we still found similar level of DNA stiffening at forces >1 pN in the force-decrease curve, but hysteresis is significantly increased, indicating moderate DNA folding (Figure 4A, blue data). Note that at forces <0.3 pN in the

force-decrease curve, the reduction in extension is due to MvaU-induced DNA folding that predominates at lower forces. These results suggest that the level of DNA stiffening caused by MvaU binding is not significantly affected by variation in KCl concentration, as force-decrease curves at forces >1 pN are largely unaffected in the range of KCl concentration tested (Figure 4A). However, the level of hysteresis between force-decrease and force-increase curves increases as KCl concentration was decreased, indicating that MvaU-induced DNA folding increases at lower KCl concentration. This indicates more folding over our experimental time scale, but it does not imply that DNA folding has reached equilibrium. The DNA folding promoted at lower KCl concentration is anticipated due to the increased electrostatic attraction between MvaU and DNA. As the osmolarity of bacteria can vary widely depending on the environment, our results suggest that MvaU can affect chromosomal DNA organization.

Similar experiments were carried out to study the effect of variation in pH, temperature, and MgCl$_2$ concentration within

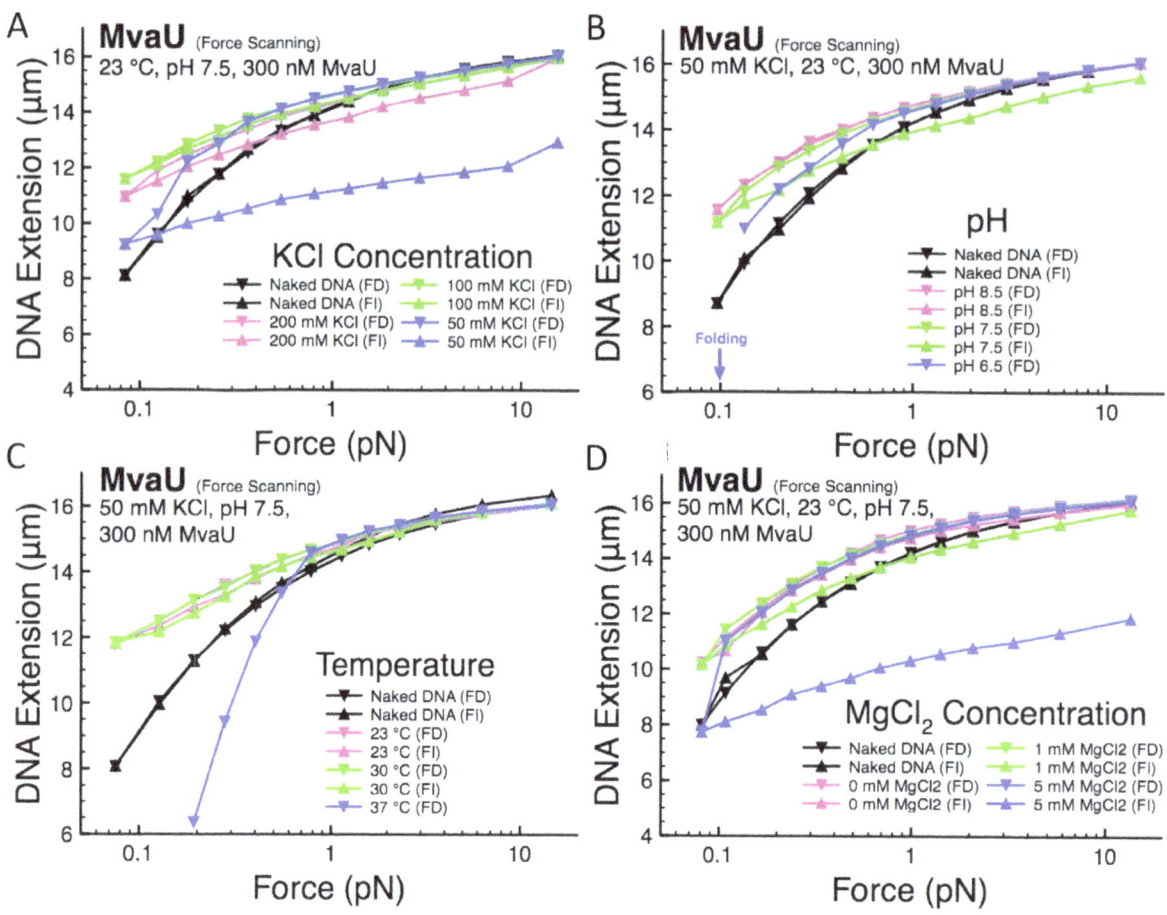

Figure 4. The effect of variation in KCl concentration, pH, temperature, and MgCl₂ concentration on MvaU-DNA. The concentration of MvaU was fixed at 300 nM MvaU at all the conditions tested. (A) Lower KCl concentration results in larger hysteresis or DNA folding. The apparent DNA stiffening is largely unaffected in 50–200 mM KCl. (B) Lower pH value results in larger hysteresis or DNA folding, while the level of DNA stiffening is largely unaffected in the pH range tested. (C) Increase in temperature leads to larger hysteresis or DNA folding, while the level of DNA stiffening is largely unaffected in the temperature range tested. (D) In the presence of MgCl₂, DNA folding is promoted as indicated by increasing amount of hysteresis as the MgCl₂ concentration was increased. The level of DNA stiffening is largely unaffected by variation in MgCl₂ concentration.

their respective physiological ranges. We examined the force-extension curves of MvaU-DNA in pH range of 6.5–8.5 (Figure 4B), temperature range of 23–37°C (Figure 4C), and MgCl₂ concentration range of 0–5 mM (Figure 4D). The results suggest that variation in these factors do not significantly affect the level of DNA stiffening, while DNA folding is facilitated at lower pH, higher temperature, or higher concentrations of MgCl₂. Although the level of hysteresis varies from one experiment to another, the general trend of increased or decreased level of DNA folding in response to variation in solution conditions is reproducible and consistent across experiments (see Figure S1C–F for additional independent experiments).

Taken together, our results show that MvaU nucleoprotein filament stiffness is largely insensitive to variation in KCl concentration, pH, temperature, and MgCl₂ concentration over the ranges tested. MvaU-induced DNA folding, however, is moderately regulated by these environmental factors.

MvaU nucleoprotein filament can protect DNA from DNase1 cleavage

It has been reported MvaU can also repress phase-variable expression of MvaT-regulated *cupA* gene cluster *in vivo*, and it is found able to complement the phenotypes of an MvaT-deficient strain [22]. It is therefore likely that both MvaT and MvaU function coordinately as gene repressor through their DNA binding [20]. Recently, the formation of rigid nucleoprotein filaments by StpA in *E. coli* and by Lsr2 in Gram-positive bacteria *Mycobacterium tuberculosis* has been found to restrict DNA accessibility [34,35]. This DNA blocking capability is proposed as the mechanism responsible for their function as gene silencers. Here we would like to examine whether MvaU can also block DNA accessibility.

We performed a single-molecule DNase1 cleavage assay to study DNA accessibility of MvaU-DNA to cleavage by DNase1 as described previously [34]. Multiple DNA tethers were stretched in the focal plane at ~7 pN to prevent folding and solution containing DNase1 was then introduced. DNA tethers were progressively cleaved, indicated by loss of detected DNA tethers (Figure 5, black data). The DNA tethers were completely cleaved by DNase1 in <5 minutes after solution containing 100 nM DNase1 in 50 mM KCl, pH 7.5, at 23°C was introduced. The rate of DNA cleavage can give us valuable information on DNA accessibility. In another independent experiment, we incubated the DNA tethers with 100 nM MvaU for 5 minutes to allow the formation of MvaU nucleoprotein filaments. Following this,

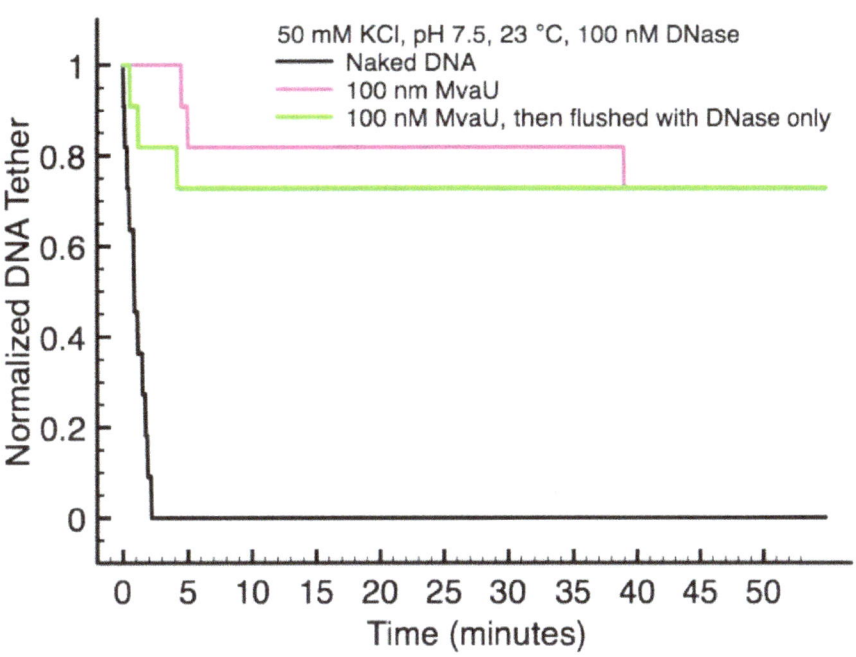

Figure 5. The formation of MvaU nucleoprotein filament effectively blocks DNase1 access to DNA. Multiple tethers of λ-DNA were completely cleaved in the presence of 100 nM DNase1 after <5 minutes of incubation (black data). In contrast, ~70% of DNA tethers remained uncut ~1 hour after the mixture of MvaU/DNase1 was introduced (magenta data), or after the buffer was changed to those containing 100 nM DNase1 without free MvaU in solution (green data).

solution containing a mixture of 100 nM MvaU and 100 nM DNase1 was introduced. We found that the formation of MvaU nucleoprotein filament and/or the presence of 100 nM MvaU in solution can significantly reduce the rate of cleavage by DNase1, as ~70% of DNA tethers were still intact for ~1 hour after DNase1 was introduced (Figure 5, magenta data).

The above DNase1 cutting experiment was conducted in the presence of 100 nM MvaU, which can also reduce the cleavage rate through competitive DNA binding. However, we speculate that the protection from DNase1 cleavage is caused by the formation of rigid nucleoprotein filaments. To test this hypothesis, we incubated the DNA tethers with 100 nM MvaU for 5 minutes to allow the formation of MvaU nucleoprotein filaments. Afterwards, we introduced solution containing 100 nM DNase1 only. Due to the absence of MvaU protein in solution, any reduction in cleavage rate can only arise from the prebound MvaU on DNA. We found that ~70% of DNA tethers were still intact for ~1 hour after solution containing 100 nM DNase1 was introduced (Figure 5, green data). Note that the protection from DNase1 cleavage is largely caused by formation of MvaU nucleoprotein filaments, as a force of ~7 pN was applied on the DNA tethers throughout the experiments to minimize DNA bridging or compaction. Under these conditions, MvaU forms rigid nucleoprotein filaments, which are remarkably stable as they do not dissociate from DNA even after ~1 hour in the absence of free MvaU protein. In the presence of DNA bridging or folding, we expect an even higher level of protection since folding may further reduce the exposed DNA site.

Overall, our single-molecule DNase1 cleavage assay reveals that the formation of MvaU nucleoprotein filaments can effectively block DNA accessibility. This finding further highlights the potential physiological importance of filament formation for their gene-silencing function.

Discussion

The organization modes of MvaU to DNA

In this work, we have elucidated the DNA organization modes of MvaU using magnetic tweezers and AFM. Our magnetic tweezers experiments revealed that MvaU can stiffen DNA and cause DNA folding at high MvaU concentrations. These results are complemented by AFM imaging experiments, which revealed the various conformations of MvaU-DNA, such as DNA bridges, MvaU nucleoprotein filaments, and compact DNA structures. The DNA stiffening found in our stretching experiments is caused by the formation of rigid MvaU nucleoprotein filaments, while the apparent DNA folding corresponds to the compact DNA structures found in the imaging experiments. Importantly, we also found that these two mechanisms are not contradictory to each other, as DNA stiffening can occur prior to DNA folding (Figure 2A, blue data). In support of this, our imaging experiments show that MvaU nucleoprotein filaments can be further compacted into thicker filaments (Figure 3C–D). We use non-specific DNA molecules in our experiments, as MvaU binds to hundreds of distinct genomic regions and we expect that the DNA organization mode of MvaU to be independent to DNA sequences [20].

The formation of DNA bridges at lower stoichiometry of 1 MvaU monomer to 10 bp suggests that MvaU can associate different segments of naked DNA molecules (Figure 3B). In addition, two or more protein-coated DNA molecules can be further compacted to form higher-order rod-like structures (Figure 3C–D). MvaT and MvaU can oligomerize and form homomeric and heteromeric complexes through their N-terminal regions [22]. Therefore, we speculate that similar to MvaT, the formation of compact structures at higher MvaU concentration is caused by concentration-dependent high-order oligomerization of MvaU in solution [22,32,33,40]. Different oligomeric conformations may correspond to different DNA binding properties. Similar to H-NS, we propose that once a DNA site is bound by an MvaU binding unit, more MvaU will oligomerize across adjacent DNA stretch, forming nucleoprotein filaments [43]. Further, MvaU may possess an additional low-affinity oligomerization domain that can mediate higher-order compaction of these nucleoprotein filaments at sufficiently high concentration.

The implication of MvaU binding on its functional role

MvaU serves as a transcriptional silencer for some genes, such as those for prophage activation, pyocyanin synthesis, fimbrial *cupA* gene expression, formation of lectin and other exoproducts, and many others [18,40]. Previously, the formation of nucleoprotein filaments has been reported for H-NS, StpA, MvaT, and Lsr2, all of which function as gene repressors [32,34,35,41]. It has been proposed that nucleoprotein filament formation may serve as the basis for gene silencing, and possibly a general gene silencing mechanism across prokaryotes [28]. In this work, we found that MvaU, which can perform some of MvaT's regulatory functions, can also form rigid nucleoprotein filaments that give rise to DNA stiffening in our stretching experiments. We also found that the formation of rigid MvaU nucleoprotein filaments restricts DNA accessibility, suggesting that the genetic information contained in the DNA is deemed inaccessible to transcription by RNA polymerase. This way, gene-silencing function can be achieved.

In addition to filament formation, we also found that MvaU can mediate high-order DNA organization at high protein concentration. Further, our data suggests that multiple environmental factors, such as KCl concentration, pH, temperature, and $MgCl_2$ concentration, can affect the level of DNA folding over physiologically relevant range. To our knowledge, there is no report on the potential role of MvaU on chromosomal DNA packaging. Our finding suggests that MvaU can effectively condense DNA and may play a role in packaging the bacterial nucleoid.

Comparison between MvaT and MvaU organization modes and their roles *in vivo*

In *Pseudomonas aeruginosa*, MvaU is a transcription regulator that shares 47% amino acid identity and 65% similarity with MvaT [22]. The resemblance of MvaT and MvaU goes beyond their structure and function, as we found striking similarities in their DNA organization modes. Both MvaT and MvaU can form rigid nucleoprotein filaments that can block access to DNA, and they can organize DNA into more complex compact structures at higher protein concentration. The nucleoprotein filaments formed by MvaT and MvaU are generally insensitive to variation in KCl concentration, pH, temperature, and $MgCl_2$ concentration, while the level of DNA folding is moderately modulated by these factors over physiological ranges [32].

The similarities in DNA organization modes of MvaT and MvaU may correspond to their predicted functional redundancy *in vivo*. In terms of DNA binding, MvaT and MvaU have nearly complete binding profile overlap in the genome, showing that they control the expression of largely the same set of genes [20]. The loss of gene encoding MvaT or MvaU leads to increase in production of the other, showing the reciprocity and cross-regulation of these two proteins [22]. The loss of both proteins, however, cannot be tolerated and causes profound effects on the cell [20,40,44]. Based on our previous work on MvaT and our current findings on MvaU, we conclude that MvaT and MvaU bind to DNA in a largely similar manner, corresponding to their *in vivo* reciprocity. They can form gene-silencing nucleoprotein filaments to regulate gene expression, and are possibly involved in

chromosomal DNA packaging. The existence of multiple H-NS paralogues in the same microorganism of *Pseudomonas* genus may be advantageous and serve to maintain a functional gene regulatory system, and therefore may be more than to simply serve as a backup system [5,20].

Finally, there are many other multiple H-NS paralogues that exist in the same microorganism [14]. It remains unclear how these seemingly redundant proteins can pose evolutionary benefits to the bacterial cell. It has been suggested that the existence of more than one H-NS-related protein may correspond to functional divergence and partition [14]. It is also possible that multiple H-NS paralogues integrate multiple environmental signals to promote higher adaptability, notably the *Pseudomonas* genus that exhibits remarkable ecological and metabolic diversity [14,45]. In addition, the ability of these H-NS paralogues to form heteromeric complexes *in vivo* through their oligomerization domain may pose additional functional significance. As MvaU bears similar DNA organization modes to MvaT, we expect that MvaT-MvaU heteromeric complexes will also exhibit similar DNA binding properties compared to its constituents. To test this, we complexed MvaT and MvaU in a test tube with ~10 mM concentration for each protein at room temperature for 4 hours to allow the two proteins to associate and reach equilibrium distribution of homomeric and heteromeric complexes. Although the formation of heteromeric complexes cannot be directly confirmed in our experiments, we reason that heteromeric complexes should predominate in the mixture as MvaT copurifies with MvaU [18,22]. Following this, we examined the DNA binding of MvaT-MvaU mixture with single-DNA stretching experiments, and we found that the mixture has similar DNA binding properties compared to MvaT and MvaU (Figure S3). The results of our studies on MvaT and MvaU give insights and advance the understanding on the nature on such intriguing bacterial regulatory system.

Supporting Information

Figure S1 Single molecule stretching of MvaU-DNA complexes. Additional independent experiments were shown to demonstrate the repeatability of the trends and consistency of the data. Note the variation on the level of hysteresis compared to the data in the main text at the same condition due to the non-equilibrium nature of DNA folding.

Figure S2 The effect of variation in KCl concentration, pH, temperature, and $MgCl_2$ concentration on naked DNA. Naked DNA shows negligible variation in its elasticity when we varied KCl concentration (A), pH (B), temperature (C), and $MgCl_2$ concentration (D). Hence, any observed changes in force-extension curves in the presence of protein due to variation in these factors can be attributed to changes in protein-DNA interaction.

Figure S3 Single-molecule stretching experiments on DNA complexed with MvaT-MvaU mixture. (A) Increasing concentration of MvaT-MvaU mixture leads to DNA stiffening, accompanied by increasing amount of DNA folding as indicated by hysteresis. (B) Time-course data of the stretching experiment in panel A, showing progressive DNA folding as force is decreased, followed by unfolding of the protein-DNA complexes as force is increased.

Acknowledgments

We thank Linda Kenney (University of Illinois at Chicago and Mechanobiology Institute Singapore) for stimulating discussions. We also thank the protein expression core facility of the Mechanobiology Institute for protein purification.

Author Contributions

Conceived and designed the experiments: JY SLD. Performed the experiments: RSW JY. Analyzed the data: RSW JY. Contributed reagents/materials/analysis tools: SC. Contributed to the writing of the manuscript: RSW JY.

References

1. Dorman C (2004) H-NS: A universal regulator for a dynamic genome. Nature Reviews Microbiology 2: 391–400.
2. Fang FC, Rimsky S (2008) New insights into transcriptional regulation by H-NS. Curr Opin Microbiol 11: 113–120.
3. Ali SS, Xia B, Liu J, Navarre WW (2012) Silencing of foreign DNA in bacteria. Curr Opin Microbiol 15: 1–7.
4. Atlung T, Ingmer H (1997) H-NS: a modulator of environmentally regulated gene expression. Mol Microbiol 24: 7–17.
5. Navarre WW, McClelland M, Libby SJ, Fang FC (2007) Silencing of xenogeneic DNA by H-NS-facilitation of lateral gene transfer in bacteria by a defense system that recognizes foreign DNA. Genes Dev 21: 1456–1471.
6. Esposito D, Petrovic A, Harris R, Ono S, Eccleston JF, et al. (2002) H-NS oligomerization domain structure reveals the mechanism for high order self-association of the intact protein. J Mol Biol 324: 841–850.
7. Smyth CP, Lundbäck T, Renzoni D, Siligardi G, Beavil R, et al. (2000) Oligomerization of the chromatin-structuring protein H-NS. Mol Microbiol 36: 962–972.
8. Ceschini S, Lupidi G, Coletta M, Pon CL, Fioretti E, et al. (2000) Multimeric self-assembly equilibria involving the histone-like protein H-NS. A thermodynamic study. J Biol Chem 275: 729–734.
9. Ueguchi C, Seto C, Suzuki T, Mizuno T (1997) Clarification of the dimerization domain and its functional significance for the Escherichia coli nucleoid protein H-NS. J Mol Biol 274: 145–151.
10. Free A, Dorman CJ (1997) The Escherichia coli stpA gene is transiently expressed during growth in rich medium and is induced in minimal medium and by stress conditions. J Bacteriol 179: 909–918.
11. Sonden B, Uhlin BE (1996) Coordinated and differential expression of histone-like proteins in Escherichia coli: regulation and function of the H-NS analog StpA. EMBO J 15: 4970–4980.
12. Zhang A, Belfort M (1992) Nucleotide sequence of a newly-identified Escherichia coli gene, stpA, encoding an H-NS-like protein. Nucleic Acids Res 20: 6735.
13. Zhang A, Rimsky S, Reaban ME, Buc H, Belfort M (1996) Escherichia coli protein analogs StpA and H-NS: regulatory loops, similar and disparate effects on nucleic acid dynamics. EMBO J 15: 1340–1349.
14. Tendeng C, Bertin P (2003) H-NS in Gram-negative bacteria: a family of multifaceted proteins. Trends in Microbiology: 511–518.
15. Bertin P, Benhabiles N, Krin E, Laurent-Winter C, Tendeng C, et al. (1999) The structural and functional organization of H-NS-like proteins is evolutionarily conserved in Gram-negative bacteria. Molecular Microbiology 31: 319–329.
16. Tendeng C, Soutourina O, Danchin A, Bertin P (2003) MvaT proteins in Pseudomonas spp.: a novel class of H-NS-like proteins. Microbiology 149: 3047–3050.
17. Diggle S, Winzer K, Lazdunski A, Williams P, Camara M (2002) Advancing the quorum in Pseudomonas aeruginosa: MvaT and the regulation of N-acylhomoserine lactone production and virulence gene expression. Journal of Bacteriology 184: 2576–2586.
18. Vallet I, Diggle S, Stacey R, Camara M, Ventre I, et al. (2004) Biofilm formation in Pseudomonas aeruginosa: Fimbrial cup gene clusters are controlled by the transcriptional regulator MvaT. Journal of Bacteriology 186: 2880–2890.
19. Westfall L, Luna A, San Francisco M, Diggle S, Worrall K, et al. (2004) The Pseudomonas aeruginosa global regulator MvaT specifically binds to the ptxS upstream region and enhances ptxS expression. Microbiology 150: 3797–3806.
20. Castang S, McManus H, Turner K, Dove S (2008) H-NS family members function coordinately in an opportunistic pathogen. Proc Natl Acad Sci U S A 105: 18947–18952.
21. Ali Azam T, Iwata A, Nishimura A, Ueda S, Ishihama A (1999) Growth phase-dependent variation in protein composition of the Escherichia coli nucleoid. J Bacteriol 181: 6361–6370.
22. Vallet-Gely I, Donovan K, Fang R, Joung J, Dove S (2005) Repression of phase-variable cup gene expression by H-NS-like proteins in Pseudomonas aeruginosa. Proc Natl Acad Sci U S A 102: 11082–11087.
23. Gordon BR, Li Y, Wang L, Sintsova A, van Bakel H, et al. (2010) Lsr2 is a nucleoid-associated protein that targets AT-rich sequences and virulence genes in Mycobacterium tuberculosis. Proc Natl Acad Sci U S A 107: 5154–5159.
24. Wolf T, Janzen W, Blum C, Schnetz K (2006) Differential dependence of StpA on H-NS in autoregulation of stpA and in regulation of bgl. J Bacteriol 188: 6728–6738.
25. Navarre WW, Porwollik S, Wang Y, McClelland M, Rosen H, et al. (2006) Selective silencing of foreign DNA with low GC content by the H-NS protein in Salmonella. Science 313: 236–238.
26. Lucchini S, Rowley G, Goldberg MD, Hurd D, Harrison M, et al. (2006) H-NS mediates the silencing of laterally acquired genes in bacteria. PLoS Pathog 2: e81.
27. Dame RT, Wyman C, Wurm R, Wagner R, Goosen N (2002) Structural basis for H-NS-mediated trapping of RNA polymerase in the open initiation complex at the rrnB P1. J Biol Chem 277: 2146–2150.
28. Lim CJ, Lee SY, Kenney LJ, Yan J (2012) Nucleoprotein filament formation is the structural basis for bacterial protein H-NS gene silencing. Sci Rep 2: 509.
29. Amit R, Oppenheim A, Stavans J (2003) Increased bending rigidity of single DNA molecules by H-NS, a temperature and osmolarity sensor. Biophysical Journal 841: 2467–2473.
30. Winardhi RS, Gulvady R, Mellies JL, Yan J (2014) Locus of Enterocyte Effacement-Encoded Regulator (Ler) of pathogenic Escherichia coli Competes Off Nucleoid Structuring Protein H-NS through Non-Cooperative DNA Binding. J Biol Chem 2890: 13739–13750.
31. Dame R, Luijsterburg M, Krin E, Bertin P, Wagner R, et al. (2005) DNA bridging: a property shared among H-NS-like proteins. Journal of Bacteriology 187: 1845–1848.
32. Winardhi RS, Fu W, Castang S, Li Y, Dove SL, et al. (2012) Higher order oligomerization is required for H-NS family member MvaT to form gene-silencing nucleoprotein filament. Nucleic Acids Res 40: 8942–8952.
33. Castang S, Dove S (2010) High-order oligomerization is required for the function of the H-NS family member MvaT in Pseudomonas aeruginosa. Molecular Microbiology 78: 916–931.
34. Lim CJ, Whang YR, Kenney LJ, Yan J (2012) Gene silencing H-NS paralogue StpA forms a rigid protein filament along DNA that blocks DNA accessibility. Nucleic Acids Res 40: 3316–3328.
35. Qu Y, Lim CJ, Whang YR, Liu J, Yan J (2013) Mechanism of DNA organization by Mycobacterium tuberculosis protein Lsr2. Nucleic Acids Res 41: 5263–5272.
36. Laurens N, Driessen RP, Heller I, Vorselen D, Noom MC, et al. (2012) Alba shapes the archaeal genome using a delicate balance of bridging and stiffening the DNA. Nat Commun 3: 1328.
37. Dame R, Wyman C, Goosen N (2000) H-NS mediated compaction of DNA visualised by atomic force microscopy. Nucleic Acids Research 28: 3504–3510.
38. Van Noort SJT, Van der Werf KO, De Grooth BG, Van Hulst NF, Greve J (1997) Height anomalies in tapping mode atomic force microscopy in air caused by adhesion. Ultramicroscopy 69: 117–127.
39. Lyubchenko YL, Oden PI, Lampner D, Lindsay SM, Dunker KA (1993) Atomic force microscopy of DNA and bacteriophage in air, water and propanol: the role of adhesion forces. Nucleic Acids Res 21: 1117–1123.
40. Li C, Wally H, Miller S, Lu C (2009) The Multifaceted Proteins MvaT and MvaU, Members of the H-NS Family, Control Arginine Metabolism, Pyocyanin Synthesis, and Prophage Activation in Pseudomonas aeruginosa PAO1. Journal of Bacteriology 191: 6211–6218.
41. Liu Y, Chen H, Kenney L, Yan J (2010) A divalent switch drives H-NS/DNA-binding conformations between stiffening and bridging modes. Genes & Development 24: 339–344.
42. Lin J, Chen H, Dröge P, Yan J (2012) Physical organization of DNA by multiple non-specific DNA-binding modes of integration host factor (IHF). PLoS One 7: e49885.
43. Bouffartigues E, Buckle M, Badaut C, Travers A, Rimsky S (2007) H-NS cooperative binding to high-affinity sites in a regulatory element results in transcriptional silencing. Nat Struct Mol Biol 14: 441–448.
44. Castang S, Dove SL (2012) Basis for the essentiality of H-NS family members in Pseudomonas aeruginosa. J Bacteriol 194: 5101–5109.
45. Renzi F, Rescalli E, Galli E, Bertoni G (2010) Identification of genes regulated by the MvaT-like paralogues TurA and TurB of Pseudomonas putida KT2440. Environmental Microbiology 12: 254–263.
46. Larkin MA, Blackshields G, Brown NP, Chenna R, McGettigan PA, et al. (2007) Clustal W and Clustal X version 2.0. Bioinformatics 23: 2947–2948.
47. Goujon M, McWilliam H, Li W, Valentin F, Squizzato S, et al. (2010) A new bioinformatics analysis tools framework at EMBL-EBI. Nucleic Acids Res 38: W695–699.
48. Combet C, Blanchet C, Geourjon C, Deléage G (2000) NPS@: network protein sequence analysis. Trends Biochem Sci 25: 147–150.

6

Structural Model of the hUbA1-UbcH10 Quaternary Complex: *In Silico* and Experimental Analysis of the Protein-Protein Interactions between E1, E2 and Ubiquitin

Stefania Correale[1❜], **Ivan de Paola**[2❜], **Carmine Marco Morgillo**[4], **Antonella Federico**[3], **Laura Zaccaro**[2], **Pierlorenzo Pallante**[3], **Aldo Galeone**[4], **Alfredo Fusco**[3], **Emilia Pedone**[2], **F. Javier Luque**[5], **Bruno Catalanotti**[4]*

1 Kedrion S.p.A., Sant 'Antimo (Na), Italy, 2 Istituto di Biostrutture e Bioimmagini, Consiglio Nazionale delle Ricerche, Napoli, Italy, 3 Istituto di Endocrinologia ed Oncologia Sperimentale Consiglio Nazionale delle Ricerche, Napoli, Italy, 4 Dipartimento di Farmacia, Università degli Studi di Napoli Federico II, Napoli, Italy, 5 Departament de Fisicoquímica and Institut de Biomedicina (IBUB), Facultat de Farmàcia, Universitat de Barcelona, Santa Coloma de Gramenet, Spain

Abstract

UbcH10 is a component of the Ubiquitin Conjugation Enzymes (Ubc; E2) involved in the ubiquitination cascade controlling the cell cycle progression, whereby ubiquitin, activated by E1, is transferred through E2 to the target protein with the involvement of E3 enzymes. In this work we propose the first three dimensional model of the tetrameric complex formed by the human UbA1 (E1), two ubiquitin molecules and UbcH10 (E2), leading to the transthiolation reaction. The 3D model was built up by using an experimentally guided incremental docking strategy that combined homology modeling, protein-protein docking and refinement by means of molecular dynamics simulations. The structural features of the *in silico* model allowed us to identify the regions that mediate the recognition between the interacting proteins, revealing the active role of the ubiquitin crosslinked to E1 in the complex formation. Finally, the role of these regions involved in the E1–E2 binding was validated by designing short peptides that specifically interfere with the binding of UbcH10, thus supporting the reliability of the proposed model and representing valuable scaffolds for the design of peptidomimetic compounds that can bind selectively to Ubcs and inhibit the ubiquitylation process in pathological disorders.

Editor: Annalisa Pastore, National Institute for Medical Research, Medical Research Council, London, United Kingdom

Funding: BC acknowledges the HPC-EUROPA2 project (projects 932 and 1200), with the support of the European Community-Research Infrastructure Action of the FP7, the Barcelona Supercomputer Center and the CINECA (ISCRA project ID HP10B81QWA and HP10CNS2K3) for the access to high performance computing resources. FJL acknowledges funding from Icrea Academia, the Spanish MCINN (SAF2011-27642), the Generalitat de Catalunya (2009SGR00249), the Xarxa de Recerca en Quimica Teorica i Computacional (XQRTC). EP and LZ acknowledge the Italian MIUR for financial support (FIRB RBAP114AMK_006). The funders had no role in study design, data collection and analysis, decision to publish, or preparation of the manuscript.

Competing Interests: The authors have the following interests. Co-author Stefania Correale Kedrion is employed by Kedrion S.p.A. Sant 'Antimo.

* Email: brucatal@unina.it

❜ These authors contributed equally to this work.

Introduction

UbcH10 is a member of the Ubiquitin Conjugation Enzymes, a component of the anaphase-promoting complex and a key regulator of cell cycle progression [1], as it induces the ubiquitination and degradation of cyclins A and B [2]. Previous studies have indicated that UbcH10 over-expression might be associated with the late stages of thyroid neoplastic transformation [3], and that high levels of UbcH10 correlate with most aggressive grade tumors in breast cancer [4]. Similar evidences have been found for several tumor types, such as ovarian [5], colorectal and brain cancers [6] and different lymphoma [7]. Moreover, in numerous cancer tissues the UbcH10 expression is relatively higher if compared with the adjacent nonmalignant tissues. All these evidences point out that the aberrant expression of UbcH10

could promote tumor expansion through dysfunction of mitotic progression, leading to deregulation of cell growth as confirmed in both thyroid [6] and breast carcinoma [8], where the interference with the UbcH10 expression significantly reduced the tumor cell proliferation. Therefore, UbcH10 appears to be a potential target for developing an anti-cancer therapy based on the suppression of its specific biological function.

A key step in the discovery of inhibitors of the UbcH10-mediated ubiquitination is the comprehension of the structural and mechanistic features that mediate the conjugation of proteins to ubiquitin (Ub), a complex process that involves a three-step cascade mechanism characterized by growing specificity ([8]; see also ref. [9] for a recent review) (Figure 1). Thus, the Ubiquitin-Activating Enzyme (UbA1, also known as E1) initiates the ubiquitination cascade by catalyzing the ATP-dependent adenylation of the Ub C-terminus (step I). The high-energy anhydride bond thus formed is attacked by the E1 active site cysteine (C632 in human UbA1), forming a thioester bond between E1 and Ub (step II). Then, Ub is transferred to the active site cysteine of an Ub-Conjugation Enzyme (denoted E2), a process promoted by the non-covalent binding of a second Ub molecule in the adenylation site (steps III and IV). Finally, Ub is conjugated to its substrate with the aid of a protein ligase (known as E3), resulting in the covalent linkage of the Ub C-terminus to the ε-amino group of a lysine in the substrate (steps V and VI). In humans, there are two E1 enzymes (UbA1 and UbA6) [10], over 30 distinct forms of E2 and about 500–1000 forms of E3, which is largely responsible for conferring specificity to ubiquitylation [11].

The preceding mechanism is common to the Ubiquitin-like proteins (Ubl), a class of signaling proteins involved in cellular homoeostasis [12]. A number of X-ray and NMR studies (reviewed in [12–14]) have examined the structural features of the recognition between Ub and Ubl (SUMO and NEDD8) with E1, while only few studies were focused on the E1–E2 interaction, including the complex between APPBP1-Uba3~NEDD8/NEDD8/MgATP/Ubc12 [13], and the construct obtained by crosslinking the catalytic cysteines of the UbA1~Ubc4/MgATP [14]. While they reveal a general preservation of the E1 structure, they have disclosed the existence of significant structural differences, particularly in the SCCH (Second Catalytic Cysteine Half-domain) and UFD (Ubiquitin Folding Domain) regions, which highlight the intrinsic flexibility of E1 for accommodating both Ub and E2. However, to the best of our knowledge, there is not a complete 3D model of the quaternary complex required for the transfer of Ub to the E2 Ubiquitin Conjugation Enzyme.

In this paper we describe a computational and experimental strategy to build up the first structural model of the transient tetrameric complex between the doubly Ub-loaded human UbA1

(hereafter denoted UbA1~Ub(T)-Ub(A)), and UbcH10, as a member of the E2 family. By combining homology modeling, protein-protein docking and molecular dynamics (MD) simulations, the structural features of the proposed model have allowed us to identify the regions that mediate the recognition between the interacting proteins. In turn, this information has been used to examine the reliability of the structural model through experimental assays performed to evaluate the binding affinities of a number of short peptides that were suitably chosen from the contact regions between interacting partners in the complex. Overall, this information can be valuable to gain insight into the specificity of the recognition between E1 and E2 partners, as well as for the design of peptidomimetic compounds that can bind selectively to E2s and inhibit the ubiquitylation process in pathological disorders.

Materials and Methods

Homology building

The amino acid sequence of human UbA1 (hUbA1) was retrieved from the National Center for Biotechnology Information (http://www.ncbi.nlm.nih.gov; accession ID P22314). To find suitable templates for homology modelling, a BLASTP [15] search was performed against the Protein Data Bank (PDB) [16]. At the beginning of this work, the search identified three templates: i) the crystal structure of mouse Ubiquitin-Activating Enzyme (PDB code 1Z7L; 2.8 Å resolution) [17], which covers 25% of the query sequence corresponding to the SCCH domain (sequence identity of 96%), ii) the crystal structure of the *Saccharomyces cerevisiae* UbA1 (scUbA1) - Ub complex (PDB code 3CMM; 2.7 Å resolution) [18], which covers 98% of the query sequence (sequence identity of 53%; similarity 71%), and iii) the NMR solution structure of a fragment of mouse UbA1 (PDB code 2V31) [19], which covers 10% of the query sequence corresponding to the FCCH region, with sequence identity of 93%. This latter

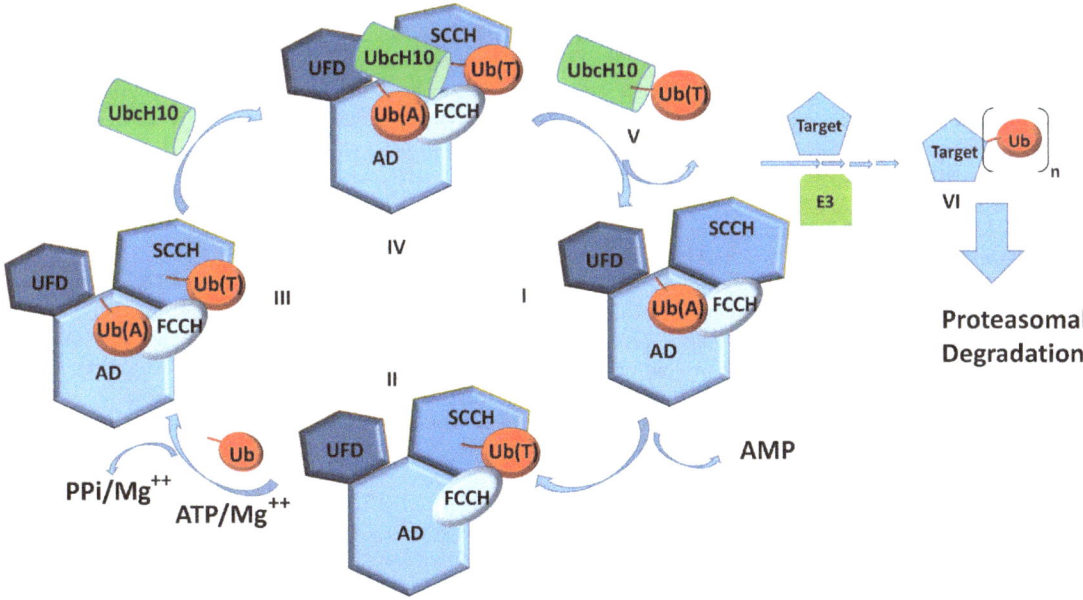

Figure 1. Ubiquitin conjugation cascade. UbA1 consists of four domains: the Adenylation domain (AD), the First Catalytic Cysteine Half-domain (FCCH), the Second Catalytic Cysteine Half-domain (SCCH) and the Ubiquitin Folding Domain (UFD). I) UbA1 catalyzes the adenylation of the Ub C-terminus in an ATP-dependent process in the AD domain; II) the activated Ub forms a thioester bond with a conserved catalytic cysteine in the SCCH domain of UbA1 [Ub(T)]; III) UbA1 is loaded with a second Ub molecule in the AD domain, followed by its C-terminal adenylation [Ub(A)]; IV) the ternary UbA1~Ub(T)-Ub(A) thioester complex recruits E2 (e.g. UbcH10); V) the thioester-linked Ub is transferred to a conserved E2 cysteine (transthioesterification); VI) E3 mediates the binding of Ub to the target lysine ε-amino groups.

structure showed that only the core region of FCCH was structured. Therefore, homology building was accomplished by using 1Z7L as template for the hUbA1 SCCH region (residues 629–884; hUbA1 numbering will be followed unless otherwise noted), and 3CMM as template for the AD, FCCH and UFD domains (residues 1–628 and 885–1057). Finally, since chains A and C in the X-ray structure 3CMM differ by a rigid-body rotation of the UFD domain, hUbA1 was modeled using the two monomers, leading then to two models hereafter designated UbA1_A and UbA1_C.

The ClustalW2 (http://www.ebi.ac.uk/Tools/msa/clustalw2/) [20] program was used for sequence alignment. The 3D structure of the target protein was modeled using SWISSMODEL [21]. The secondary structure of the target protein was assigned using DSSP [22]. Coordinates for two loops with undetermined coordinates in the UbA1 template structure (residues 812–824 and 964–969) were built up using the loop building ProMod tool [23] by scanning through the loop database in SWISSMODEL. The models were refined on the basis of energy minimization by GROMOS96 [24] and the models were validated for the 3D–1D profile with VERIFY3D [25], non-bonded interactions with ERRAT2 [26] and stereochemical qualities with PROCHECK [27] and WHATCHECK [28]. The comparison of the final model with the recently released structure of *Schizosaccharomyces Pombe* UbA1 (spUbA1; PDB code 4II3) revealed similar homology parameters with hUbA1 (covered sequence 94%; sequence identity of 54%; similarity 70%) and a RMSD for the backbone atoms of 1.6 Å, thus confirming the reliability of the model.

General strategy for docking calculations

The 3D model of the quaternary complex between hUbA1, two Ub molecules and UbcH10 was determined by using an experimentally-guided incremental strategy that relies on the building and refinement of models for the dimeric and trimeric complexes. Thus, we first explored the recognition between hUbA1 and Ub, leading to the UbA1~Ub(T) complex (Ub(T) stands for Ub bound to E1 through a thioester bond). Next, this model was used to build up the ternary UbA1~Ub(T)-Ub(A) system (Ub(A) denotes Ub bound to the AD domain). Finally, this model was the starting point for assembling the quaternary complex, UbA1~Ub(T)-Ub(A)/UbcH10. To this end, we adopted a computational approach that combines protein-protein docking, guided by the available structural information, and subsequent refinement through MD simulations (see below).

In order to generate the structural models, two docking programs were used: HADDOCK [29] and RosettaDock [30]. HADDOCK uses experimentally derived data, in conjunction with the available structures, to carry out flexible data-driven docking of proteins. Residues that are known to be implicated in the protein-protein recognition are designated active and are used to introduce suitable restraints to drive the docking process (i.e, the so called ambiguous interactions restraints; AIRs). HADDOCK expert interface was used to generate a reasonable rough complex, which was subsequently refined with the HADDOCK refinement interface.

To assess the initial orientation of the interacting partners in order to check the suitability of the restraints to be imposed in HADDOCK calculations (i.e., the extension and solvent accessibility of the region comprising passive residues, which are solvent-exposed residues that surround the active ones) the mutual complementarity of the interacting partners was first explored by superposing the structures of Uba1 and Ub(T) in the X-ray structure of the APPBP1-Uba3~NEDD8-NEDD8-Ubc12 complex [13] (PDB entry 2NVU). A list of the restraints used in calculations is given in Supporting Information (Table S1).

Finally, the RosettaDock server performs a local docking searching for conformations near the starting 3D structure in order to find the optimal fit between the partners. It was then used to calibrate the models derived from HADDOCK.

The UbA1~Ub(T)-Ub(A)-UbcH10 complex

Following the incremental docking strategy, the dimeric UbA1~Ub(T) system was generated using as input structures the previously generated hUbA1_A and hUbA1_C models, and the NMR structure of human Ub (PDB ID 2K6D) [32]. For HADDOCK calculations, the active residues were only those involved in covalent interactions, i.e. UbA1 Cys632 and Ub Gly76, and passive residues were defined as neighboring residues in a range of 8.5 Å from the active ones. Residues in the Ub tail (residues 70–76) and in the loop above the catalytic cysteine, whose coordinates were undetermined in template structures (residues 803–819), were set as fully flexible during all stages of the docking protocol. Since RosettaDock accepts a maximum of 600 residues, docking was performed using a truncated form of UbA1 that retains the residues pertaining to the interaction domain. Taking into account that RosettaDock allows the sliding of proteins around 8 Å, a binding region that includes residues 216–296 and 627–888 was defined.

The ternary complex was generated taking into account experimental information taken from the PDB structure 3CMM, in which Ub is bound to the AD domain of scUbA1. In HADDOCK calculations the active residues were those known to participate in the binding between UbA1 (Arg239, Asp576, Tyr600) and Ub (Asp32, Arg72, Gly75, Gly76). Passive residues were automatically defined as neighbors in a range of 8.5 Å from active residues. Besides the Ub tail, full flexibility was also given to residues of the UbA1 crossover loop (residues 592–630) to facilitate the accommodation of the Ub tail.

Finally, to build up the 3D model of the quaternary complex, the UbcH10 structure was taken from PDB ID 1I7K. Let us note that this structure is functionally active even though it lacks the first 30 residues at the N-terminus [33]. Note also that Ser114 in the crystal structure was mutated to Cys to restore the native sequence. In order to enhance the sampling in predicting the quaternary complex, four starting structures of the UbA1~Ub(T)-Ub(A) complex were generated by combining the two UbA1 models (UbA1_A and UbA1_C) with two orientations of Ub in the UbA1~Ub(T)complex (denoted Ha and Hb; see below). Hence, a total of four ternary models were used to build up the 3D structure of the quaternary complex. Active residues in HADDOCK calculations comprised those involved in E1–E2 interactions on the basis of mutagenesis studies [13,31,41–43]: Glu1037, Asp1047 and Glu1049 in UbA1 (numbering for the UbA1-Ub2 complex), and Lys33′ and Gln37′ in UbcH10. Moreover, the two catalytic cysteine residues (Cys632 in UbA1 and Cys114′ for UbcH10) involved in the transthiolation process were also treated as active residues in order to guide the complex formation. Passive residues were defined as neighbors in a range of 8.5 Å from active residues. Residues of the UbA1 crossover loop (residues 592–630) and the Ub tail (residues 70–76) were also flexible. Each ternary model was docked twice with UbcH10 structure yielding a total of 80 clusters.

Molecular Dynamics

MD simulations were performed to refine the different complexes. To this end, each complex was immersed in a pre-equilibrated octahedral box of TIP3P water molecules, and the system was neutralized. The final systems contained between 93000 and 99000 atoms. All simulations were performed with the parmm99SB force field [34]. The thioester bond between Ub(T)

Gly76 and UbA1 Cys362 was manually added, and suitable force field parameters were derived using $CH_3CH_2SCOCH_3$ as a representative model. The AMP position was derived from the ATP molecule as found in the PDB structure 2NVU. To this end, the AD domain of the UbA1~Ub(T)-Ub(A) model was superimposed to the AD domain of NAE1/UbA3. On the other hand, the phosphodiester bond between Ub(A) Gly76 and AMP was manually added, and the force field parameters for the phosphodiester linkage between UbA1 Cys632 and Ub(T) Gly76 were derived using $CH_3OP(O)_2OCOCH_3$ as a model system.

For each complex the geometry was minimized in four steps, which involve: i) water molecules and counterions (3000 steps of steepest descent and 7000 steps of conjugate gradient), ii) hydrogen atoms in the protein (500 steps of steepest descent and 4500 steps of conjugate gradient), iii) then, hydrogen atoms, water molecules and counterions (3500 steps of steepest descent and 11500 steps of conjugate gradient, and iv) finally the whole system (2500 steps of steepest descent and 8500 steps of conjugate gradient). Thermalization of the system was performed in four steps of 60 ps, increasing the temperature from 50 to 298 K. Concomitantly, the atoms that define the protein backbone were restrained during thermalization using a variable restraining force. Thus, a force constant of 20 kcal·mol^{-1} Å^{-2} was used in the first stage of the thermalization and was subsequently decreased by increments of 5 kcal·mol^{-1} Å^{-2} in the next stages. Then, an additional step of 250 ps was performed in order to equilibrate the system density at constant pressure (1 bar) and temperature (298 K). Finally, an extended trajectory was run using a time step of 2 fs. SHAKE was used for those bonds containing hydrogen atoms in conjunction with periodic boundary conditions at constant pressure and temperature, particle mesh Ewald for the treatment of long range electrostatic interactions, and a cutoff of 10 Å for nonbonded interactions. The structural analysis was performed using in-house software and standard codes of Amber 12.

Steered Molecular Dynamics and refinement of the final complex

Comparison of the final MD structures and the recently solved X-ray structure of Uba1 in complex with Ubc4 (PDB entry 4II2; [14]) showed that the loop masking the hUbA1 catalytic cysteine (Cys-cap loop) prevented a close packing between UbcH10 and the ternary complex. Accordingly, the protein-protein interface was refined by means of steered molecular dynamics (SMD) simulations, which were set up using Amber 12. To this end, the Cys-cap loop (residues 801–825) was deleted and capping groups were added to the newly formed terminals. The distance between the sulfur atom of the UbcH10 catalytic cysteine (C114) and the carbon atom of the terminal carboxy group of Ub(T) was constrained to 3 Å in 4 steps: i) from the initial distance (9.4 Å) to 7 Å in 0.5 ns with a force constant of 5 kcal/mol; ii) from 7 to 4 Å in 1.5 ns with a force constant of 5 kcal/mol; iii) from 4 to 3 Å in 2 ns with a force constant of 10 kcal/mol; iv) and finally from 3 to 2.5 Å in 4 ns with a force constant of 20 kcal/mol. At the end, the system was rebuilt by adding the removed Cys-cap loop (UbA1 residues 801–825), equilibrated with suitable constraints in order to relax the residues in the Cys-cap loop, and finally subjected to an unrestrained MD (50 ns) simulation.

Binding free energy evaluation and virtual alanine scanning

Binding free energies of the docking solutions and sampled in MD simulations were estimated using the Solvated Interaction Energy (SIE) method [35], as implemented in the Sietraj program

[36]. Analysis of the MD trajectory was carried out by calculating SIE on a 0.1 ns interval at the end of the trajectory. On the other hand, the contribution of specific residues to the binding between interacting proteins was examined by using alanine scanning [37,38].

GST-UbcH10 preparation for experimental binding assays

The PCR product was cloned into pGEX-4T1 expression vector (GE Healthcare), leading to a protein with a cleavable N-terminal GST tag (GST-UbcH10). E. coli BL21 (DE3) RP strain was transformed with the recombinant plasmid for GST-UbcH10. Overnight cultures were used to inoculate 500 ml LB medium containing 50 µg/ml ampicillin, and protein induction was performed by the addition of 1 mM IPTG at 22°C when an OD$_{600}$ value of 0.7 was reached. After approximately 16 h the cells were harvested and the proteins were isolated by sonicating cell pellets resuspended in 30 ml PBS1X in the presence of an EDTA free protease inhibitor cocktail (Roche Diagnostics). The crude cell extract was cleared by centrifugation at 18000 rpm and the supernatant was loaded onto a 1 ml GST-trap column connected to AKTA FPLC system (GE-Healthcare) equilibrated with binding buffer PBS1X. After washing with ten volumes of binding buffer, a single elution step was performed with 50 mM TrisHCl, 10 mM reduced glutathione. The fractions containing GST-UbcH10 were pooled and extensively dialyzed against PBS1X at 4°C. The homogeneity of the protein was tested by SDS-PAGE and mass spectrometry.

Peptides synthesis

A series of peptides chosen to mimic specific regions of the protein-protein interface (L1, L2, U1 U2, S1 and S2), as well as the L2- scrambled (ScrL2) and U1-scrambled (ScrU1) peptides were obtained by Fmoc solid-phase strategy. To mimic the fragment within the parent protein, the N- and C-terminus were acetylated and amidated, respectively. The syntheses were carried out with Novasyn TGR resin (substitution 0.25 mmol g^{-1}). Coupling reactions were performed by using 10 equiv of Fmoc protected amino acids activated in situ with HBTU (9.8 equiv)/ HOBt (9.8 equiv)/DIPEA (20 equiv) in DMF for 1 h. Fmoc protecting group was removed by treatment with 30% piperidine in DMF two times for 10 min. Before the cleavage from the resin, all peptides were acetylated or biotinylated at the N-terminus to obtain the corresponding derivatives. The acetylation reaction was carried out two times for 10 min using a solution of acetic anhydride (0.5 M)/DIPEA (0.15 M)/HOBt (0.125 M) in DMF. Biotinylated peptides were obtained using a solution of N-(+)-biotinyl-6-aminocaproic acid (2 equiv)/PyBop (2 equiv)/DIPEA (4 equiv) in DMF overnight. All peptides were cleaved off the resin by treatment with a mixture of TFA/H$_2$O/ethanedithiol (EDT)/ triisopropylsilane (TIS) (94:2.5:2.5:1v/v/v/v) for 3 h at room temperature. The resins were filtered and the crude peptides were precipitated with diethyl ether, dissolved in a H$_2$O/CH$_3$CN (1:1 v/v) solution and lyophilized.

L1, L2, L2-scrambled, U2, S1 and S2 peptides were purified by preparative RP-HPLC on a Shimadzu system equipped with the UV-Vis detector SPD10A using a Phenomenex Jupiter Proteo column (21.2×250 mm; 4 µm; 90 Å) and a linear gradient of H$_2$O (0.1% TFA)/CH$_3$CN (0.1% TFA) from 5 to 70% of CH$_3$CN (0.1% TFA) in 20 min at flow rate of 5 ml/min. U1 and U1-scrambled peptides were dissolved in H$_2$O/CH$_3$CN solution with TCEP to avoid S-S bridge formation and purified using a linear gradient of ammonium formate buffer 0.1 M (pH = 7.0) and ammonium formate buffer/CH$_3$CN 0.1 M (pH = 7.0) (1:1 v/v) from 20 to 65% of ammonium formate buffer/CH$_3$CN 0.1 M (pH = 7.0) in 25 min at flow rate of 5 mL/min. The collected fractions containing the peptides were lyophilized. The identity and purity of peptides were assessed by an ESI-LC-MS instrument (ThermoFinnigan, NY, USA) equipped with a diode array detector combined with an electrospray ion source and ion trap mass analyzer using a Phenomenex Jupiter Proteo column (150×2 mm; 4 µm; 90 Å) and a linear gradient of H_2O (0.1% TFA)/CH_3CN (0.1% TFA) from 5 to 70% of CH_3CN (0.1% TFA) in 15 min at flow rate of 200 µl/min for L1 and L2 peptides and from 20 to 80% of CH_3CN (0.1% TFA) in 15 min for U1 peptides.

ELISA assay

Wells were coated overnight at 4°C with 100 µg/ml GST-UbcH10 in PBS 1X, 1 mM TCEP, in the presence of an EDTA free protease inhibitor cocktail (Roche Diagnostics). Binding step was performed with different concentrations of biotinylated peptides L1, L2, ScrL2, U1, ScrU1, S2 (2.2, 11, 22, 44, 108 µM) in PBS 1X (with 1 mM TCEP for U1 and ScrU1). A blocking solution 1% BSA in PBS 1X, 0.05% Tween-20 was used. Washes were executed with PBS 1X, 0.05% Tween-20. To verify the interaction a 1:10000 dilution of horseradish peroxidase-conjugated streptavidin (Sigma Aldrich) in 0.3% BSA, PBS 1X was incubated for 1 hour. The colorimetric reaction was carried out with SIGMA-FAST OPD reagent (Sigma Aldrich), according to the manufacturer's instructions. Finally, readings were performed at 490 nm on a Model680 MicroplateReader (Bio-Rad, Hercules, CA-USA), and data were recorded by Microplate Manager 5.2 program and elaborated by GraphPad Prism program. Negative control experiments with the fusion tag GST in coating were performed in the same conditions described above.

Results and Discussion

In order to determine the 3D model of the tetrameric complex responsible for UbcH10 transthiolation and identify the regions involved in protein recognition, we have first built the trimeric complex formed by UbA1 with two Ubs, one covalently bound at UbA1 Cys632 (Ub(T)) through a thioester bond (indicated with the symbol ~) and the other non-covalently bound at the AD site (Ub(A)) following an incremental docking procedure that follows the series of events leading to the quaternary system (Figure 1). The model of the quaternary complex was then experimentally validated by competitive binding assays using a series of peptides chosen for their contribution to the protein-protein interface in the 3D model (see below).

Homology modeling of hUbA1

The structural model of hUbA1 was built up by using 1Z7L and 3CMM structures as templates for the hUbA1 SCCH region (1Z7L) and for the AD, FCCH and UFD domains (3CMM), respectively. Moreover, the two conformations of S. cerevisiae UbA1 (scUbA1) found in the X-ray structure 3CMM were considered, leading to 3D models named UbA1_A and UbA1_C (see Materials and Methods). The quality of the models was checked by considering a number of structural features, including stereochemical properties, the compatibility between the amino acid sequence and the environment of amino acid side chains, and the patterns of non-bonded interactions (see Table 1). The Ramachandran plots for the two UbA1 models showed that around 98% of the total residues were located within the allowed regions (88% in the most favored ones), and only 3 (UbA1_A) or 4 (UbA1_C) residues were found in disallowed regions (0.3%) (Figure S1). The global PROCHECK G-factor for UbA1_A and

UbA1_C was −0.08 and −0.15, respectively, indicating that the two structures are acceptable, because the recommended value must be greater than −0.50. On the other hand, the VERIFY3D scores above the threshold of 0.2 (86.7% and 90% for UbA1_A and UbA1_C, respectively) also indicated good local structural environments. Finally, the ERRAT2 analysis, which examines the quality of non-bonded interactions, yielded an estimate above 95%, indicating that the two models exhibit interresidue contacts that compare well with the patterns observed in high-resolution structures.

Taking into account the similar scores obtained for the two models and their structural resemblance (RMSD = 1.2 Å), MD refinement was accomplished only for UbA1_C. A stable structure was obtained after the first 5 ns of the trajectory (Figure S2; A). The increase in the RMSD was mainly due to structural rearrangements of the domains present in UbA1, leading to an average displacement of *ca.* 6 Å. Nevertheless, the structure of each domain was very stable along the trajectory, as demonstrated by the stability of the RMSD of the single domains (Figure S2; B).

hUbA1~Ub(T) and hUbA1~Ub(T)-Ub(A) complexes

To build up the tetrameric complex between hUba1, Ub(T), Ub(A) and UbcH10, we first modeled the hUbA1~Ub(T) thioester complex, which was subsequently used to dock a second Ub molecule in the AD domain. Modeling the binding mode of Ub(T) is challenged by the lack of structural and biochemical information about this interaction, and by the covalent linkage of Ub, which is an unusual feature in protein-protein docking. To this end, a multistep strategy that included the use of two protein-protein docking webservers, HADDOCK and RosettaDock, in order to disclose the non-covalent interfaces between the E1 and Ub(T), was adopted. Accordingly, we first docked Ub to hUba1 using HADDOCK by restraining the contact between Cys632 (UbA1) and Gly76 (Ub). Among the 9 clusters that embody the 200 best structures yielded by HADDOCK (Table S2), solutions were chosen on the basis of four criteria: i) the distance from the sulfur atom of Cys632 and the carboxylic oxygen of Gly76, ii) the total score, iii) the total number of poses, and iv) the buried surface area. The selected poses lead to a distance lower than 3.8 Å, and are characterized by a high score, a large number of poses, and a large burial of surface area (see Table S2). These poses (denoted Ha and Hb) mainly differ in the orientation of Ub relative to the SCCH domain (Figure 2-A). In the lowest energy solution (Ha), Ub forms contacts with SCCH, mainly through ionic and polar interactions, and FCCH, primarily through hydrophobic interactions via the Ile44 patch, which is known to be involved in other non-covalent interactions of Ub, such as in the recognition of UbcH5c, UEV and GLUE domains [39]. In the second pose (Hb), Ub only showed polar contacts between residues in the Ub tail with the SCCH domain.

The two poses were then checked using RosettaDock. The best ranked solution turned out to be very similar to the best HADDOCK solution (Ha), as noted in a RMSD of 0.82 Å. In contrast, calculations started from pose Hb yielded solutions that showed significant structural differences with regard to the initial structure. Therefore, due to the structural consistency of pose Ha, it was chosen as a model of the hUbA1~Ub(T) complex and subsequently refined by MD simulations, which led to a stable trajectory after the first 2 ns (see Figure S2-A, C). The refined structure supports the hydrophobic contacts between residues Leu8, Ile44, Val70 and Leu73, which interact with FCCH residues Y286, Met223, Val277, Leu178 and Thr233. The hydrophobic interactions involving the Ile44 patch were also

Table 1. Structural models of hUbA1.

Protein	Procheck[1]				Errat2	Verify3D
	Most favoured	**Allowed regions**	**Generously allowed**	**Disallowed**	**%**	**%(avg>0.2)**
CMM_A	780 (88.1%)	98 (11.1%)	1 (0.1%)	6 (0.7%)	91.8	88.0
hUbA1_A	781 (87.9%)	93 (10.5%)	12 (1.3%)	3 (0.3%)	95.8	86.7
CMM_C	787 (88.3%)	90 (10.1%)	7 (0.8%)	7 (0.8%)	96.4	94.0
hUbA1_C	789 (88.6%)	95 (10.7%)	3 (0.3%)	4 (0.4%)	95.6	90.0

Validation results for the lowest energy models of hUbA1, compared with the corresponding templates.
[1]Number of residues in the region (the percentage is given in brackets).

reinforced by ionic interactions between Arg42 (Ub(T)) and Asp236 and between Arg74 (Ub(T)) and Asp811 (Figure 2-B).

The trimeric complex was obtained through docking of Ub to the AD site and subsequent MD refinement, which led to a stable trajectory after the first 2 ns (see Figure S2-A, D). The 3D structures closely resembled the X-ray template 3CMM (RMSD of 1.1 Å; Figure 3A). Three different interfaces might be identified: i) the loop pocket defined by hUbA1 residues Tyr618, Ser621, Glu626, Arg515, Asn512, Asn516 and Arg551 interacting with Ub(A) tail residues Arg72, Arg42 and Arg74 and AMP (Figure 3, B); ii) an hydrophobic patch formed by the Ub residues Leu8, Ile44 and Val70 that form contacts with an hydrophobic area in the hUbA1 AD region formed by residues Phe933 and Phe926

(Figure 3-C); and iii) the polar interface formed by Ub(T) residues Thr9, Lys11, Thr12, Asp3 interacting with the FCCH region, mainly with residues Glu243, Arg239 and Asn212 (Figure 3-C). Moreover, interactions between Ub(T)-Lys48 and Asp920 and Glu938, not present in the 3CMM structure, were also found.

Finally, during the submission of the article, a novel structure of the scUbA1 loaded with two ubiquitin molecules was released in the PBD with the name 4NNJ [40]. The superimposition of the model of hUbA1-Ub(T)-Ub(A) complex obtained with our incremental strategy to the crystal structure (chains C,D,E) leads to an rmsd of 1.5 Å (determined for the Cα carbon atoms), and showed a position of Ub(T) very similar to the pose C_Ha selected

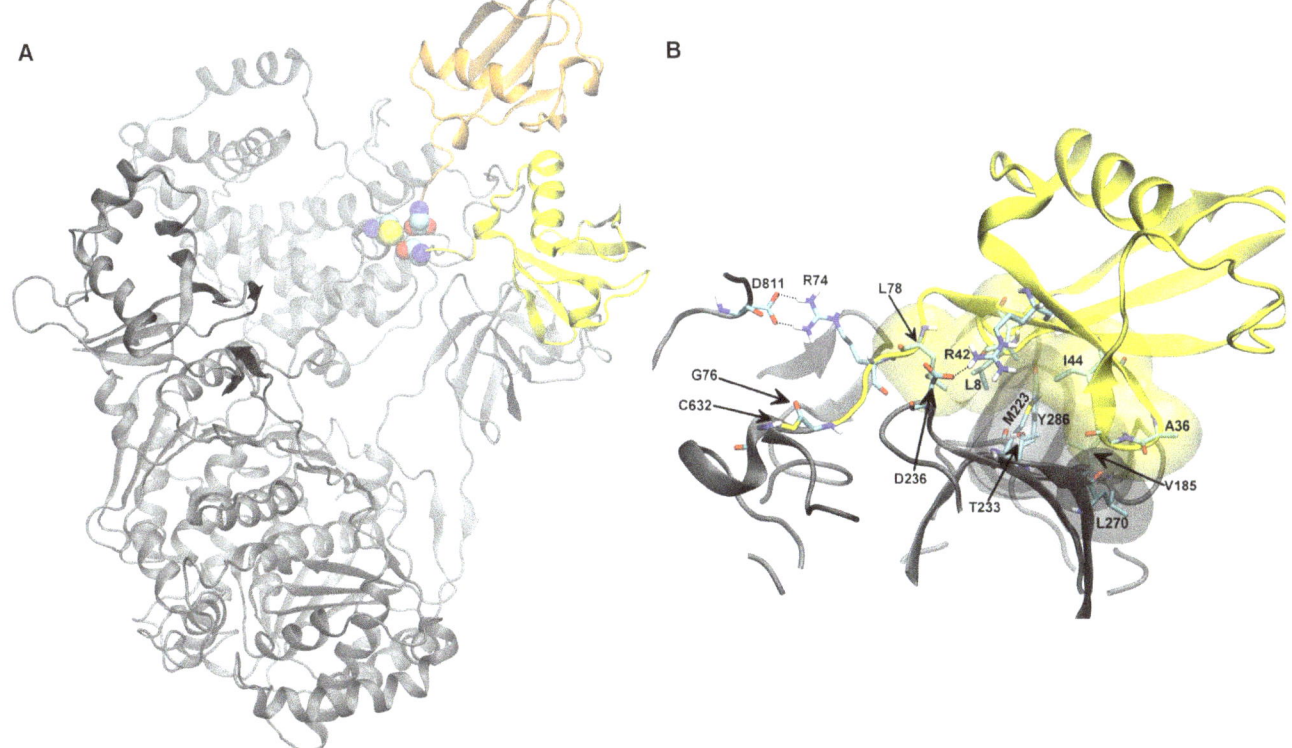

Figure 2. Model of hUbA1-Ub(T) complex. A) Comparison of the best two binding modes of Ub resulting from HADDOCK calculation, Ha and Hb, shown in yellow and orange, respectively. Terminal Ub-Gly76 and catalytic UbA1-Cys632 are highlighted in spheres coloured by atom type. B) Detail of hUbA1~Ub(T) interactions in the lowest energy MD frame (time 9.6 ns). Apolar hydrogens were omitted for sake of clarity. Colour code: hUbA1, grey; Ub(T), yellow; van der Waals interactions are highlighted with transparent Connolly surfaces. Carbons are in cyan, nitrogen in blue, oxygens in red, sulphur in yellow and hydrogens in white.

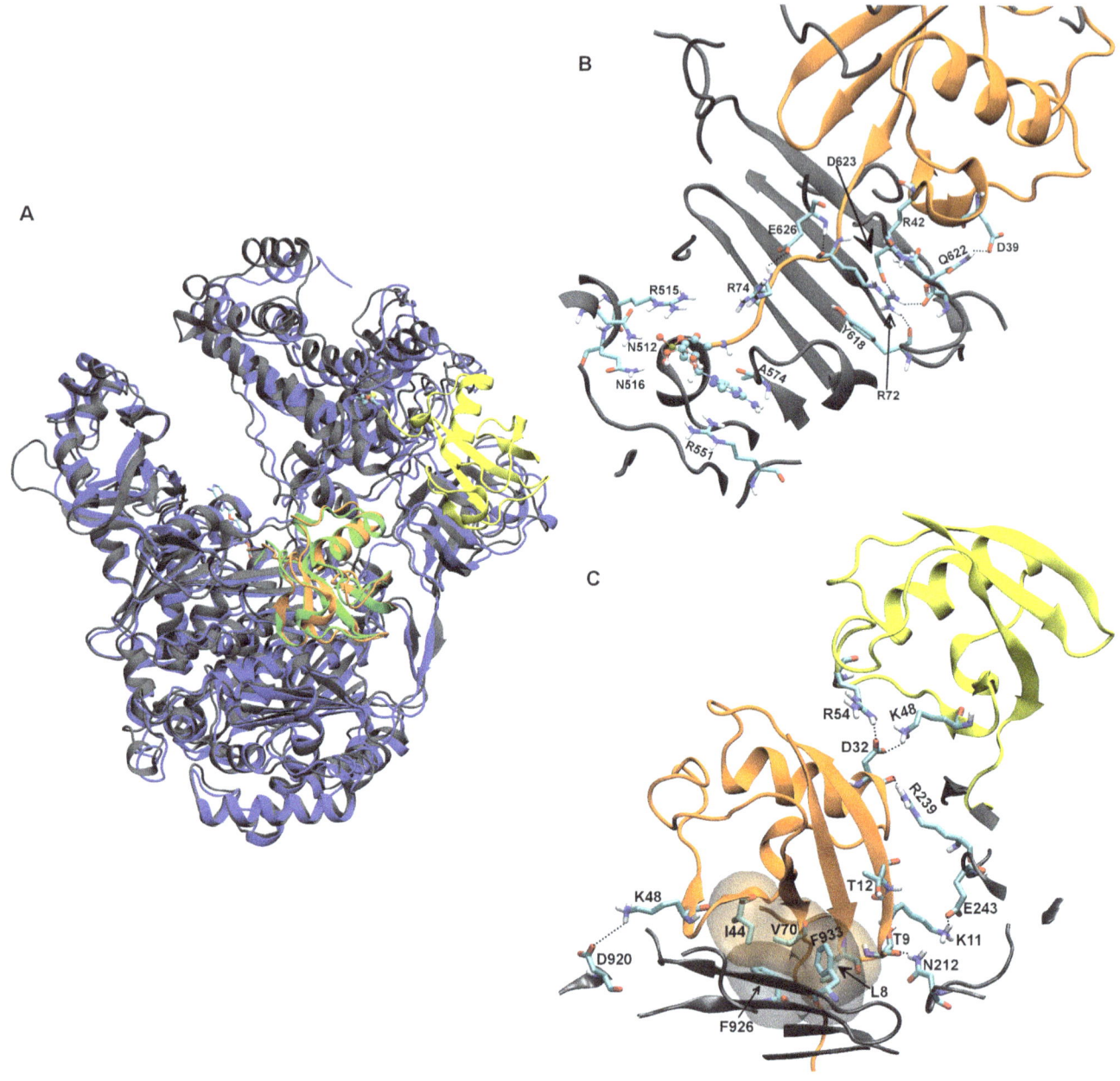

Figure 3. Model of hUbA1-Ub(T)-Ub(A) complex. A) Superimposition of the hUbA1~Ub(T)-Ub(A) model C_Ha on the crystal structure 3CMM_C; B) Detail of the main interactions of Ub(A) and AMP in the hUbA1 loop pocket; C) Detail of the main interactions between Ub(A) and hUbA1: hydrophobic and polar interface. AMP is highlighted in CPK. Apolar hydrogens were omitted for sake of clarity. Colour code: hUbA1, grey; Ub(T) yellow; Ub(A), orange; scUbA1, blue; scUb(A), green.

in our calculations (Figure S3), thus supporting the reliability of our model.

hUbA1~Ub(T)-Ub(A)-UbcH10 complex

The hUbA1~Ub(T)-Ub(A) model was the starting point for the docking with UbcH10. In order to make a more exhaustive sampling, up to four different starting points were considered for docking calculations (see Materials and Methods). HADDOCK calculations yielded 80 clusters. The results were filtered by selecting only poses where the distance from the sulphur atom of the UbcH10 catalytic cysteine (Cys114) to the carbonyl carbon of the C-terminal Gly in Ub(T) was lower than 20 Å, considering this

limit as indicative of the side of UbcH10 facing the SCCH region. Moreover, this criterion is consistent with the distance between the Cys residues involved in the transthiolation reaction in the crystal structure 2NVU, representing the tetrameric complex of the NEDD8 system. Only six clusters satisfied the distance cutoff. Among these clusters, Ub(T) adopted the Ha binding mode in five cases, and the Hb arrangement was found in a single case. This suggests that the Ha binding mode position was better suited to accommodate the E2 partner within the E1 groove with the catalytic cysteines facing each other. Table 2 shows the distinct poses ranked according to HADDOCK score as well as to the binding free energy of the complex calculated with the SIE method using the snapshots collected in the MD simulation. A_Ha_L2 emerges as the best pose according to HADDOCK score and SIE binding affinity. Both HADDOCK and SIE scores are consistent in suggesting C_Ha_R and C_Ha_L as feasible poses. The structures of these complexes differ by around 5.5 Å relative to A_Ha_L2. The A_Ha_L1 and A_Ha_R poses were structurally similar to A_Ha_L2 (rmsd of 3.3 Å) and Ca_Ha_R (rmsd of 4.3 Å), respectively. However, SIE calculations predict that they are less stabilized compared to A_Ha_L1 and A_Ha_R. Finally, C_Hb_R was rejected due to their low energetic score.

It is experimentally known that the N-terminal helix and β1β2 loop of E2 are directly involved in the formation of the complex [13,31,41–43]. In particular, mutational and structural studies disclosed the main role of two basic residues, conserved in the E2 family (positions 33 and 37 in UbcH10 numbering), in assisting the binding to E1. We have therefore examined the role played in the selected models by i) the conserved acidic residues of the UFD region of hUbA1 (i.e. Glu1037, Asp1047 and Glu1049) and ii) the conserved basic residues of the N-terminal helix of UbcH10 (i.e. Lys33′ and Gln37′). It is worth noting that while a basic residue in position 33 is conserved in all the members of the E2 family, position 37 shows a higher variation, albeit basic or polar residues are generally found in this position. To this end, the best three solutions (A_Ha_L2, C_Ha_R and C_Ha_L) were subjected to a virtual alanine scanning analysis in order to evaluate the contribution of these residues to the E1–E2 interaction. Even though the results (Table 3) did not show significant interactions (defined as $\Delta\Delta G \geq -0.5$ kcal/mol) with Gln37′, the best three models showed a significant contribution to the binding of at least one residue from the N-terminal helix and one residue from the UFD region of UbA1. For the sake of comparison, no significant contribution was found for the mutations in the N-terminal helix for poses A_Ha_L1 and A_Ha_R. In fact, only a single mutation in hUbA1 (Asp1047→Ala) was found to lead to a significant destabilization. This finding, together with the structural resemblance to A_Ha_L2 and Ca_Ha_R and the lower SIE binding free energy (see above), led to their exclusion from further refinements.

The three models were further refined by running a series of 50 ns MD simulations, and the binding free energy was determined from SIE calculations performed for the snapshots sampled in the last four 10 ns windows. The results consistently showed that the best binding affinity was obtained for model C_Ha_R (-26.6 ± 0.2 kcal/mol), it being more favorable by 6 and 9 kcal/mol compared to A_Ha_L (-20.2 ± 1.4 kcal/mol) and C_Ha_L (-17.2 ± 1.5 kcal/mol) models. On the basis of the preceding findings, the C_Ha_R model was further refined by extending the MD simulation to 500 ns. The analysis of the trajectory revealed a progressive stabilization of the complex,

leading to a binding affinity close to -31 kcal/mol in the last 250 ns (Figure 4-A). The alanine scanning analysis also demonstrated that the residues known to be critical to E1–E2 complex formation contributed significantly to the protein-protein interaction with the only exception of Gln37′ (Figure 4-B). During the MD run we observed a change in hUbA1 associated to the rotation in opposite directions of the UFD and SCCH domains with respect to the AD domain (by 20° and of 13°, respectively, as calculated with DynDom [44]). This conformational change caused the widening of the groove defined by the three domains, thus allowing a closer contact between hUbA1 and UbcH10, leading to an increase of the interaction surface (Table S3) and the gradual decrease of the distance between the UbcH10 catalytic cysteine and the Ub(T) C-terminal glycine crosslinked to UbA1-Cys632 until it stabilised at around 8 Å (Figure S4). Analysis of the last 50 ns of the trajectory revealed the presence of two main interaction surfaces, which involve contacts between i) UbcH10 helix H1 and β1β2 loop and hUbA1 UFD domain, and ii) hUbA1 Cys-cap loop and Ub(T) (Figure 5).

Final refinement of the tetrameric complex

Although MD simulations led to a progressive refinement of the quaternary complex, the distance between residues Cys114 in UbcH10 and the terminal glycine of the crosslinked Ub(T) was still too large ($\sim$8 Å; Figure S4) as to mimic a state that resembles the catalytic arrangement of the interacting proteins. Inspection of the final MD structure showed that a closer approach between hUbA1 and UbcH10 was prevented by the Cys-cap loop, which retained the orientation found in the PDB template 3CMM. In contrast, in the available structure of the E1–E2 complex (PDB structure 4II2) the Cys-cap loop is not assigned, thus suggesting a large flexibility in the covalent construct that mimics the thioester crosslinking event. We have therefore forced the approach of UbcH10 by using steering forces applied on the sulphur of the UbcH10 Cys114 toward the carbonyl group of the crosslinked Ub(T) C-terminal glycine, after manual removing of the Cys-cap loop. SMD simulations allowed us to reduce the distance between those atoms from 8 Å to 3.2 Å in 8 ns. After loop reconstruction, the final structure was refined in a 50 ns MD simulation, leading to a stable trajectory (Figure S5). This approach led to a closer fitting of UbcH10 into the groove defined by the UFD and SCCH domains, increasing the total interaction surface, especially between hUbcA1 SCCH and the UbcH10 region around the catalytic cysteine, in better agreement with the crystal structure of the E1–E2 crosslinked construct (Table S3). Moreover, SIE calculations revealed an increase of the binding energy to -42.7 kcal/mol.

Comparison of the refined model with the recently reported X-ray crystallographic structure of the trimeric complex of *S. pombe*

Table 2. Comparison of structural and energy data for selected docking results of the quaternary complex.

Docking	UbcH10 S-C114 HUbA1 S-C632	UbcH10 S-C114 Ub(T) C-ter G76	Haddock Score	SIE ΔG
A_Ha_L2	15,2	17,6	-147.7 ± 11.2	-11.0
C_Ha_R	14,6	14,5	-126.7 ± 5.4	$-9,6$
C_Ha_L	18,1	17,4	-125.0 ± 5.4	$-9,2$
A_Ha_L1	16,9	15,1	-118.9 ± 7.3	-8.0
A_Ha_R	11,3	11,8	-125.0 ± 10.0	$-7,6$
C_Hb_R	16,7	18,5	-96.6 ± 8.3	$-3,6$

The structural data include the distance (Å) from the sulphur of the UbcH10 Cys114 to the hUbA1 Cys632, and to the C-terminal Gly76 of Ub(T). The energy data report the score of the docked structures obtained from HADDOCK and from SIE (kcal/mol).

Table 3. Alanine scanning.

Docking	ΔΔG hUbA1			ΔΔG UbcH10	
	E1037A	D1047A	E1049A	K33A	N37A
A_Ha_L2	0.0	−0.8	0.0	−0.8	−0.4
C_Ha_R	−0.2	−0.4	−1.5	−0.7	0.0
C_Ha_L	−0.3	−1.3	0.1	−1.0	−0.1
A_Ha_L1	0,0	−1,2	0,0	−0,2	−0,2
A_Ha_R	−0,2	−0,6	0,0	−0,3	0,1
C_Hb_R	−0,4	0,2	−0,6	−0,7	0,0

Results of the virtual alanine scanning (ΔΔG; kcal/mol) due to the mutation to Ala of residues Glu1037, Asp1047 and Glu1049 in hUbA1 and Lys33′ and Gln37′ in UbcH10 are reported.

Uba1-Ub-Ubc4 (PDB ID: 4II2) lends support to the theoretical 3D model of the quaternary complex. Thus, after deletion of the E2 partners (UbcH10 and Ubc4) and the additional Ub present in the quaternary complex, superposition of the backbone Cα carbon atoms leads to a positional rmsd of 2.5 Å, which indicates the similar structural arrangement of the AD, SCCH and UFD domains in the two complexes (see also Figure S6). Furthermore, retention of the E2 partners in the superposed structures leads to an rmsd value of 2.6 Å, thus suggesting a similar arrangement in the trimeric and quaternary complexes.

The analysis of the snapshots sampled in the last 20 ns of the trajectory allowed us to identify key interactions in the complex, which involve three interfaces: i) the contacts between the hUbA1 UFD domain and the UbcH10 helix H1 and β1β2 loop, ii) the interaction between the hUbA1 SCCH domain and Ub(T) with the region surrounding the UbcH10 Cys114′, involving residues from the 3–10 helix and helices H2 and H3, and iii) the contacts between the hUbA1 crossing loop and Ub(A) with UbcH10.

The first interface (Figure 6) comprises the UbcH10 residues Lys33′ and Gln37′, which are experimentally known to be critical for the interaction between E1 and E2 [41–43]: Lys33′ interacts with Asp1042 and with Ser1044, and Gln37′ is hydrogen-bonded to the backbone oxygen of Cys1040 and the hydroxyl group of Thr988 (Figure 6-F). Moreover, hydrogen bonds were also formed between the side chains of Gln36′ and Asp1042, between Tyr91

and Asp1047, and between the N-terminal Pro30′ with Glu1049 (Figure 6-F). Finally, the ionic interactions were supplemented by hydrophobic contacts involving hUbA1 residues Met989, Val994, Met996, Phe1000, Phe1001, and UbcH10 residues Leu42′, Pro54′, Leu59′ and Phe60′ (Figure 6-E).

Interactions between the hUbA1 SCCH domain and the region surrounding the E2 catalytic cysteine were mainly characterized by a number of ionic and polar interactions between residues from H3 and H4 helices of UbcH10 (Glu154′, Lys164′, Lys172′ and Tyr165′) and residues from the hUbA1 Cys-cap loop (Gln728, Lys806, Glu813, Asp811 and Asp822) (Figure 6-B), while the region around the UbcH10 catalytic cysteine, including residues in the 3–10 helix, were involved in interactions with the residues around hUbA1 Cys632 and Ub(T). In particular, two main clusters of interactions are formed: i) the first, mainly based on hydrophobic interactions, between the UbcH10 helix H3 (Pro147′, Ala153′ and T150′) with the UbA1 coiled stretch between H24 and H25 (Phe637, Phe729 and Phe741), also supported by hydrogen bonds between Asn728 with His151′ and Thr150′ (Figure 6-C); ii) the second mainly involves residues closer to the catalytic cysteines, such as the ionic contact between the UbcH10 Asp145′ with UbA1-Lys746 and Ub(T)-Arg 72, and interactions between charged residues in the UbcH10 3–10 helix (Asp116′, Asp120′ and Lys121′) with Ub(T) residues (Gln40,

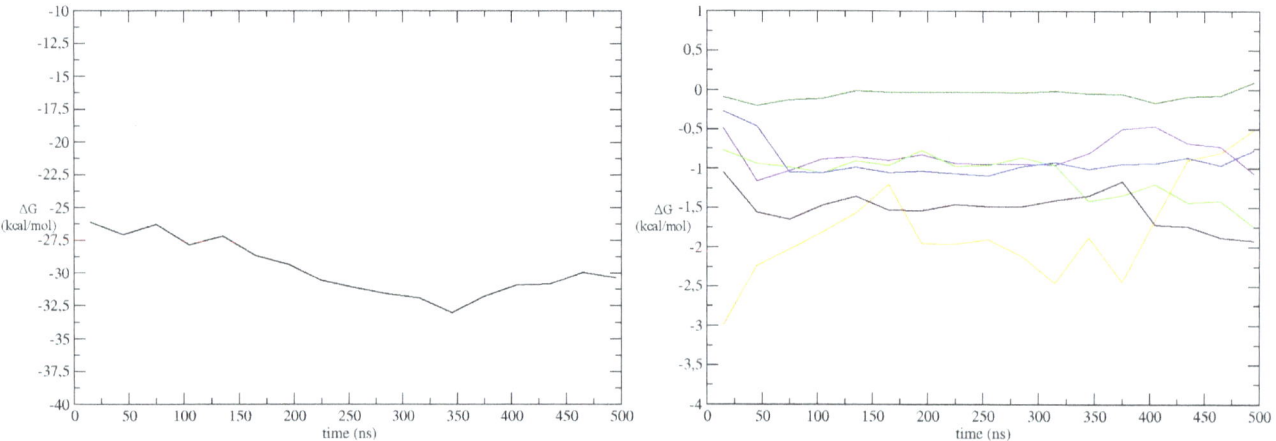

Figure 4. Energetic analysis. A) SIE values (kcal/mol) determined for the E1-E2 interaction along the trajectory (averaged for 20 ns windows). B) Contribution of key residues as derived from alanine scanning in UbcH10 N-terminal helix and the hUbA1 UFD region during the MD simulation of the model C_Ha_R. Colours: Glu1037, orange; Asp1047, violet; Glu1049, light green; Lys33′, bordeaux: Gln36′, blue; Gln37′, dark green.

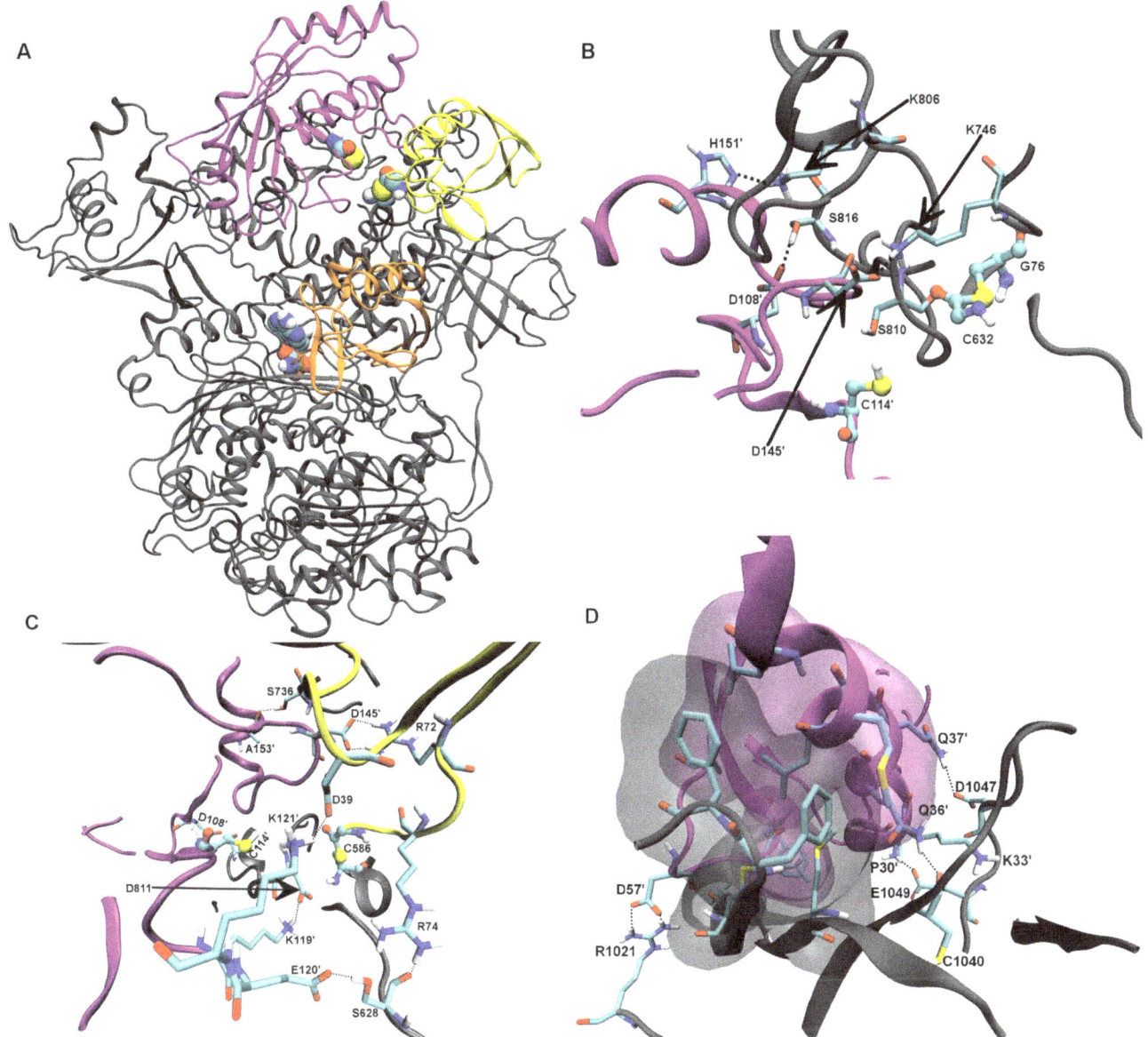

Figure 5. Model of the hUbA1-Ub(T)-Ub(A)-UbcH10 complex. A) Average structure of last 20 ns of MD simulation of the C_Ha_R model. The catalytic cysteines, the thioester bond and AMP were highlighted in spheres. Detail of interactions between B) UbcH10 and SCCH region, C) UbcH10 and Ub(T) and D) UbcH10 and UFD region. Catalytic cysteins and the thioester bond were highlighted in CPK. Apolar hydrogens were omitted for the sake of clarity. Colour code: hUbA1, grey; scUbA1, blue; Ub(T) yellow; Ub(A), orange; UbcH10, violet. The van der Waals interactions are highlighted with transparent Connolly surfaces.

Arg74, Asp39) and with the hUbA1 FCCH domain (Glu237) (Figure 6-D).

These findings demonstrated that the crosslinked Ub plays a key role in the transthiolation intermediate with UbcH10. In particular, MD simulations highlighted that the approach of the catalytic cysteines induced a rotation of 25° of Ub(T) with respect to hUbA1 and a rearrangement of the Ub(T) pattern of interactions showed in models lacking E2 (Figure 7). In particular, in absence of E2 Arg74 was hydrogen-bonded to Cys-cap residues (Asp811 and Gln812), while in the final model it was involved in ionic interactions with UbcH10-Glu120′ and Asp116′, and with Glu237, bearing to the hUbA1 FCCH domain. These data support the hypothesis that products of the transthiolation reaction might be released upon a process involving the rearrangement of

the Ub(T) binding to E1, driven by the charged residues in the region surrounding the catalytic cysteine of E2. Finally, we also observed some interactions in the loop region of hUbA1, in particular hydrogen bonds between the side chain of Asn622 with the backbone of UbcH10-Tyr 91′ (Figure 6-F), and between the backbone of Ser628 and the side chain of Glu120′ (Figure 6-D). A representative snapshot of the 3D model is available as supplemental PDB file.

Peptide affinity assays

On the basis of the 3D model of the quaternary complex, we have designed six peptides as molecular probes in order to calibrate their ability to interfere the binding of UbcH10. This

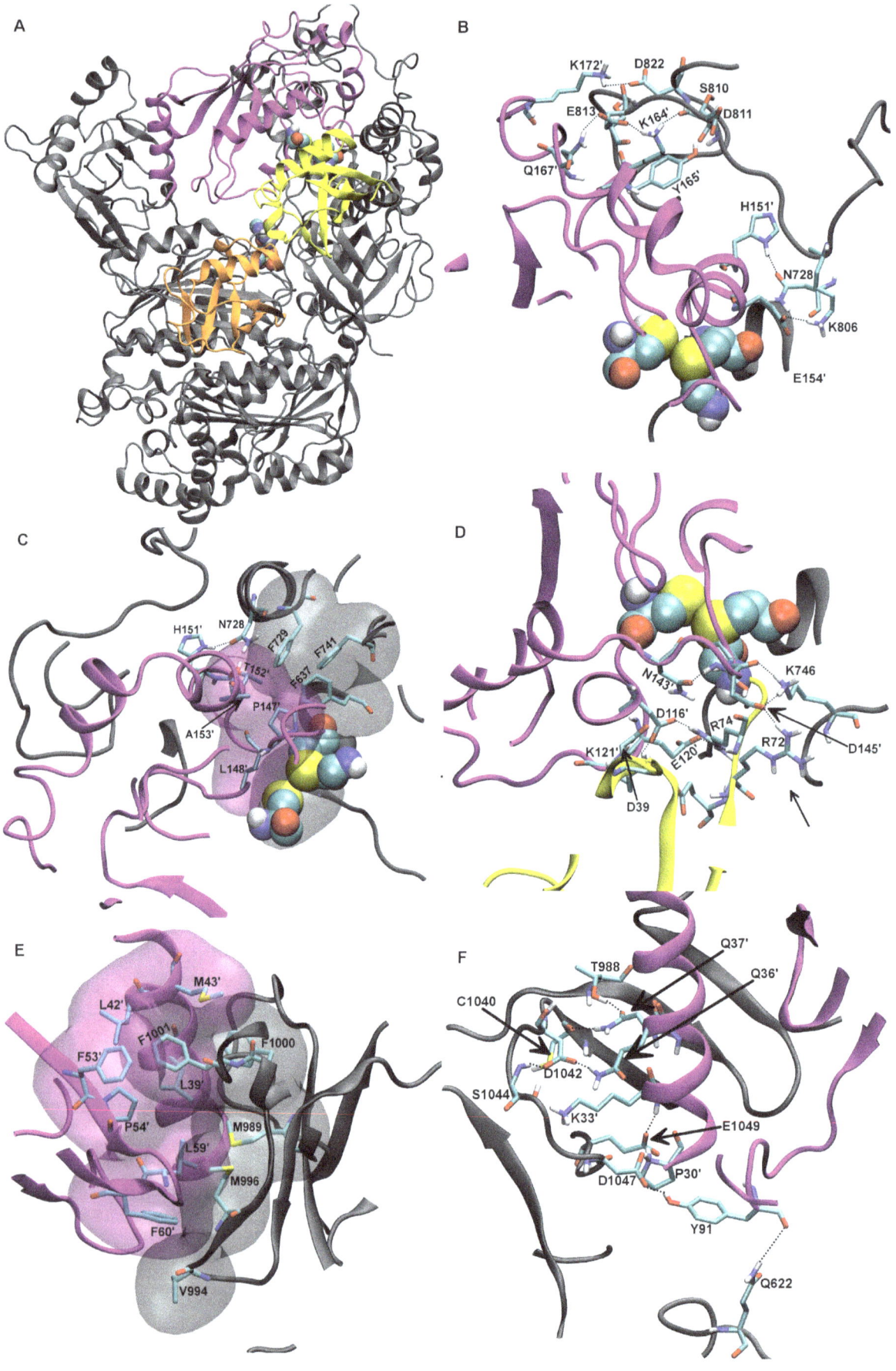

Figure 6. Final refined model of the tetrameric complex. A) Average structure of the last 20 ns of MD simulation of the model after SMD. B) Detail of the UbcH10-Cys-cap loop interactions. C) Detail of UbcH10-Cys region involved in hydrophobic interactions; D) Detail of the UbcH10-Cys region involved in polar interactions; E) Detail of the hydrophobic interactions between hUbA1 UFD and UbcH10. F) Detail of the polar interactions between hUbA1 UFD and UbcH10. Colour code: hUbA1, grey; Ub(T) yellow; Ub(A), orange; UbcH10, violet. Catalytic cysteins were highlighted in spheres. Apolar hydrogens were omitted for the sake of clarity. The van der Waals interactions are highlighted with transparent Connolly surfaces.

strategy was motivated by two main reasons. First, the identification of short peptides that mediate protein-protein interaction seemed *a priori* effective for disrupting the protein-protein recognition and binding. While other strategies, i.e. introduction of specific mutations, may also be envisaged, it is unclear whether single-point mutants might lead to a significant destabilization of the complex or even to impede the formation of the quaternary complex. Second, since our ultimate goal is the design of compounds that might disrupt the ubiquitilation process, testing a series of suitably chosen short peptides represents a valuable proof-of-concept for supporting the potential therapeutic effect of peptidomimetics. Specifically, the peptides were designed to examine the capability of hUbA1 stretches that contribute to the protein-protein interface with UbcH10 (Figure S7). In particular, we have designed two peptides per interface, which will be denoted S for SCCH region, L for cross loop, and U for UFD (Table 4). In the UFD region peptides U1 and U2 were selected to test the relevance of the acidic residues in mediating the binding of the UbcH10 H1 helix. In the SCCH region peptide S1 was chosen to test the role of the Cys-cap in binding UbcH10, while peptide S2, corresponding to helix H19 in the SCCH region, was designed as negative control, since the 3D model revealed the lack of any interaction in the complex. Finally, peptides L1 and L2 were chosen to explore the role of the cross loop region in assisting the interaction with UbcH10. All the peptides were synthesized as biotinylated derivatives by solid phase method and purified by RP-

HPLC. Unfortunately, S1 and U2 were insoluble and so not testable in binding assays. The ability of the soluble peptides to bind recombinant GST-UbcH10 was checked by ELISA, utilizing GST as control (data not shown).

The best results (Table 4) were obtained with U1 and L2, which were found to bind UbcH10 with an apparent K_D of about 10 and 20 μM, respectively. In order to confirm that the binding of U1 and L2 peptides was sequence-dependent, two scrambled peptides were synthesized, ScrU1 and ScrL2. The results demonstrated that these peptides exhibited a very poor binding, much weaker than U1 and L2, which might then be considered indicative of native protein-protein interactions. In particular the good affinity showed by U1 allowed us to validate the role of the acidic residues of the UFD region in binding E2, thus giving confidence to our 3D model. The U1 peptide, indeed, contained D1047 and E1049, two of the three acidic residues involved in the hUbA1 UFD-UbcH10 H1 interface. Unfortunately, the low solubility of U2 did not allow us to verify the role of E1037, which is the third residue proven to be involved in the interaction by mutagenesis studies. Similarly, the results obtained for L2 support the role of Gln622 in assisting the interaction of the crossover loop with UbcH10, in agreement with the 3D model. The low affinity showed by L1 peptide, which contains Gln622 at the N-terminus side of the sequence, might be indicative of the importance of flanking residues in L2 binding. Finally, the results obtained from the SCCH peptides allowed us to exclude a role in the binding of the helix region corresponding to S2, as expected for this peptide, which was designed as negative control.

Overall, the results support the involvement of the selected peptides in mediating the protein-protein interactions in the hUbA1~Ub(T)-Ub(A)-UbcH10, which in turn reinforces the reliability of the 3D model built up for the quaternary complex between E1, E2 and Ub partners. On the other hand, they also demonstrate the feasibility of interfering the formation of the complex, which paves the way to the structure-based design of peptidomimetics for UbcH10-related anticancer strategies.

Conclusions

We have simulated the dynamic process associated to the formation of the complex leading to the transthiolation reaction between doubly loaded hUbA1 and UbcH10. The formation of the complex takes place through protein-protein interactions in three main interfaces: i) the first between the hUbA1 UFD domain and the UbcH10 helix H1 and β1β2 loop, ii) the second formed by the hUbA1 SCCH domain and Ub(T) with the region surrounding the UbcH10 Cys114', involving residues from the 3–10 helix and helices H2 and H3, and iii) the third between the hUbA1 crossing loop and Ub(A) with UbcH10. The involvement of these regions has been supported by the ELISA assays performed for a series of short peptides that encompass the residues that mediate the interaction between UbA1, UbcH10 and the two Ubs. In particular, peptides U1 and L2, pertaining to the UBA1 UFD domain and to the UbA1 loop, have been able to interfere the assembly of the E1–E2 complex. The availability of this structural model should facilitate the understanding of the structural details of the ubiquitination cascade, to rationalize the details of the recognition between E1 and E2 partners, and finally to facilitate

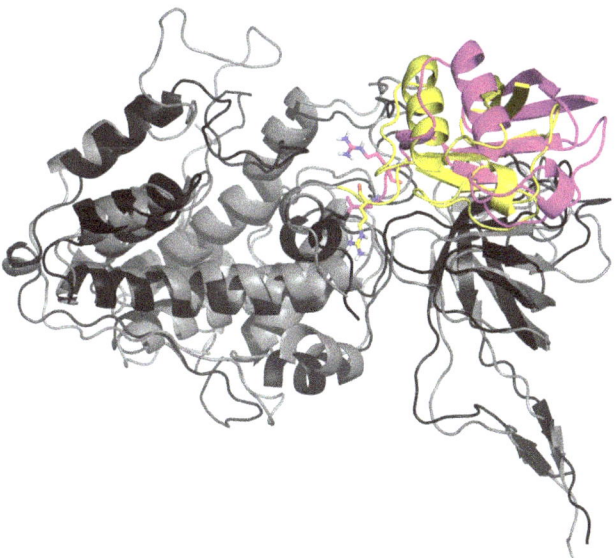

Figure 7. Rotation of the Ub(T) induced by UbcH10 interaction in the final quaternary model with respect to the UbA1~Ub(T) model. Superimposition was made on the Cα atoms of the SCCH and FCCH domains. Colour code: Final quaternary complex model: hUbA1, grey; UbA1~Ub(T) model: UbA1, black; Ub(T), magenta; Arg74 in the two models is shown as licorice. Apolar hydrogens were omitted for sake of clarity.

Table 4. Sequence and binding assays results of the designed peptides.

UbA1 region	peptide	sequence	K_D (μM)
UFD	U1	1038-LCCNDESGEDVEV-1050	10
	U2	1036-LELCCNDESGEDV-1048	Not soluble
SCCH	S1	807-IHVSDQELQSA-817	Not soluble
	S2	649-RDEFEGLFKQPAEN-662	No binding
LOOP	L1	621-SQDPPEKSIPI-631	No binding
	L2	615-TESYSSSQDPPEK-627	20
Scrambles	ScrU1	ELNCDVEVEGSDC	>100
	ScrL2	SDPSKTSEYQPSE	>80

Numbering is referred to the human UbA1 sequence. K_D values are calculated by Elisa assays.

the design of peptidomimetics or small size compounds able to interfere with the formation of the E1–E2 complex, which might be valuable to open new strategies against tumorigenic processes.

Supporting Information

Figure S1   Comparison of the Ramachandran plot of the models A (A) and C (C) with the corresponding conformations of the template 3CMM (B and D respectively).

Figure S2   A: Time evolution (ns) of the RMSD (Å) of the UbA1_C model: hUbA1 apo (red), hUbA1~Ub(T) (yellow), hUbA1~Ub(T)-Ub(A) (green) and hUbA1~Ub(T)-Ub(A)-UbcH10 (black). B, C and D: Structural preservation of the structure of each region. of: hUbA1 apo (B), hUbA1-Ub(T) (C) and UbA1~Ub(T)-Ub(A) (D), AD (black), UFD (green), SCCH (red) and FCCH (blue).

Figure S3   Superposition of the backbone for the X-ray structure 4NNJ and the 3D model of the ternary complex. Front and side views are shown in the left and right pictures, respectively. Colour code: hUbA1, grey; Ub(T) yellow; Ub(A), orange; scUbA1, magenta scUb(A), green, scUb(A), Cyan.

Figure S4   Analysis of the distance between the sulphur atom of the UbcH10 Cys114 and the carbonyl group of the crosslinked Ub(T) terminal glycine during the 500 ns unconstrained MD.

Figure S5   Analysis of the distance between the sulphur of the UbcH10 Cys114 and the carbonyl group of the crosslinked Ub(T) C-terminal glycine during the 50 ns

unconstrained MD of the final model obtained after SMD.

Figure S6   Superposition of the UbA1 and Ub(A) backbone for the X-ray structure 4II2 (grey) and the 3D model of the quaternary complex (Green). Front and rear views are shown in the left and right pictures, respectively.

Figure S7   Strategy of peptide design, highlights of the hUbA1 regions used to design the peptides. Colour code: hUbA1, grey; Ub(T) yellow; Ub(A), orange; UbcH10, violet; S1 and S2 red; U1 and U2 green; L1 and L2 blue.

Table S1   List of the active residues used in each docking step.

Table S2   Results from HADDOCK calculations performed for the dimeric hUbA1 and Ub(T) complex. Clusters of poses are given ordered by total score. The best models are highlighted in bold.

Table S3   Time evolution of interaction surface ($Å^2$) for selected domains in hUbA1.

File S1   Atomic Coordinates of a representative snapshot of the 3D model.

Author Contributions

Conceived and designed the experiments: LZ A. Federico PP EP FJL BC. Performed the experiments: SC IdP A. Federico CMM BC. Analyzed the data: LZ PP AG A. Fusco EP FJL BC. Wrote the paper: EP FJL BC.

References

1. Rape M, Kirschner MW (2004) Autonomous regulation of the anaphase-promoting complex couples mitosis to S-phase entry. Nature 432: 588–595.
2. De Gramont A, Ganier O, Cohen-Fix O (2006) Before and after the spindle assembly checkpoint—an APC/C point of view. Cell Cycle 5: 2168–2171.
3. Pallante P, Berlingieri MT, Troncone G, Kruhoffer M, Orntoft TF, et al. (2005) UbcH10 overexpression may represent a marker of anaplastic thyroid carcinomas. Br J Cancer 93: 464–471.
4. Fujita T, Liu W, Doihara H, Date H, Wan Y (2008) Dissection of the APCCdh1-Skp2 cascade in breast cancer. Clin Cancer Res 14: 1966–1975.
5. Berlingieri MT, Pallante P, Guida M, Nappi C, Masciullo V, et al. (2007) UbcH10 expression may be a useful tool in the prognosis of ovarian carcinomas. Oncogene 26: 2136–2140.
6. Donato G, Iofrida G, Lavano A, Volpentesta G, Signorelli F, et al. (2008) Analysis of UbcH10 expression represents a useful tool for the diagnosis and therapy of astrocytic tumors. Clin Neuropathol 27: 219–223.
7. Troncone G, Guerriero E, Pallante P, Berlingieri MT, Ferraro A, et al. (2009) UbcH10 expression in human lymphomas. Histopathology 54: 731–740.
8. Haas AL, Warms JV, Hershko A, Rose IA (1982) Ubiquitin-activating enzyme. Mechanism and role in protein-ubiquitin conjugation. J Biol Chem 257: 2543–2548.
9. Kleiger G, Mayor T (2014) Perilous journey: a tour of the ubiquitin-proteasome system. Trends Cell Biol 24: 352–359.
10. Jin J, Li X, Gygi SP, Harper JW (2007) Dual E1 activation systems for ubiquitin differentially regulate E2 enzyme charging. Nature 447: 1135–1138.

11. Fang S, Weissman AM (2004) A field guide to ubiquitylation. Cell Mol Life Sci 61: 1546–1561

12. Schulman BA, Harper JW (2009) Ubiquitin-like protein activation by E1 enzymes: the apex for downstream signalling pathways. Nat Rev Mol Cell Biol 10: 319–331.

13. Huang DT, Hunt HW, Zhuang M, Ohi MD, Holton JM, et al. (2007) Basis for a ubiquitin-like protein thioester switch toggling E1-E2 affinity. Nature 445: 394–398.

14. Olsen SK, Lima CD (2013) Structure of a ubiquitin E1-E2 complex: Insights to E1-E2 thioesther transfer. Mol Cell 49: 884–896.

15. Altschul SF, Gish W, Miller W, Myers EW, Lipman DJ (1990) Basic local alignment search tool. J Mol Biol 215: 403–410.

16. Berman HM, Westbrook J, Feng Z, Gilliland G, Bhat TN, et al. (2000) The protein data bank. Nuc Acids Res 28: 235–242.

17. Szczepanowski RH, Filipek R, Bochtler M (2005) Crystal structure of a fragment of mouse ubiquitin-activating enzyme. J Biol Chem 280: 22006–22011.

18. Lee I, Schindelin H (2008) Structural insights into E1-catalyzed ubiquitin activation and transfer to conjugating enzymes. Cell 134: 268–278.

19. Jaremko L, Jaremko M, Wojciechowski W, Filipek R, Szczepanowski RH, et al. (2006) NMR assignment of a structurally uncharacterised fragment of recombinant mouse ubiquitin-activating enzyme. J Biomol NMR 36 (Suppl 1): 43.

20. Thompson JD, Higgins DG, Gibson TJ (1994) CLUSTAL W: improving the sensitivity of progressive multiple sequence alignment through sequence weighting, position-specific gap penalties and weight matrix choice. Nuc Acids Res 22: 4673–4680.

21. Bordoli L1, Kiefer F, Arnold K, Benkert P, Battey J, et al. (2009) Protein structure homology modeling using SWISS-MODEL workspace. Nat Protoc. 4(1): 1–13.

22. Kabsch W, Sander C (1983) Biopolymers 22: 2577–2637.

23. Peitsch MC (1996) ProMod and Swiss-Model: Internet-based tools for automated comparative protein modelling. Biochem Soc Trans 24: 274–279.

24. Van Gunsteren WF, Berendsen HJC (1990) Computer simulation of molecular dynamics: Methodology, applications, and perspective in Chemistr Angew. Chem Int Ed. 29: 992–1023.

25. Luthy R, Bowie JU, Eisenberg D (1992) Assessment of protein models with three-dimensional profiles. Nature 356: 83–85.

26. Colovos C, Yeates TO (1993) Verification of protein structures: patterns of nonbonded atomic interactions. Prot Sci 2: 1511–1519.

27. Laskowski RA, MacArthur MW, Moss DS, Thornton JM (1993) PROCHECK: a program to check the stereochemical quality of protein structures. J Appl Cryst 26: 283–291.

28. Hooft RWW, Vriend G, Sander C, Abola EE (1996) Errors in protein structure. Nature 381: 272.

29. de Vries SJ, van Dijk M, Bonvin AM (2010) The HADDOCK web server for data-driven biomolecular docking. Nat Protoc 5: 883–897.

30. Lyskov S, Gray JJ (2008) The RosettaDock server for local protein-protein docking. Nuc Acids Res 36: W233–W238.

31. Tokgöz Z, Siepmann TJ, Streich F Jr, Kumar B, Klein JM, et al. (2012) E1-E2 interactions in ubiquitin and Nedd8 ligation pathways. J Biol Chem 287: 15512–15522.

32. Bezsonova I, Bruce MC, Wiesner S, Lin H, Rotin D, et al. (2008) Interactions between the three CIN85 SH3 domains and ubiquitin: implications for CIN85 ubiquitination. Biochemistry 47: 8937–8949.

33. Lin Y, Hwang WC, Basavappa R (2002) Structural and Functional Analysis of the Human Mitotic-specific Ubiquitin-conjugating Enzyme, UbcH10. J Biol Chem 277: 21913–21921.

34. Hornak V, Abel R, Okur A, Strockbine B, Roitberg A, et al. (2006) Comparison of multiple Amber force fields and development of improved protein backbone parameters. Proteins 65: 712–725.

35. Naïm M, Bhat S, Rankin KN, Dennis S, Chowdhury SF, et al. (2007) Solvated interaction energy (SIE) for scoring protein-ligand binding affinities. 1. Exploring the parameter space. J Chem Inf Model 47: 122–133.

36. Cui Q, Sulea T, Schrag JD, Munger C, Hung MN, et al. (2008) Molecular Dynamics and Solvated Interaction Energy Studies of Protein-Protein Interactions: the MP1-p14 Scaffolding Complex. J Mol Biol 379: 787–802.

37. Kortemme T, Kim DE, Baker D (2004) Computational alanine scanning of protein-protein interfaces. Sci STKE 219: pl2

38. Huo S, Massova I, Kollman PA (2002) Computational alanine scanning of the 1:1 human growth hormone-receptor complex. J Comput Chem 23: 15–27.

39. Dikic I, Wakatsuki S, Walters KJ (2009) Ubiquitin-binding domains - from structures to functions. Nat Rev Mol Cell Biol 10: 659–71.

40. Schäfer A, Kuhn M, Schindelin H (2014) Structure of the ubiquitin-activating enzyme loaded with two ubiquitin molecules. Acta Cryst D70, 1311–1320

41. Pitluk ZW, McDonough M, Sangan P, Gonda DK (1995) Novel CDC34 (UBC3) ubiquitin-conjugating enzyme mutants obtained by charge-to-alanine scanning mutagenesis. Mol Cell Biol 15: 1210–1219.

42. Sullivan ML, Vierstra RD (1991) Cloning of a 16-kDa ubiquitin carrier protein from wheat and Arabidopsis thaliana. Identification of functional domains by in vitro mutagenesis. J Biol Chem 266: 23878–23885.

43. Bencsath KP, Podgorski MS, Pagala VR, Slaughter CA, Schulman BA (2002) Identification of a multifunctional binding site on ubc9p required for smt3p conjugation. J Biol Chem 277: 47938–47945.

44. Poornam G, Matsumoto A, Ishida H, Hayward S (2009) A method for the analysis of domain movements in large biomolecular complexes. Proteins 76: 201–221.

7

Modulating the Structure of EGFR with UV Light: New Possibilities in Cancer Therapy

Manuel Correia[1,9], **Viruthachalam Thiagarajan**[2,3], **Isabel Coutinho**[2], **Gnana Prakash Gajula**[2], **Steffen B. Petersen**[4,5], **Maria Teresa Neves-Petersen**[2*,9]

1 Department of Physics and Nanotechnology, Aalborg University, Aalborg, Denmark, 2 BioPhotonics Group, Department of Nanomedicine, International Iberian Nanotechnology Laboratory (INL), Braga, Portugal, 3 School of Chemistry, Bharathidasan University, Tiruchirappalli, India, 4 Department of Health Science and Technology, Aalborg University, Aalborg, Denmark, 5 The Institute for Lasers, Photonics and Biophotonics, University at Buffalo, The State University of New York, New York, United States of America

Abstract

The epidermal growth factor receptor (EGFR) is a member of the ErbB family of receptor tyrosine kinases. EGFR is activated upon binding to e.g. epidermal growth factor (EGF), leading to cell survival, proliferation and migration. EGFR overactivation is associated with tumor progression. We have previously shown that low dose UVB illumination of cancer cells overexpressing EGFR prior to adding EGF halted the EGFR signaling pathway. We here show that UVB illumination of the extracellular domain of EGFR (sEGFR) induces protein conformational changes, disulphide bridge breakage and formation of tryptophan and tyrosine photoproducts such as dityrosine, N-formylkynurenine and kynurenine. Fluorescence spectroscopy, circular dichroism and thermal studies confirm the occurrence of conformational changes. An immunoassay has confirmed that UVB light induces structural changes in the EGF binding site. A monoclonal antibody which competes with EGF for binding sEGFR was used. We report clear evidence that UVB light induces structural changes in EGFR that impairs the correct binding of an EGFR specific antibody that competes with EGF for binding EGFR, confirming that the 3D structure of the EGFR binding domain suffered conformational changes upon UV illumination. The irradiance used is in the same order of magnitude as the integrated intensity in the solar UVB range. The new photonic technology disables a key receptor and is most likely applicable to the treatment of various types of cancer, alone or in combination with other therapies.

Editor: Federico Quaini, University-Hospital of Parma, Italy

Funding: MC acknowledges the support from "Fundaço para a Ciência e Tecnologia" (FCT) for the PhD grant (SFRH/BD/61012/2009) supported by "Programa Operacional Potencial Humano" (POPH) in the framework of "Quadro de Referência Estratégica Nacional" (QREN) and co-financed by the European Social Fund ("Fundo Social Europeu", FSE). The funders had no role in study design, data collection and analysis, decision to publish, or preparation of the manuscript.

Competing Interests: The authors have declared that no competing interests exist.

* Email: nevespetersen@gmail.com

9 These authors contributed equally to this work.

Introduction

The ErbB family of receptor tyrosine kinases (RTKs) plays a key role in regulating normal cellular processes such as cell survival, proliferation and migration [1,2] and have a critical role in the development and progression of cancers [3]. The epidermal growth factor receptor (EGFR; ErbB1) is a member of this family [4]. EGFR binding to ligands such as epidermal growth factor (EGF) or transforming growth factor alpha (TGF-α) leads to receptor dimerization and to the activation of the intracellular tyrosine kinase domain [1,2]. The extracellular domain of EGFR (sEGFR, soluble extracellular region of EGFR) comprises 4 subdomains: 2 large homologous binding domains (I and III), and 2 homologous furin-like cysteine rich domains (II and IV). Domains I, II and III have been found to be directly involved in ligand binding and dimer formation that precede the mechanism of signal transduction by RTKs [1,5,6]. Cancer progression has been correlated with the increase in the number of EGFR molecules on the cell surface [7]. High expression of EGFR is generally associated with invasion, metastasis, late-stage disease, chemotherapy resistance, hormonal therapy resistance and poor general therapeutic outcome. EGFR overexpression has been found to be a strong prognostic indicator in head and neck, ovarian, cervical, bladder and oesophageal cancers, a modest prognostic indicator in breast, colorectal, gastric and endometrial cancer and a weak prognostic indicator in non-small-cell lung cancer [7]. EGFR is the target of many chemotherapeutical approaches because EGFR activation results in cell signaling cascades that promote tumor growth. Inhibition of EGFR function is therefore a rational treatment approach. Typical chemotherapeutical agents are EGFR tyrosine kinase inhibitors which compete with ATP at the intracellular tyrosine kinase domain, and monoclonal antibodies (mAbs) that prevent ligand-binding or receptor dimerization. Blocking the binding of EGF to EGFR can abolish cancer proliferation, invasion, metastasis, angiogenesis and inhibition of apoptosis [8].

We have previously reported that UVB illumination (280 nm, 0.35 W/m^2 for 30 min) of cancer cells overexpressing EGFR led to the arrest of the EGFR signaling pathway [9]. Proof-of-concept has been documented on cell lines A431 (human epidermoid carcinoma cells) and Cal39 (derived from human vulva squamous cell carcinoma cells). The irradiance used was lower than the total UVB solar irradiance [10]. UVB prevented EGF induced activation of EGFR, abolishing phosphorylation of the EGFR intracellular domain and of other key downstream signaling proteins such as AKT (Protein Kinase B) and the mitogen activated protein kinases (ERK1 and 2). AKT plays a key role in e.g. glucose metabolism, apoptosis, cell proliferation, transcription and cell migration. AKT is involved in cellular survival pathways by inhibiting apoptotic processes [11–16]. The ERK kinases act in a signaling cascade that regulates cellular processes such as proliferation, differentiation, and cell cycle progression [17].

One of the possible reasons for the observed UV light induced arrest of the EGFR signaling pathway is the UVB induced photochemistry, leading to conformational changes in EGFR which most likely prevent the correct binding to EGF. Our previous work on UVB induced photochemistry in proteins (wavelengths used 275–295 nm) [18–22] supports this hypothesis. UVB excitation of aromatic residues in proteins leads to the disruption of disulphide bridges and to the formation of photoproducts, such as N-formylkynurenine (NFK), kynurenine (Kyn) [23,24] and dityrosine (DT) [25–27] (see Figure 1). Such reactions will most likely induce structural changes in proteins which may impair their activity [22]. Proteins that are rich in aromatic residues and disulphide bridges are likely to have their structure considerably impaired upon prolonged UVB excitation. sEGFR is extremely rich in disulphide bridges when compared to the natural average abundance of disulphide bridges in protein structures of the size of sEGFR, as will be shown in the results section. The % of disulphide bridges in sEGFR is approximately 13 times higher than expected for a protein of its size. The expected average results have been previously reported by our group after a comprehensive analyses of the features of the disulphide bridges present in 131 non-homologous single chain protein structures [28]. Furthermore, the extracellular domain of sEGFR has 40 aromatic residues and 25 disulphide bridges per monomer and many of the aromatic residues are in close proximity of disulphide bridges (see Figure 2). Therefore, it is likely that UV illumination of this protein will lead to disulphide bridge breakage and to photoproducts. This hypothesis will be tested in our present paper.

Whereas the dangers of overexposure to sunlight have been well publicized, less attention has been given to the health benefits of UV-exposure. The effects of UV light on biomolecules, cells and tissues will depend on the energy delivered per unit area. UVB (280–315 nm) radiation is the primary contributor to vitamin D production, which has a protective effect in colon, prostate, and breast cancer [29,30]. Exposure to sunlight in low doses has also been linked to improved cancer survival rates [31,32]. UV light has been used to successfully treat rickets, psoriasis, lupus vulgaris, vitiligo, atopic dermatitis and localized scleroderma and jaundice (World Health Organization. The known health effects of UV. *Ultrav. Radiat. INTERSUN Program.* - *FAQ* at http://www. who.int/uv/faq/uvhealtfac/en/index1.html). There is no direct and conclusive evidence to suggest an increased risk of skin cancer from UVB treatments for psoriasis if radiation doses are respected. UV light is currently being used to treat cutaneous T-cell lymphoma (American Cancer Society, 2011 at http://www. cancer.org). This therapy has been improved by the Food and Drug Administration (FDA, USA). The protective effects of regular weekend sun exposure have been reported, particularly in the case of limb tumors [33]. Melanoma is more frequent among

people with indoor occupations than among people having outdoor activities (without sunburn) such as farmers, fishermen, and kids that play outdoors [34,35]. Skin cancer is still on the increase but this is a result of exposure to UV light both by acute overdosing (causing sunburn) and lifelong cumulative exposure [36]. It is mainly the UVB fraction of solar radiation that gives erythema, melanogenesis (melanin production, which acts as a sunscreen protecting against DNA photodamage and erythema), vitamin D synthesis, and non-melanoma skin cancer, while melanoma is likely to be caused by UVA [37]. Jones et al (1987) found that among the UVA wavelengths, 365 nm had the highest mutagenic effect [38]. UVA fluency rates are several times higher for sunbeds than in the case of direct solar radiation [39]. The above examples document that the health effects of UV exposure are dose and wavelength dependent.

The goal of the present work is to investigate if a low dose of UVB light can induce structural changes in the EGF binding site of EGFR that could impair the correct binding of molecules, such a specific antibody that is known to compete with EGF for EGFR binding. If observed, this will give us insight into why UVB illumination (280 nm) of cancer cells overexpressing EGFR led to the arrest of the EGFR signaling pathway [9]. The UVB irradiance used is in the same order of magnitude as the total irradiance of solar UVB. Fluorescence spectroscopy and circular dichroism studies were carried out in order to detect UVB induced conformational changes. Thermal unfolding studies have been done in order to determine the melting point of the protein prior to and after illumination. Light induced breakage of disulphide bridges has been quantified and the formation of Trp/Tyr photoproducts has been detected. A binding immunoassay was carried out in order to infer the effects of UVB illumination on the structure of the EGF binding site. Our study confirms that low dose UVB (280 nm) illumination of sEGFR induced structural changes in the EGFR binding domain that impaired the binding of a specific antibody known to compete with EGF for EGFR binding to that particular domain. We address the beneficial effects of UV light, its present application in the treatment of diseases and the potential of our putative new photonic therapy for the treatment of localized tumors associated with the overexpression of UV sensitive cellular receptors.

Results

Three-dimensional structure of monomeric and dimeric sEGFR

In Figure 2A is displayed the 3D structure of monomeric sEGFR (1ivo.pdb, chain A) bound to Epidermal Growth Factor (EGF, in red). sEGFR contains 25 SS bridges, 6 Trp residues, 16 Tyr residues, and 18 Phe residues, displayed as CPK: SS bridges in yellow, Trp in green, Tyr in violet (Tyr246 and Tyr251 in pink) and Phe in cyan. Only 18 out of 25 SS bridges, 5 out of 6 Trp residues, 13 out of 16 Tyr residues and 17 out of 18 Phe residues are displayed since some residues are missing in the pdb file. Several aromatic residues are located in close spatial proximity of SS bonds.

In Figure 2B is displayed the 3D structure of the EGFR dimer (1ivo.pdb, chains A and B) formed upon binding EGF (in red). The dimer interface is rich in disulphide bridges and aromatic residues (Fig. 2B). Residues Tyr246 and Tyr251 (Figure 2B, in pink) from one chain, are intertwined with the same residues from the other chain. A zoom into one of the chains is displayed in Figure 2C and distances between aromatic residues and disulphide bridges are shown. EGF docking to EGFR is displayed in Figure 2D (EGFR monomer). The interaction between EGFR and EGF appears to

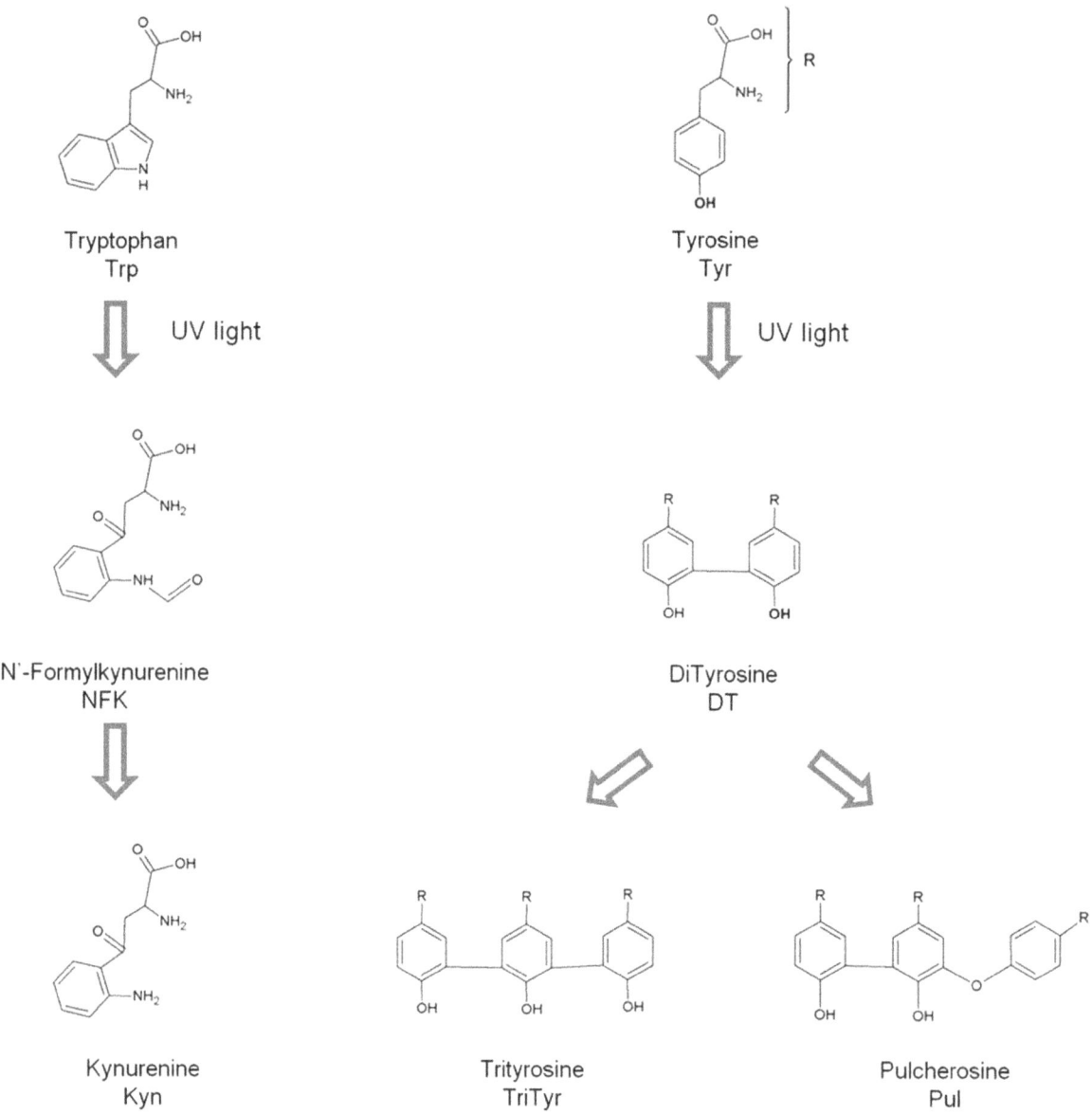

Figure 1. Typical photoproducts generated upon UV illumination of the aromatic residues Trp and Tyr.

be strongly dependent on Van der Waals interactions. Some of those interactions are π-π interactions, such as the interaction between Phe357 (in cyan) of sEGFR and Tyr13 (in violet) of EGF (5 Å).

Protein Bioinformatics

In Figure 3 is displayed the fraction of disulphide bridges present in sEGFR and the dependence of the average fraction of disulphide bridges on the protein chain length [28]. The expected average fraction of disulphide bridges for proteins with sequence length of 600–650 residues is ~0.3%, while in sEGFR (622 aa) it is ~4%, confirming that this protein is exceedingly rich in disulphide bridges. The expected average results has been previously reported by our group after doing a comprehensive analyses of the features of the disulphide bridges present in 131 non-homologous single chain protein structures [28].

Steady State Fluorescence

The fluorescence emission intensity of sEGFR at 350 nm decreases upon continuous 280 nm excitation (Figure 4A). The kinetic trace is best fitted by a bi-exponential model (Figure 4A and methods). The root mean square error R^2 was 0.99774. The values recovered from the fitting for C1 and C2 were 367240.44±1182.23 and 208449.95±1054.90, respectively. The rate constants of fluorescence emission intensity decrease k1 and k2 fitted values were 117.63±0.71 min-1 and 12.20±0.10 min-1, respectively.

Emission Spectra (excitation 280 nm and 295 nm). Emission spectra of sEGFR were recorded upon 280 nm excitation before and after continuous 280 nm illumination (15 min, 30 min, 45 min and 75 min at 2.5 W/m²) (not shown). A decrease in the intensity of the fluorescence emission at ~337 nm, where Trp and Tyr emit, is observed due to continuous illumination. The emission intensity at 337 nm decreases by 43%

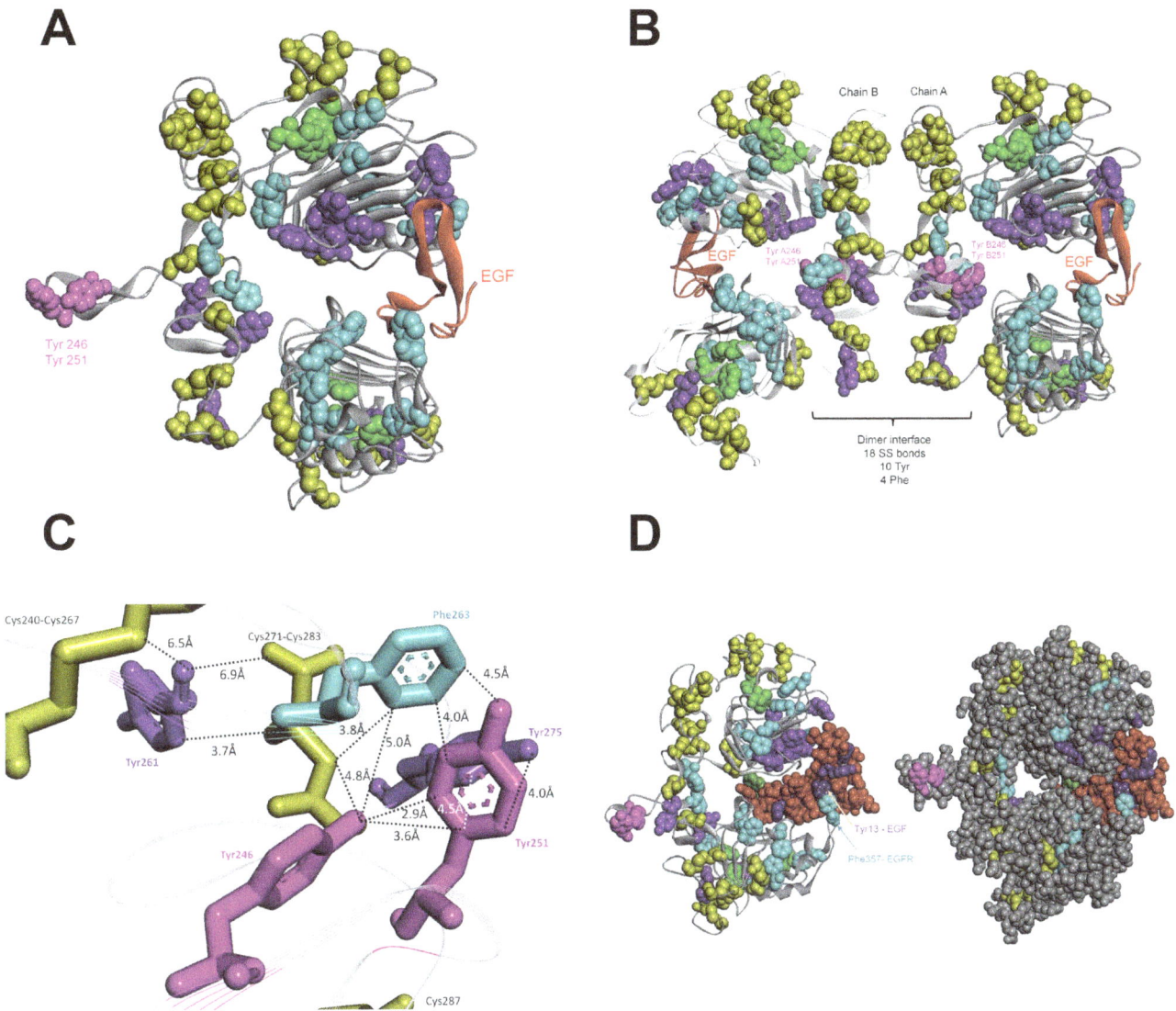

Figure 2. Structural analysis of the EGFR extracellular domain and interactions with EGF upon dimerization. (A) Crystal structure of EGFR extracellular domain (1ivo.pdb, chain A) [1]. The disulphide bridges and aromatic residues atoms are displayed as CPK: SS bridges in yellow, Trp in green, Tyr in violet (Try 246 and Tyr 251 are displayed in pink) and Phe in cyan. Only 18 out of 25 SS bridges, 5 out of 6 Trp residues, 13 out of 16 Tyr residues and 17 out of 18 Phe residues are displayed since some residues are missing in the pdb file. (B) Crystal structure of a 1:1 complex between human EGF and the dimeric form of EGFR extracellular domain (1ivo.pdb, chain A and B and 2 EGF molecules [1]). EGF is displayed in red. (C) Zoom into interface region present in the dimeric of EGFR (extracellular domains). Distances among some of the aromatic residues and the nearby disulphide bridges are also displayed. (D) Crystal structure of a 1:1 complex between human EGF and the dimeric form of EGFR extracellular domain. (1ivo.pdb, chain A and B and 2 EGF molecules [1]) EGF is shown in red. In the right panel all atoms are displayed as CPK. EGF docking to sEGFR involves extensive non-covalent contacts.

and 74% after 15 min and 75 min of continuous 280 nm illumination, respectively. The wavelength of the most intense peak centered at 337 nm remains constant.

The emission spectra of sEGFR recorded upon 295 nm excitation before and after of continuous 280 nm illumination (15 min, 30 min, 45 min and 75 min) are shown in Figure 4B. A decrease in the intensity of the fluorescence emission at ~337 nm, where Trp emits, is observed. The emission intensity of sEGFR at 337 nm decreases by 42% and 77% after 15 min and 75 min of continuous 280 nm illumination, respectively. The wavelength of the most intense peak centered at 337 nm is observed to red-shift to 343 nm after 75 min of illumination.

Excitation Spectra (emission 334 nm). In Figure 4C is displayed the excitation spectra of sEGFR with fluorescence emission fixed at 334 nm, before and after continuous 280 nm illumination of the protein for different time periods: 15 min, 30 min, 45 min, 60 min and 75 min. Trp and Tyr absorption contribute to this spectrum. Continuous 280 nm illumination leads to a progressive decrease in the fluorescence excitation intensity. After 15 min and 75 min of illumination, the excitation intensity at 282 nm decreases by 40% and 71%, respectively. The correspondent normalized excitation spectra (not shown) show no shift in the wavelength at maximum excitation intensity (~282 nm).

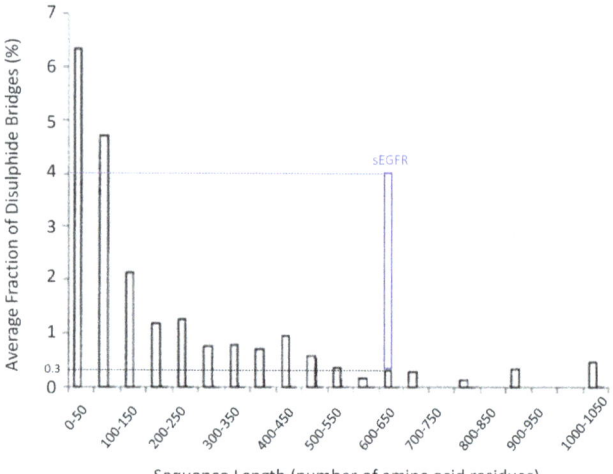

Figure 3. Dependence of the average fraction of disulphide bridges on the protein chain length. The fraction of disulphide bridges was calculated as the number of disulphide bridges found in a protein divided by the protein sequence length (number of amino acids) x 100. The fraction of disulphide bridges present in the monomeric extracellular domain of EGFR is displayed.

Fluorescence based protein thermal unfolding studies. The fluorescence emission intensity at 330 nm (exc. 295 nm) of a fresh sEGFR sample (non-illum.) and illuminated sEGFR sample (75 min at 280 nm) was monitored from 4°C to 90°C and from 90°C to 4°C (see Fig. 5). For both samples the fluorescence emission intensity decreases upon heating. A transition between 65–75°C is observed for the non-illuminated sEGFR. Data has been fitted using a modified Boltzmann function (see methods). The root mean square error for the fitting R^2 was 0.99774. The values recovered from the fitting for $A1$, $B1$, $A2$, $B2$ and dx were 1.06202 ± 0.0014, 0.45482 ± 0.02255, $-0.00426 \pm 5.41\text{E-04}$, $-0.00296 \pm 2.72\text{E-04}$ and 1.74 ± 0.17, respectively. The recovered parameter $x0$ ($69.7° \pm 0.24°C$ C) corresponds to the mid-point of transition, i.e. to the melting temperature of the protein. This value is in agreement with the inflection point values obtained from the second derivative of the fitted data (not shown). The second derivative is zero (inflection point) within the interval 68.3–69.2°C. Such transition is not observed for the illuminated sample. Furthermore, no transition is observed for both samples when cooling the protein from 90°C to 4°C.

Time Resolved Fluorescence. The decay times (τ_i) and pre-exponential factors (f_i) recovered from the time resolved intensity decays for the control sEGFR sample (75 min in the dark, negative control, NC) and the illuminated sEGFR sample (illum. for 75 min

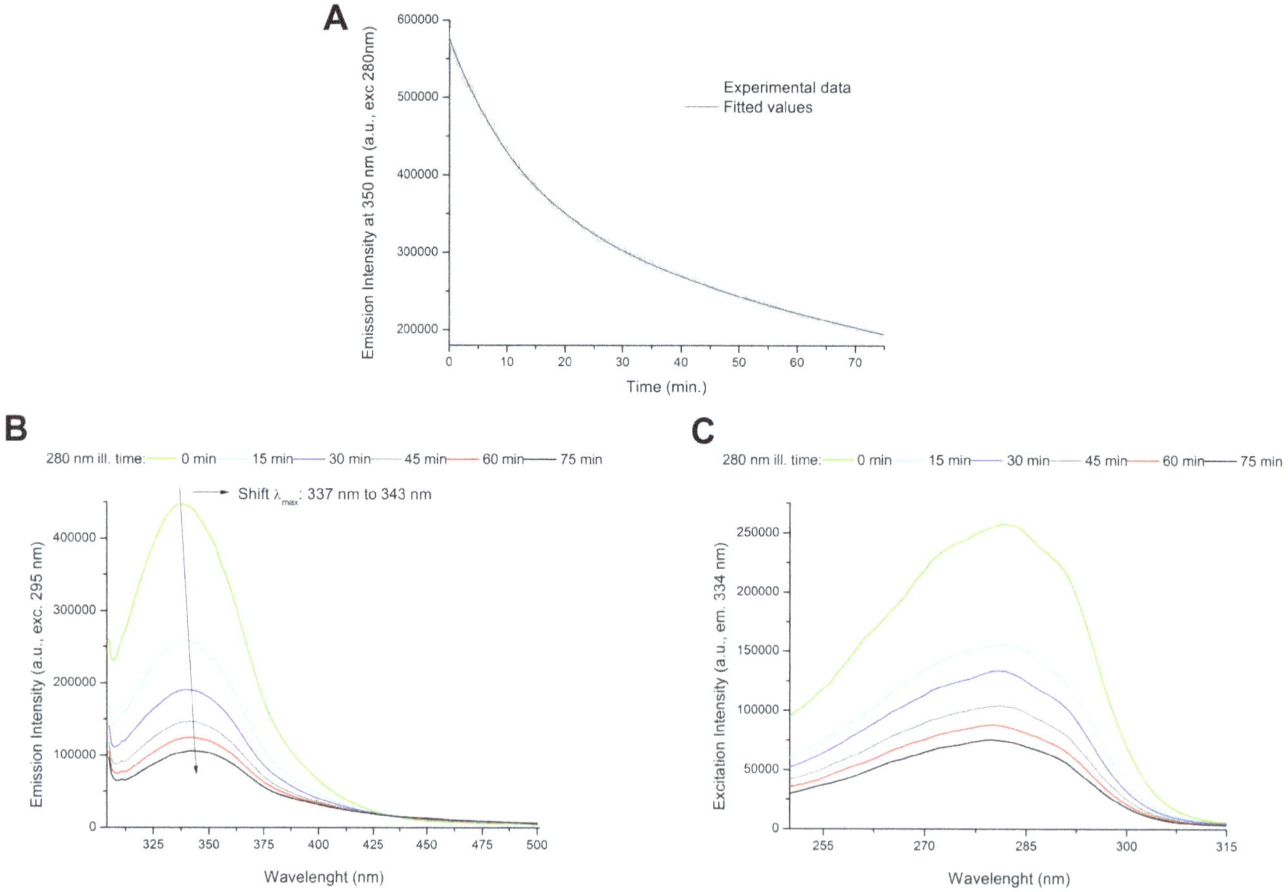

A

B

Shift λ_{max}: 337 nm to 343 nm

C

Figure 4. Effects of UV illumination on the intrinsic fluorescence emission of human extracellular EGFR. (A) Fluorescence emission intensity kinetics at 350 nm of human extracellular EGFR (sEGFR) sample during 75 min UV illumination at 280 nm. (B) Fluorescence emission spectra of sEGFR samples obtained upon 295 nm excitation after 280 nm illumination for 0 min, 15 min, 30 min, 45 min, 60 min and 75 min. (C) Fluorescence excitation spectra of sEGFR samples recorded with emission fixed at 334 nm excitation after 280 nm illumination for 0 min, 15 min, 30 min, 45 min, 60 min and 75 min.

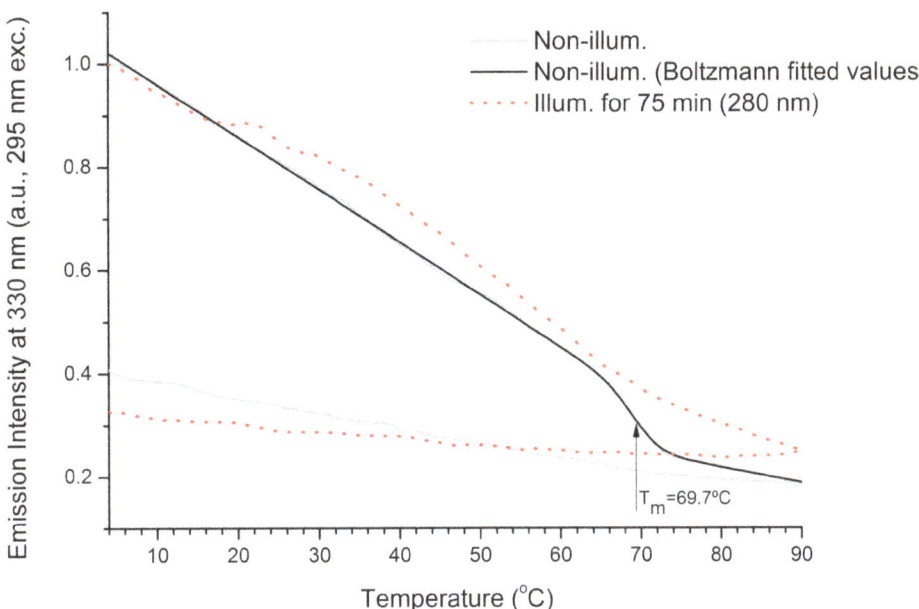

Figure 5. Temperature dependent fluorescence emission intensity at 330 nm upon 295 nm excitation. Non illuminated (non illum.) and UV illuminated (illum. for 75 min) human extracellular EGFR (sEGFR) samples were heated from 4 °C to 90 °C. Fitted values (non-illuminated sample) were obtained using a modified Boltzmann function (see Results section). The transition mid-point recovered from the fitting was at 69.72±0.24°C.

at 280 nm) at pH 7.5 are given in Table 1. Figure 6 shows the experimental data, the fit and residuals assuming three lifetime decays for sEGFR control sample (NC). The value of χ^2 dropped significantly when progressing from a one lifetime component fit to two and to three components. The fluorescence mean lifetime of sEGFR is shown in Table 1. Non-illuminated sEGFR displays three lifetimes: 1.04 ns, 2.74 ns and 6.23 ns with intensity fractions of 25.6%, 53.1% and 21.1%, respectively. Upon illumination, the contribution of the shortest lived 1.04 ns species increased from 25.6% to 30.6%, the contribution of the 2.74 ns species was kept constant and the contribution of the longer lived species decreased from 21.1% to 16.4%. Furthermore, the lifetime of the longer lived component increased from 6.2 ns to 7.0 ns while the lifetime of the other two components remained constant.

Excitation Spectra (emission 400 nm). In Figure 7A are displayed the excitation spectra (emission at 400 nm) obtained before and after continuous 280 nm illumination (0 min, 15 min, 30 min, 45 min, 62 min and 75 min). Spectra were recorded in order to verify the presence of NFK, dityrosine and Kyn (Fig. 1). In Table 2 are displayed their absorbance and fluorescence emission properties. At 0 min, one excitation peak is observed centered at ~281 nm. Its intensity decays upon UV illumination of sEGFR, decreasing 52% after 75 min. However, illumination of sEGFR leads to an increase of fluorescence excitation above 305 nm. After 75 min, fluorescent excitation intensity increased 115% at 315 nm, where dityrosine maximally absorbs (Abs^{max} at 316 nm, [40], see Table 2), 149% at 322 nm, where NFK maximally absorbs (Abs^{max} at 321 nm, [23], see Table 2) and 173% at 360 nm, where Kyn maximally absorbs (Abs^{max} at 360 nm, [23], see Table 2).

Emission Spectra (excitation 320 nm). In order to verify the formation of dityrosine (DT) and NFK, emission spectra were obtained upon 320 nm excitation before and after 280 nm illumination (Fig. 7B). An increase in fluorescence emission intensity at 400–405 nm (exc. 320 nm) is observed. The fluorescence emission maximum is centered ~400 nm, corresponding to

the emission maximum of DT (Em^{max} at 400–409 nm, [25,40], see Table 2) and to a spectral region where NFK can also emit (for NFK, Em^{max} at 400–440 nm [23], Table 2). After 75 min, fluorescence emission intensity at 400 nm increases by 129%. A small emission peak at ~368 nm is observed in the spectra

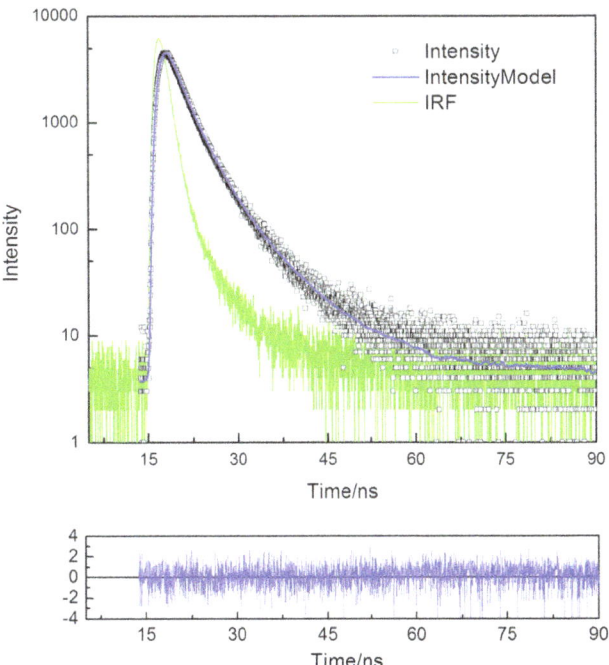

Figure 6. Time-resolved fluorescence measurements. Time-resolved intensity decay data, fitting curve and residuals obtained using the ISS routine for the control sEGFR sample (kept in the dark for 75 min, negative control, NC).

Table 1. Lifetimes (τ_i) and pre-exponential values (α_i) for sEGFR (negative control and UV illuminated samples) at pH 7.5 obtained by a nonlinear fit (see Figure 6 for example, negative control sEGFR) using the ISS software.

Lifetime (ns)		Intensity fraction		Pre-exponential	
sEGFR Dark for 75 min (negative control, NC)					
τ_1	1.04±0.06	f_1	0.258±0.02	α_1	24.8
τ_2	2.74±0.1	f_2	0.531±0.01	α_2	19.4
τ_3	6.23±0.2	f_3	0.211 Fixed	α_3	3.38
$<\tau>_{f\ weighed}$	3.04				
sEGFR Illum. for 75 min (at 280 nm)					
τ_1	1.02±0.04	f_1	0.306±0.02	α_1	30
τ_2	2.88±0.1	f_2	0.53±0.01	α_2	18.4
τ_3	7.0±0.2	f_3	0.164 Fixed	α_3	2.34
$<\tau>_{f\ weighed}$	2.99				

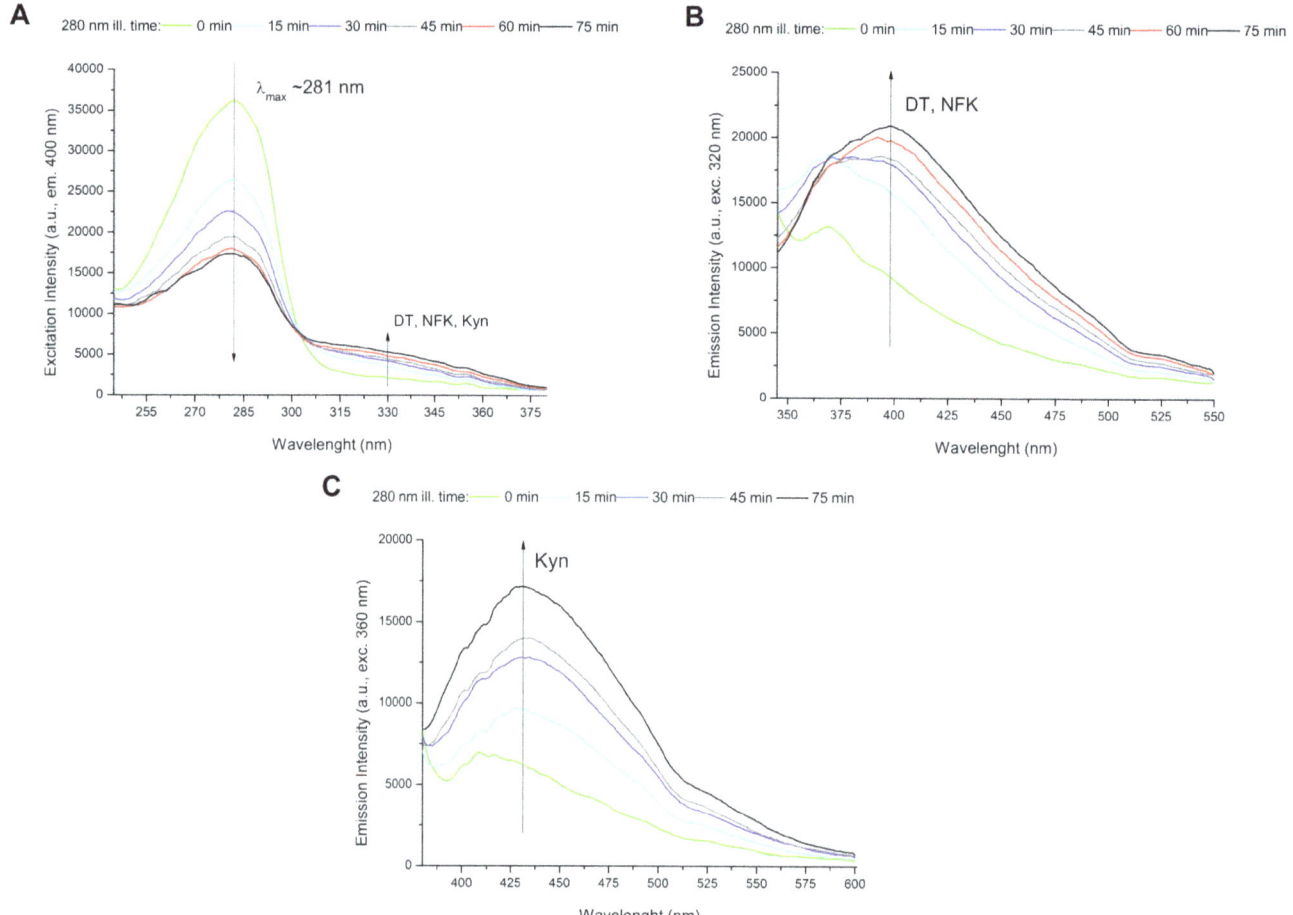

Figure 7. Formation of new fluorescence species and changes in thermal unfolding of the human extracellular EGFR upon UV illumination. (A) Fluorescence emission spectra of human extracellular EGFR (sEGFR) samples recorded upon 320 nm excitation after 280 nm illumination for 0 min, 15 min, 30 min, 45 min, 60 min and 75 min. (B) Fluorescence emission spectra of sEGFR recorded upon 360 nm excitation after 280 nm illumination for 0 min, 15 min, 30 min, 45 min and 75 min. (C) Fluorescence excitation spectra of sEGFR obtained with fluorescence emission fixed at 400 nm after 280 nm illumination for 0 min, 15 min, 30 min, 45 min, 60 min and 75 min.

Table 2. Absorption and fluorescence spectral characteristics of N-formylkynurenine (NFK), dityrosine (DT) and kynurenine (Kyn) [23,25,40].

	Absorption (λ_{max}, nm)	Fluor. Em. (λ_{max}, nm)
NKF	261, 322	400–440[a]
DT	284, 316	400–409
Kyn	258, 360	434–480[a]

[a]Depending on solvent characteristics (e.g. polarity).

obtained at 0 min and 15 min of illumination due to the Raman contribution of water. Raman spectral corrections did not completely remove this peak.

Emission Spectra (excitation 360 nm). In order to verify the presence of kynurenine (Kyn) (Abs^{max} at 360 nm, [23], Table 2) in sEGFR, emission spectra were obtained upon 360 nm excitation before and after 15 min, 30 min, 45 min, 60 min and 75 min of 280 nm illumination (Fig. 7C). Kyn absorbs maximally at 360–365 nm and fluoresces maximally at ~434–480 nm ([23], Table 2). Continuous UVB illumination of sEGFR leads to an increase of fluorescence centered at ~430–435 nm. The fluorescence intensity at 430 nm increased 53% and 172 after 15 min and 75 min illumination, respectively.

Thiol group's quantification

In order to confirm that UV illumination of sEGFR has led to SS disruption, the concentration of solvent accessible thiol groups has been determined with the Ellman's assay for a control sEGFR sample kept in the dark for 75 min (negative control, NC) and for a sample previously illuminated at 280 nm for 75 min. The detected concentration of free thiol groups is 2.9 fold higher in the illuminated sample (Fig. 8). Free thiol groups in illuminated sEGFR is ~1 µM.

Circular dichroism studies

In Figure 9A are displayed the CD spectra of fresh sEGFR (non-illum.) and illuminated sEGFR (75 min at 280 nm). The far-UV CD spectrum of the non-illuminated sEGFR displays some of the classical far-UV features of protein secondary structure, with the presence of a double minimum at 208–210 nm and 220–

222 nm, characteristic of α-helical content. β-sheet structural content (characteristic minimum at ~218 nm) may also contribute for the second minimum at 220–222 nm [41]. Prolonged excitation with UVB light leads to a decrease in the ellipticity intensity at 205–225 nm and to a spectral shift. The ellipticity at 207.5 nm and at 220 nm has decreased by 9% and 20%, respectively. Furthermore, the first negative peak has shifted from 207.5 nm to 203 nm.

Circular dichroism based protein thermal unfolding studies. The ellipticity intensity at 220 nm of fresh sEGFR (non illum.) and of illuminated sEGFR (75 min at 280 nm) was continuously monitored from 4°C to 90°C (Fig. 9B). For both samples the ellipticity intensity at 220 nm decreases upon heating. A transition with mid-point between 60–70 °C is observed for the non-illuminated sEGFR sample. Data has been fitted by a Boltzmann function (see methods). The root mean square error for the fitting R^2 was 0.99921. The values recovered from the fitting for $A1$, $A2$ and dx were -0.96 ± 9.97E-4, -0.83 ± 0.002, and 2.59 ± 0.08, respectively. The temperature of mid-transition $x0$ fitted value was 64.70 ± 0.07 °C, corresponding to the melting temperature of the protein. This value is in agreement with the value recovered by fluorescence spectroscopy displayed in Figure 5. Such transition is not observed for the illuminated sample.

EGFR binding immunoassay

A binding immunoassay was used to indirectly access the effects of UV illumination on the structure of the sEGFR binding site to EGF/TGF-α. Results displayed in Figure 10 show that non-illuminated sEGFR binds LA1 anti-EGFR (lanes "No-UV", fresh sample; and "NC", negative control, sample kept in the dark for 75 min). sEGFR sample illuminated with 280 nm light for 75 min no longer binds LA1 anti-EGFR, confirmed by the complete disappearance of the sEGFR band (Figure 10, lanes "UV"). Two sets of duplicate samples were analysed. Signal intensity profiles along the protein bands are shown. The intensity observed in the regions where illuminated sEGFR was present ("UV" lanes) is within the observed noise level (noise intensity from ~0.02 to 0.2).

Discussion

UVB excitation of aromatic residues leads to the formation of photoproducts. Tryptophan may form tryptophanyl cation radical, N-formylkynurenine (NKF) and kynurenine (Kyn) (Fig. 1). NFK and Kyn are of particular importance as they are photosensitizers that can generate reactive oxygen species (ROS) upon UV absorption [42], further contributing to protein structural damage. Tyrosine residues are known to be converted into e.g. tyrosil radicals, dityrosine, trityrosine and pulcherosine (Fig. 1). These reactions are likely to lead to changes or to complete loss of protein structure and function [22,43]. UVB excitation also leads to

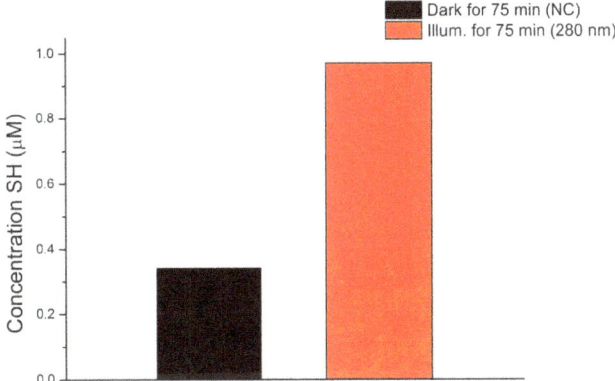

Figure 8. Increase in the concentration of free thiol groups detected in sEGFR upon UV illumination for 75 min. The control sample of sEGFR was kept in the dark for 75 min (negative control, NC). Free thiol groups have been detected using the Ellmann's assay.

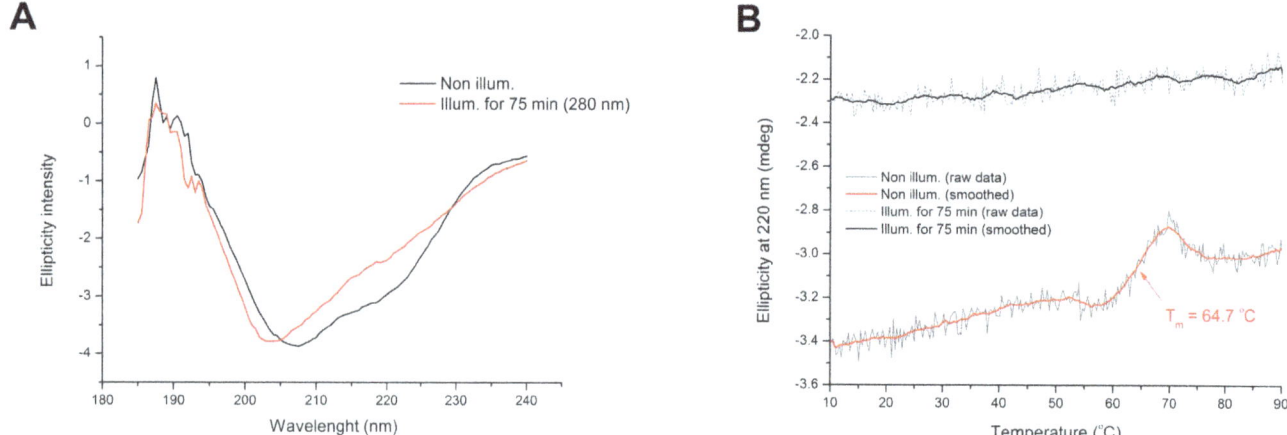

Figure 9. Effect of UVB illumination on the far UV circular dichroism features of sEGFR. (A) Far UV CD spectra were recorded for a fresh sEGFR solution (non illum.) and a sEGFR sample that was previously illuminated at 280 nm (illum. for 75 min). (B) Circular dichroim thermal unfolding curves of fresh sEGFR (non-illum.) and UVB illuminated sEGFR (illum. for 75 min) were obtained upon heating from 4 °C to 90 °C (1 °C.min^{-1}). The ellipticity signal was constantly monitored at 220 nm. The melting temperature of non illuminated sEGFR, which corresponds to the transition mid-point of the curve, was recovered upon fitting the curve with a Boltzmann function (see methods).

electron ejection from the side chains of aromatic residues [20,24,44–46]. The electron can be captured by disulphide bridges, leading to the formation of a transient disulphide electron adduct RSSR$^{\cdot-}$ (see schemes 1–8, [49]), which upon dissociation will form free thiol groups (scheme 9, [47]). Nature has kept aromatic residues in spatial close proximity to disulphide bridges

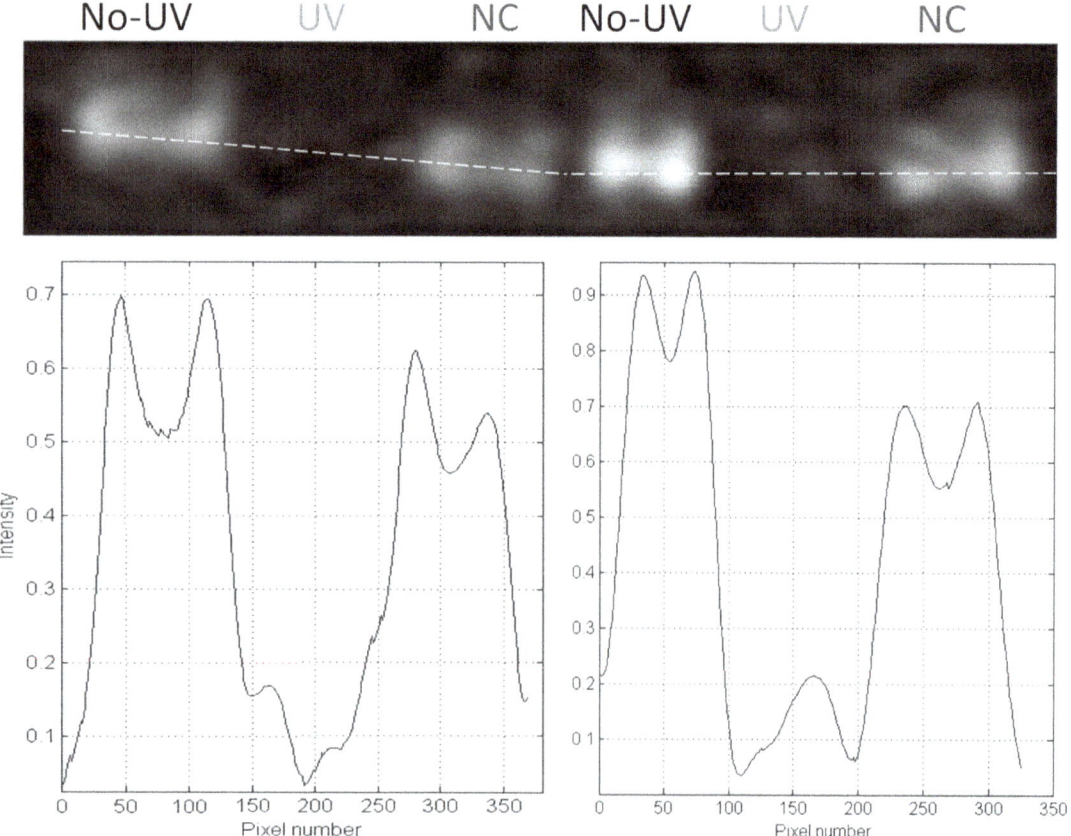

Figure 10. UV illumination of sEGFR prevents binding of anti-EGFR neutralizer antibody LA1 to its target antigen. In each well, exactly 1.4 μg of non illuminated ("No-UV"), UV illuminated for 75 min ("UV") and negative control ("NC") sEGFR samples was loaded. Samples loaded on well 1–3 are duplicates of samples 4–6, but were treated independently after UV illumination. The intensity profile along the wells shows that signal is observed in the wells with non-illuminated protein but no signal is present in the wells containing UV illuminated protein samples.

(SS) in proteins [19,28], making the disruption of SS a likely event upon UV excitation. The schemes below summarize some of the photoproducts formed which contribute to SS disruption, leading to conformation changes that can lead to loss of protein function. For further details please see referenced literature [21,22,24].

$$^1\text{Tyr-OH} + h\nu \rightarrow {}^1\text{Tyr-OH}^{\bullet+} + e_{aq}^- \tag{1}$$

$$^1\text{Trp} + h\nu \rightarrow {}^1\text{Trp}^{\bullet+} + e_{aq}^- \tag{2}$$

$$^1\text{Tyr-OH} + h\nu \rightarrow {}^3\text{Tyr-OH} \tag{3}$$

$$^1\text{Tyr-O}^- + h\nu \rightarrow {}^1\text{Tyr-O}^{\bullet} + e_{aq}^- \tag{4}$$

$$^1\text{Trp} + h\nu \rightarrow {}^1\text{Trp}^* \tag{5}$$

$$^1\text{Trp}^* \rightarrow {}^3\text{Trp} \tag{6}$$

$$e_{aq}^- + \text{RSSR} \rightarrow \text{RSSR}^{\bullet-} \tag{7}$$

$$^3\text{Trp} + \text{RSSR} \rightarrow \text{Trp}^{\bullet+} + \text{RSSR}^{\bullet-} \tag{8}$$

$$\text{RSSR}^{\bullet-} \Leftrightarrow \text{RS}^{\bullet} + \text{RS}^- \tag{9}$$

Proteins rich in aromatic residues and disulphide bridges are most vulnerable to photochemistry and damage. sEGFR is such a protein. It has a total of 6 Trp, 16 Tyr, 18 Phe residues and 25 disulphide bridges (Figure 2A). Interestingly, aromatic residues and disulphide bridges play a critical structural role at the dimer interface (Figures 2B, 2C, and 2D) indicating that UVB induced photochemistry will most likely impair EGFR dimerization. The close proximity between aromatic residues and disulphide bridges (Figure 2C) promotes electron transfer from the aromatic residues to the bridges, leading to their disruption. UV-induced protein inactivation involves fast and short-range electron transfer between photoexcited aromatic residues and nearby disulphide bridges [20,44,45,48]. Zhi Li et al. [49] showed that fast electron transfer is consistent with direct electron transfer between triplet tryptophan and a nearby disulphide bridge, leading to RSSR•⁻ and likely bridge breakage (schemes 8 and 9). As displayed in Figure 3, proteins as large as sEGFR (600–650 aa) are observed to have an average fraction of disulphide bridges no larger than ~0.3%. This is presumably due to the stabilizing effect of the large hydrophobic core. A short protein has a much smaller hydrophobic core and depends to a larger extent on the structural stability provided by disulphide bridges. An example of this is insulin, a molecule formed by two polypeptide chains containing 21 and 30 residues, respectively, and that has 3 disulphide bridges (SS average fraction of 5.9%). Two disulphide bridges are the only covalent links between the chains. sEGFR displays an average SS fraction of ~4% compared to the expected ~0.3%, which means that disulphide bridges in sEGFR are ~13 times more abundant than

expected for a protein of this length. Due to the fact that disulphide bridges are good acceptors of the electrons produced upon UV excitation of the protein, several disulphide bridges might be disrupted during illumination, compromising the structure of sEGFR. Data confirms that UVB illumination of sEGFR leads to the disruption of disulphide bridges (Fig. 8).

Receptors that are not as rich in aromatic residues and disulphide bridges as EGFR is are known be less sensitive of UV light. Huang et al. have reported that PDGF receptors did not appear to be involved in the cells response to UVC illumination, contrary to what they observed with EGFR [50]. They observed that PDGF-B was able to stimulate the tyrosine phosphorylation of the UV illuminated receptor, confirming that the 3D structure of the PDGFR binding domain remained intact after UV illumination. PDGFR lower sensitivity to UV light compared to EGFR might be due to a considerably lower number of disulphide bridges (3 per monomer) and Trp residues (2 per monomer), and lower number of aromatic residues in close spatial proximity to disulphide bridges.

Our previous work on the UV illumination effects on native and W69A mutant cutinase (Trp-depleted mutant) with 280 nm and 295-nm light also documents the crucial interplay between the aromatic residue and the disulphide bridge [51]. Cutinase has a single Trp residue in very close spatial proximity to a disulphide bridge. 280 nm and 295 nm excitation of theTrp residue is known the lead to the disruption of the disulphide bridge mediated by an electron transfer process [20]. This mechanism is severely impaired in the mutant cutinase. In our present study, the ultimate control control would be a sEGFR mutant that did not have aromatic residues and/or disulphide bridges in its structure or at least that did not have aromatic residues in close spatial proximity of disulphide bridges. Those mutants are not available for the simple reason that so many residues would have to be mutated and most certainly the structure of such mutant would no longer be the same as the native structure of the extracellular domain of EGFR. Therefore, studies done with EGFR depleted mutant cells are important in asserting the important of EGFR in UV mediated responses. Evidence is found in literature that strengthens the knowledge that most of the signal pathways involved in UV-induced processes are thought to originate at plasma membrane receptors such as EGFR. Zhang et al. showed that UVA stimulation of the epidermal growth factor receptor (EGFR) may lead to activation of kinases (p70^{S6K}/p90^{RSK}) through phosphatidyl isositol (PI)-3 kinase and extracellular receptor-activated kinases (ERKs) [52]. Evidence is provided that phosphorylation and activation of p70^{S6K}/p90^{RSK} induced by UVA were prevented in Egfr-/- cells (in which basal EGFR expression and its tyrosine phosphorylation induced by UVA were absent) and were also markedly inhibited by EGFR-specific tyrosine kinase inhibitors. Hence, data suggest that UVA-induced p70^{S6K}/p90^{RSK} signalling activation is initiated by EGFR-dependent pathways. Furthermore, Xu et al. reported that B82 mouse L cells devoid of EGFR confirmed the key role of EGFR in UVB-dependent signal transduction [53].

Data displayed in Figures 7A, 7B and 7C confirms that UVB illumination of sEGFR leads to the progressive conversion of Trp and Tyr residues into species such as NKF, Kyn and DT. The longer the illumination time the larger such conversion is. Kyn formation is very clear since it absorbs at 360 nm [23] (Table 2). Figure 4B confirms that UVB induces conformational changes in sEGFR due to the observed spectral shift observed upon illumination (solvent relaxation effect). Such red shift is due to the fact that the Trp pool in this protein became more solvent accessible upon illumination. Temperature dependent fluorescence studies (Fig. 5) confirm that illumination of sEGFR induces conformation changes. Prior to illumination a clear thermal transition around 69.7°C was observed. Such transition is due to conformation changes that happen in the Trp moiety of the protein and is associated with changes in solvent accessibility of these residues. The transition is absent if the protein previously has been illuminated for 75 min. This is correlated with the observation that UV illumination renders Trp residues in sEGFR more solvent accessible (Fig. 4B). The cooling experiments have shown that the sEGFR could not refold into its native 3D structure in our *in vitro* experiments. It is also likely that protein aggregation could have occurred upon heating the protein, contributing also to preventing protein refolding. However, the typical formation of a turbid solution associated with protein aggregation was not observed. CD data also confirms that UVB induced protein conformational changes leading to the loss of secondary structural content. As can be seen in Figure 9A, UVB illumination lead to the loss of ellipticity intensity at 207.5 nm (9%) and 220 nm (20%), characteristic of α-helical content and possibly β-sheet structural content. It is likely that the large amount of disulphide bridges in the protein has prevented further loss of ellipticity. The temperature dependent CD data (Figure 9B) shows that non-illuminated sEGFR has a melting temperature around 64.7°C. This value is in good agreement with the value recovered from the temperature dependent fluorescence studies (69.7°C). After 75 min illumination at 280 nm, the observed transition disappears, confirming that UVB has induced protein denaturation. The observed structural changes induced in sEGFR as well as the observed photochemistry are likely causes for the observed changes in fluorescence lifetime distribution in sEGFR (Table 1). Conformational changes that alter the environment surrounding the aromatic residues will most likely induce changes in their lifetimes' distribution. The protein immunoassays displayed in Figure 10 confirm that the UVB illumination of sEGFR induces structural changes in the EGFR binding site to EGF/TGF-α. After illumination, sEGFR no longer binds LA1 anti-EGFR antibody, confirmed by the complete disappearance of the band corresponding to sEGFR (Figure 10). The monoclonal antibody used competes with EGF and TGF-α for binding to EGFR [54]. This demonstrates that UV-illumination of sEGFR compromises the 3D structure of the EGF binding site in the protein. EGF docking to sEGFR is dependent on extensive non-covalent and Van der Vaals interactions between the two molecules, including π-π interactions (see Results, Fig. 2D). It is likely that UV induces conformational changes, disrupting the native tertiary fold that promotes sEGFR-EGF contacts.

The mechanisms through which EGF binds and induces receptor dimerization are not fully understood and the standing model is based on structural studies [1,55,56]. Inactive EGFR is kept in its closed conformation via interactions between two sEGFR disulphide rich domains (II and IV) [56]. EGF binding occurs primarily through interactions with domains I and III [1,55]. Ligand binding requires a change in the relative positions of domains I and III and dimerization occurs upon interaction of subdomains II of two EGFR monomers [1,5,55,56]. It is believed that domain IV has a role in high affinity binding and signal transduction [6]. Considering that the mechanisms of EGF binding and posterior EGFR dimerization can involve all four domains of sEGFR, it is not surprising that UV induced conformational changes will most likely impair binding to EGF. UV induced SS disruption in domain II will likely impair correct EGFR dimerization, but may also affect EGF binding as it is involved in the change of the relative position of domains I and III. Furthermore, domains I and III also contain disulphide bonds,

though in lower extent, and SS breakage in these regions can also impair EGF binding. It has also been reported by several groups that EGFR exists as a preformed dimer on the cell-surface [57]. Also in this case, UV induced photochemistry and consequent structural changes in the EGF binding site will most likely impair EGF binding.

Our data confirms that low dose UVB leads to conformational changes in sEGFR, impairing its ability to bind an EGFR specific antibody that competes with EGF for binding EGFR, confirming that the 3D structure of the EGFR binding domain suffered conformational changes upon UV illumination. The present molecular level *in vitro* studies allow us to predict that UV light will most likely also change the structure/function of the extracellular domain of EGFR when present in the cell surface of cancer cells overexpressing EGFR, halting this way EGF-EGFR activation and EGFR dependent key metabolic pathways. Our most recent studies on the UVB (280 nm) illumination of lung cancer cells overexpressing EGFR confirm our predictions (paper in preparation). The present data also supports our previously publish results showing that low dose UVB illumination of cancer cells overexpressing EGFR (A431 and Cal39) led to the arrest of the EGFR signaling pathway [9]. The irradiance used in the present study (2.5 $W.m^{-2}$) and in the previous study (0.35 $W.m^{-2}$) is in the same order of magnitude or lower, respectively, than the total irradiance of sunlight in the UVB region, reported to be 1.75 $W.m^{-2}$ in summer and 0.4 $W.m^{-2}$ in winter (below 313 nm) [10]. The total amount of energy given to the protein solution after 75 min illumination at 280 nm is 90 mJ. This energy is lower than the limit values recommended by the British Photodermatology Group (1000 J, Psoriasis and Psoriatic Arthritis Alliance (PAPAA), 2008, available online at http://www.nhs.uk/ipgmedia/national/psoriasis) in order to prevent cancer. We envision that low dose UVB light can be used as a new photonic therapeutical approach used in order to stop the development of localized cancer, which cells overexpress EGFR or another receptor which structure will be labile to UV light. The treatment could be easily applicable to epidermal skin cancer because UVB light penetrates the skin down to 150–200 μm [37] (online information from the Department of Dermatology School of Medicine, University of California, San Francisco. UV Radiation. at http://www.dermatology.ucsf.edu/skincancer/General/prevention/UV_Radiation.aspx). The thickness of the stratum corneum of our skin is 10–20 μm, the thickness of the epidermis can vary from 50 to 150 μm [37] (online information from the Department of Dermatology School of Medicine, University of California, San Francisco. UV Radiation. at http://www.dermatology.ucsf.edu/skincancer/General/prevention/UV_Radiation.aspx) while the thickness of the dermis varies from 30 to 300 μm (Brannon, H. Skin Anatomy. at http://dermatology.about.com/cs/skinanatomy/a/anatomy.htm). If the localized tumor is located deeper location, UVB light could be delivered to those locations my means of an optical fiber or generated via multiphoton excitation using e.g. IR light. This possible new photonic therapy can also be applied during a surgical intervention.

The role of UV light on the activation or deactivation of EGFR and in promoting cell death or cell survivel remains controversial. We have previously described the attenuation of EGFR signaling as detected by the phosphorylation status of key downstream molecules i.e. AKT and the mitogen activated protein kinases ERK1 and 2 [9]. In response to UV (280 nm) illumination phosphorylation of AKT and ERK1/2 is not detectable upon activation with EGF. Given the observed upregulation of p21WAF1 in our previous work [9], which is a cyclin dependent kinase inhibitor, we must assume that UV illumination causes the cells to arrest the cell cycle in G1 phase, which would be beneficial in inhibiting the proliferative potential of EGFR overexpressing cells. Adachi S. et al have also recently shown that UV irradiation can induce evasion of colon cancer cells from stimulation of epidermal growth factor [58]. They report that UV (254 nm) caused inhibition of cell survival and proliferation, concurrently inducing the decrease in cell surface EGFR and subsequently its degradation. Furthermore, the same team has reported that UV (254 nm) significantly inhibited platelet-derived growth factor PDGF-BB-induced phosphorylation of Akt on pancreatic cancer cells [59]. Several reports exist, describing how UV light can activate the EGF receptor hence activating the AKT and MAPK pathway (43–50 from ref. 9). These observations are in contrast to our results. Wan et al. have reported that UVB irradiation stimulated PI 3-kinase activity in human skin *in vivo* [60]. UV also stimulated phosphorylation of the downstream AKT effectors, S6 kinase and BAD. Inhibitors of EGFR and PI 3-kinase blocked UV-induced phosphorylation of BAD, suggesting that EGFR mediates UV-activated cell survival pathway. They concluded that both positive and negative roles for UV activation of the PI 3-K/AKT pathway in human skin can be envisioned. The PI 3-K/AKT pathway likely plays a critical role in balancing UV-induced apoptotic signals, thereby preventing widespread skin cell death. Conversely UV activation of the PI 3-K/AKT pathway may enhance survival of mutated cells, thereby promoting skin cancer, as has been found in several other types of cancer. Cao et al. [61] have concluded that UV (>290 nm, UVB/UVA2 source peaking we believe at 445 nm) induces multiple signalling pathways mediated by EGFR trans-activation leading to possible maturation, apoptosis and survival, and EGFR activation protects against UV-induced apoptosis in cultured mouse dendritic cells. Authors also report that most of the UV responses are mediated by the production of reactive oxygen species (ROS) and can be blocked by antioxidants. UVC is reported to induce rapid and transient expression of early growth response-1 gene (Egr-1) encoding a transcription factor that plays a role in cell survival. UVC irradiation causes tyrosine phosphorylation of EGFR in mouse NIH 3T3 fibroblasts and HC11 mouse mammary cells [50]. Possible reasons for the reported discrepancies could be found in the wavelength used in the experiments (from 254 nm to 445 nm), in the illumination power per unit of illuminated area and in the type of cells used (mutated vs non-mutated cells). Unfortunately these values are not accurately reported in all papers in order to make comparisons possible.

The present paper has provided unequivocal evidence for a UVB induced structural modification in EFGR, resulting in structural changes in the EGF binding site and loss of function. Interestingly, these changes were observed at a photonic power level approximately in the same order of magnitude as the integrated power in the solar UVB range. Therefore, it is reasonable to believe that UVB possesses therapeutic properties, especially towards skin and other superficial types of cancer. Clearly there must be a threshold irradiance level and illumination time, below which UVB light can be used for cancer treatment. Our most recent studies on the on the UV (280 nm) effects on EGF-EGFR activation of A549 lung carcinoma cells overexpressing EGFR (paper in preparation) confirm the hypothesis raised in this study that 280 nm light will most likely also change the structure/function of the extracellular domain of EGFR when present in the cell surface of cancer cells overexpressing EGFR, preventing this way EGF-EGFR binding and activation.

Materials and Methods

Structure Analysis

The crystallography data used for the display of the 3D protein structure (Figure 2) was extracted from 1ivo.pdb (extracellular domain of human epidermal growth factor receptor complexed with EGF, 3.3 Å resolution [1]). Accelrys Discovery Studio Visualizer 3.5 was used for displaying the 3D structure. Distances between protein residues were obtained by using the monitor tool in the program to determine the distance between atoms in the 3D structure. The fraction of disulphide bridges was calculated as the number of disulphide bridges found in a protein divided by the protein sequence length (number of amino acids) x 100 (Figure 3). The pdb dataset used in order to display the dependence of the average fraction of disulphide bridges on the protein chain length has been published previously by our group [28].

Fluorescence Studies

Fluorescence studies were carried out in order to analyze the effects of 280 nm excitation on the structure of the extracellular domain of human EGFR (sEGFR). sEGFR was purchased from Speed Biosystems (YCP1031). This protein was expressed in human embryonic kidney (HEK) 293 cells. The amino acid sequence corresponds to sEGFR (Leu25-Ser645) with a C-terminal polyhistidine (His) tag. The protein was dissolved directly in 10 mM Sodium Phosphate Buffer (NaPB) at pH 7.5 (stock solutions). Milli-Q water with conductivity below 0.2 μS.cm^{-1} was used. sEGFR concentration was determined by Abs^{280nm} using a molar extinction coefficient of 60,000 M^{-1}cm^{-1} estimated using the bioinformatic tool ProtParam (Expasy, [62], entry: sequence of extracellular domain of EGFR (Leu25-Ser645) with a C-terminal polyhistidine (His) tag). Stock solutions were used within 2-3 days after dissolution in buffer and was kept desiccated and protected from light at 4°C. Before each experiment the stock solution was diluted to 3.9 μM. Unless stated otherwise, this was the concentration used in all the fluorescence studies.

Steady state studies

UVB illumination of sEGFR was carried out in a ChronosBH spectrometer (ISS) with a T-configuration, using a 300-W Xenon arc lamp coupled to a monochromator. Excitation and emission slits were set to 4 nm and 8 nm, respectively. Lamp power at the sample location was 89 μW (at 280 nm). The illumination spot was 0.35 cm^2. Irradiance was 2.5 W.m^{-2}. The temperature of the solution was kept at 20°C using a Peltier element at the cuvette holder location. One hundred L of sEGFR stock solution was placed in a quartz cuvette (2 mm×10 mm×6 mm excitation volume) and excited at 280 nm during a maximum time of 75 min. Samples were illuminated through the 2 mm×6 mm window and fluorescence collected from the 10 mm×6 mm window. A fresh sample was used for each illumination session. Emission spectra were acquired with 280 nm, 295 nm, 320 nm and 360 nm excitation. Excitation spectra were recorded with the emission fixed at 320 nm, 334 nm, 350 nm, and 400 nm. Each of the previous spectra was recorded before illumination (0 min), and after 15 min, 30 min, 45 min, 60 min and 75 min of illumination. The same emission and excitation spectra were acquired for the buffer. The emission and excitation intensity values obtained were corrected in real-time for oscillations in the intensity of the excitation lamp. Buffer Raman signal was subtracted from each emission spectrum. The 350 nm fluorescence emission intensity upon continuous 280 nm excitation was recorded during 75 min of illumination.

Fluorescence based protein thermal unfolding studies. Trp emission is sensitive to the extent of solvent accessibility. The more solvent accessible Trp is the more red shifted its fluorescence emission will be. Therefore, Trp emission is usually used as a probe for protein conformational changes and can be used to determine the melting temperature of the protein.

The 330 nm fluorescence emission intensity (exc. 295 nm) of a fresh sEGFR sample (3.49 µM) and of a 280 nm illuminated for 75 min sEGFR sample (3.07 µM) was monitored from 4 °C to 90 °C. The sEGFR sample was continuously illuminated at 280 nm and as it was previously described. The heating rate was 1°C/min. A point was acquired every minute. After reaching 90 °C the protein samples were cooled from 90 °C to 4 °C at a cooling rate of 1 °C/min and the fluorescence intensity at 330 nm was monitored (exc. 295 nm).

Time resolved studies. A sEGFR sample was continuously illuminated at 280 nm during 75 min. The experimental set-up and parameters used for UVB illumination were the same as previously described. As negative control, a freshly prepared sEGFR sample was kept in the dark for 75 min. After UV illumination or dark period (negative control), the sample was kept in the cuvette and lifetime measurements were carried out at magic angle conditions using TCSPC method. A 283 nm light emitting diode from ISS (FWHM = 9 nm) was used to excite the samples. The fluorescence emission at magic angle (54.7°) was counted by a GaAs detector (Hamamatsu H7422P-40), and a 300 nm long-pass filter (Semrock) was employed in order to stray light. A solution of Ludox in Millipore water was used as a reference sample. The instrument response function for this system is 180 ps.

Detection of thiol groups' concentration formed upon UV illumination of sEGFR

Preparation of sEGFR solutions and buffers was carried as described in the previous section. One hundred L of sEGFR work solution (3.9 µM) was placed in a quartz macro cuvette (1 cm path length) and illuminated at 280 nm during 75 min. 280 nm illumination was carried out as described in the previous section. In parallel, 100 µL of the same sEGFR work solution was kept in the dark for 75 min (negative control sample, dark NC). Ellman's assay was used in order to detect free thiol groups formed upon UVB induced disruption of disulphide bridges in sEGFR [19,21,22,63]. Ellman's reagent, 5,5′-dithiobis-2-nitrobenzoic acid (DTNB) was purchased from Molecular Probes (product D8451, Life Technologies, Naerum, Denmark). One hundred mM stock solution was prepared in DMSO and stored at 4°C. After illumination, 150 µL of 2 times diluted illuminated or control sEGFR solution was mixed with an excess of DTNB (1.5 µL of 100 mM stock solution). The molar ratio DTNB/sEGFR was 5.13. Four minutes after mixing the two components (sample kept in the dark), the absorbance intensity at 412 nm was measured in a UV/Visible spectrophotometer (Shimadzu Corporation, 3. Kanda-Nishikicho 1-chome, Chiyoda-ku, Tokyo 101–8448, Japan), using a 1 cm path length quartz cuvette. The product of the reaction is the 2-nitro-5-thiobenzoate ion (TNB^{2-}), which absorbs at 412 nm. Abs_{412nm} is proportional to the amount of thiol groups present in solution. The concentration of thiol groups was determined using an extinction molar coefficient for TNB^{2-} of 14150 $M^{-1}.cm^{-1}$ at 412 nm [63].

Circular dichroism studies

The circular dichroism spectrum of a fresh sEGFR sample and of a 280 nm illuminated for 75 min sEGFR sample was acquired in order to monitor the UVB loss of secondary structural elements.

Preparation of sEGFR solutions and buffers was carried as previously described. One hundred µL of sEGFR work solution (3.9 µM) was placed in a quartz macro cuvette (1 cm path length) and illuminated at 280 nm during 75 min. The experimental set-up, conditions and parameters used during the illumination procedure were the same as previously described. Immediately after the illumination, 200 µL of 3 times diluted illuminated sEGFR solution was placed in a quartz microcuvette with a path length of 0.1 cm and cooled to 4°C using a Peltier element at the cuvette holder's location. Afterwards, a far-UV CD spectrum was recorded. Temperature was kept constant at 4°C during the measurement. The same was done for a fresh sEGFR sample. Far-UV CD spectra (190–240 nm) were acquired using the following parameters: 1.0 nm band width, resolution 0.5 nm, 4 accumulations, scan speed 20 nm/min, sensitivity high, 16 s response time. Far-UV CD spectrum was also acquired for the buffer. The buffer signal was subtracted from all spectra. The measurements were carried out on a JASCO J-815 CD spectrometer (JASCO Corporation, Ishikawa-cho Hachioji-shi, Tokyo, Japan).

Circular dichroism based protein thermal unfolding studies. The ellipticity intensity at 220 nm of a fresh sEGFR sample and of a 280 nm illuminated for 75 min sEGFR sample (both solutions 3.9 µM) was continuously monitored from 4°C to 90°C. The experimental set-up, conditions and parameters used during the illumination procedure were the same as described in the previous sections. The heating rate was 1 °C/min. A point was acquired every minute. Far-UV CD spectra (190–240 nm) were acquired at 4°C (before heating) and at 90°C. Afterwards the protein samples were cooled from 90°C to 4°C at a cooling rate of 1 °C/min and the ellipticity intensity at 220 nm was monitored every minute. At the end, far-UV CD spectra (190–240 nm) were acquired at 4°C. Far-UV CD spectra (190–240 nm) were recorded using the parameters described in the previous section.

Immunoassay

A binding immunoassay was carried out in order to analyze the effects of 75 min of UV illumination at 280 nm on the structure of the EGF-sEGFR binding site. The primary mouse monoclonal antibody anti-EGFR neutralizer antibody LA1 from Milipore (05–101), which competes with EGF and TGF-α for binding EGFR [54] was used.

For each experiment, 1.4 µg of fresh non-illuminated protein sample, 1.4 µg of a 280 nm illuminated sample for 75 min and 1.4 µg of a protein sample that has been in the dark for 75 min (negative control, NC) were analyzed by Western blot. The experimental set-up, conditions and parameters used during UV illumination procedure were the same as described in the previous sections. Samples were resolved by reduced sodium dodecyl sulfate polyacrylamide gel electrophoresis (SDS-PAGE), 7.5% polyacrylamide from Bio-Rad (4561023) and transferred to a 0.2 µm nitrocellulose membrane from Whatman (10402495). Then, the membrane was probed with 11000 dilution of LA1 primary antibody followed by incubation with 15000 dilution of goat anti-mouse horseradish peroxidase (HRP)-conjugated secondary antibody from Santa Cruz Biotech (SC-2005). Immune complexes were visualized on nitrocellulose by enzyme-linked enhanced chemiluminescence (ECL) from GE Healthcare (RPN2232) and detected using the CCD camera of G:Box chemi XT4 controlled by Genesys software from Syngene. Band intensities were quantified using the gel analysis software GeneTools from Syngene. The immunoassay was carried out twice.

Data Analysis

All data analysis, plotting and fitting procedures were done using Origin 8.1.

Emission Spectra and Excitation Spectra. Emission Spectra (280 nm, 295 nm, 320 nm, and 360 nm excitation) were first smoothed using a 12 points adjacent averaging. Excitation Spectra (emission fixed at 334 nm and 400 nm excitation) were smoothed using a 5 points adjacent averaging. All fluorescence spectra obtained were first Raman corrected by subtracting the spectra recorded for the buffer in solution. Normalized emission and excitation spectra were obtained by dividing each data point by the maximum intensity value in each spectrum.

Fitting Procedures. *Fluorescence emission kinetic trace at 350 nm upon 280 nm exc*: The fluorescence emission kinetic trace (em. 350 nm, Fig. 4A) was fitted using a bi-exponential function $F(t) = y_0 + C1.e^{-k1.t} + C2.e^{-k2.t}$. $F(t)$ is the fluorescence emission intensity at 350 nm (a.u.) at 280 nm excitation time t (min), y_0, $C1$ and $C2$ are constants and $k1$ *and* $k2$ is the rate constant of fluorescence emission intensity decrease (min^{-1}). y_0 value was fixed to 0.

Fluorescence based protein thermal unfolding studies: The fluorescence emission intensity thermal curves (Fig. 5) were first smoothed using a 5 points adjacent averaging. The fluorescence emission intensity values were afterwards normalized by dividing each value by the initial 330 nm emission intensity value. The curve corresponding to the non-illuminated sEGFR sample (non illum.) was then fitted using a modified Boltzmann function:

$$y = A2 + B2.x + \frac{(A1 + B1.x - A2 + B2.x)}{\left(1 + \exp\left(\frac{x - x0}{dx}\right)\right)}$$

where y (a.u.) is the 330 nm fluorescence emission intensity at temperature x (°C), $A1$, $B1$, $A2$, $B2$ and dx are constants. The fitting parameter $x0$ (°C) corresponds to the inflection point of the Boltzmann curve and the corresponding temperature is the temperature of mid-transition determined upon probing Trp fluorescence emission.

Fluorescence lifetimes: The fluorescence decay was analyzed by an ISS routine based on the Marquardt least-squares minimization. The governing equations for the time-resolved intensity decay data were assumed to be a sum of discrete exponentials as in:

$$F(t) = \sum_i \alpha_i . \exp(-t/\tau_i)$$

where $F(t)$ is the intensity decay, α_i is the amplitude (pre-exponential factor), τ_i the fluorescence lifetime of the i-th discrete component, and $\sum \alpha_i = 1.0$.

The fractional intensity fi of each decay time is given by:

$$f_i = \frac{\alpha_i \tau_i}{\sum_i \alpha_i \tau_i}$$

and the mean lifetime is:

$$\langle \tau \rangle = \sum_i f_i \tau_i$$

Circular dichroism based protein thermal unfolding studies. Only the data values obtained above 10°C were then considered for analysis. The ellipticity intensity values were first smoothed using a 12 points adjacent averaging. The data values were then normalized by dividing each ellipticity intensity value by the initial ellipticity intensity value (i.e. first data point above 10°C). The curve corresponding to the non-illuminated sEGFR sample (non illum.) was then fitted using a Boltzmann function:

$$y = A2 \frac{(A1 - A2)}{\left(1 + \exp\left(\frac{x - x0}{dx}\right)\right)}$$

where y (a.u.) is the 220 nm ellipticity intensity values at the temperature x (°C), $A1$, $A2$, and dx are constants. The fitting parameter $x0$ (°C) is the inflection point of the Boltzmann curve and its value corresponds to the temperature of mid-transition of the curve. The fitting was done for the interval 57.12–69.48°C, which comprises the thermal transition part of the curve.

Author Contributions

Conceived and designed the experiments: MC VT IC GPG SBP MTNP. Performed the experiments: MC VT IC. Analyzed the data: MC VT IC SBP MTNP. Contributed reagents/materials/analysis tools: MC VT IC GPG SBP MTNP. Wrote the paper: MC VT IC SBP MTNP.

References

1. Ogiso H, Ishitani R, Nureki O, Fukai S, Yamanaka M, et al. (2002) Crystal structure of the complex of human epidermal growth factor and receptor extracellular domains. Cell 110: 775–787.
2. Han W, Lo HW (2012) Landscape of EGFR signaling network in human cancers: biology and therapeutic response in relation to receptor subcellular locations. Cancer Lett 318: 124–134.
3. Suzuki S, Dobashi Y, Sakurai H, Nishikawa K, Hanawa M, et al. (2005) Protein overexpression and gene amplification of epidermal growth factor receptor in nonsmall cell lung carcinomas. An immunohistochemical and fluorescence in situ hybridization study. Cancer 103: 1265–1273.
4. Yarden Y (2001) The EGFR family and its ligands in human cancer. Signalling mechanisms and therapeutic opportunities. Eur J Cancer 37 Suppl 4: S3–8.
5. Liu P, Cleveland TE, Bouyain S, Byrne PO, Longo PA, et al. (2012) A single ligand is sufficient to activate EGFR dimers. Proc Natl Acad Sci U S A 109: 10861–10866.
6. Macdonald J, Li Z, Su W, Pike IJ (2006) The membrane proximal disulfides of the EGF receptor extracellular domain are required for high affinity binding and signal transduction but do not play a role in the localization of the receptor to lipid rafts. Biochim Biophys Acta 1763: 870–878.
7. Nicholson RI, Gee JM, Harper ME (2001) EGFR and cancer prognosis. Eur J Cancer 37 Suppl 4: S9–15.
8. Noonberg SB, Benz CC (2000) Tyrosine kinase inhibitors targeted to the epidermal growth factor receptor subfamily: role as anticancer agents. Drugs 59: 753–767.
9. Olsen BB, Neves-Petersen MT, Klitgaard S, Issinger OG, Petersen SB (2007) UV light blocks EGFR signalling in human cancer cell lines. Int J Oncol 30: 181–185.
10. Pirie A (1971) Formation of N'-formylkynurenine in proteins from lens and other sources by exposure to sunlight. Biochem J 125: 203–208.
11. Casanova ML, Larcher F, Casanova B, Murillas R, Fernández-Aceñero MJ, et al. (2002) A critical role for ras-mediated, epidermal growth factor receptor-dependent angiogenesis in mouse skin carcinogenesis. Cancer Res 62: 3402–3407.
12. Kiguchi K, Beltrán L, Rupp T, DiGiovanni J (1998) Altered expression of epidermal growth factor receptor ligands in tumor promoter-treated mouse epidermis and in primary mouse skin tumors induced by an initiation-promotion protocol. Mol Carcinog 22: 73–83.
13. Sibilia M, Fleischmann A, Behrens A, Stingl L, Carroll J, et al. (2000) The EGF receptor provides an essential survival signal for SOS-dependent skin tumor development. Cell 102: 211–220.
14. Segrelles C, Ruiz S, Perez P, Murga C, Santos M, et al. (2002) Functional roles of Akt signaling in mouse skin tumorigenesis. Oncogene 21: 53–64.

15. Budiyanto A, Bito T, Kunisada M, Ashida M, Ichihashi M, et al. (2003) Inhibition of the epidermal growth factor receptor suppresses telomerase activity in HSC-1 human cutaneous squamous cell carcinoma cells. J Invest Dermatol 121: 1088–1094.

16. Kim YN, Dam P, Bertics PJ (2002) Caveolin-1 phosphorylation in human squamous and epidermoid carcinoma cells: dependence on ErbB1 expression and Src activation. Exp Cell Res 280: 134–147.

17. Roux PP, Blenis J (2004) ERK and p38 MAPK-activated protein kinases: a family of protein kinases with diverse biological functions. Microbiol Mol Biol Rev 68: 320–344.

18. Neves-Petersen MT, Petersen S, Gajula GP (2012) UV Light Effects on Proteins: From Photochemistry to Nanomedicine. In: Saha S (Dr., editor.Molecular Photochemistry - Various Aspects.InTech. pp.125–158.

19. Neves-Petersen MT, Gryczynski Z, Lakowicz J, Fojan P, Pedersen S, et al. (2002) High probability of disrupting a disulphide bridge mediated by an endogenous excited tryptophan residue. Protein Sci 11: 588–600.

20. Neves-Petersen MT, Klitgaard S, Pascher T, Skovsen E, Polivka T, et al. (2009) Flash photolysis of cutinase: identification and decay kinetics of transient intermediates formed upon UV excitation of aromatic residues. Biophys J 97: 211–226.

21. Correia M, Neves-Petersen MT, Parracino A, di Gennaro AK, Petersen SB (2012) Photophysics, photochemistry and energetics of UV light induced disulphide bridge disruption in apo-α-lactalbumin. J Fluoresc 22: 323–337.

22. Correia M, Neves-Petersen MT, Jeppesen PB, Gregersen S, Petersen SB (2012) UV-light exposure of insulin: pharmaceutical implications upon covalent insulin dityrosine dimerization and disulphide bond photolysis. PLoS One 7: e50733.

23. Fukunaga Y, Katsuragi Y, Izumi T, Sakiyama F (1982) Fluorescence characteristics of kynurenine and N'-formylkynurenine. Their use as reporters of the environment of tryptophan 62 in hen egg-white lysozyme. J Biochem 92: 129–141.

24. Kerwin BA, Remmele RL (2007) Protect from light: photodegradation and protein biologics. 96: 1468–1479.

25. Malencik DA, Anderson SR (2003) Dityrosine as a product of oxidative stress and fluorescent probe. Amino Acids 25: 233–247.

26. Malencik DA, Anderson SR (1987) Dityrosine formation in calmodulin. Biochemistry 26: 695–704.

27. Giulivi C, Traaseth NJ, Davies KJ (2003) Tyrosine oxidation products: analysis and biological relevance. Amino Acids 25: 227–232.

28. Petersen MT, Jonson PH, Petersen SB (1999) Amino acid neighbours and detailed conformational analysis of cysteines in proteins. Protein Eng 12: 535–548.

29. Grant WB (2002) An estimate of premature cancer mortality in the U.S. due to inadequate doses of solar ultraviolet-B radiation. Cancer 94: 1867–1875.

30. Banwell CM, Singh R, Stewart PM, Uskokovic MR, Campbell MJ (2003) Antiproliferative signalling by 1,25(OH)2D3 in prostate and breast cancer is suppressed by a mechanism involving histone deacetylation. Recent Results Cancer Res 164: 83–98.

31. Gorham ED, Garland CF, Garland FC, Grant WB, Mohr SB, et al. (2005) Vitamin D and prevention of colorectal cancer. J Steroid Biochem Mol Biol 97: 179–194.

32. Garland CF, Garland FC, Gorham ED, Lipkin M, Newmark H, et al. (2006) The role of vitamin D in cancer prevention. Am J Public Health 96: 252–261.

33. Newton-Bishop JA, Chang YM, Elliott F, Chan M, Leake S, et al. (2011) Relationship between sun exposure and melanoma risk for tumours in different body sites in a large case-control study in a temperate climate. Eur J Cancer 47: 732–741.

34. Pukkala E, Martinsen JI, Lynge E, Gunnarsdottir HK, Sparén P, et al. (2009) Occupation and cancer - follow-up of 15 million people in five Nordic countries. Acta Oncol 48: 646–790.

35. Kaskel P, Sander S, Kron M, Kind P, Peter RU, et al. (2001) Outdoor activities in childhood: a protective factor for cutaneous melanoma? Results of a case-control study in 271 matched pairs. Br J Dermatol 145: 602–609.

36. Webb AR, Engelsen O (2006) Calculated ultraviolet exposure levels for a healthy vitamin D status. Photochem Photobiol 82: 1697–1703.

37. Moan J (2001) Visible Light and UV radiation. In: Brune D, Hellborg R, Persson B, Paakkonen R, editors. Radiation at Home, Outdoors and in the Workplace. Oslo: Scandinavian Publisher. pp.69–85.

38. Jones CA, Huberman E, Cunningham ML, Peak MJ (1987) Mutagenesis and cytotoxicity in human epithelial cells by far- and near-ultraviolet radiations: action spectra. Radiat Res 110: 244–254.

39. Nilsen LTN, Aalerud TN, Hannevik M, Veierod MB (2011) UVB and UVA irradiances from indoor tanning devices. Photochem Photobiol Sci 10: 1129–1136.

40. Jacob JS, Cistola DP, Hsu FF, Muzaffar S, Mueller DM, et al. (1996) Human phagocytes employ the myeloperoxidase-hydrogen peroxide system to synthesize dityrosine, trityrosine, pulcherosine, and isodityrosine by a tyrosyl radical-dependent pathway. J Biol Chem 271: 19950–19956.

41. Greenfield NJ (2006) Using circular dichroism spectra to estimate protein secondary structure. Nat Protoc 1: 2876–2890.

42. Finley EL, Dillon J, Crouch RK, Schey KL (1998) Identification of tryptophan oxidation products in bovine alpha-crystallin. Protein Sci 7: 2391–2397.

43. Finley EL, Busman M, Dillon J, Crouch RK, Schey KL (1997) Identification of photooxidation sites in bovine alpha-crystallin. Photochem Photobiol 66: 635–641.

44. Bent DV, Hayon E (1975) Excited state chemistry of aromatic amino acids and related peptides. I. Tyrosine. J Am Chem Soc 97: 2599–2606.

45. Bent DV, Hayon E (1975) Excited state chemistry of aromatic amino acids and related peptides. III. Tryptophan. J Am Chem Soc 97: 2612–2619.

46. Creed D (1984) The photophysics and photochemistry of the near-uv absorbing amino acids–I. Tryptophan and its simple derivatives. Photochem Photobiol 39: 537–562.

47. Hoffman MZ, Hayon E (1972) One-electron reduction of the disulfide linkage in aqueous solution. Formation, protonation, and decay kinetics of the RSSR-radical. J Am Chem Soc 94: 7950–7957.

48. Bent DV, Hayon E (1975) Excited state chemistry of aromatic amino acids and related peptides. II. Phenylalanine. J Am Chem Soc 97: 2606–2612.

49. Li Z, Lee WE, Galley WC (1989) Distance dependence of the tryptophan-disulfide interaction at the triplet level from pulsed phosphorescence studies on a model system. Biophys J 56: 361–367.

50. Huang RP, Wu JX, Fan Y, Adamson ED (1996) UV Activates Growth Factor Receptors via Reactive Oxygen Intermediates. The Journal of Cell Biology 133 (1): 311–220.

51. Neves-Petersen MT, Snabe T, Klitgaard S, Duroux M, Petersen SB (2006) Photonic activation of disulfide bridges achieves oriented protein immobilization on biosensor surfaces. Protein Science 15: 343–351.

52. Zhang Y, Dong Z, Dode AM, Ma WY, Chen, Dong Z (2001) Induction of EGFR-Dependent and EGFR-Independent Signaling Pathways by Ultraviolet A Irradiation. DNA and Cell Biology 20(12): 769–777.

53. Xu Y, Voorhees JJ, Fisher GJ (2006) Epidermal growth factor receptor is a critical mediator of ultraviolet B irradiation-induced signal transduction in immortalized human keratinocyte HaCaT cells. The American Journal of Pathology 169(3): 823–830.

54. Kawamoto T, Kishimoto K, Takahashi K, Matsumura T, Sato JD, et al. (1992) Polymorphonuclear leukocytes-mediated lysis of A431 cells induced by IgG1 mouse anti-epidermal growth factor receptor monoclonal antibodies. In Vitro Cell Dev Biol 28A: 782–786.

55. Garrett TPJ, McKern NM, Lou M, Elleman TC, Adams TE, et al. (2002) Crystal structure of a truncated epidermal growth factor receptor extracellular domain bound to transforming growth factor alpha. Cell 110: 763–773.

56. Ferguson KM, Berger MB, Mendrola JM, Cho HS, Leahy DJ, et al. (2003) EGF activates its receptor by removing interactions that autoinhibit ectodomain dimerization. Mol Cell 11: 507–517.

57. Tao RH, Maruyama IN (2008) All EGF(ErbB) receptors have preformed homo- and heterodimeric structures in living cells. Journal of Cell Science 121: 3207–3217.

58. Adachi S, Yasuda I, Nakashima M, Yamauchi T, Kawaguchi J (2011) Ultraviolet Irradiation Can Induce Evasion of Colon Cancer Cells from Stimulation of Epidermal Growth Factor. The Journal of Biological Chemistry 286: 26178–26187.

59. Kawaguchi J, Adachi S, Yasuda I, Yamauchi T, Yoshioka T, et al. (2012) UVC irradiation suppresses platelet-derived growth factor-BB-induced migration in human pancreatic cancer cells. Oncology Reports 27(4): 935–939.

60. Wan YS, Wang ZQ, Shao Y, Voorhees JJ, Fisher GF (2001) Ultraviolet irradiation activates PI 3-kinase/AKT survival pathway via EGF receptors in human skin in vivo. International Journal of Oncology 18(3): 461–466.61.

61. Cao C, Lu S, Jiang Q, Wang WJ, Song X, et al. (2008) EGFR activation confers protections against UV-induced apoptosis in response to UV radiation in cultured mouse skin dendritic cells. Cell Signal. 20(10): 1830–1838.

62. Gasteiger E, Hoogland C, Gattiker A, Duvaud S, Wilkins MR, et al. (2005) Protein Identification and Analysis Tools on the ExPASy Server. In: Walker JM, editor. The Proteomics Protocols Handbook. Humana Press. pp.571–607.

63. Riener CK, Kada G, Gruber HJ (2002) Quick measurement of protein sulfhydryls with Ellman's reagent and with 4,4'-dithiodipyridine. Anal Bioanal Chem 373: 266–276.

8

Chronic Beryllium Disease: Revealing the Role of Beryllium Ion and Small Peptides Binding to HLA-DP2

Marharyta Petukh[1]*, **Bohua Wu**[2], **Shannon Stefl**[1], **Nick Smith**[1], **David Hyde-Volpe**[3], **Li Wang**[1], **Emil Alexov**[1]

1 Computational Biophysics and Bioinformatics, Physics Department, Clemson University, Clemson, South Carolina, United States of America, **2** School of Nursing, Clemson University, Clemson, South Carolina, United States of America, **3** Department of Chemistry, Clemson University, Clemson, South Carolina, United States of America

Abstract

Chronic Beryllium (Be) Disease (CBD) is a granulomatous disorder that predominantly affects the lung. The CBD is caused by Be exposure of individuals carrying the HLA-DP2 protein of the major histocompatibility complex class II (MHCII). While the involvement of Be in the development of CBD is obvious and the binding site and the sequence of Be and peptide binding were recently experimentally revealed [1], the interplay between induced conformational changes and the changes of the peptide binding affinity in presence of Be were not investigated. Here we carry out *in silico* modeling and predict the Be binding to be within the acidic pocket (Glu26, Glu68 and Glu69) present on the HLA-DP2 protein in accordance with the experimental work [1]. In addition, the modeling indicates that the Be ion binds to the HLA-DP2 before the corresponding peptide is able to bind to it. Further analysis of the MD generated trajectories reveals that in the presence of the Be ion in the binding pocket of HLA-DP2, all the different types of peptides induce very similar conformational changes, but their binding affinities are quite different. Since these conformational changes are distinctly different from the changes caused by peptides normally found in the cell in the absence of Be, it can be speculated that CBD can be caused by any peptide in presence of Be ion. However, the affinities of peptides for Be loaded HLA-DP2 were found to depend of their amino acid composition and the peptides carrying acidic group at positions 4 and 7 are among the strongest binders. Thus, it is proposed that CBD is caused by the exposure of Be of an individual carrying the HLA-DP2*0201 allele and that the binding of Be to HLA-DP2 protein alters the conformational and ionization properties of HLA-DP2 such that the binding of a peptide triggers a wrong signaling cascade.

Editor: Gernot Zissel, University Medical Center Freiburg, Germany

Funding: The work of MP, SS, NS, LW and EA was supported by a grant from NIH, NIGMS grant number R01GM093937. The funder had no role in study design, data collection and analysis, decision to publish, or preparation of the manuscript.

Competing Interests: The authors have declared that no competing interests exist.

* Email: mpetukh@g.clemson.edu

Introduction

Chronic Beryllium Disease (CBD) is a pulmonary granulomatous disorder caused by an immune reaction when individuals are exposed to beryllium (Be) [2]. About 18% of people who are exposed to Be in the workplace may develop CBD depending on a number of risk factors such as their genetic susceptibility, the duration, the concentration of the Be exposure, and their smoking habits [3,4,5,6,7]. Out of these risk factors, the genetic susceptibility of the individuals is shown to be the dominant contributor in the development of CBD [2,8,9]. Understanding the molecular mechanism of CBD would help prevention and treatment of disease [10,11,12,13,14].

The Major Histocompatibility Complex II (MHC II) controls the activation of the immune system in response to foreign microorganisms. The first component in this process is the binding of small peptides (that are part of an antigen) to the MHC II. The second component is the conformational change of MHC II induced by the peptide binding which makes the MHC II complex "recognizable" by T-cell receptors along with the subsequent activation of T-cells (TCs) [9,15,16]. Different conformational changes result in the activation of different TCs and thus lead to different immune responses. These conformational changes are caused by the binding of variety of peptides to alleles of MHC II. In addition, the associations of MHC II alleles with various diseases have been previously reported [17,18,19].

Of particular interest for this study is one of the MHCII molecules, the HLA-DP2 (coded by DPA1*0103, DPB1*0201), which has been shown to be associated with development of CBD [3,20,21]. By analyzing various HLA-DP molecules of the MHCII and focusing on the amino acid sequence forming the peptide binding pocket, it was demonstrated that approximately 85% of CBD patients have a glutamic acid residue at position 69 in the β-chain (βGlu69) of HLA-DP [20,22]. In other HLA-DP proteins, position 69 is taken by a lysine (Lys) amino acid. Because the side chain of Glu residue is shorter than of Lys, it can be speculated that the binding pocket of HLA-DP2 is wider comparing with the other HLA-DP.

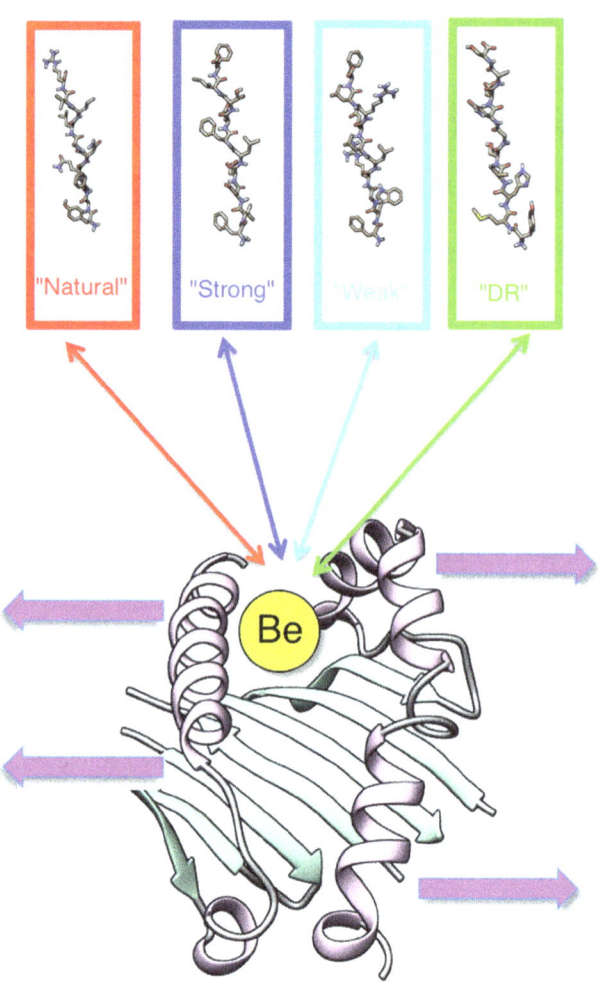

Figure 1. Graphical depiction of our investigation of the effects of the Be ion on the complex made up of a HLA-DP2 protein, the Be ion, and a small peptide. We investigated four types of small peptides ("Natural", "Strong", "Weak", and "DR") and two binding scenarios: (1) the ion bound to a small peptide and (2) the ion bound to the HLA-DP2 protein. Effect predictions include binding affinity, conformation changes of the peptide binding pocket, pKa shifts of titratable groups of the protein upon peptide and/or the Be ion binding.

The selectivity of HLA-DP2 towards different peptides is typically analyzed by reviewing the corresponding binding pockets for each amino acid of the peptide. Typically these pockets are labeled according to the residues in the peptide, i.e. pocket p1 refers to the protein pocket where the side chain of the first amino acid of peptide will be accommodated. Some of these pockets are very specific, others are not. Thus, the crystal structure of HLA-DP2 with a self-peptide derived from the HLA-DR α-chain (pDRA) [23] reveals that HLA-DP2 has four pockets which are able to bind variety of amino acids [4,23,24,25]. Thus, the p1 pocket prefers hydrophilic residues, p2 can accommodate positively charged residues, p6 could be occupied with aromatic residues, and p9 might accept larger residues. A molecular docking approach discovered that p1, p2 and p6 peptide positions are the most important for binding to the HLA-DP2 protein [24,26]. Comparing HLA-DP molecules with the HLA-DP2, the replacement of Lys69 with Glu69 (pointing toward peptide position 7, p7) and the other two glutamic acid residues, βGlu68 and βGlu26

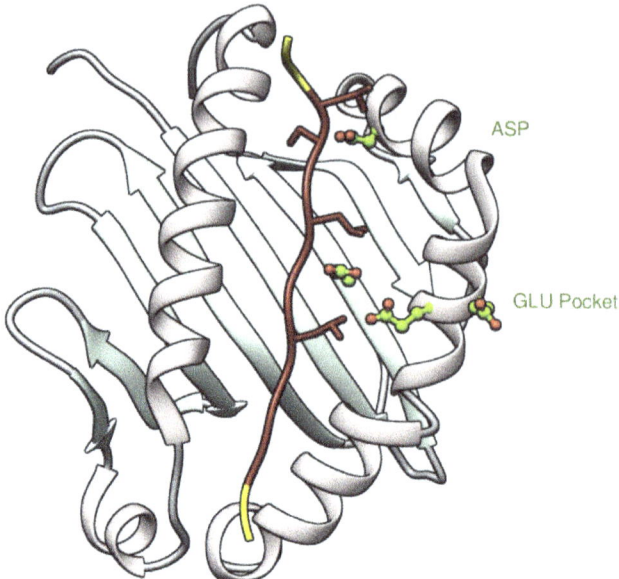

Figure 2. Illustration of the two investigated binding scenarios. In the top panel (A) the Be ion is in complex with the protein and then the peptides were added, peptide→(Be+protein) scenario; and in the bottom panel (B) the Be ion is in complex with the peptides and then the protein is added, (Be+peptide)→protein scenario.

(pointing toward p4 and p6), on HLA-DP2, results in an acidic pocket that might potentially accommodate the Be ion [1,23,26]. The existence of such an acidic pocket, providing a strong negative potential, was the reason for the speculation that even an endogenous self-peptide with Be can bind to HLA-DP2 and be recognized by the T cells [26]. Thus a recent *ex vivo* study demonstrated that some endogenous self-peptides are able to bind with HLA-DP2 in the presence of Be [26]. When these peptides were exposed to the HLA-DP2 protein, Be-responsive TC receptors were able to recognize those peptides and stimulate the accumulation of cytokines by TCs. Another study examined 40 human peptides for their ability to stimulate the secretion of interleukin-2 (IL-2), a type of cytokine signaling molecule secreted mainly by CD4+ and CD8+ T-cell, *ex vivo*, and 11 of them were found to initiate production of IL-2 by TCs in the presence of the Be ion [26].

Although, in multiple papers the Be ion is suggested to bind directly to HLA-DP2 [27,28], only recently it was experimentally shown [1] that Be binds first and then the peptide binds to the HLA-DR2-Be complex. However, the reason for this sequence of events is not clear and here we provide plausible explanation based on binding energy calculations. In addition, it is not clear if the sequence of the binding events causes different conformational changes of the HLA-DR2-Be-peptide complex. In this work, we investigate *in silico* both scenarios: (a) peptides which bind to the protein with the Be ion placed inside the protein pocket which induces conformational changes in HLA-DP2 that are necessary for its recognition by TCs; and (b) the Be ion binds to the peptide first and thus changes the peptide-binding specificity and affinity to HLA-DP2 resulting in conformational changes in the HLA-DP2 that are necessary for its recognition by TCs. This is done by analyzing the changes in binding affinity and conformational changes of the protein upon binding the four sets of small peptides (Figure 1): peptides that are known to cause the activation of TC receptors in normal immune response; peptides that prefer to bind to HLA-DR but not HLA-DP; and peptides that are known to bind to the protein in the presence of the Be ion and induce the production of high/low concentration of inflammation cytokines in TCs that are the cause of the autoimmune disease [26]. It is anticipated that by comparing the effects of each set of peptides we will be able to reveal additional details of molecular mechanism of CBD along the finding of recent experimental work [1].

Methods

This section describes the selection of peptides to be investigated and the methods for the 3D structure modeling, the Be ion binding site prediction, the molecular dynamics (MD) simulations, the geometry analysis, the binding free energy and pKa calculations. In each of the cases, two scenarios will be investigated: (a) the peptide binding to the (Be-HLA-DP2) complex and (b) the Be bound first to the peptide resulting in a (Be-peptide) complex, then the Be-peptide complex binds to the HLA-DP2 protein (see Figure 2, panels A and B respectively).

Selection of peptides to be investigated

To investigate the specificity of peptides to HLA-DP2, four different classes of peptides were selected (Figure 1): (1) 10 peptides that bind to the HLA-DP2 were selected from the 40 human peptides based on their ability to stimulate the immune response detected experimentally as an increase of the IL-2 secretion by more than 15 pg/ml in the Be-specific T cell hybridomas AV22 and by 45 pg/ml in the Be-specific T cell hybridomas AV9, which were termed "strong" peptides [26]; (2) 10 additional peptides were selected from the same 40 human peptides which were also found to increase the IL-2 secretion but not as much as the peptides in the "strong" class (the increase of the IL-2 secretion was less than 6 pg/ml in both Be-specific T cell hybridomas AV22 and AV9); these were termed "weak" peptides; (3) 10 peptides which were bound to HLA-DP2 and known to trigger the immune system reaction without the Be ion were chosen and were termed "natural" peptides [5]; and (4) 10 peptides that naturally bind to HLA-DR which were randomly selected from syfpeithi database (http://www.syfpeithi.de) and were termed "DR" peptides. This last class was created because the HLA-DR structure is similar to the HLA-DP2 structure [6], which provides the opportunity to compare the physical properties of peptides binding to HLA-DR and HLA-DP2. All peptides in each class were cut to a length of 10 amino acids. In investigating the molecular mechanism of CBD, it is anticipated that peptides belonging to different classes may have a different affinity and may induce different conformational changes. For example, peptides from the "strong" class may have the highest affinity to bind to HLA-DP2 in the presence of Be and induce conformational and pKa changes distinctly different from peptides in the "natural" and "DR" classes. The binding of peptides in the "weak" class may cause effects similar to the "strong" peptides, but the magnitude of these changes is expected to be smaller. Alternatively, it can be envisioned that the presence of the Be ion in combination with the HLA-DP2 molecule is the leading factor and any peptide will cause CBD.

Structure modeling

All of the above selected peptides are of various lengths which could introduce artifacts into the computational modeling. So, this is why the lengths need to be made the same. Since the experimentally studied peptides (the "strong" and "weak" classes) are of length 10 [23,26], the other "natural" and "DR" peptides were also trimmed to a 10 amino acid length. The trimming was

guided by preserving the interactions of peptide residues with the key amino acids in the HLA-DP2 protein [9,23,26].

The crystal structure of HLA-DP2 was retrieved from the Protein Data Bank (PDB) (http://www.rcsb.org), PDB ID: 3LQZ. It contains an α chain, a β chain, two ligand chemical components, and it shows HLA-DP2 crystalized with a bound natural peptide of 15 amino acids. We chose to include the 10 residues of the crystallized peptide that were located completely within peptide binding pocket of HLA-DP2 (shown in red in Figure 2): peptide residue numbers 1-10 (sequence FHYLPFLPST), to be consistent with Ref. [1]. Our decision of where to cut the peptide was reinforced based on the interaction of the peptide with the key residues in the protein. We predict that the positively charged Be ion will bind to the protein near the negatively charged pocket formed by the three glutamic acid residues in the protein (shown in green in Figure 2) at positions 68, 69, and 26 [20,23,26], in accordance with recent experimental work [1]. Using the crystallographic structure, it was found that the peptide residue numbers 4 and 7 are in close proximity to the putative Be binding site and because of this, these two residues are considered to be the key positions in the peptide. When choosing where to cut the other "natural" and "DR" peptides, several considerations were made. First, we wanted to keep the key peptide positions, p4 and p7, and secondly we wanted to keep the peptide residue number 10 as well. The reason for keeping peptide residue number p10 is the following. In the HLA-DP2 protein there is another key residue, an aspartic acid, at position 55 (also shown in green in Figure 2) that is expected to form hydrogen bonds with the peptides or the T-cell receptor. In the crystallized natural peptide, possible interactions between the peptide and this aspartic acid corresponds to a polar tyrosine at peptide amino acid p10, making this position also a key residue in the peptide. Thus, the final selection was made to consider the peptide segment from residues 1-10 as a common 10 amino acid length segment for the modeling of all the peptides.

The 3D structures of the "strong", "weak", and other "natural" peptides are not available in neither free form nor bound to HLA-DP2 and therefore have to be modeled (note that the structure of HLA-DP2-Be-peptide complex [1] was not available prior the manuscript was submitted for review). We used the peptide in PDB ID: 3LQZ as a template and have built models for the rest of "natural", "strong", and "weak" peptides bound to HLA-DP2. In doing so, several considerations were made (all alignments are available in Material S1). The alignment of the other natural peptides to the 10 amino acid positions of the crystallized peptide template were based mainly on maintaining the biophysical characteristics of key peptide positions 4, 7, and 10. Additionally, we also tried to maintain the biophysical properties of the remaining 7 positions. For example, at positions 1 and 6 in the crystallized peptide, there are hydrophobic phenylalanine residues. Thus, in other peptides, we tried to conserve the biophysical characteristics of these positions by aligning the other peptides to have hydrophobic residues at positions 1 and 6. Similar considerations were made for the alignment of the "strong" and "weak" peptides to the template peptide.

The alignment and cutting of the peptides that bind to HLA-DR ("DR" peptides) was done in a similar fashion. We started with the crystallized structure of HLA-DR with a bound peptide (PDB ID:1KG0). We then used the crystallized DR peptide as a template for aligning and cutting other "DR" peptides. Because we intended to use these peptides as a control (they do not normally bind to HLA-DP2), we first aligned the peptide crystallized with DR to the same peptide template used for the "natural" peptides (taken from PDB ID: 3LQZ) to increase the likelihood of a peptide that naturally bind to HLA-DR to bind to HLA-DP2. We did the alignment in the same way (aligning residue positions 4, 7, 10, and then other positions) and cut the crystallized HLA-DR peptide to include the amino acids 1-10 (sequence YVKQNTLKLA). (The fact that the DR-peptide was also cut to include positions 1-10 was coincidental; the alignment was based entirely on biophysical characteristics). After the "DR" peptide was aligned to and cut based on the HLA-DP2 peptide, we then used it as a template for cutting the other 10 "DR" peptides. The alignment was done in the same fashion as stated before for the HLA-DP2 natural peptides. To create the three-dimensional structures of the peptides, we used the crystallized peptide as a structural template and mutated each residue individually via side chain replacement using VMD [29]. Once all of the peptide structure files were created, we then created additional structure files containing the ion, protein, and peptide to represent the two different binding processes. For the binding processes, we first identified the appropriate binding site of the Be ion on both the HLA-DP2 protein and the peptides.

Placement of the Be ion on the surface of HLA-DP2 and the peptides

To investigate the sequence of forming the (Be-HLA-DP2-peptide) complex, we modeled two different scenarios (Fig 3). This is done for computational purposes to compare the energy components and conformational changes, while not addressing questions if this occurs intracellularly or on the cell surface.

(a)   the Be ion is first bound to the HLA-DP2 protein forming a (Be-HLA-DP2) complex, and then the peptides bind to the (Be-HLA-DP2) complex (in text described as "peptide→(Be+ protein)" scenario); This requires predicting the Be binding site on HLA-DP2 in the absence of the peptides.

(b)   the Be is first bound to the peptide forming a (Be-peptide) complex, that binds to the HLA-DP2 protein (in text described as "(Be+peptide)→HLA-DP2" scenario). This requires predicting the Be binding site on the free peptides.

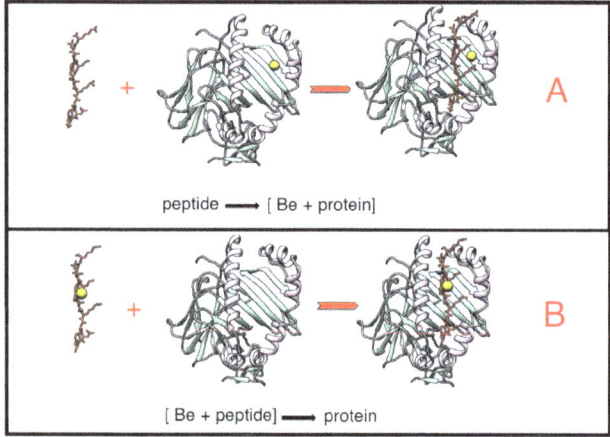

peptide ⟶ [Be + protein]

[Be + peptide] ⟶ protein

Figure 3. The HLA-DP2 binding pocket in complex with the crystallized natural peptide of 15 residues. The 10 residues of the peptide used as a cutting template are shown in red with important residues having extended structure. The removed residues are in yellow. Key residues in HLA-DP2 are shown in green; the aspartic acid is labeled as *ASP* and the three glutamic acid residues are labeled as *GLU Pocket*.

It can be expected that the Be ion binds to the HLA-DP2 or to the peptide surface non-specifically, and that the electrostatic force favors this binding. To determine the position of the positively charged Be ion on both the surface of the HLA-DP2 protein and on the exterior of the small peptides that bind to the HLA-DP2 protein, we utilized the BION webserver (http://compbio. clemson.edu/bion_server/) [30]. The algorithm implemented in BION relies on a DelPhi [31] generated potential map in conjunction with an in-house clustering algorithm [32]. The predictions take into account the magnitude of the electrostatic potential at selected surface-bound grid points and the biophysical properties of the ions such as radius and charge. The representative grid points are sorted in descending order (by absolute value) of the potential and the position of a given point within this list is termed Rank. Only the first ranked plausible ion position was used for further analysis. Note that Be ion binding sites for each unbound peptide were independently predicted and are slightly different depending on peptide sequence and conformation.

Placement of peptides onto the protein

Once the Be was placed onto either the protein or the peptide, we then placed each of the 40 peptides separately into the HLA-DP2 protein using Chimera [33]. Each peptide was placed in such a way as to be as similar as possible to the placement of the natural peptide found in the PDB file. We used the coordinates of the crystallized HLA-DP2 peptide found in the PDB file as the initial position of all the other peptides ("natural", "strong", "weak", and "DR") when binding them to the HLA-DP2 protein. When there were clashes between the side chains of the peptide and the side chains of the HLA-DP2, we manually adjusted the position of the entire peptide in order to very slightly lift the peptide out of the peptide binding pocket of HLA-DP2 or tilt the peptide within the pocket until the side chains did not overlap. In the case where the peptide side chains overlapped with the ion radius, side chain rotamers defined in the Chimera rotamer library were used to eliminate the overlap. All structures are available for download from http://figshare.com/articles/Chronic_Beryllium_Disease_ Revealing_the_role_of_beryllium_ion_and_small_peptides_binding _to_HLA_DP2/1147422.

Energy minimization and MD simulations

Once the structures were built, the protein, peptides and Be ion were relaxed and simulated using NAMD [34]. Each case was split by chain and protonated using the Charmm22 force field. These protonated chains were then recombined and an explicit water sphere was created around the complex using VMD with the TIP3P water model [29]. This sphere was created with a padding of 15 Å around the protein. To avoid initial structural clashes this structure was first relaxed by NAMD software [34] for 5000 steps via its conjugate gradient algorithm, and then, the minimized structure was simulated at a temperature of 298K for 1 nanosecond using 2 femtosecond time steps. The simulation used spherical boundary conditions to contain the complex and Langevin dynamics to incorporate random forces and damping of motion into the run. Every 100 steps, a snapshot of the protein was exported to a DCD trajectory file that was later unzipped into multiple PDB snapshots.

RMSD and pocket distance analysis

Once the results were obtained from the MD simulation, the snapshots of each of the cases were utilized for the screening of conformational changes of the Be binding pocket of the protein as a function of time. These changes in HLA-DP2 pocket due to the

Table 1. Selected distances between residues forming peptide-binding pocket of HLA-DP2.

Distance label	Residue 1	Residue 2
1	Arg76.A	Glu50.B
2	Leu70.A	Tyr58.B
3	Ile65.A	Asp64.B
4	Leu60.A	Met76.B
5	Ala56.A	Leu83.B

binding of small peptides and the Be ion were analyzed in two ways:

1. The change in RMSD of residues 50–86 of chain B (Figure 4, labeled in dark purple). Our preliminary data suggests that among two helices (one formed by chain A residues, and another by chain B residues) that form the peptide-binding pocket of HLA-DP2, the helix that is formed by chain B residues experiences the primary conformational transformation upon interacting with the "natural" peptides (not presented).

2. In order to depict the process of the opening-closing of the pocket, we observed the change in five different distances between pairs of residues between the helices forming the peptide-binding pocket of HLA-DP2 (Table 1, Figure 4, labeled in red).

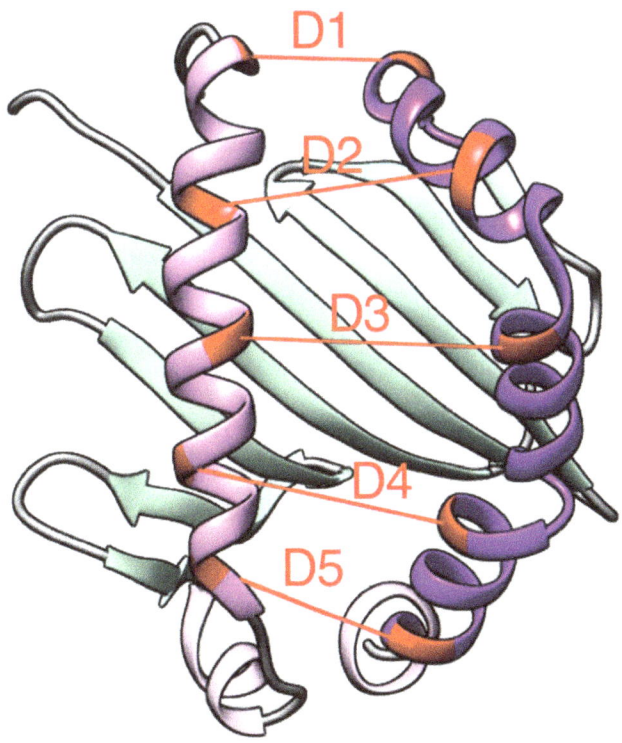

Figure 4. Cartoon representation of HLA-DP2 protein (PDB ID 3LQZ). Peptide binding pocket is shown in light and dark purple. Residues taken into account for distance calculations are in red. Change in RMSD due to peptide/Be binding was analyzed for residues in dark purple. Illustration was made with Chimera software [33].

For statistical significance, we used the peptide class representative RMSD and the distances between five pairs of residues that were calculated as the median RMSD values, and as the mean of distance values respectively within each case in the group.

Binding energy calculations

The binding free energy of the energetically-minimized protein complex was calculated via the following [35,36,37,38]:

$$\Delta G = G_{AB} - G_A - G_B \qquad (1)$$

where G_{AB} is the free energy of the complex; G_A and G_B are energies of A and B parts of the complex. Only the electrostatic – Coulombic (EE) and polar component of solvation energy (SP) – were calculated using the Delphi software (scale was 1 Å per grid, perfil 70%, dielectric constants for protein and water were 1 and 80 respectively) [31] and van der Waals (VE) energies, calculated with NAMD [34] (we used the data from one step of the minimization of complex and each of its partners), were taken into account in order to calculate the free energy of the complex and each of its parts:

$$G = EE + SP + VE \qquad (2)$$

Such an approach, with slight variations, was previously applied by us to model the changes of the binding free energy [35,36,37,38] and a variation of this method was benchmarked against experimental data [39]. For "(Be+peptide)→protein" scenario (Be ion bound first to the peptide), the binding energy was calculated between the protein (G_A) and the complex of the peptide and the Be (G_B). For the "peptide→(Be+protein)" scenario (Be attached to the protein first), we analyzed the binding affinity between the protein (G_A) and the Be-peptide complex (G_B). We also calculated the binding free energy between peptides (G_A) and HLA-DP2 protein (G_B) when modeled without the Be ion. In both cases, G_{AB} refers to the energy of the entire complex Be-HLA-DP2-peptide. It should be pointed out that the current implementation of eq. (2) does not use weighted coefficients and thus overestimates the energy changes, as outlined in another study of protein folding free energy [40]. Due to this, the results obtained using eq. (1) are used for ranking only, not to infer absolute binding free energy changes.

pKa calculations

We also performed pKa calculations for the ionizable groups within these complex structures and investigated the effects of the presence of the Be ion. The pKa values were obtained with Multi-Conformer-Continuum-Electrostatics (MCCE) program [41,42,43]. MCCE calculates the equilibrium state of each conformation and the charged state of the ionizable residues. It simulates the   the complex recognition by TC receptors. As the manuscript was

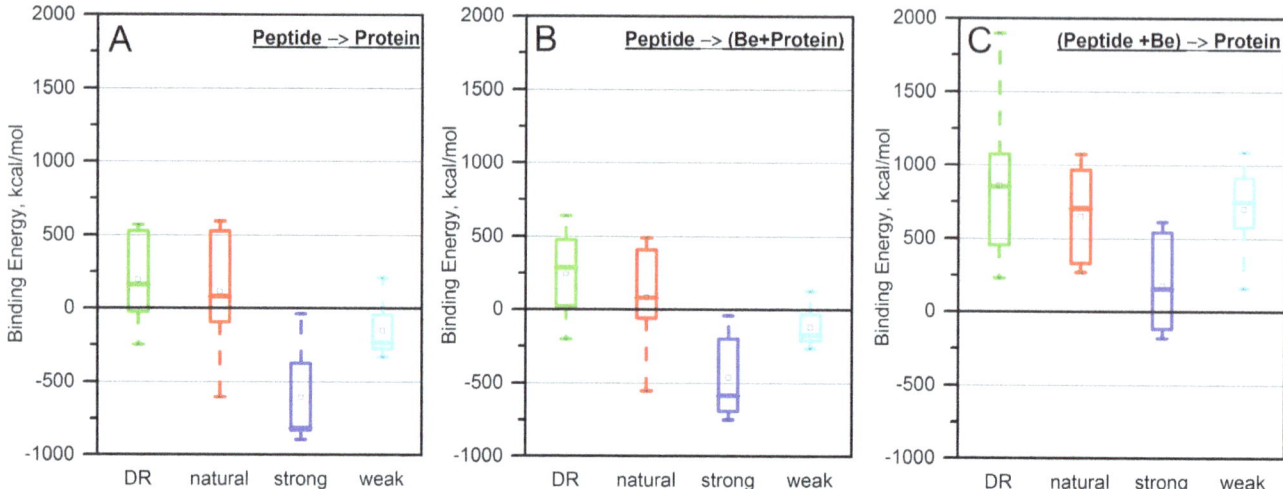

Figure 5. Binding free energy between A) peptides and HLA-DP2 protein without the Be ion; B) peptide and complex of HLA-DP protein with the Be ion in case of peptide→(Be+protein) scenario; C) complex of peptides and the Be ion with the protein for (Be+peptide)→protein scenario.

conformational and ionization changes in a Monte Carlo procedure and couples the protonation events with the conformational changes. The calculation was done as a function of pH with an internal dielectric constant of 4.0. The Be ion topology file was created with the parameters: radius = 1.53 Å, charge = +2.0e, and reference energy = −18.430 Kcal/mol.

The pKa calculations were performed on 40 bound complex structures in absence of Be ion (protein+peptide) along with 80 structures with the Be ion (40 structures of the peptides bound to the Be-HLA-DP2 complex and 40 structures of HLA-DP2 to bind with the Be-peptide complex). MCCE predicted the pKa value of the ionizable residue of each structure, and the pKa shifts were analyzed and discussed, with emphasis on the role of the Be ion and the binding.

Results and Discussion

At normal conditions, MHCII controls the activation of the immune system in response to foreign microorganisms. Its proteins bind a very specific range of exogenous peptides to the peptide-binding pocket and undergo some conformational changes that are recognizable by various TC receptors. However, in patients with CBD, high concentration of Be ions was shown to influence HLA-DP2 activity and resulting in the initiation of "false" signal cascades that cause the development of the chronic autoimmune disease. In order to investigate the molecular mechanism of the HLA-DP2 malfunction in the presence of Be ions, we investigate two scenarios: peptide→(Be+protein) and (Be+peptide)→protein, along with the binding of four different sets of peptides ("natural", "DR", "strong", and "weak", see Methods section for details). The goal is to analyze the changes in the key processes of the initial immune response associated with HLA-DP2 (peptide binding and subsequent protein conformational changes) for all cases. It is important to emphasize that the binding site of TC receptors is on the same side of the peptide-binding pocket of HLA-DP2, and thus the "false" signal cascade initiation depends mostly on the conformational change of the protein, and the strength of peptides binding to the protein might be less important for the process of

under review, this hypothesis was experimentally confirmed and it was shown that TC receptor recognizes the conformational changes in the surface of DP2-peptide complex induced by the internally bound Be ion [1].

Binding affinity of small peptides to HLA-DP2 protein

If the immune response is related to the binding affinity of the corresponding peptides to HLA-DP2, then one can expect that the peptides in the "strong" set should have the strongest affinity in the presence of the Be ion, and this affinity should be similar to the affinity of the "natural" set of peptides in the absence of the Be ion. The Be ion should favor binding of peptides in the "weak" set to the protein as well, however, the binding affinity should be smaller than for peptides in the "strong" set. Furthermore one expects the "DR" peptides to have the lowest or no affinity to HLA-DP2, in both the presence and in the absence of the Be ion. Although the binding affinity may not be a key factor, as indicated above, the binding free energy estimations may address the question of which scenario is more energetically favorable, the scenario "peptide→(Be+protein)" or "(Be+peptide)→protein".

Figure 5 shows the calculated binding free energy (eq.1) after 5000 steps of minimization. The grouped box chart was created based on the data for all of the peptides within the corresponding set ("strong", "weak", "natural", and "DR") and based on the model for placing the Be ion in the complex ("peptide→(Be+ protein)" and "(Be+peptide)→protein").

The immediate observation can be made by comparing Figure 5.B and C: The scenario "(Be+peptide)→protein" is less energetically favorable than the "peptide→(Be+protein)" scenario. This is due to several reasons. The peptides themselves, including "strong" and "weak" peptides, do not have an acidic cluster, i.e. do not have a binding pocket with a strong negative potential which can attract positively charged Be ions. In addition, free peptides are expected to be quite flexible and therefore are unable to keep tightly bound ions.

Comparing Figure 5.A (binding without the Be ion) and Figure 5.B (binding with the Be ion), no significant changes are

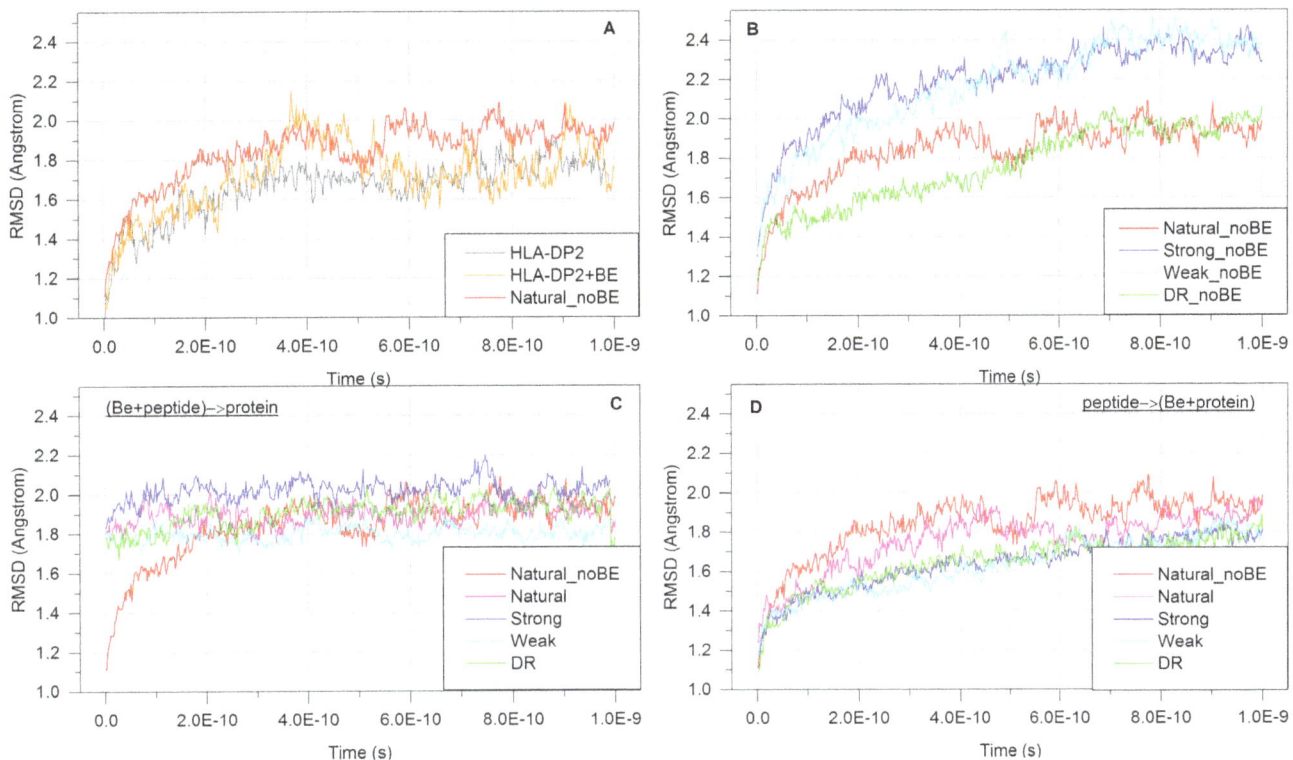

Figure 6. Change in RMSD of one of the walls of the peptide-binding pocket of HLA-DP2 during MD simulations in presence of peptides and/or the Be ion.

predicted for each of the peptides, however, "strong" peptides are predicted to have the highest affinity. The reason for this prediction will be discussed in the *pKa* section below. Here, we simply mention that the formation of the acidic cluster of three Glu residues affects the protonation states of these Glu residues in both the free HLA-DP2 and in the bound HLA-DP2 protein. These changes are not taken into account in the protocol that calculates the binding free energy. This was done to study the binding, conformational, and ionization changes separately.

Further analysis of Figure 5 reveals that the binding affinities are ranked in accordance with our expectations: the peptides in the "strong" set are the strongest binders, followed by "weak", "natural", and "DR" peptides. As we have indicated above, the binding free energies are for ranking only and it is quite possible that "DR" peptides do not bind at all to HLA-DP2.

Conformational changes in HLA-DP2 upon small peptides and/or Be ion binding

The binding of the peptide and the Be ion to HLA-DP2 should trigger conformational changes that are distinctly different from the conformational changes induced by "natural" peptides in the absence of Be. This is what will initiate the "false" signaling cascade and is the cause of CBD. We assume that these conformational changes should be caused by distinct different conformational changes within the peptide binding pocket as well. Thus, 1 ns MD simulations were conducted and later analyzed to observe the changes in the RMSD of one of the "walls" that form the peptide-binding pocket (helix 50–86.B) and the width of the binding pocket via the monitoring of five selected distances between the CA atoms of the residues that form this pocket

(Figure 4). The distance and RMSD fluctuations were analyzed for the last 0.4 ns.

1. RMSD analysis. Because the initial conformation of HLA-DP2 is the same for all cases (before the energy minimization steps), it allowed us to monitor how it will change over a period of time while binding different types of peptides and/or the Be ion. Here, we are focusing on the changes associated with helix 50–86.B (residues 50–86 of chain B), which forms one of the walls of the binding pocket and is expected to play an important role in the peptide and Be binding by HLA-DP2 and is expected to undergo conformational changes which are different from the changes observed for "natural" peptides in the absence of the Be ion.

In order to have a reference to compare with, the initial simulations were done on the unbound HLA-DP2, on HLA-DP2 with natural peptide, and on HLA-DP2 with the bound Be ion (Figure 6.A). One can see that in the unbound HLA-DP2, the helix 50–86.B undergoes some conformational fluctuations and reaches saturation at about 1.8 Å. The same simulations that were performed with "natural" peptides result in larger fluctuations, reaching RMSD saturation at about 2.0 Å. The simulations done on HLA-DP2 with the bound Be indicate that Be increases the magnitude of the fluctuations (±0.1 Å), but the saturation RMSD is almost identical to the HLA-DP2 without Be. Based on these results, it can be speculated that the binding of "natural" peptides to the HLA-DP2 protein causes the helix 50–86.B to deviate further from its crystallographic position and thus to open the peptide binding side. These changes can be recognized by specific TC receptors and trigger the "normal" cascade of signals.

Figure 6.B shows the RMSD without the Be ion present. It can be seen that "natural" and "DR" peptide binding causes almost the same RMSD on the HLA-DP2 complex, while the binding of "strong" and "weak" peptides in absence of Be causes a much

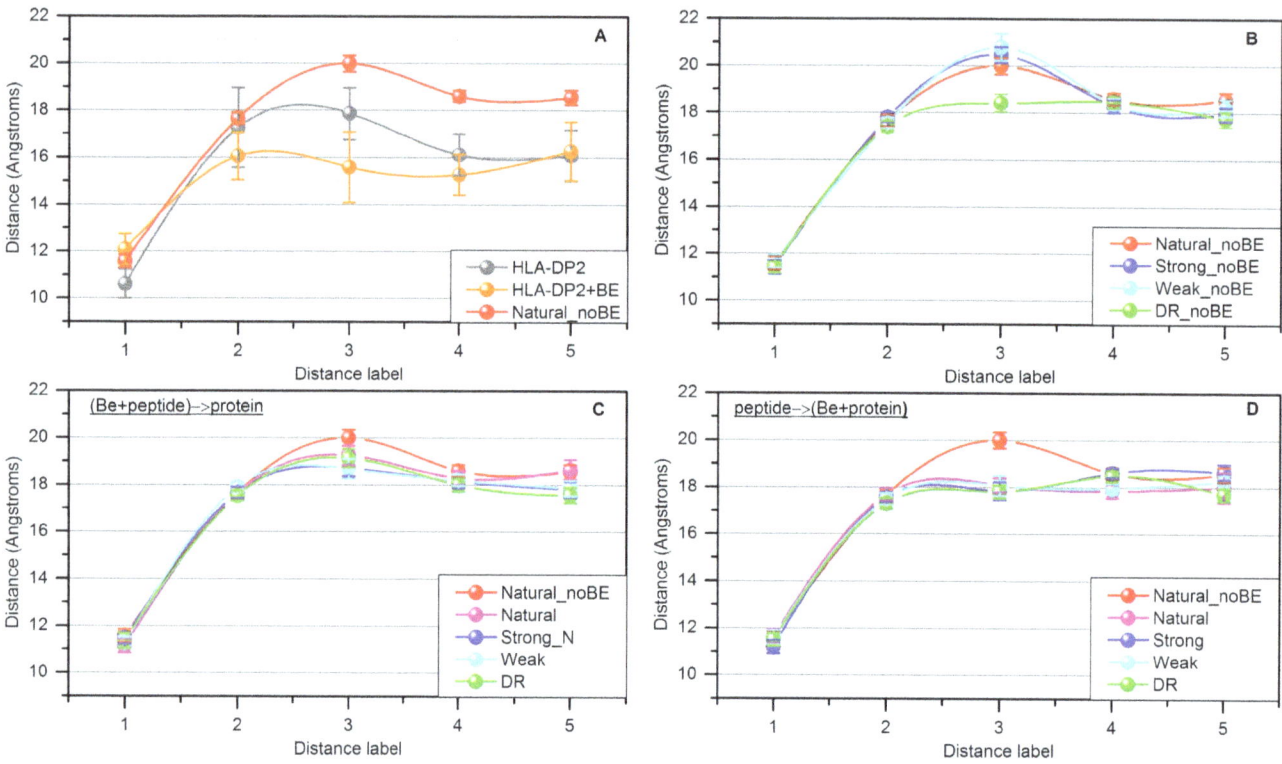

Figure 7. Average distances between the CA atoms of residues forming peptide-binding pocket. Error bar illustrated as ± two values of standard deviation.

larger RMSD of helix 50–86.B. The direction of the change, the increase in the RMSD upon the binding of the peptides, is the same as for binding "natural" peptides to HLA-DP2, and perhaps will initiate the same immune cascade.

The simulations performed with all the sets of peptides ("DR", "natural", "strong", and "weak") with the Be ion in the scenario "(Be+peptide)→protein" indicate that there is practically no difference between the RMSD of the complexes (Figure 6.C). This is similar to the RMSD induced by the Be ion alone and is in agreement with the binding free energy calculations (Figure 5). Indeed, all peptides within scenario "(Be+peptide)→protein" were calculated to bind much more weakly than "normal" peptides in the absence of the Be ion. This is the reason why the RMSD fluctuations are very similar for all of the peptides to the RMSD of Be-HLA-DP2, simply because the type of peptide does not matter.

If the same procedure is applied to the structures built under the scenario "peptide→(Be+protein)", the effect is opposite (Figure 6.D). The binding of all peptides onto Be-HLA-DP2 complex reduces the RMSD associated with helix 50–86.B. This response is just opposite to the increase of RMSD of helix 50–86.B upon binding of "natural" peptides in absence of the Be ion. The most striking observation is that, in presence of the Be ion, peptides from the "strong", "weak", and "DR" sets induce practically the same conformational changes in helix 50–86.B: reduce both the level of saturation and fluctuations compared to HLA-DP2 bound to the "natural" peptides. The same is valid for "natural" peptides to lesser extent. It can be viewed as the binding is rigidifying the pocket. Since this change is completely opposite to the change induced by the binding of 'natural' peptides in the absence of Be, it can be speculated that it might trigger a different signaling cascade.

2. Analyzing the changes in the width of peptide-binding pocket. To analyze the change in the size of peptide-binding pocket upon the binding of the Be ion and peptides in the four different sets, we monitored the change in the distances between five selected positions within the helices that form the "walls" of the pocket (see Figure 4). The results are shown in Figure 6 where the position between CA atoms for each of the 5 pairs of residues (see Figure 4, Table 1) averaged for the last 0.4 ns of MD simulations within the corresponding set are plotted. To demonstrate the distance fluctuations, we also included two standard deviations of the average data as error bars on the graphs.

As in the previous paragraph, we begin the modeling with simulations of peptide-free HLA-DP2, along with HLA-DP2 loaded with a "natural" peptide and with a Be ion. The results are shown in Figure 7.A. It can be seen that in the case of free HLA-DP2, the distance between the helices forming the gate of the binding pocket fluctuates at each positions by about 4 Å. Adding the Be ion narrows the gate at positions D3–D5, while the fluctuations remain almost the same as in the free HLA-DP2. However, when the "natural" peptides are simulated with HLA-DP2, the trend is the opposite: opening the pocket at positions D3–D5 while reducing the fluctuations. Combined with the observations made in the previous section, these results indicate that the binding of natural peptides to HLA-DP2 makes the binding pocket wider while rigidifying it.

Figure 7.B shows the results for the four sets of peptides in the absence of the Be ion. The calculated distances between the helices forming the gate of the binding pocket are almost identical for "natural", "strong", and "weak" peptide sets. At first, this may be considered as a contradiction of the observation made about the RMSD of helix 50–86.B in the previous section. However, the analysis of the other helix forming the gate indicates that the

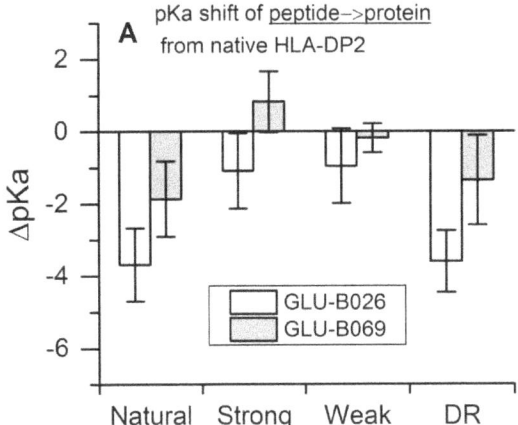

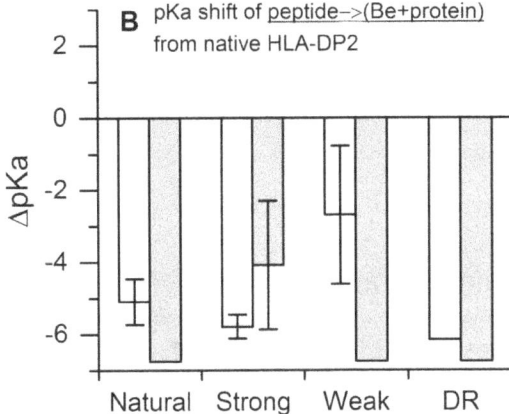

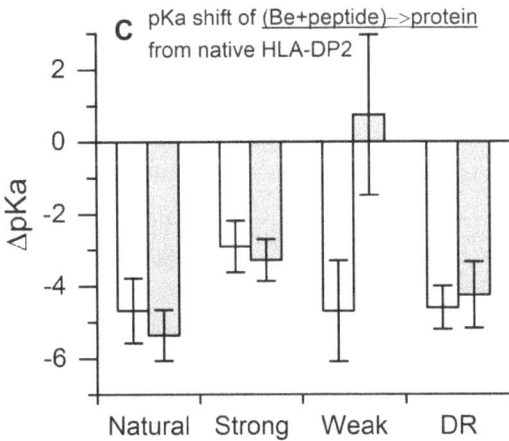

Figure 8. The pKa shift of two ionizable residues βGlu26 and βGlu69 on (A) HLA-DP2 binds to peptides (Four types as "natural", "strong", "weak" and "DR", the same to the followings) (B) peptides bind to the complex (Be+protein) (C) (Be+peptides) complex binds to HLA-DP2 protein from pKa of native unbond protein. The pKa value was calculated as the average of 10 structures for each type of peptides.

"strong" and "weak" peptides in the absence of Be only deform the pocket without changing its width. This indicates that the presence of the Be ion may be an important factor for regulating

the conformational changes of the gate of the peptide binding pocket.

The simulations of the complexes with HLA-DP2 loaded with the four sets of peptides with the Be ion using structures delivered via the "(Be+peptide)→protein" scenario are shown in Figure 7.C. For comparison, the figure also shows the results for the HLA-DP2 with the "natural" peptides without Be. The data shows that there is no difference in the size of the pocket between each of the sets of peptides with the Be ion and the "natural" peptide without the ion. The same similarities were shown previously while analyzing the binding affinity (Figure 5.B) and the conformational changes of helix 50–86.B (Figure 6.C).

The same calculations for the "peptide→(Be+protein)" scenario (Fig 7.D) show that all sets of peptides bound to the HLA-DP2 cause the same conformational changes. This change is distinctly smaller at the distance D3 as compared to HLA-DP2 loaded with "natural" peptides. In contrast, the length at distance markers D4 and D5 is slightly longer in comparison with the HLA-DP2 protein in complex with the "natural" peptides. This is a distinctive signature of the binding effects caused by binding of each of the peptides onto HLA-DP2 already loaded with the Be ion. These conformational changes are significantly different from the conformational changes induced by the binding of "natural" peptides in the absence of the Be ion and perhaps will trigger a different signaling cascade.

pKa calculation of ionizable residues due to small peptides and/or Be ion binding to HLA-DP2

Binding frequently causes ionization changes, typically referred to as proton uptake/release [44,45] and thus affects the electrostatic component of the binding free energy [46]. To investigate the effects of small peptides and the Be ion binding to HLA-DP2 on the ionization states of titratable groups, we carried out a pKa calculation using the MCCE program [41,42,43]. The pKa's of all ionizable residues of HLA-DP2 were calculated, and the protonation changes caused by the peptide binding, especially for the residues within the pocket, were analyzed. As described in the methods section, 120 structures were subjected to the pKa calculations. 40 of them were the HLA-DP2 bound with a peptide in the absence of the Be ion and 40 of them were the protein-Be complex bound to a small peptide and the rest were the protein bound to the Be-peptide complex. They were also divided into 4 groups by the peptide categories: "DR", "natural", "strong", and "weak". Also, for each group, we obtained the average pKa values of 10 structures along with the standard error and the calculated pKa shift of them from the native HLA-DP2 protein (PDB ID: 3LZM) (see Figure 8).

We found two acidic residues, βGlu26 and βGlu69, always experience a large pKa change that is caused by the peptide or Be ion binding, although several other ionizable residues also showed slight pKa shifts. Figure 8 shows the pKa shifts for βGlu26 (blue bar) and βGlu69 (red bar) as HLA-DP2 binds four different types of peptides in presence/absence of Be ion. The pKa values for βGlu26 and βGlu69 on free HLA-DP2 are 6.2 and 6.8 respectively, i.e. they are almost neutral at pH = 7. We found that in the absence of the Be ion, the pKa shift that was caused by the binding of most of the types of peptides is less than 2.0 for βGlu26 and βGlu69 although the pKa shift for βGlu26 is about 3.5 for the "natural" and "DR" peptides. However, when the Be ion participates in the binding, both pKa shifts increase significantly and this results in deprotonation (both glutamic acids become fully charged). Especially when the Be ion binds with HLA-DP2 first and then the complex binds to the peptide, most of the pKa's were lowered more than 5.0, which indicates that the glutamic acids are fully deprotonated due to the Be ion. This is another observation in favor of the hypothesis that the Be ion is crucial for the HLA-DP2 response to the binding and the nature of the binding peptide is a secondary factor. However, since the binding affinity strongly differs among peptides, it is expected that "strong" peptides will dominate.

Conclusions

Three questions were addressed in this study: (a) Where does the Be ion bind?; (b) What is the sequence of the events for forming the Be-HLA-DP2-peptide complex?; and (c) What is the specific "signature" of the effects induced by CBD-causing peptides and Be?

The putative Be position was predicted by the BION server [30] and it was found to be in the previously suggested acidic pocket composed of three acidic residues, Glu26, Glu68 and Glu69 [20,23,26]. The predicted Be position was tested in MD stimulations both with and without peptides and it was observed that the Be ion stays within the pocket and does not drift away within 1 ns of simulations. As the manuscript was under review, the Be position was experimentally determined [1] and our predicted Be position was in close proximity to the experimental one (~3.2 Å).

Our calculations show that for the scenario "(Be+peptide)→ protein" the conformational changes of the HLA-DP2 protein induced by the peptides with the attached Be ion are identical to those caused by "natural" peptides alone. This indicates that if this scenario occurs in CBD development, then the common TC receptors should recognize the complex and exert the "normal" immune response. However, the possibility of this scenario is negligibly small because of a) the decreased binding affinity between (Be+pepide) and protein and b) small peptides are very flexible in the free state (unbound state) and holding the Be ion is quite unlikely. In addition, since the binding energies for all four sets of peptides in the scenario "(Be+peptide)→protein" were much worse when compared with the "peptide→(Be+protein)" scenario,

it was concluded that the "peptide→(Be+protein)" scenario is more likely to occur in CBD.

Comparing the calculated effects across the four sets of peptides, it can be suggested that the specific "signature" of CBD is complex. The key player is the Be ion, which, when binds to the acidic pocket of HLA-DP2, changes the conformational properties of the peptide binding pocket. The peptide affinity is unaffected by the Be presence, but once the peptides bind to the pocket pre-loaded with the Be ion, they all induce similar conformational and ionization changes. These changes are distinctively different from the changes caused by the binding of "natural" peptides to the HLA-DP2 in absence of Be and, thus, are expected to trigger a completely different cascade of immune reactions which then leads to CBD. This conclusion is supported by the experimental results suggesting that in patients with CBD, the Be-specific CD4+T cells (Vβ5+CD4+T cell [47]) are responsible for the complex recognition and immune response propagation [48,49,50]. Furthermore, recent experimental work suggests that Be-specific T cells recognize the peptide-Be-HLA-DP2 complex via an unconventional docking topology [51], which in our work is referred to as a different conformational change of HLA-DP2 in CBD as compared with the normal immune response. Thus, we hypothesize that the peptide binding to the Be doped HLA-DP2 causes CBD development as recently confirmed experimentally [1].

Author Contributions

Conceived and designed the experiments: MP BW SS NS DHV LW EA. Performed the experiments: MP BW SS NS DHV LW. Analyzed the data: MP BW SS NS DHV LW EA. Contributed reagents/materials/analysis tools: MP BW SS NS DHV LW EA. Wrote the paper: MP BW SS NS DHV LW EA.

References

1. Clayton GM, Wang Y, Crawford F, Novikov A, Wimberly BT, et al. (2014) Structural Basis of Chronic Beryllium Disease: Linking Allergic Hypersensitivity and Autoimmunity. Cell 158: 132–142.
2. Van Dyke MV, Martyny JW, Mroz MM, Silveira LJ, Strand M, et al. (2011) Exposure and genetics increase risk of beryllium sensitisation and chronic beryllium disease in the nuclear weapons industry. Occupational and environmental medicine 68: 842–848.
3. Silveira LJ, McCanlies EC, Fingerlin TE, Van Dyke MV, Mroz MM, et al. (2012) Chronic beryllium disease, HLA-DPB1, and the DP peptide binding groove. The Journal of Immunology 189: 4014–4023.
4. Bowerman NA, Falta MT, Mack DG, Kappler JW, Fontenot AP (2011) Mutagenesis of beryllium-specific TCRs suggests an unusual binding topology for antigen recognition. The Journal of Immunology 187: 3694–3703.
5. Díaz G, Cañas B, Vazquez J, Nombela C, Arroyo J (2005) Characterization of natural peptide ligands from HLA-DP2: new insights into HLA-DP peptide-binding motifs. Immunogenetics 56: 754–759.
6. Wang Y, Dai S (2013) Structural basis of metal hypersensitivity. Immunologic research 55: 83–90.
7. Welch LS, Ringen K, Dement J, Bingham E, Quinn P, et al. (2013) Beryllium disease among construction trade workers at department of Energy nuclear sites. American journal of industrial medicine 56: 1125–1136.
8. Van Dyke MV, Martyny JW, Mroz MM, Silveira LJ, Strand M, et al. (2011) Risk of chronic beryllium disease by HLA-DPB1 E69 genotype and beryllium exposure in nuclear workers. American journal of respiratory and critical care medicine 183: 1680–1688.
9. Dai S, Falta MT, Bowerman NA, McKee AS, Fontenot AP (2013) T cell recognition of beryllium. Current opinion in immunology 25: 775–780.
10. Thomas CA, Deubner DC, Stanton ML, Kreiss K, Schuler CR (2013) Long-Term Efficacy of a Program to Prevent Beryllium Disease. American journal of industrial medicine 56: 733–741.

11. Mayer A, Brazile W, Erb S, Barker E, Miller C, et al. (2013) Developing Effective Health and Safety Training Materials for Workers in Beryllium-Using Industries. Journal of Occupational and Environmental Medicine 55: 746–751.
12. Salvator H, Gille T, Hervé A, Bron C, Lamberto C, et al. (2013) Chronic beryllium disease: azathioprine as a possible alternative to corticosteroid treatment. European Respiratory Journal 41: 234–236.
13. Maier LA, Barkes BQ, Mroz M, Rossman MD, Barnard J, et al. (2012) Infliximab therapy modulates an antigen-specific immune response in chronic beryllium disease. Respiratory medicine 106: 1810–1813.
14. Seidler A, Euler U, Müller-Quernheim J, Gaede K, Latza U, et al. (2012) Systematic review: progression of beryllium sensitization to chronic beryllium disease. Occupational medicine 62: 506–513.
15. Chain JL, Martin AK, Mack DG, Maier LA, Palmer BE, et al. (2013) Impaired function of ctla-4 in the lungs of patients with chronic beryllium disease contributes to persistent inflammation. The Journal of Immunology 191: 1648–1656.
16. Bowerman NA, Falta MT, Mack DG, Wehrmann F, Crawford F, et al. (2014) Identification of Multiple Public TCR Repertoires in Chronic Beryllium Disease. The Journal of Immunology: 1400007.
17. Yucesoy B, Johnson VJ, Lummus ZL, Kashon ML, Rao M, et al. (2014) Genetic Variants in the Major Histocompatibility Complex Class I and Class II Genes Are Associated With Diisocyanate-Induced Asthma. Journal of Occupational and Environmental Medicine 56: 382–387.
18. Morris D, Fernando M, Taylor K, Chung S, Nititham J, et al. (2014) MHC associations with clinical and autoantibody manifestations in European SLE. Genes and immunity.
19. Pimentel-Santos F, Matos M, Ligeiro D, Mourão A, Ribeiro C, et al. (2013) HLA alleles and HLA-B27 haplotypes associated with susceptibility and severity of ankylosing spondylitis in a Portuguese population. Tissue Antigens 82: 374–379.
20. Richeldi L, Sorrentino R, Saltini C (1993) HLA-DPB1 glutamate 69: a genetic marker of beryllium disease. Science 262: 242–244.

21. Fontenot AP, Torres M, Marshall WH, Newman LS, Kotzin BL (2000) Beryllium presentation to CD4+ T cells underlies disease-susceptibility HLA-DP alleles in chronic beryllium disease. Proc Natl Acad Sci U S A 97: 12717–12722.

22. Bill JR, Mack DG, Falta MT, Maier LA, Sullivan AK, et al. (2005) Beryllium presentation to CD4+ T cells is dependent on a single amino acid residue of the MHC class II beta-chain. J Immunol 175: 7029–7037.

23. Dai S, Murphy GA, Crawford F, Mack DG, Falta MT, et al. (2010) Crystal structure of HLA-DP2 and implications for chronic beryllium disease. Proc Natl Acad Sci U S A 107: 7425–7430.

24. Patronov A, Dimitrov I, Flower DR, Doytchinova I (2011) Peptide binding prediction for the human class II MHC allele HLA-DP2: a molecular docking approach. BMC structural biology 11: 32.

25. Doytchinova I, Petkov P, Dimitrov I, Atanasova M, Flower DR (2011) HLA-DP2 binding prediction by molecular dynamics simulations. Protein Science 20: 1918–1928.

26. Falta MT, Pinilla C, Mack DG, Tinega AN, Crawford F, et al. (2013) Identification of beryllium-dependent peptides recognized by CD4+ T cells in chronic beryllium disease. The Journal of experimental medicine 210: 1403–1418.

27. Amicosante M, Sanarico N, Berretta F, Arroyo J, Lombardi G, et al. (2001) Beryllium binding to HLA-DP molecule carrying the marker of susceptibility to berylliosis glutamate β69. Human immunology 62: 686–693.

28. Scott BL, Wang Z, Marrone BL, Sauer NN (2003) Potential binding modes of beryllium with the class II major histocompatibility complex HLA-DP: a combined theoretical and structural database study. Journal of inorganic biochemistry 94: 5–13.

29. Humphrey W, Dalke A, Schulten K (1996) VMD: visual molecular dynamics. J Mol Graph 14: 33–38, 27–38.

30. Petukh M, Kimmet T, Alexov E (2013) BION web server: predicting non-specifically bound surface ions. Bioinformatics 29: 805–806.

31. Li C, Petukh M, Li L, Alexov E (2013) Continuous development of schemes for parallel computing of the electrostatics in biological systems: Implementation in DelPhi. Journal of computational chemistry.

32. Petukh M, Zhenirovskyy M, Li C, Li L, Wang L, et al. (2012) Predicting Nonspecific Ion Binding Using DelPhi. Biophys J 102: 2885–2893.

33. Pettersen EF, Goddard TD, Huang CC, Couch GS, Greenblatt DM, et al. (2004) UCSF Chimera–a visualization system for exploratory research and analysis. J Comput Chem 25: 1605–1612.

34. Phillips JC, Braun R, Wang W, Gumbart J, Tajkhorshid E, et al. (2005) Scalable molecular dynamics with NAMD. Journal of computational chemistry 26: 1781–1802.

35. Teng S, Madej T, Panchenko A, Alexov E (2009) Modeling effects of human single nucleotide polymorphisms on protein-protein interactions. Biophysical journal 96: 2178–2188.

36. Zhang Z, Teng S, Wang L, Schwartz CE, Alexov E (2010) Computational analysis of missense mutations causing Snyder-Robinson syndrome. Human mutation 31: 1043–1049.

37. Zhang Z, Miteva MA, Wang L, Alexov E (2012) Analyzing effects of naturally occurring missense mutations. Computational and mathematical methods in medicine 2012.

38. Nishi H, Tyagi M, Teng S, Shoemaker BA, Hashimoto K, et al. (2013) Cancer missense mutations alter binding properties of proteins and their interaction networks. PLoS One 8: e66273.

39. Li M, Petukh M, Alexov E, Panchenko AR (2014) Predicting the Impact of Missense Mutations on Protein-Protein Binding Affinity. Journal of Chemical Theory and Computation.

40. Zhang Z, Wang L, Gao Y, Zhang J, Zhenirovskyy M, et al. (2012) Predicting folding free energy changes upon single point mutations. Bioinformatics 28: 664–671.

41. Alexov E, Gunner M (1997) Incorporating protein conformational flexibility into the calculation of pH-dependent protein properties. Biophysical journal 72: 2075–2093.

42. Alexov E, Gunner M (1999) Calculated protein and proton motions coupled to electron transfer: electron transfer from QA-to QB in bacterial photosynthetic reaction centers. Biochemistry 38: 8253–8270.

43. Georgescu RE, Alexov EG, Gunner MR (2002) Combining conformational flexibility and continuum electrostatics for calculating pK(a)s in proteins. Biophys J 83: 1731–1748.

44. Onufriev AV, Alexov E (2013) Protonation and pK changes in protein–ligand binding. Quarterly reviews of biophysics 46: 181–209.

45. Petukh M, Stefl S, Alexov E (2013) The Role of Protonation States in Ligand-Receptor Recognition and Binding. Current pharmaceutical design 19: 4182.

46. Zhang Z, Witham S, Alexov E (2011) On the role of electrostatics in protein-protein interactions. Phys Biol 8: 035001.

47. Bowerman N, Falta M, Mack D, Crawford F, Kappler J, et al. (2013) Characterizing the T cell receptor repertoire of beryllium-responsive CD4 T cells (P5021). The Journal of Immunology 190: 110.111.

48. Fontenot AP, Falta MT, Freed BM, Newman LS, Kotzin BL (1999) Identification of pathogenic T cells in patients with beryllium-induced lung disease. J Immunol 163: 1019–1026.

49. Fontenot AP, Keizer TS, McCleskey M, Mack DG, Meza-Romero R, et al. (2006) Recombinant HLA-DP2 binds beryllium and tolerizes beryllium-specific pathogenic CD4+ T cells. The Journal of Immunology 177: 3874–3883.

50. Martin AK, Mack DG, Falta MT, Mroz MM, Newman LS, et al. (2011) Beryllium-specific CD4+ T cells in blood as a biomarker of disease progression. Journal of allergy and clinical immunology 128: 1100-1106. e1105.

51. Bowerman N, Falta M, Mack D, Kappler J, Fontenot A (2011) Beryllium-specific T cells adopt an unusual binding topology for antigen recognition. The Journal of Immunology 186: 100.125.

9

Protein Thermal Stability Enhancement by Designing Salt Bridges: A Combined Computational and Experimental Study

Chi-Wen Lee[1][9][¶], **Hsiu-Jung Wang**[2][9][¶], **Jenn-Kang Hwang**[1]*, **Ching-Ping Tseng**[2]*

1 Institute of Bioinformatics and Systems Biology, College of Biological Science and Technology, National Chiao Tung University, Hsinchu, Taiwan, Republic of China,
2 Department of Biological Science and Technology, College of Biological Science and Technology, National Chiao Tung University, Hsinchu, Taiwan, Republic of China

Abstract

Protein thermal stability is an important factor considered in medical and industrial applications. Many structural characteristics related to protein thermal stability have been elucidated, and increasing salt bridges is considered as one of the most efficient strategies to increase protein thermal stability. However, the accurate simulation of salt bridges remains difficult. In this study, a novel method for salt-bridge design was proposed based on the statistical analysis of 10,556 surface salt bridges on 6,493 X-ray protein structures. These salt bridges were first categorized based on pairing residues, secondary structure locations, and Cα–Cα distances. Pairing preferences generalized from statistical analysis were used to construct a salt-bridge pair index and utilized in a weighted electrostatic attraction model to find the effective pairings for designing salt bridges. The model was also coupled with B-factor, weighted contact number, relative solvent accessibility, and conservation prescreening to determine the residues appropriate for the thermal adaptive design of salt bridges. According to our method, eight putative salt-bridges were designed on a mesophilic β-glucosidase and 24 variants were constructed to verify the predictions. Six putative salt-bridges leaded to the increase of the enzyme thermal stability. A significant increase in melting temperature of 8.8, 4.8, 3.7, 1.3, 1.2, and 0.7°C of the putative salt-bridges N437K–D49, E96R–D28, E96K–D28, S440K–E70, T231K–D388, and Q277E–D282 was detected, respectively. Reversing the polarity of T231K–D388 to T231D–D388K resulted in a further increase in melting temperatures by 3.6°C, which may be caused by the transformation of an intra-subunit electrostatic interaction into an inter-subunit one depending on the local environment. The combination of the thermostable variants (N437K, E96R, T231D and D388K) generated a melting temperature increase of 15.7°C. Thus, this study demonstrated a novel method for the thermal adaptive design of salt bridges through inference of suitable positions and substitutions.

Editor: Chandra Verma, Bioinformatics Institute, Singapore

Funding: This work was supported by the National Science Council, Taiwan (NSC-98-2321-B-009-002). The funders had no role in study design, data collection and analysis, decision to publish, or preparation of the manuscript.

Competing Interests: The authors have declared that no competing interests exist.

* Email: jkhwang@cc.nctu.edu.tw (JKH); cpts@cc.nctu.edu.tw (CPT)

9 These authors contributed equally to this work.

¶ These authors are first authors on this work.

Introduction

The need for thermostable proteins has greatly promoted protein engineering in chemical, biotechnological, and food industries [1,2]. In thermal processes, the use of thermostable proteins exhibits several advantages, including increased productivity, reduced resource depletion, and prevention of microbial contamination. Several studies have been conducted to identify the factors associated with thermal adaptation [3–11]. For instance, the thermal stability of proteins is affected by hydrophobicity, packing density [3,4], number of disulfide bonds [5], strength of electrostatic interactions [6,7], length of surface loops [4], conformational rigidity [8,9], amino acid coupling patterns [10], and local structural entropy [11]. These factors not only provide information regarding thermal resistance but pose a great challenge on the development of thermostable proteins.

Electrostatic interaction is one of the most relevant factors in determining protein thermal stability [6,7,12,13]. Furthermore, an increased number of surface charges and salt bridges are observed in thermophilic proteins, in which they contribute to conformational specificity, thermal stability, and oligomerization [14–16]. Therefore, increasing electrostatic attraction on protein surfaces was suggested to enhance protein thermal stability [17–19]. These stabilizing effects often originate from the increase in protein rigidity and decrease in free energy [20]. The suitable substitutions on protein surfaces can prevent the disruption of conformational changes that occur in a well-packed buried core [6,21]. Thus, the design of salt bridges on the protein surface is a promising strategy to enhance thermal stability.

Salt bridges in proteins are often formed when two oppositely charged groups interact; in particular, the two charged groups include the cationic ammonium ion of a basic amino acid residue and the anionic carboxylate ion of an acidic amino acid residue spanning a gap of <4 Å. Salt bridges are rarely isolated protein structures; instead, these structures are an integrated part of complex interaction networks that often occur in thermophilic proteins [22,23]. Therefore, specific geometric conformations are critical in determining the stabilization effect of salt bridges [14,24]. These geometric conformations depend on the orientation of the side-chains of positively charged residues (e.g. Lysine and Arginine) with respect to the oppositely charged residues (e.g. Aspartate and Glutamate), solvent accessibility, structural context, and local environment, among others [24,25]. However, rational design of salt bridges to enhance protein thermal stability remains a difficult task because the rules for predicting the positions that result in the effective replacement by charged residues to form salt bridges are still unclear.

B-factor represents the atomic displacement parameters that are obtained from X-ray crystallographic data. This parameter indicates the blurring of atomic electron densities with respect to their equilibrium positions because of thermal motion and positional disorder [26]. Residues with high B-factors often exhibit high flexibility and low thermal stability. Reetz and his group [9] developed an approach coupled with an evolutionary strategy to obtain thermostable proteins by using saturation mutagenesis at the high B-factor residues. Besides, packing density is another important factor that affects protein thermal stability. Our previous studies have shown that the weighted contact number (WCN) can reflect the packing density and is highly correlated with the B-factor [27,28]. The interaction of two opposite-charge groups to form a salt bridge enhances the rigidity of the connection between two secondary structures where the salt bridge is located, which results to a decrease in local flexibility and an increase in packing density. Thus, protein thermal stability is enhanced [20,29,30]. In summary, B-factor and WCN of an amino acid residue are indicators that can be used to design stable salt bridges.

In this study, we developed a method with a scoring system to enhance protein thermal stability based on a salt-bridge design on protein surface. The surface residues, which qualify the criteria of B-factor, WCN and other parameters, were selected. Substitutions that exhibit a high probability of forming a bridge with the surrounding charged residues were computed for each selected position. For example, the polar residues Ser, Thr, Cys, Asn, and Gln are located near negatively charged residues with a high probability of forming salt bridges. Thus, these residues can be mutated to positively charged amino acids, such as Lys or Arg, to form new salt bridges. As a result, electrostatic stabilization is increased. Afterwards, the electrostatic stabilizations from the substitutions (Lys, Arg, Asp, and Glu) of each selected position were compared with the scores that were calculated by using our scoring model.

The production of cellulase required to degrade lignocellulosic substrates constitute a great portion of the total cost of bio-alcohol production [31]. The use of thermostable cellulases in cellulose degradation offers advantages such as reducing the enzyme dosage because of the extension of the enzyme half-life and the capacity to increase the activity at high temperatures [32]. The β-glucosidase A (BglA, EC 3.2.1.21, PDB code: 1BGA [33]) from *Bacillus polymyxa* is a member of family 1 glycosyl hydrolases. The BglA catalyzes the hydrolysis of β-1,4-glucosidic bonds of cellobiose, which is the last step in cellulose degradation. In addition, this enzyme is mesophilic and exhibits a homo-octameric structure [(β/α)$_8$] in the native state. Therefore, the BglA can be used as a polymeric model for the salt-bridge design. In this study, our method was applied on the BglA, and six pairs of putative salt-bridges were successfully designed. The experimental results suggested that our method can identify novel salt bridges stabilized by native electrostatic environments to increase protein thermal stability.

Materials and Methods

Protein data set for determining geometric preferences of salt bridges

To determine the specific geometric features of salt bridges, a protein structure set was obtained from the Pisces web server with the following criteria: resolution <2.5 Å; $R<0.25$; sequence similarities between any pair protein <25%; and chain length ranging from 60 to 800 residues [34]. The data set comprised of 6,493 X-ray protein structures (Text S1).

Residue pairs of salt bridges

A salt bridge is formed between oppositely charged residues, such as Arg (R) or Lys (K) and Asp (D) or Glu (E). In this study, His (H) was excluded because this amino acid exhibits an ambiguous protonation state at pH 7.0. Therefore, the following four types of attractive interactions were considered: K–D; K–E; R–D; and R–E. The salt bridge is defined based on the following statements; let atom A be $N^{\eta 1}$ or $N^{\eta 2}$ of R or N^{ξ} of K, and let atom B be $O^{\delta 1}$ or $O^{\delta 2}$ of D or $O^{\varepsilon 1}$ or $O^{\varepsilon 2}$ of E. A distance between atoms A and B less than 4 Å indicates the formation of a salt bridge between A and B. According to this definition and the considered types of residue pairings, 32,096 salt bridges were isolated from the 6,493 protein structures in the data set.

Relative solvent accessibility and secondary structure locations of salt bridges

The 32,096 salt bridges were classified into buried or exposed state according to the Relative Solvent Accessibility (RSA) of the salt-bridge residues. RSA is the ratio of the solvent exposed surface area of a residue [35]. To calculate the RSA of a salt-bridge residue, the residue's solvent accessibility (ASA) obtained from DSSP program was divided by a reference value of surface area for the given amino acid (X) in an extended Gly–X–Gly peptide. Salt bridges were considered at the exposed state if the RSA of one residue was >35% and that of the other residue was >25% [36,37]. After that, the exposed salt-bridges were further categorized according to the located secondary structure and other parameters. The following types of secondary structures are defined in the DSSP program: H, G, I, B, E, T, S, and U [35]. For simplicity, these structures were categorized as helix (H = "H, G, and I"), sheet (S = "B and E"), and coil (C = "T, S, and U").

Geometric orientation of salt bridges

The Cα and Cβ atomic coordinates of two residues (i, j) of a salt bridge were used to determine the angles (θ_1, θ_2). Two angles of a salt bridge $(i–j)$, $\angle C\beta_1 C\alpha_1 C\alpha_2$ (θ_1) and $\angle C\beta_2 C\alpha_2 C\alpha_1$ (θ_2), were used to describe the geometric interaction of two oppositely charged groups according to the relative orientation of the two residues' Cα–Cβ vectors (Figure 1A).

Salt-bridge pair index of weighted electrostatic attraction model

To find the effective pairings for a salt-bridge design, we developed a weighted electrostatic attraction model (E_i), which is a scoring function, to measure the attractive interactions of residue i in a protein structure. E_i is defined as follows:

$$E_i = -w_{ij} \frac{Q(A_i)Q(A_j)}{r_{ij}} \tag{1}$$

where j represents the residues within a radius of 15 Å of i; A_i and A_j are the amino acids of residues i and j, respectively; $Q(A)$ is an

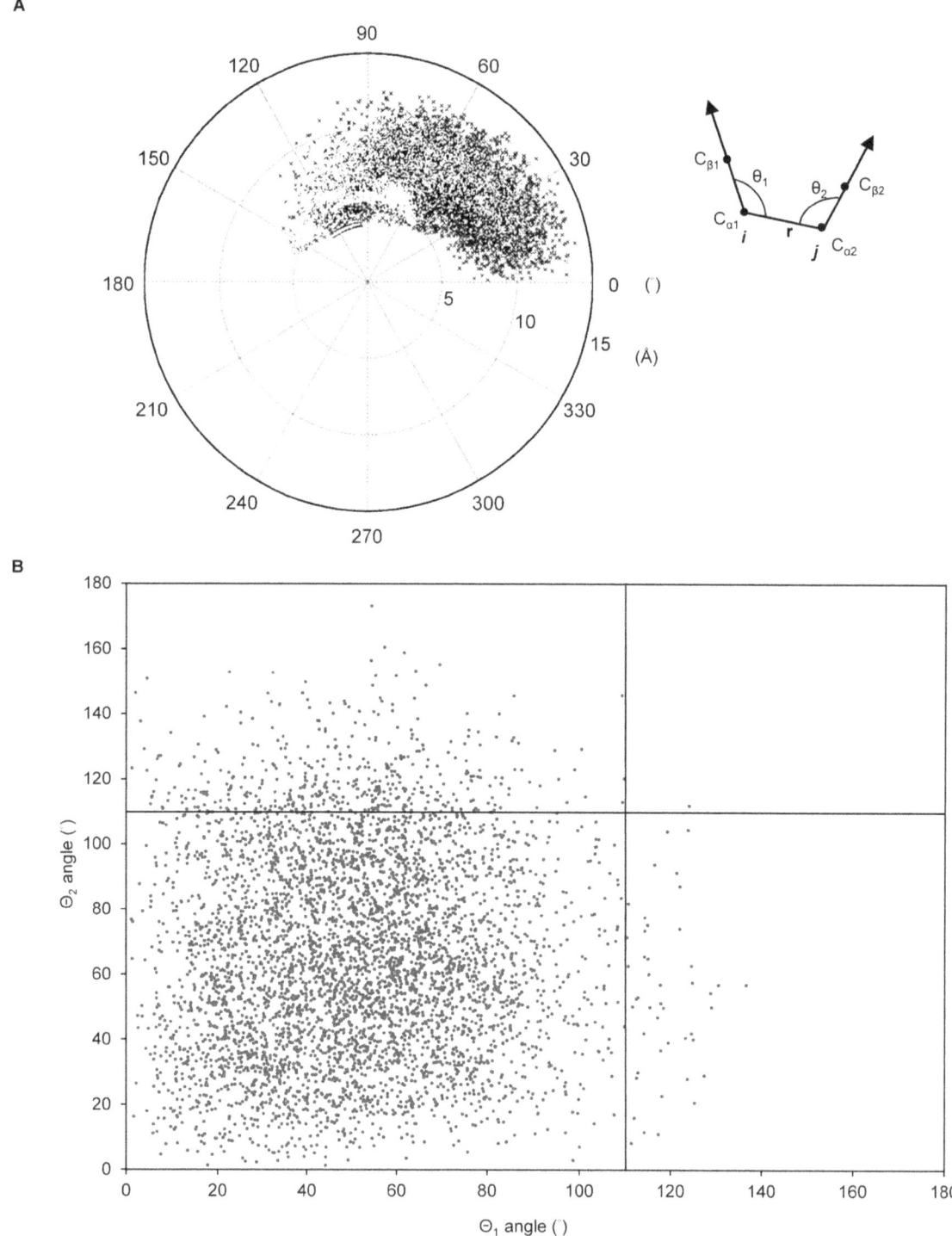

Figure 1. Spatial orientation and Cα–Cα distances of salt bridges on protein surfaces. (A) Statitical analysis of angles (θ_1, θ_2) of 10,556 salt bridges on the surfaces of 6,493 X-ray protein structures. Two angles of a salt bridge (*i–j*), $\angle C\beta_1 C\alpha_1 C\alpha_2$ (θ_1) and $\angle C\beta_2 C\alpha_2 C\alpha_1$ (θ_2), were used to describe the charge group interaction based on the relative orientation of the two residues' Cα–Cβ vectors as indicated. All of the angles are in the range of 0° to 180° (θ_1 and θ_2 color in black and gray). The length of radius are corresponding to Cα–Cα distance (Å). (B) The scatter plot shows the angles (θ_1, θ_2) of the salt bridges at a backbone distance >7 Å, which are restrained within 0° to 110°.

electrostatic charge function that depends on the charge of the amino acid: $Q(A) = 1$ if $A = $ Lys or Arg; $Q(A) = -1$ if $A = $ Asp or Glu; and $Q(A) = 0$ if $A = $ other types of amino acids. w_{ij} is an empirical parameter generated by the statistical analysis of 10,556

surface salt bridges in the data set. These salt bridges were first classified based on the following parameter set: (1) Cα–Cα distance (r_{ij}) of a salt bridge; (2) oppositely charged amino acid types (i.e., A_i and A_j); and (3) the secondary structures of residues *i* and *j* locations. A w_{ij} set for each type of salt bridges depends on the frequency of occurrence in the 10,556 surface salt bridges. Therefore, these w_{ij} values corresponding to different frequency of specific types of salt bridges could be seen as a salt-bridge pair index to help determine the designable salt bridges with a high probability of formation. In addition, since the repulsive interactions were not taken into consideration, the frequency of repulsive pairs were not analyzed: $w_{ij} = 0$, if $Q(A_i)Q(A_j) = 1$.

Basic rationale of our approach

To design salt bridges on a protein by using our approach, the following procedures were performed. The residues with the following characteristics were isolated: RSA >35%; Z-score of B-factor (z_B-factor) >0; Z-score of rWCN (z_rWCN) >0; and conservation <4. The idea of this pre-filtering was to select the potential residues that can be designed to improve protein thermal stability. Note that only polar residues (Ser, Thr, Cys, Asn, and Gln) and charged residues (Lys, Arg, Asp, and Glu) were considered because polar residues are scarcer in thermophilic proteins compared with their homologous counterparts [38]. The WCN of an atom is the sum of the inverse squares of the distances between the atom and other atoms; rWCN is the reciprocal of WCN [27,28]. The rWCN value was utilized in the selection of potential residues for designing salt bridges. The rWCN and B-factor of a residue were computed using those values of the Cα atom only. The values were further normalized to their respective Z-scores (z_B-factor and z_rWCN). The normalization was calculated as $Z = (u - \bar{u})/\sigma_u$, where $\bar{u}$ and σ_u are the mean and standard deviation of u. The value of u is either the B-factor or the rWCN value. The evolutionary conservation of a residue was also taken into consideration because low conservation residue may indicate less importance in determining the structural conformation and function of proteins [39]. Therefore, low conservation residues can be mutated to design salt bridges without affecting protein folding and activity. The conservation scores obtained from the Consurf data set were used in the selection [39]. Each position fit all of the above criteria was then independently assumed as K, R, D, and E and scored by Eq. (1) to determine the most efficient pairing of each position with one of the surrounding charge residues within a radius of 15 Å. If a multimeric protein complex was obtained, the residues across the interface of another inter-subunit within a radius of 15 Å were also considered. Before the scoring, the surrounding charge residues within 15-Å radius of each position were located. Therefore, the putative pairs of each position can be obtained. If the putative pairs of a position were all at an interval of <5 residues of the primary sequence, they were not scored because the pairings may not be influential for increasing thermal stability. Besides, to ensure that the side chains of the two pairing residues were oriented toward each other, the angles of one pair were restricted to a limited degree. The angular criteria were established from the statistical analysis of the surface salt bridges in the data set and discussed in the results and discussion section. Only the pairs that fit the angular criteria were scored by Eq. (1). The most efficient type of charged amino acid for each position to form the highest-scoring salt bridge was identified to perform site-directed mutagenesis.

Site-directed mutagenesis and Protein overexpression

A plasmid encoding the BglA (pBglA) was used as a site-directed mutagenesis template. Fifteen pBglA plasmid amplification cycles were performed with the primers containing the mutated nucleotides and KOD DNA polymerase (Takara Inc.). The amplified material was digested with Dpn I and the digested mixtures were used to transform Escherichia coli DH5α. Kana-mycin-resistant colonies were then isolated on a LB plate (MDBio Inc.) containing 25 µg mL^{-1} kanamycin. The mutated plasmids were extracted and used to transform E. coli BL21 Star (DE3) for protein overexpression. Cells were grown in LB broth containing kanamycin (25 µg mL^{-1}) at 37°C with shaking to optical density at 600°nm (OD$_{600}$) of 0.4. Isopropyl β-D-1-thiogalactopyranoside (IPTG) of 0.5 mM was added to induce the BglA overexpression overnight at 25°C.

Protein purification

Cell destruction was performed using an Avestin Emulsiflex Homogenizer EF-C3 (Utek Process) under 20,000 psi pressure for five cycles. Insoluble cell debris was removed by centrifugation at $16,000 \times g$ for 15°min at 4°C. The supernatant was passed through a HisTrap column (GE Healthcare Inc.). After the column was washed with a phosphate binding buffer (10 mL, 20°mM) containing imidazole (20 mM), elution was performed using 20°mM phosphate buffer containing imidazole (100 mM). The elution buffer was replaced with 20°mM PBS buffer by using a HiTrap desalting column (GE Healthcare Inc.). His-tag was removed by incubating the purified protein with thrombin (GE Healthcare Inc.) at 20°C for 10°h. The resulting sample was reloaded on the HisTrap column. The eluted protein was collected and concentrated using a 10,000 MWCO centrifugal filter (Millipore Inc.). The concentrated protein was purified by gel filtration (HiLoad 26/60 Superdex 200 prep-grade, GE Healthcare Inc.) using citrate phosphate buffer (50 mM, pH 7) to elute 400 kDa BglA.

Thermal denaturation and kinetic studies of BglA and mutants

The thermal stability and kinetics of the purified BglA and mutants were measured. The melting temperature (T_m) was determined by differential scanning calorimeter (VP-DSC, GE Healthcare Inc.) with protein samples (2.0 mg/mL) in a solution of 50°mM citrate phosphate buffer at pH 7.0. The scanning temperature ranged from 20°C to 70°C at a heating rate of 1°C/min. The resulting data were fitted to non-2-state model. 4-Nitrophenyl β-D-glucopyranoside (pNPG; Sigma) was used as an artificial substrate for detecting β-glucosidase activity. The kinetic assays were conducted at 37°C for 5.5 minutes in 50°mM citrate phosphate buffer, pH 7.0, using 1 µg.800 µl^{-1} enzyme and 0.02 − 3 mM pNPG, modified based on the previously described method [40]. The results were fit to the Michaelis-Menten equation for determination of the K_m and k_{cat} by the software of GraphPad Prism 5.

Results and Discussion

Geometry preferences of salt bridges

In total, the data set consisted of 32,096 salt bridges from 6,493 X-ray protein structures. To analyze the specific geometry of salt bridges on the protein surface, we further classified these salt bridges as a buried or exposed state based on the RSA value (Table 1). The analysis revealed that the K–E pairs favored the formation of exposed salt bridges, whereas R–D pairs were more common in buried salt bridges. The result may suggest that stability of K–E pairings were higher on protein surfaces than in buried regions. Hence, the analyses were consistent with those in previous reports, in which K and E are more common on the surface of thermophilic proteins than on mesophilic homologues [38]. The 10,556 salt bridges at the exposed state were regarded as surface salt bridges and used for the following statistical analysis (Table 1).

Table 1. Relative solvent accessibility of salt bridges in 6,493 X-ray protein structures.

Location	Salt-bridge pair				Total
(RSA)	K–D	K–E	R–D	R–E	
All	8040	9479	7240	7337	32096
(≥0%)[a]	25.00%	29.50%	22.60%	22.90%	
Buried	657	525	750	730	2662
(<9%)[a]	24.70%	19.70%	28.20%	27.40%	
Exposed	2947	3764	1742	2103	10556
(>35%, >25%)[b]	27.90%	35.70%	16.50%	19.90%	

Abbreviation: K, Lysine; R, Arginine; D, Asparagine; E, Glutamine; and RSA: Relative solvent accessibility.
[a]RSA of both residues of a salt bridge are indicated.
[b]RSA of one residue of a salt bridge is >25% and the other is >35%.

Figure 1A shows the $C\alpha$–$C\alpha$ distance and angular distributions of the 10,556 surface salt bridges. The angles of the surface salt bridges ranged from 0° to 180°, which could be used as an angle constraint. The longest $C\alpha$–$C\alpha$ distance was observed in the R–E pair in PDB 1HBN (R535–E39 coil–helix at 14.235 Å, θ_1:27.16°, and θ_2:9.45°), whereas the shortest was observed in the K–D pair in PDB 1D2T (K90–D91 coil–coil at 3.719 Å, θ_1:97.08°, and θ_2:119.25°). The results suggested that the range of possible interaction in a salt-bridge extended to 15 Å of $C\alpha$–$C\alpha$ distance approximately. Hence, we set a 15-Å radius of a position as the considered area in the search for the surrounding charge residues. Figure 1B shows the angles (θ_1, θ_2) of the salt bridges with $C\alpha$–$C\alpha$ distance >7 Å. Based on statistical analysis, the two angles (θ_1, θ_2) of 89.7% salt bridges were both restrained within 0° to 110° when the $C\alpha$–$C\alpha$ distance (r) was >7 Å. For the remaining 10.2% salt bridges, when one of the angles was >110°, then the second angle should be <110°, with only one exception. Therefore, we set the angular restraint of <110° for at least one angle (θ_1 or θ_2) if the $C\alpha$–$C\alpha$ distance is >7 Å to ensure the close proximity of the ionic groups of the salt bridges to each other.

We further classified the surface salt bridges based on the secondary structure locations. Figure 2 shows the frequency distributions of K–D, K–E, R–D, and R–E salt-bridges corresponding to different secondary structure pairs. The highest frequency occurred at the helix-helix structure for all pairings of the salt bridges except the R–D pairings. Thus, such salt bridge formation at the helix-helix structure may efficiently stabilize protein structures. Based on previous studies, the formation of R–E salt-bridge at the helix–helix structure may account for much of the stabilization of certain thermophilic proteins [41]. To determine the exact value standing for the preference of salt bridge formation of a particular type of ion pair, we constructed a matrix of empirical weights (w_{ij}) based on the frequency of each type of salt bridges. To calculate the frequency of each type of salt bridges, the 10,556 surface salt-bridge pairs were first classified based on pairing residues, $C\alpha$–$C\alpha$ distances, and secondary structure locations. Table S1 shows the empirical weight of each salt bridge type depending on the frequency. The weight matrix was then used in the weighted electrostatic attraction model to score each pairings. Finally, the angular constraints and the weighted electrostatic attraction model could be used in the last steps of identifying the pairings and the substitutions after the selection of potential positions for salt-bridge design (Figure 3).

Potential pairings in intra- and inter-subunits

The BglA obtained from *B. polymyxa* was used as a polymeric model for the design of inter- and intra-salt bridges to enhance thermal stability. Polar and charge residues were considered as the pre-filtering criterion in the isolation of potential positions for different reasons. Polar residues are reported to be scarcer in thermophilic proteins, and charged residues can be selected for charge optimization; for example, to avoid electrostatic repulsion [38,42]. Proteins from thermophiles and hyper-thermophiles exhibit more frequently networked salt bridges than proteins from the mesophilic counterparts [43]. Hence, increasing the thermal stability of proteins by optimization of charge–charge interactions is a good strategy for an evolutionary solution. Besides, the criteria of RSA >35%; z_B-factor >0; z_rWCN >0; and conservation < 4 were also used as pre-filters for isolation of residues on the surface of a protein, conformational fluctuation above the average, packing density below the average, and low conservation, respectively [27,28,36,39]. The characteristics used in the pre-filtering of all the 447 BglA residues are shown in Table S2. Thirty four positions that fit the pre-filtering criteria in the BglA were isolated (Pre-filters: STCNQ/KRDE: 185, RSA >35%: 67, z_B-factor >0:50, z_rWCN >0:49, and conservation <4:34) as shown in Table S3. To determine the potential pairings for each 34 positions, we located the charge residues within a radius of 15 Å of the position. The putative pairs of the 34 positions were separately identified as in Table S4. Only 11 positions that could be paired with the surrounded residues were found according to the angular constrains of θ_1 and θ_2 generated from the statistical analysis of the surface salt-bridges (Figure 1) and at least one pair for each position that met the sequence separation criteria (≥5 residues). The putative pairs for each 11 positions were separately identified and scored using Eq. (1). To demonstrate how to identify the charge amino acids for substitutions at the 11 positions, Table S5 shows the theoretical scores of the putative pairings between each of the 11 positions and their surrounded charge residues. The predicted pair equivalent to the highest-score pair of each position was the pair with the highest preference for salt-bridge formation. The suggested amino acids for mutations were the one that fit the predicted pair at each position. For example, the suggested mutations would be E96K because E96K–D28 is the predicted pair of E96 position. However, if a position's highest-scoring pair exhibited a short sequence separation (i.e. <5 residues), the corresponding mutation was not adopted. Moreover, an alternative charged substitution was performed when that substitution for alternative pair formation (i.e., sequence separation ≥5 residues)

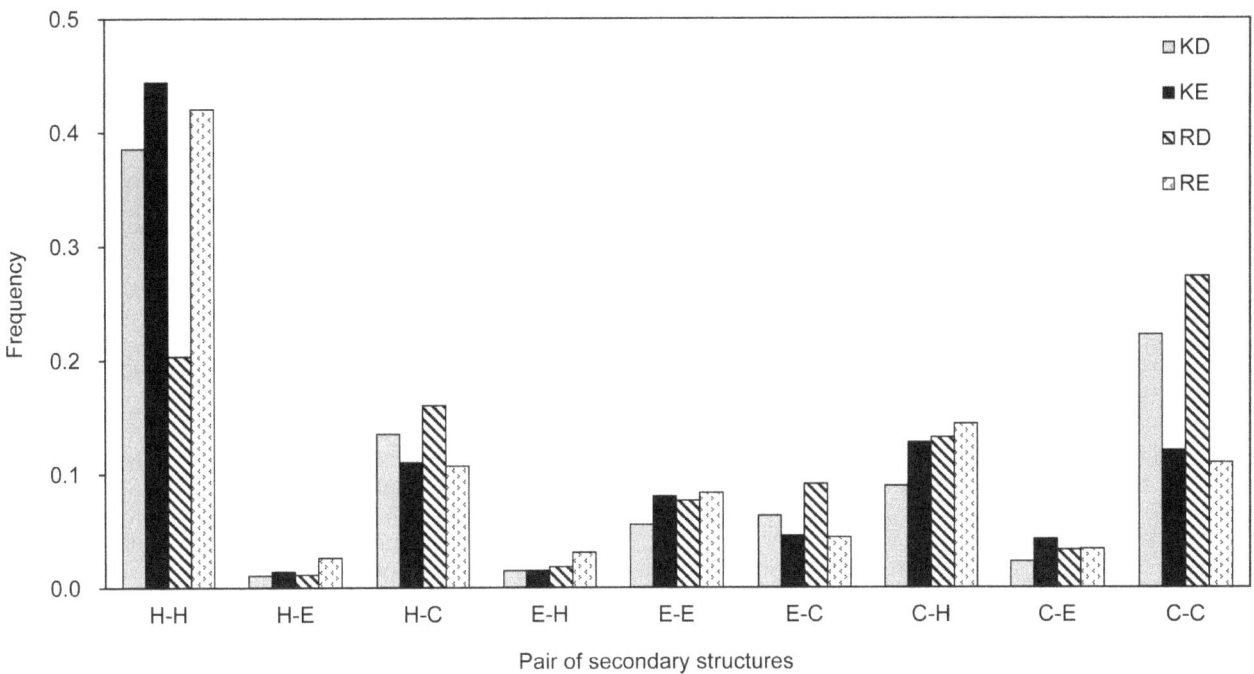

Figure 2. Frequency distributions of surface salt bridges at different pairs of secondary structures. Four types of salt-bridge pairings, Arg/Asp (R/D), Arg/Glu (R/E), Lys/Asp (K/D), and Lys/Glu (K/E), were considered in the statistical analysis.

was scored higher (i.e., higher preference for forming the salt bridge) than the same substitution for inefficient pair formation (i.e., sequence separation <5 residues). The highest-scoring pair was inferred when two or more of such alternative pairs were found. One example in Table S5 was the S440E–R436 pair. The pair exhibits a pair interval <5 residues of the primary sequence and was the highest-scoring pair for the R436 position. Moreover, the S440R–E70 pair's score was higher than that of the S440R–

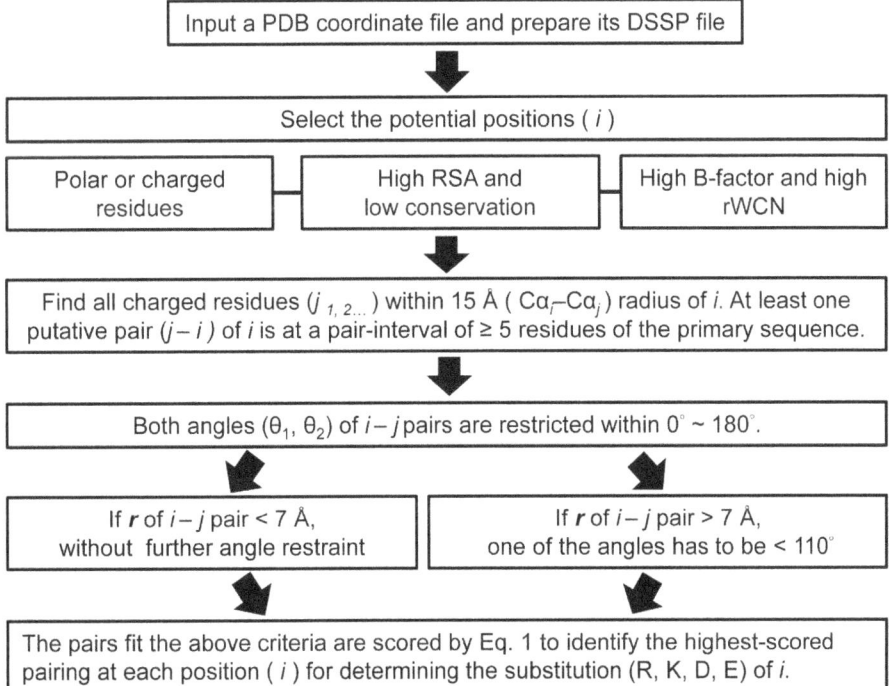

Figure 3. Flowchart of salt-bridge design in this study. Identification of potential positions and mutation types for a given protein structure are demonstrated.

R436 pair. Therefore, the suggested S440 mutation was S440R for the S440R–E70 pair formation.

Six of the eleven positions in Table S5 (i.e., N47, N88, Q95, T273, E360, and N437) exhibited their highest-scoring pairs to be the pairs with a short sequence separation (<5 residues). Moreover, no alternative substitutions were found for these positions except for the N437 position. Besides, two positions, E96 and Q277, each was found with two suggested mutations for two predicted pairs that showed equal scores (Table S5). Hence, eight mutations for the eight predicted pairs consisted of two inter-subunit pairs (i.e., T231K–D388 and N437K–D49) and six intra-subunit pairs (i.e., E96K–D28, E96R–D28, Q141K–E148, Q277D–R137, Q277E–R137, and S440K–E70) were constructed to perform the experimental thermal stability test. In addition, to validate the pre-filtering criteria set for the salt-bridge design, the Q216 was used as a negative control position because the RSA, z_B-factor and z_rWCN values of Q216 are not qualified but close to the criteria. The z_B-factor and z_rWCN of the Q216 are close but less than 0 and the RSA is less than 35% as shown in Table 2. The D289 was the qualified charge residues to bridge with the Q216R (Cα–Cα distance: 5.52 Å; scores of D289–Q216K/R/D/E: 0.04/0.07/0/0; θ_1 and θ_2 of Q216R–D289 are 60.36°°and 11.61°) as shown in Table S5. Therefore, the T_m resulted from the Q216R–D289 pair was expected not to be higher than that of the wild type enzyme. Figure 4 shows the locations of the eight predicted pairs and Q216R–D289 pair in the BglA structure. Two residues (T231, N437) at the A/D and A/E interfaces and four residues (E96, Q141, Q277, and S440) of the intra-subunit pairs, as well as Q216, were subjected to site-directed mutagenesis. The eight suggested mutations of the six positions (E96, Q141, T231, Q277, N437, and S440) were E96K, E96R, Q141K, T231K, Q277D, Q277E, N437K, and S440K. The structural character-istics of the seven positions (E96, Q141, T231, Q277, N437, S440, and Q216) are summarized in Table 2. All the six positions (E96, Q141, T231, Q277, N437, and S440) qualify the following characteristics of the pre-filtering criteria: polar or charged residue, RSA >35%, z_B-factor >0, z_rWCN >0, and conser-vation <4. Figure S1 shows the z_rWCN and the z_B-factor profiles of 1BGA: chain A, which demonstrate that the z_rWCN and z_B-factor of the six positions are higher than the average. On the contrary, the Q216, used as a negative control position, exhibits the z_B-factor and z_rWCN slightly lower than the average. The profiles also demonstrate that Q216 is a relatively stable and buried position around the local sequence region.

Thermal stability and kinetics of BglA mutants

The ΔT_m of the enzymes was obtained from DSC experiments, in which the ΔT_m values and the melting curves of the wild type and the mutants are compared in Table 3 and Figure S2, respectively. The irreversible denaturation of the enzymes could cause the uneven baselines of the melting curves (Figure S2). Therefore, the melting temperatures described below were apparent melting temperatures. Six of the eight suggested mutants showed greater stability than that of the wild type based on the ΔT_m value (i.e., E96K, E96R, T231K, Q277E, N437K, and S440K) while the two mutants (Q141K and Q277D) showed similar Tm as that of the wild type. The ΔT_m caused by the six suggested mutations were 3.7, 4.8, 1.2, 0.7, 8.8, and 1.3°C (E96K, E96R, T231K, Q277E, N437K, and S440K, respectively). Each predicted position that successfully increased the Tm of the BglA through the suggested mutations was individually mutated to all the four charged amino acids, K, R, D, and E, via site-directed mutagenesis. These additional mutations were conducted to further confirm the predicted accuracy and investigate the relation

between the increase of melting temperature (ΔT_m) and the predicted score of each mutant. The ΔT_m values resulted from the suggested charged mutations that leaded to increased T_m and the same type of charge mutation at each of the E96, T231, Q277, N437, and S440 positions were higher than that of the opposite type of charge mutations except the Q277D and Q277R mutations (Table 3). The latter two mutations displayed a similar Tm. These findings suggested that the prediction of the pairings according to the scores calculated by Eq. (1) of the K, R, D, and E substitutions at each five positions were in agreement with the ΔT_m results. Besides, the k_{cat}/K_m values of the thermostable mutants were also measured as in Table 3. Moreover, the mechanisms of thermal adaptation of each position (E96, Q216, T231, N437, Q277, and S440) were varied and discussed in the following sections.

Case 1: E96, reversing the negatively charged residue

The E96 residue exhibits high B-factor and rWCN values (z_B-factor: 2.46, z_rWCN: 0.81), which indicate a large fluctuation and a low packing density, respectively (Table 2). Based on Table S5, E96 in a helix revealed the highest probability of forming an intra-molecular salt bridge with a neighboring D28 residue in a loop structure by reversing from negatively to positively charged K or R residues. The D28 residue is also in a flexible region as indicated by the z_B-factor. Therefore, the bridging of K96 or R96 with D28 could stabilize two flexible regions to increase protein rigidity. In previous studies, the E96K mutation was also detected by random mutation [44]. The mutant was crystallized to demonstrate the formation of a link between K96 and D28 [45]. This crystalline structure also indicates a twist conformation caused by E96K mutation, where the K96 side chain turns to the D28 side chain. We also found that ΔT_m caused by E96R mutation was even higher than that of E96K mutation. The E96R mutant showed a ΔT_m of 4.8°C compared with 3.1°C of E96K (Table 3). On the contrary, when a negatively charged Aspartate was introduced at E96, the T_m of the BglA dropped 1°C. Thus, thermal adaptation occurred more effectively in this environment than in the original environment in the event of reversal of a negatively charged residue (E96) to its positively charged residues (K96 or R96). Previous research has also mentioned that charge reversal of a residue could convert a destabilizing effect to a stabilizing effect because charge reversal alters the core of the interactions; as a result, such destabilized interactions are stabilized [42]. A four-fold increase of k_{cat}/K_m value was found in E96K mutant, whereas that of E96R mutant remained almost unchanged.

Case 2: Q216, with low B-factor and WCN at a slightly buried state

The Q216 residue at a slightly buried state exhibits lower RSA, B-factor, and rWCN values (RSA: 0.16, z_B-factor: −0.14, and z_rWCN: −0.19) than that prescribed in the pre-filtering criteria (Table 2); the surrounded D289 could be paired with the Q216R according to the scores calculated by Eq. (1). Thus, Q216R–D289 was used as a negative control pair. The ΔT_m of Q216 mutants were lower or similar than that of the wild type enzyme, which was consistent with our predictions (Table 3). Thermal stability of the Q216D and Q216K variants showed 0.1 and −0.5°C of ΔT_m while a large decrease in T_m of Q216E and Q216R (−2.2 and −3.5°C) mutants were found, which can be explained by the repulsion that arised from the larger side chains of E and R than that of D and K. The large decrease in T_m caused by the Q216E and Q216R mutations could be expected from the slightly buried and crowded position indicated by the low RSA and z_rWCN of

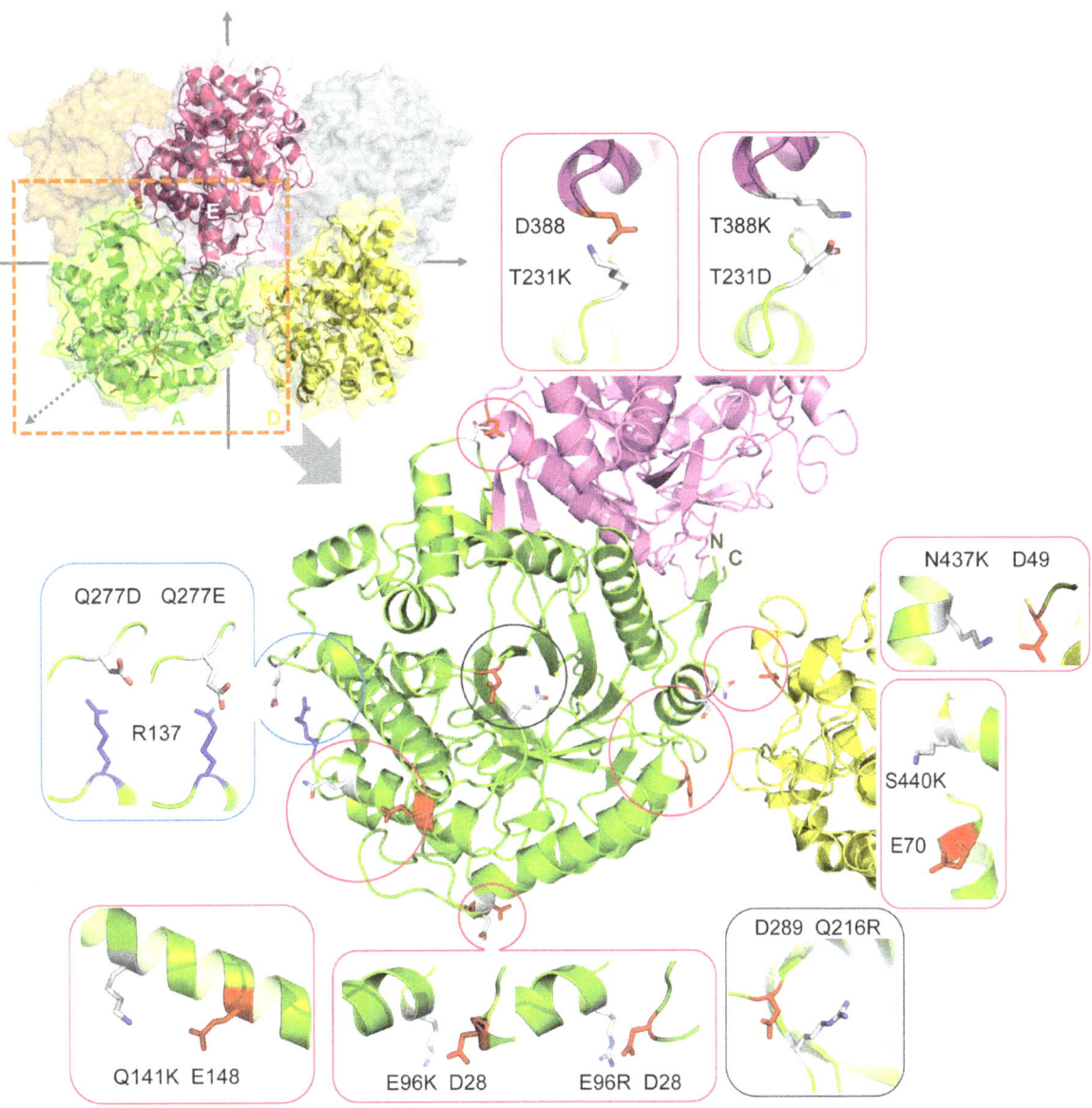

Figure 4. Locations of eight predicted pairs and one control pair in the octameric structure (PDB: 1BGA). A, D, E subunits are represented in green, yellow, and purple, respectively. The N437–D49 pair is between subunits A and D; the T231–D388 pair is between subunits A and E; whereas E96–D28, Q141–E148, Q277–R137, S440–E70 and control pair Q216-D289 are intra-subunit pairs. The predictive pairs as well as T231D–D388K and Q216R–D289 pairs were simulated by Pymol program. Each pair is magnified in an independent window showing secondary structure elements and the paired residues (red for negatively charged residues and blue for positively charged residues). The neighboring charge residue of each position (E96, Q141, T231, Q277, N437, S440, and Q216) suggested the oppositely charged substitutions for forming salt bridges.

Q216. Therefore, the ΔT_m of the Q216R mutant was in agreement with the pre-filtering criteria set for the isolation of designable positions. Bleicher et°al. have proposed that the existence of salt bridges in the hydrophobic core of a mesophilic laminarinase (versus its absence in a hyperthermophilic laminarinase) may be responsible for the differences in the thermal stability between the two enzymes [46]. In the mesophilic enzyme, the existence of charge-charge interactions permeating the hydrophobic core of the enzyme actually destabilizes the structure by facilitating water penetration into hydrophobic cavity. These findings could support the ΔT_m result caused by the putative pairing of Q216R–D289 in this study.

Case 3: T231, polarity reversal of a putative electrostatic pair

The T231 residue locates at a coiled structure between two subunits of A and E subunits shows a large rWCN value (z_rWCN: 3.28) in Table 2. When an exposed polar T231 residue was mutated into a positively charged residue (T231K), T_m was slightly increased (ΔT_m: 1.2°C), whereas the T231R mutant showed a similar thermal stability as that of the wild type enzyme (Table 3). On the contrary, the T_m of T231D and T231E mutants was 3.8 and 3.2°C lower than that of the wild type. Originally, K231 or R231 was expected to form a salt bridge with D388 across the interface (Table S5). We suspected that the small

Table 2. Characteristics of positions in BglA for mutations.

Position	SS	RSA	z_B-factor	z_rWCN	Conservation	Location (subunit)
E96	H	0.46	2.46	0.81	3	Intra (A)
Q141	H	0.70	0.16	0.11	3	Intra (A)
Q277	C	0.52	1.12	1.96	2	Intra (A)
S440	H	0.54	0.43	0.66	2	Intra (A)
T231	C	0.96	0.91	3.28	1	Inter (A/E)
N437	H	0.54	0.29	0.53	1	Inter (A/D)
Q216	E	0.16	−0.14	−0.19	4	Intra (A)

Secondary structure (SS: H, helix; E, beta-sheet; C, coil), relative solvent accessibility (RSA), Z-scores of B-factor (z_B-factor) and reciprocal of weighted contact number (z_rWCN), conservation score, and inter- or intra- subunit location are indicated.

increase in T_m caused by T231K mutation was triggered by an intra-subunit electrostatic interaction instead of an inter-subunit electrostatic interaction. Further considering the local environment, we found that the positively charged residue K231 could stabilize the negatively charged environment and interact with one of the three negatively charged residues (E233, E234, and D235) from the neighboring helix of the same subunit at Cα–Cα distances of 6.6, 8.5 and 7.0 Å, respectively. The three negatively charged residues are also closely located to 231 position in the primary sequence. Hence, these negatively charged residues

Table 3. ΔT_m and kinetic parameters of wild-type BglA and mutant enzymes.

No.	Mutants	ΔT_m (°C)	k_{cat} (sec^{-1})	K_m (mM)	k_{cat}/K_m (sec^{-1} mM^{-1})
0	Wild-type	0.0[a]	44.73±0.81	1.21±0.03	36.89±0.38
1	E96K[b]	3.7	28.42±0.15	0.19±0.01	149.00±4.97
	E96R[b]	4.8	21.19±0.50	0.47±0.02	44.81±1.08
	E96D	−1	n.d.	n.d.	n.d.
2	Q141K[b]	−0.3	n.d.	n.d.	n.d.
3	Q216K	−0.5	30.62±0.54	0.47±0.04	64.90±4.90
	Q216R	−3.5	26.02±0.12	0.46±0.00	56.87±0.19
	Q216D	0.1	n.d.	n.d.	n.d.
	Q216E	−2.2	n.d.	n.d.	n.d.
4	Q277K	0.4	n.d.	n.d.	n.d.
	Q277R	0.0	n.d.	n.d.	n.d.
	Q277D[b]	−0.1	1.04±0.02	18.06±0.29	17.41±0.12
	Q277E[b]	0.7	0.85±0.31	15.68±3.50	19.00±2.72
5	T231K[b]	1.2	36.32±1.36	0.61±0.07	59.31±4.96
	T231R	−0.3	n.d.	n.d.	n.d.
	T231D	−3.8	n.d.	n.d.	n.d.
	T231E	−3.2	n.d.	n.d.	n.d.
	T231D/D388K	4.8	32.76±0.76	0.19±0.01	174.90±9.48
6	N437K[b]	8.8	34.85±0.18	0.50±0.01	69.74±1.45
	N437R	5.3	10.27±0.51	0.40±0.06	25.57±2.39
	N437D	−0.4	n.d.	n.d.	n.d.
	N437E	−0.4	n.d.	n.d.	n.d.
7	S440K[b]	1.3	1.63±0.03	24.39±0.30	14.99±0.14
	S440R	2.4	3.96±0.32	31.78±1.90	8.04±0.17
	S440D	−0.7	n.d.	n.d.	n.d.
	S440E	0.5	n.d.	n.d.	n.d.
8	E96R/T231D/D388K/N437K	15.7	10.27±0.51	0.40±0.06	25.57±2.39

[a]T_m of wild-type protein was detected to be 38.5°C.
[b]Suggested mutation.
n.d.: Not determined.

could interfere the expected inter-subunit attraction (T231K–D388) by drawing the positive charge of K231 side chain away from the negative charge of D388 side chain in the opposite direction. Therefore, we hypothesized that the reversal of the charges of the two residues, that is, K231 and D388 into D231 and K388, is more likely to promote the salt bridge formation at the interface due to the interferences are dismissed. Thus, the double mutant (T231D–D388K) was constructed to investigate the hypothesis, which resulted in a large increase in T_m of 4.8°C. This change was significantly greater than that of T231K mutation (Table 3). The results suggested that the negatively charged residues closely located at the helix next to residue D231, may hinder the flexibility of D231 and drive the side chain of D231 to rotate to cross the interface, which formed a bridge with the newly reverse-charged variant K388 at another subunit. Therefore, the T_m of T231K and T231D–D388K mutants may also potentially demonstrate that an inter-subunit electrostatic interaction could increase protein thermal stability more efficiently than an intra-subunit interaction of polymeric proteins. A five-fold increase in the k_{cat}/K_m value caused by the T231D–D388K mutation was observed, whereas the k_{cat}/K_m value of T231K remained almost unchanged (Table 3). The results suggested that the inter-subunit interaction may have influenced the enzyme kinetics of the BglA.

Case 4: N437, elongating the side chain across the interface

The N437 residue locates in a helix structure was predicted to form an inter-subunit salt bridge with the D49 residue in a coil structure between the A and D subunits by the N437K substitution (Table S5). The N437K and N437R mutations resulted in an increase in T_m of 8.8 and 5.3°C, respectively (Table 3). The N437K mutation was also found as a thermostable mutation by random mutation in a previous study [44]. Based on the BglA structure [33], there is no direct interaction between the N437 residue of subunit A and D49 of subunit D. In addition, the distance between the amide group of the N437 side chain and the carboxyl group of the D49 side chain is 4.5 Å which is longer than 4 Å required for the salt-bridge formation. Therefore, the substitution of R437 or K437 for N437 may shorten the distance by elongating the length of the residue side chain. Therefore, the putative salt bridge between K437 or R437 and D49 may form across the interface by extending the range of the side chain interaction. By contrast, the T_m values of N437D and N437E mutants were lower than that of the wild type. During the BglA purification process, the protein gel-filtration data revealed that the N437D and N437E mutants both eluted differently sized tetramers from the native octamer (data not shown). Hence, this position was suggested to function crucially in protein oligomerization.

Case 5: Q277 and S440, introducing intra-subunit pairings at long distances

The Q277 in a coil structure was predicted to form an intra-subunit salt bridge with the R137 in a helix structure through the Q277D or Q277E mutation (Table S5). The predicted pairs (Q277D–R137 and Q277E–R137) exhibit a 10.96 Å Cα–Cα distance. The Q277E mutant showed a higher Tm (ΔT_m: 0.7°C) than that of the Q277D mutant (ΔT_m: −0.1°C) (Table 3). Hence, the results suggested that instead of aspartate, a longer glutamate side-chain was more suitable to reach the R137 side chain.

The S440 and E70 residues at a long interval (Cα–Cα: 10.31 Å) were also predicted to interact with each other through the S440K

mutation (Table S5). Both the S440K and S440R mutations resulted in increased ΔT_m values (i.e., 1.3 and 2.4°C, respectively) (Table 3). However, both mutations caused a 0.5-fold drop of $k$cat/Km. A decrease of $k$cat/Km caused by the Q277E and Q277D mutations was also observed.

Case 6: Combining the thermostable mutations

The combination of the E96R, T231D–D388K, and N437K mutations with each showed an increased $k$cat/Km value compared with that of the wild type provided a maximum ΔT_m of approximately 15.7°C (Table 3). The large increase in T_m observed in the combined mutant emphasized the importance of increasing the number of salt bridges in enhancing protein thermal stability [38]. The k_{cat}/K_m value was only slightly less than that of the wild type at 37°C.

Previous simulation studies of mesophilic and hyper-thermophilic proteins have shown that protein unfolding may be initiated at sites prone to large thermal fluctuations; an example of such sites include the loop region in rubredoxin [47,48], whose flexibility allows water to access and unfold the protein structure. In the present study, the novel salt bridges were constructed within the flexible region to minimize fluctuations; hence, the development of resistance to high temperatures was promoted [49]. Previous reports have also indicated the relevance of high rigidity as a factor that influences the integrity of the native folded structure by increasing the packing density on a protein surface [27–29], which is consistent with the strategy of constructing salt bridges at positions of low contact number and high solvent accessibility in this study.

The two mutation pairs, T231D–D388K and N437K–D49, in the A/E and A/D inter-subunit could increase multimeric stabilization. The ΔTm of T231D–D388K mutant was 4.8°C, and that of N437K–D49 was 8.8°C higher than that of the wild type. According to the previous studies, inter-subunit interactions are known to preserve oligomer structures and increase resistance protein thermal stability [50,51]. Besides, the k_{cat}/K_m values of these two mutants were approximately five- and two-fold greater than that of the wild-type, respectively (Table 3). A similar observation was reported that the inter-subunit salt bridge of quinone reductase obtained from *Bacillus subtilis* affects not only protein oligomerization but also catalytic activity [51].

In the present study, the weighted electrostatic attraction model could be used to calculate the interaction between two oppositely charged residues to identify the preferred pairings and the possible substitution in a salt-bridge design. The most stable electrostatic pair at each of the three identified positions increased the BglA thermal stability by charge addition, charge reversal, and polarity reversal. The computational prediction scores clearly identified the positively or negatively charged substitutions with increased T_m, which suggested that the model could be useful in the design of thermostable proteins.

Accordingly, various rational techniques have been employed to improve protein thermal stability, including consensus-guided mutagenesis and the comparison of thermophiles and mesophilic homologues [52–55]. Most of these methods adopt the evolutionary analysis of multiple sequence alignment to substitute the conserved residue for the non-conserved ones at an aligned position or to substitute a residue from thermophiles for a residue from mesophiles. The bioinformatics method used in this study managed to recognize the potential charge substitutions for the low conservation residues from the surrounding electrostatic environment.

Conclusion

Our approach recognized effective residues and suitable substitutions for designing salt bridges to improve protein thermal stability. Statistical analysis of 10,556 surface salt bridges on 6,493 X-ray protein structures was performed to reveal the preferred geometric characteristics of specific types of salt bridges. The designed weight matrix was used in a weighted electrostatic attraction model to identify the effective pairings. We successfully designed electrostatic pairs at five positions on a mesophilic β-glucosidase, which was experimentally verified. These individual electrostatic pairs may also provide insights into the thermal adaptation exhibited by salt bridges. Combining three electrostatic pairs generated an increase of 15.7°C in T_m. Thus, the method varied from the existing popular thought and could provide a simple alternative to increase the thermal stability of proteins.

Supporting Information

Figure S1   rWCN (black line) and B-factor (dotted line) profiles of 1BGA. A. The positions for mutations (i.e., E96, Q141, T231, Q277, N437, S440 and Q216) are indicated with hollow circles and diamonds corresponding to their z_rWCN and z_B-factor values, respectively. If two symbols are overlapped, only circle is indicated. The residues in the interface regions are indicated by black (A/D) and gray (A/E) thick lines on the horizontal axis. rWCN and B-factor values are normalized to their respective Z-scores (z_rWCN and z_B-factor values).

Figure S2   Melting curves of BglA and mutant proteins. The mutants showed in the legend are listed in the order of their Tm values from high to low. The melting curves of (A) T231, Q277, (B) E96, N437, S440, (C) Q216, and Q141 mutants as well as the mutant containing the three thermostable pairs (E96R–D28, N437K–D49, and T231D–D388K) were compared with that of the wild type.

Table S1   Empirical weight index of salt-bridge pairing preferences. These weight values (w_{ij}) corresponds to the frequency of the specific types of salt bridges in a total of 10,556 surface salt bridges classified by residue pairs, Cα–Cα distances, and secondary structure locations. The sum of all the values equals to 100.

Table S2   Characteristics of 447 residues of BglA for the pre-filtering. The parameters demanded in the pre-filtering contains the structural factors (RSA, B-factor, rWCN) and the conservation score obtained from obtained from the Consurf server.

Table S3   A total of 34 residues fit the pre-filtering criteria in BglA. The listed 34 residues fit the pre-filtering criteria as follows: polar or charge amino acids (STCNQ/KRDE), RSA>35%, z_bf>0, z_rWCN>0, and sequence conservation <4.

Table S4   Identification of potential pairs for each 34 qualified positions. The positions that possess at least one pairing qualified the angular and sequence separation constrains could be identified.

Table S5   Prediction of the charge-charge interactions between each 11 positions and their surrounding residues by scoring each interaction with Eq. (1). The predicted pair equivalent to the highest-score pair of each position; the suggested amino acids for mutations were the one that fit the predicted pair at each position. If a position's highest-scored pair is a $i–j$ <5 pair, then the corresponding mutation is not suggested. An alternative charged substitution is adopted if that substitution for the alternative pair ($i–j \geq 5$) formation was scored higher than the same substitution for the $i–j$ <5 pairs. The highest-scored pair is inferred if two or more of such alternative pairs are found.

Text S1   Data set for determining geometric preferences of salt-bridges. The data set contains 32,096 salt bridges isolated from 6,493 protein structures. Four types of salt-bridge pairs were considered: K–D; K–E; R–D; and R–E.

Author Contributions

Conceived and designed the experiments: CWL HJW JKH CPT. Performed the experiments: CWL HJW. Analyzed the data: CWL HJW JKH CPT. Contributed reagents/materials/analysis tools: JKH CPT. Wrote the paper: CWL HJW.

References

1. Demirjian DC, Moirs-Varas F, Cassidy CS (2001) Enzymes from extremophiles. Curr. Opin. Chem. Biol. 5: 144–151.
2. Rothschild LJ, Mancinelli RL (2001) Life in extreme environments. Nature 409: 1092–1101.
3. Delboni LF, Mande SC, Turley S, Hol WGJ, Rentier-Delrue F, et al. (1995) Crystal structure of recombinant triosephosphate isomerase from bacillus stearothermophilus. An analysis of potential thermostability factors in six isomerases with known three-dimensional structures points to the importance of hydrophobic interactions. Protein Sci 4: 2594–2604.
4. Chang C, Park BC, Lee DS, Suh SW (1999) Crystal structures of thermostable xylose isomerases from Thermus caldophilus and Thermus thermophilus: possible structural determinants of thermostability. J Mol Biol 288: 623–634.
5. Rosato V, Pucello N, Giuliano G (2002) Evidence for cysteine clustering in thermophilic proteomes. Trends Genet 18: 278–281.
6. Xiao L, Honig B (1999) Electrostatic contributions to the stability of hyperthermophilic proteins. J Mol Biol 289: 1435–1444.
7. Brian N. Dominy HMCLB, III (2004) An electrostatic basis for the stability of thermophilic proteins. Proteins: Structure, Function, and Bioinformatics 57: 128–141.
8. Jaenicke R (2000) Do ultrastable proteins from hyperthermophiles have high or low conformational rigidity? Proc Natl Acad Sci U S A 97: 2962–2964.
9. Reetz MT, Carballeira JD, Vogel A (2006) Iterative saturation mutagenesis on the basis of B factors as a strategy for increasing protein thermostability. Angew Chem Int Ed Engl 45: 7745–7751.
10. Liang HK, Huang CM, Ko MT, Hwang JK (2005) Amino acid coupling patterns in thermophilic proteins. Proteins 59: 58–63.
11. Chan CH, Liang HK, Hsiao NW, Ko MT, Lyu PC, et al. (2004) Relationship between local structural entropy and protein thermostability. Proteins: Structure Function and Bioinformatics 57: 684–691.
12. Alsop E, Silver M, Livesay DR (2003) Optimized electrostatic surfaces parallel increased thermostability: a structural bioinformatic analysis. Protein Eng 16: 871–874.
13. Zhou H-X, Dong F (2003) Electrostatic Contributions to the Stability of a Thermophilic Cold Shock Protein. Biophys J 84(4): 2216–2222.
14. Kumar S, Nussinov R (1999) Salt bridge stability in monomeric proteins. J Mol Biol 293: 1241–1255.
15. Bosshard HR, Marti DN, Jelesarov I (2004) Protein stabilization by salt bridges: concepts, experimental approaches and clarification of some misunderstandings. J Mol Recognit 17: 1–16.
16. Donald JE, Kulp DW, DeGrado WF (2011) Salt bridges: Geometrically specific, designable interactions. Proteins: Structure, Function, and Bioinformatics 79: 898–915.

17. Gribenko AV, Patel MM, Liu J, McCallum SA, Wang C, et al. (2009) Rational stabilization of enzymes by computational redesign of surface charge-charge interactions. Proc Natl Acad Sci U S A 106: 2601–2606.

18. Schweiker KL, Zarrine-Afsar A, Davidson AR, Makhatadze GI (2007) Computational design of the Fyn SH3 domain with increased stability through optimization of surface charge-charge interactions. Protein Sci 16: 2694–2702.

19. Vetriani C, Maeder DL, Tolliday N, Yip KS, Stillman TJ, et al. (1998) Protein thermostability above 100 degreesC: a key role for ionic interactions. Proc Natl Acad Sci U S A 95: 12300–12305.

20. Lam SY, Yeung RCY, Yu T-H, Sze K-H, Wong K-B (2011) A Rigidifying Salt-Bridge Favors the Activity of Thermophilic Enzyme at High Temperatures at the Expense of Low-Temperature Activity. PLOS Biol 9: e1001027.

21. Elcock AH (1998) The stability of salt bridges at high temperatures: implications for hyperthermophilic proteins. Journal of Molecular Biology 284: 489–502.

22. Rovo P, Farkas V, Hegyi O, Szolomajer-Csikos O, Toth GK, et al. (2011) Cooperativity network of Trp-cage miniproteins: probing salt-bridges. J Pept Sci 17: 610–619.

23. Luke KA, Higgins CL, Wittung-Stafshede P (2007) Thermodynamic stability and folding of proteins from hyperthermophilic organisms. FEBS J 274: 4023–4033.

24. Kumar S, Nussinov R (2002) Relationship between Ion Pair Geometries and Electrostatic Strengths in Proteins. Biophys J 83: 1595–1612.

25. Kumar S, Nussinov R (2000) Fluctuations between stabilizing and destabilizing electrostatic contributions of ion pairs in conformers of the c-Myc-Max leucine zipper. Proteins: Structure Function and Genetics 41: 485–497.

26. Parthasarathy S, Murthy MR (2000) Protein thermal stability: insights from atomic displacement parameters (B values). Protein Eng 13: 9–13.

27. Lin CP, Huang SW, Lai YL, Yen SC, Shih CH, et al. (2008) Deriving protein dynamical properties from weighted protein contact number. Proteins 72: 929–935.

28. Lu C-H, Huang S-W, Lai Y-L, Lin C-P, Shih C-H, et al. (2008) On the relationship between the protein structure and protein dynamics. Proteins 72: 625–634.

29. Rader AJ (2010) Thermostability in rubredoxin and its relationship to mechanical rigidity. Physical Biology 7: 016002.

30. Radestock S, Gohlke H (2011) Protein rigidity and thermophilic adaptation. Proteins 79: 1089–1108.

31. El-Bondkly AM, El-Gendy MM (2012) Cellulase production from agricultural residues by recombinant fusant strain of a fungal endophyte of the marine sponge Latrunculia corticata for production of ethanol. Antonie Van Leeuwenhoek 101: 331–346.

32. Zamost BL, Nielsen HK, Starnes RL (1991) Thermostable Enzymes for Industrial Applications. Journal of Industrial Microbiology 8: 71–81.

33. Sanz-Aparicio J, Hermoso JA, Martinez-Ripoll M, Lequerica JL, Polaina J (1998) Crystal structure of beta-glucosidase A from Bacillus polymyxa: insights into the catalytic activity in family 1 glycosyl hydrolases. J Mol Biol 275: 491–502.

34. Wang G, Dunbrack RL (2003) PISCES: a protein sequence culling server. Bioinformatics 19: 1589–1591.

35. Kabsch W, Sander C (1983) Dictionary of protein secondary structure: Pattern recognition of hydrogen-bonded and geometrical features. Biopolymers 22: 2577–2637.

36. Qin SB, He Y, Pan XM (2005) Predicting protein secondary structure and solvent accessibility with an improved multiple linear regression method. Proteins-Structure Function and Bioinformatics 61: 473–480.

37. Burkhard Rost, Chris Sander (1994) Conservation and prediction of solvent accessibility in protein families. Proteins: Structure, Function, and Genetics 20: 216–226.

38. Fukuchi S, Nishikawa K (2001) Protein surface amino acid compositions distinctively differ between thermophilic and mesophilic bacteria. J Mol Biol 309: 835–843.

39. Armon A, Graur D, Ben-Tal N (2001) ConSurf: an algorithmic tool for the identification of functional regions in proteins by surface mapping of phylogenetic information. J Mol Biol 307: 447–463.

40. Gonzalez-Candelas L, Aristoy MC, Polaina J, Flors A (1989) Cloning and characterization of two genes from Bacillus polymyxa expressing beta-glucosidase activity in Escherichia coli. Appl Environ Microbiol 55: 3173–3177.

41. Olson CA, Spek EJ, Shi Z, Vologodskii A, Kallenbach NR (2001) Cooperative helix stabilization by complex Arg-Glu salt bridges. Proteins 44: 123–132.

42. Loladze VV, Ibarra-Molero B, Sanchez-Ruiz JM, Makhatadze GI (1999) Engineering a thermostable protein via optimization of charge-charge interactions on the protein surface. Biochemistry 38: 16419–16423.

43. Jelesarov I, Karshikoff A (2009) Defining the role of salt bridges in protein stability. Methods Mol Biol 490: 227–260.

44. Gonzalez-Blasco G, Sanz-Aparicio J, Gonzalez B, Hermoso JA, Polaina J (2000) Directed Evolution of beta -Glucosidase A from Paenibacillus polymyxa to Thermal Resistance. J Biol Chem 275: 13708–13712.

45. Sanz-Aparicio J, Hermoso JA, Martinez-Ripoll M, Gonzalez B, Lopez-Camacho C, et al. (1998) Structural basis of increased resistance to thermal denaturation induced by single amino acid substitution in the sequence of beta-glucosidase A from Bacillus polymyxa. Proteins 33: 567–576.

46. Bleicher L, Prates ET, Gomes TC, Silveira RL, Nascimento AS, et al. (2011) Molecular basis of the thermostability and thermophilicity of laminarinases: X-ray structure of the hyperthermostable laminarinase from Rhodothermus marinus and molecular dynamics simulations. J Phys Chem B 115: 7940–7949.

47. Daggett V, Levitt M (1992) A model of the molten globule state from molecular dynamics simulations. Proc Natl Acad Sci U S A 89: 5142–5146.

48. Lazaridis T, Lee I, Karplus M (1997) Dynamics and unfolding pathways of a hyperthermophilic and a mesophilic rubredoxin. Protein Sci: a publication of the Protein Society 6: 2589–2605.

49. Cavagnero S, Debe DA, Zhou ZH, Adams MW, Chan SI (1998) Kinetic role of electrostatic interactions in the unfolding of hyperthermophilic and mesophilic rubredoxins. Biochemistry 37: 3369–3376.

50. Tanaka Y, Tsumoto K, Yasutake Y, Umetsu M, Yao M, et al. (2004) How oligomerization contributes to the thermostability of an archaeon protein. Protein L-isoaspartyl-O-methyltransferase from Sulfolobus tokodaii. J Biol Chem 279: 32957–32967.

51. Binter A, Staunig N, Jelesarov I, Lohner K, Palfey BA, et al. (2009) A single intersubunit salt bridge affects oligomerization and catalytic activity in a bacterial quinone reductase. FEBS Journal 276: 5263–5274.

52. Bogin O, Levin I, Hacham Y, Tel-Or S, Peretz M, et al. (2002) Structural basis for the enhanced thermal stability of alcohol dehydrogenase mutants from the mesophilic bacterium Clostridium beijerinckii: contribution of salt bridging. Protein Sci 11: 2561–2574.

53. Charbonneau DM, Beauregard M (2013) Role of Key Salt Bridges in Thermostability of G. thermodenitrificans EstGtA2: Distinctive Patterns within the New Bacterial Lipolytic Enzyme Family XV. PLOS One 8: e76675.

54. Sandgren M, Gualfetti PJ, Shaw A, Gross LS, Saldajeno M, et al. (2003) Comparison of family 12 glycoside hydrolases and recruited substitutions important for thermal stability. Protein Sci 12: 848–860.

55. Anbar M, Gul O, Lamed R, Sezerman UO, Bayer EA (2012) Improved thermostability of Clostridium thermocellum endoglucanase Cel8A by using consensus-guided mutagenesis. Appl Environ Microbiol 78: 3458–3464.

10

Development of Quantitative Proteomics Using iTRAQ Based on the Immunological Response of *Galleria mellonella* Larvae Challenged with *Fusarium oxysporum* Microconidia

Amalia Muñoz-Gómez[1,2,3]*****, **Mauricio Corredor**[2], **Alfonso Benítez-Páez**[3], **Carlos Peláez**[1]

1 Grupo Interdisciplinario de Estudios Moleculares (GIEM), Instituto de Química, Universidad de Antioquia, Medellín, Antioquia, Colombia, **2** Genetic and Biochemistry of Microorganisms group (GEBIOMIC), Instituto de Biología, Universidad de Antioquia, Medellín, Antioquia, Colombia, **3** Bioinformatic Analysis Group (GABi), Centro de Investigación y Desarrollo en Biotecnología, CIDBIO, Bogotá, Distrito Capital, Colombia

Abstract

Galleria mellonella has emerged as a potential invertebrate model for scrutinizing innate immunity. Larvae are easy to handle in host-pathogen assays. We undertook proteomics research in order to understand immune response in a heterologous host when challenged with microconidia of *Fusarium oxysporum*. The aim of this study was to investigate hemolymph proteins that were differentially expressed between control and immunized larvae sets, tested with *F. oxysporum* at two temperatures. The iTRAQ approach allowed us to observe the effects of immune challenges in a lucid and robust manner, identifying more than 50 proteins, 17 of them probably involved in the immune response. Changes in protein expression were statistically significant, especially when temperature was increased because this was notoriously affected by *F. oxysporum* 10^4 or 10^6 microconidia/mL. Some proteins were up-regulated upon immune fungal microconidia challenge when temperature changed from 25 to 37°C. After analysis of identified proteins by bioinformatics and meta-analysis, results revealed that they were involved in transport, immune response, storage, oxide-reduction and catabolism: 20 from *G. mellonella*, 20 from the Lepidoptera species and 19 spread across bacteria, protista, fungi and animal species. Among these, 13 proteins and 2 peptides were examined for their immune expression, and the hypothetical 3D structures of 2 well-known proteins, unannotated for *G. mellonella*, i.e., actin and CREBP, were resolved using peptides matched with *Bombyx mori* and *Danaus plexippus*, respectively. The main conclusion in this study was that iTRAQ tool constitutes a consistent method to detect proteins associated with the innate immune system of *G. mellonella* in response to infection caused by *F. oxysporum*. In addition, iTRAQ was a reliable quantitative proteomic approach to detect and quantify the expression levels of immune system proteins and peptides, in particular, it was found that 10^4 microconidia/mL at 37°C over expressed many more proteins than other treatments.

Editor: Eleftherios Mylonakis, Warren Alpert Medical School of Brown University, United States of America

Funding: This work was supported by the Sustainability Strategy 2011-2012 from University of Antioquia for Interdisciplinary Group of Molecular Studies (GIEM Grupo Interdisciplinario de Estudios Moleculares, Universidad de Antioquia). The funders had no role in study design, data collection and analysis, decision to publish, or preparation of the manuscript.

Competing Interests: The authors have declared that no competing interests exist.

* Email: amunoz@cidbio.org

Introduction

Galleria mellonella larvae offer considerable advantages as an infection and virulence model because they are simple to handle and can be studied in large numbers while carrying out cost-effective experiments [1]. Most importantly, and unlike many other invertebrate models such as *Drosophila melanogaster* and Caenorhabditis elegans species, experiments can be performed at 37°C, an optimal temperature for the vast majority of human pathogens [2]. The remarkable advantages of this model also draw upon the innate immune response of *G. mellonella* larvae, which share a high degree of homology with mammalian organisms [3].

Furthermore, *G. mellonella* has been widely used as a heterologous host for several fungal pathogens, including *Candida*

albicans, *Cryptococcus neoformans* and *Aspergillus fumigatus* [4]. A positive correlation has been observed in pathogenicity using these yeasts or fungi since this insect model is comparable to mammalian models [5]. For example, *C. neoformans* can proliferate in the *G. mellonella* hemocoel, leading to the eventual death of the caterpillar. Therefore, it is unquestionable that *G. mellonella* should emerge as a model of host defense against fungal infection [6].

The hemocytes in the cellular hemolymph produce a robust oxidative burst, which performs similarly to phagocytic cells in response to bacterial infections [7]. This feature, among other conserved aspects of the anti-microbial process, might explain the positive correlation between data obtained from *Galleria* and mice infections [8] with either prokaryotic or eukaryotic pathogens.

Therefore, *G. mellonella* emerges as a model that has since been successfully applied to assess the virulence in a variety of bacteria, including the Burkholderia cepacia complex [9], Enterococcus faecium [10], *Francisella tularensis* [11], Legionella pneumophila [12], Listeria monocytogenes [13], Pseudomona aeruginosa [8], [14] and various corynebacteria [15]. Additionally, virulence in *G. mellonella* has been assessed using various fungi, as mentioned above, A. fumigatus [16], C. albicans [17] and C. neoformans [6] and evaluated with some viruses, such as baculovirus [18] and parvovirus [19].

Fusarium oxysporum is a trans-kingdom pathogen and is well known for producing harmful secondary metabolites, called mycotoxins, that cause several diseases in humans, animals, insects, and even plants [20], [21]. This mold is the causal agent of vascular wilt disease, which affects a wide range of plant species as tomato, but also can produce disseminated infections in human beings and immunodepressed mice. The *Fusarium* species are an ideal model for a comparative analysis of fungal virulence in plant and animal hosts because they are capable of manifesting a multitude of clinical infections [20]. Few studies have been developed to use the larvae of *G. mellonella* as a heterologous host for fusaria. *G. mellonella* is able to resist high concentrations of *F. oxysporum* microconidia. When conidia are injected into the hemocoel of this Lepidopteran system, both clinical and environmental isolates of the fungus are able to kill the larvae at 37°C, although the killing occurs rapidly when incubated at 30°C [21].

Regarding the immune response of insects against fungal infections, the insect hemolymph plays a critical role, acting as a means of transporting cells, proteins-peptides, oxygen, hormones, nutrients, and metabolites associated with the immune system [22], [23]. Immunological studies of insects represent an important challenge to understand the molecular mechanisms of the innate immune response in the animal kingdom. Moreover, the innate immune system is well known as the primary defense against pathogen or competitors' attacks because it is widespread in all three major kingdoms of life [24].

Multiplexed high-throughput proteomics strategies provide an integrated and broader view of biological regulatory networks and pathways [25]. Protein activity is a crucial task that depends on the interactions and modifications of antagonistic and synergistic proteins [25]. iTRAQ is suitable for exploratory molecular studies of the pathogenic mechanisms and pathophysiology of diseases [26].

This study likely represents one of the first reports that uses iTRAQ to detect *G. mellonella* proteins after a fungal challenge because the main obstacle to the development of a computational approach with iTRAQ is the unavailability of *Galleria*'s genome, which makes protein identification problematic. Our proposal in this experimental and computational biology work was to apply an alternative approach, matching up the isolated iTRAQ peptides with the few *G. mellonella* proteins in databases and even with those from related butterflies or other organisms that have been previously reported in databases. In this work, we implemented a proteomic iTRAQ approach to detect and quantify protein expression patterns in *G. mellonella* hemolymph following challenge with a sub-lethal infection of *F. oxysporum* microconidia in order to induce a response of the innate immune system.

Materials and Methods

Maintenance of biological samples

Sixth-instar larvae of the wax moth *G. mellonella* were reared in our laboratory and fed an artificial diet (5.54% dry yeast, 22.17% wheat bran, 11.09% yellow corn flour, 11.09% powdered milk, 5.54% wheat germ, 17.65% honey, 22.70% glycerol, 3.88% beeswax and 0.033% formaldehyde solution). Larvae with a body weight between 230 and 330 mg were used. Caterpillars were harvested for treatments with the best quality control (color, size and vigor). Additionally, a strain of *F. oxysporum* (Foxy-GIEM isolated from soil) was grown in PDA (Oxoid, Ltd., Basingstoke, England) and maintained at 25°C for 14 days at intervals of light and darkness (16 h light/8 h dark).

F. oxysporum microconidia preparation

A surfactant solution of 0.1% Tween-80 was added to isolate the conidia from the mycelium, using Drigalski spatula for scraping. The resulting suspension was filtered through sterile gauze and collected in a sterilized screw-cap tube. The volume was increased to 10 mL with 0.1% Tween-80. The microconidia suspension was vortexed three times for 20 seconds each. The concentration of the suspension was determined with the Neubauer–improved counting chamber (Paul Marienfeld GmbH & Co. KG, Lauda-Königshofen, Germany) and adjusted to $1 \times 10^4 - 1 \times 10^8$ microconidia/mL.

Larval survival test and lethal concentration of microconidia

G. mellonella larvae were arranged into 10 groups, randomly selected during the last instar of the larval development and inoculated with conidia (1×10^4, 1×10^5, 1×10^6, 1×10^7 and 1×10^8 microconidia/mL) by injecting 20 μL of suspension into the hemocoel through the last pro-leg [27]. Before injection, the area was disinfected with an alcohol swab (70%). A 1 mL disposable syringe was used with a 31G×8.0 mm needle (BD Ultra-Fine, Becton Dickinson, Franklin Lakes, NJ, USA). Larvae were stored in the dark and incubated in Petri dishes at 25 and 37°C. The experiment was prepared with two control groups: the first group represented the control injection process, regarding the physical trauma into hemocoel of larvae [6] inoculated with 20 μL of 0.1% Tween-80, and the second control group received no injection. Mortality was assessed after 48 hours based on the brown color of the cuticle (melanization) and the absence of movement in response to stimulation. All experiments were performed in triplicate. Each experiment was classified as sublethal microconidia concentrations when approximately 80–100% of the larvae survived.

Statistics for the survival treatment of *G. mellonella*

To check if any assumption has been violated on the statistically significant difference between microconidia treatments and temperature, the hypothesis for testing the difference in proportions was established. To assess whether there was a relation of statistical significance between treatment and temperature. The hypothesis for testing a difference in the means of survival when temperature changed was performed for each microconidia concentration (1×10^4, 1×10^5, 1×10^6 microconidia/mL and means of survival at 25 and 37°C). The p-values were calculated using R software ($\alpha \leq 0.5$).

Infection of larvae with microconidial sublethal concentrations

G. mellonella larvae, in groups of 10, were inoculated with conidia by injecting 20 μL of the suspension into the hemocoel through the last pro-leg with the sub-lethal concentrations selected, 1×10^4 and 1×10^6 microconidia/mL [27]. Two groups of 10 larvae were arranged in triplicate; one set was incubated at 25°C, and the other set was incubated at 37°C in Petri dishes after being kept in the dark for 48 hours. To ensure that larvae did not die by different treatments such as a single puncture injection or incubation at 37°C or with 20 μL of 0.1% Tween-80 surfactant solution, these processes were included as part of the treatments as controls [6]. The mortality rate was recorded 2 days post-injection. Mortality was assessed based on the brown cuticle coloration (melanization) and the absence of movement in response to stimulation. All experiments were performed in triplicate. The same statistical analysis was performed to know if there was a statistical significance. The same hypothesis for testing the difference in the means of survival was used when temperature changed for each microconidia concentration (1×10^4 and 1×10^6 microconidia/mL and means of survival at at 25 and 37°C). The p-values were calculated using R software ($\alpha \leq 0.5$).

Collection and preparation of cell-free hemolymph

Prior to hemolymph collection, insects were anesthetized by chilling for 5 min at 8°C, and the abdominal surface was disinfected with a 70% (v/v) ethanol solution. Hemolymph samples were obtained by puncturing the larval abdomen with a sterile insulin needle. The outflowing hemolymph was immediately collected into sterile, chilled Eppendorf tubes containing 0.150 mM phenylthiourea [28] to prevent melanization. The cell-free hemolymph was obtained by centrifugation at 1606 g for 5 min and then for 15 min at 4°C to pellet the cell debris. The pooled supernatants were stored at -20°C [29].

Preparation and quantification of proteins

The protein concentration was determined with the Bradford method using bovine serum albumin (BSA) as a standard [29], [30]. The Quick Start Bradford protein assay kit (Bio-Rad Laboratories, Inc. USA) was used following the manufacturer's instructions.

Identification and characterization of peptides by iTRAQ

The sample preparation for LC/MS/MS first used 100 μg of protein from each hemolymph sample, challenged or unchallenged with microconidia. The samples were processed using the protocol of Applied Biosystems iTRAQ 8-plex [31].

Labeling of free-cell hemolymph sample

The hemolymph sample from larvae challenged with 1×10^4 microconidia/mL of *F. oxysporum* and incubated at 25°C for 48 h was labeled as iTRAQ-113. The hemolymph of larvae injected with 0.1% Tween-80 and incubated at 25°C for 48 h was labeled as iTRAQ-114. The hemolymph sample from larvae challenged with 1×10^4 microconidia/mL and incubated at 37°C for 48 h was labeled as iTRAQ-115. The hemolymph from larvae without any injection and incubated at 25°C for 48 h was labeled as iTRAQ-116. The sample from larvae challenged with 1×10^6 microconidia/mL and incubated at 37°C for 48 h was labeled as iTRAQ-117. The sample from larvae injected with 0.1% Tween-80 and incubated at 37°C for 48 h was labeled as iTRAQ-118. Treatments are summarized in table 1.

Triple TOF mass spectrometry

The multiplexed isobaric chemical tagging reaction was elaborated in the YPED proteomics laboratory, Yale Cancer Center Mass Spectrometry Resource & W.M. Keck Foundation Biotechnology Resource Laboratory. Peptides were separated on a Waters nanoACQUITY system (75 μm x 150 mm BEH C18 eluted at 500 nL/min) via MS analysis on an AB Sciex 5600 Triple TOF mass spectrometer. The iTRAQ allowed the multiplexing of two to six protein samples and produced identical MS/MS sequencing ions for all six versions of the same derivatized tryptic peptide. Quantitation was achieved by comparing the peak areas and the resultant peak ratios for the six MS/MS reporter ions, which range from 113 to 118 Da (119 and 121 were not used for this experiment). The mass spectrometric analysis was visualized using the Peakview software and quantified with ProteinPilot (AB Sciex, MA, USA).

Statistics and bioinformatics analyses

The LC/MS/MS raw data were analyzed using the Mascot Distiller software, version 2.2 (Matrix Science, United Kingdom), to match the peptides with a protein database using the first default parameter. The protein ratio type was set to median, the normalization method was median ratio, no outlier removal was chosen and the peptide threshold was set to at least homology. Mascot search engine with ProteinPilot software was utilized. The search for proteins was performed using the non-redundant NCBInr_20121109 (National Center for Biotechnology Information) database. Proteins matching two or more peptides were considered a positive identification, but trypsin peptides, the enzyme that digests proteins, were excluded from analysis. Protein concentrations were determined using the ProteinPilot software, version 4.0 (AB Sciex, MA, USA) to infer concentrations based on the amounts of peptides assigned. ProteinPilot software, which employs the Paragon search algorithm, was used for peptide matching, protein identification and relative protein quantitation. The ProteinPilot Descriptive Statistics Template (PDST) tool automatically generates a wealth of important information from data-intensive proteomics experiments. Protein expression ratios were calculated from the pair-wise comparison of two iTRAQ channels. For each ratio, the iTRAQ peak area was corrected for Observed Bias Correlation. The general level of confidence was 66% for the protein hits. The percent of confidence was expressed in ProtScore units. The levels of confidence were 99% (2.0), 95% (1.3), 90% (1.0) and 66% (0.47), and the ProtScore units are presented in parentheses. The results file data was stored in YPED proteomics database, School of Medicine, WM Keck foundation, Yale University. The P-value (Probability that the deviation from unity is by chance) of ratios was calculated (biased and non-biased). Identified Lepidoptera proteins were compared with *G. mellonella* peptides using pairwise alignment with the Blast algorithm version BLAST 2.2.29 [32] and multiple alignments with Muscle version 3.2 [33]. Protein data were analyzed in Uniprot, Pfam, Gene Ontology and KEGG to identify the biological processes associated with immunological functions.

Validation of iTRAQ protein expression test with RT-PCR

In order to validate protein expression detected by iTRAQ, we compared these results with Real-Time-PCR (RT-PCR). Four larvae were previously treated with 10^6 microconidia/mL and four were untreated, both at 37°C. RNA was extracted from hemolymph and cDNA synthesis was elaborated with Fermentas Thermo Scientific kit, following manufacturer's instructions. Three genes were selected randomly from iTRAQ list. RT-PCRs were developed with specific primers and the reactions were set on the Applied Biosystem device following classic quantitative PCR (qPCR) protocol. The genes selected were: serpin, cationic protein 8p and 26 kDa ferritin subunit (see table S4). The housekeeping genes were Ribosomal protein S7e, lipocalin and fungal protease inhibitor. The RT-PCR amplification data were compared with iTRAQ protein expression data using Log2 value of fold-changes. Both iTRAQ and RT-PCR were correlated statistically. The chi-square goodness of fit analysis was performed to determine whether there was correspondence between the iTRAQ and

Table 1. Treatments with microconidia concentrations and temperatures.

iTRAQ 8-plex tags	Microconidia concentration and temperatures	Volume recovered
113	Hemolymph challenge with *F.oxysporum* 10^4 microconidia/mL at 25°C	6 mL
114	Hemolymph injected only with Tween-80 (0.1% v/v) at 25°C	6 mL
115	Hemolymph challenge with *F.oxysporum* 10^4 microconidia/mL at 37°C	6 mL
116	Hemolymph untreated at 25°C	6 mL
117	Hemolymph challenge with *F.oxysporum* 10^6 microconidia/mL at 37°C	6 mL
118	Hemolymph injected only with Tween-80 (0.1% v/v) at 37°C	6 mL

Sample tags for iTRAQ analysis. For each treatment or pool around 6 mL were recovered. The protein concentration was calculated by Bradford assay.

qPCR data obtained for the indicated proteins. The p-values greater than 0.05 indicate that both sets of data match.

Secondary and tertiary structure of proteins

Peptides of two unannotated proteins for *G. mellonella* that had an adequate peptide coverage (more than 50%) were used to be aligned with other protein sequences. Peptides of each protein were reanalyzed by Blast version 2.2.29 [32] and the alignments were elaborated using the Muscle version 3.2 [33] and ClustalW version 2.0 algorithms [34]. Secondary and tertiary structures were defined using Phyre version 2.0 and i-Tasser version 3.0 servers on line and the standalone package. The files generated were visualized in UCSF Chimera version 1.8. The mutated amino acids were used to generate tertiary structures that were aligned with the PBD structure database. Additionally, the tertiary structures were aligned using structural comparison and 3D matching in Chimera. Hypothetical structures were analyzed and reported [35], [36].

Results

Survivor assay and selection of non-lethal dose from microconidia

To characterize the interaction between *G. mellonella* and *F. oxysporum*, we first reproduced the infection model to investigate the biological processes, such as protein transport, defense and nodulation, protein and lipid condensation and, finally, insect immune response during fungal infection. For this purpose, we inoculated larvae with different microconidia doses at two different temperatures. We noted that larvae treated with 10^4, 10^5, 10^6 microconidia/mL at 25°C survived, whereas the mortality increased when microconidia concentration was higher (figure 1). The difference in means test obtained a p-value <0.01 for all treatments (10^4, 10^5, 10^6 microconidia/mL concentrations), indicating that the survival mean of each treatment differs with temperature. In other words, mortality and survival are correlated and the same correlation would be obtained working with mortality rate (Data in Table S2). The experiment was repeated with new larvae with 10^4 and 10^6 microconidia/mL using the same statistical treatment. Once again, similar results were obtained (data in table S2). That is, at a fixed concentration, the average surviving larvae at 25°C is different from mean survival of larvae at 37°C, the latter being higher for both concentrations (Data in Table S3).

The best larvae survival rate throughout the study (~99%) was observed at the 10^4 microconidia/mL concentration at 37°C (figure 1). An increase in the survival was also observed at 37°C and was greater than at 25°C after treatments with 10^5 and 10^6 microconidia/mL. When concentrations of 10^7 or 10^8 microco-

nidia/mL were used, every larva died in both temperatures. Using 10^6 microconidia/mL at 25°C produced a low number of survivors and data were not included in figure 1.

To identify the proteins obtained by larvae challenged with *F. oxysporum*, we selected two microconidia concentrations where 10^4 microconidia/mL produced a lower mortality than 10^6 microconidia/mL did. This last concentration, despite the mortality, allowed us to retrieve a suitable number of larvae survivors for subsequent experiments.

Protein identification inside larval hemolymph infected with fungi

A total of 374 peptides corresponding to 59 different proteins were identified in each treatment made by iTRAQ 8 plex. Regarding Lepidoptera, iTRAQ 8 plex identified 351 peptides dispersed among 40 proteins (table 2). Among these, 47 proteins were identified with a level of confidence ≥99%, and 54 proteins were identified with a confidence level ≥95%. From this set of proteins identified, 20 correspond to *G. mellonella* and 20 to the following Lepidoptera order: *Danaus plexippus* (4), *Bombyx mori* (3), *Corcyra cephalonica* (2), *Hyphantria cunea* (1), *Papilio polytes* (1), *Manduca sexta* (1), *Sesamia nonagrioides* (1), *Pieris rapae* (1), *Arcte modesta* (1), *Chilo suppressalis* (1), *Papilio xuthus* (1), *Samia ricini* (1), *Choristoneura fumiferana* (1), *Spodoptera exigua* (1) (table 2). The remaining proteins were identified as belonging to other Insecta orders, other invertebrates, bacteria, protozoan and fungi (table S1). Notably, more than 350 peptides were assigned to Lepidopteran proteins (table 2).

Among the 59 proteins identified by iTRAQ, 17 showed good coverage with two or more peptides, as summarized in figure 2. Among these, the minimal peptide coverage was 59.85, corresponding to a cellular retinoic acid binding protein with 3 peptides; the maximum coverage was arylphorin with 340 labeled peptides detected (peptides in plot ratios, figure 2), corresponding to 48 different peptides with 83.9 coverage (figure 2). We selected those 17 proteins, not only for their coverage, but also for their excellent detection level in LC/MS/MS with different treatments. Of those proteins, 15 correspond to *G. mellonella*, 1 to *B. mori* (actin) and 1 to *D. plexippus* (cellular retinoic acid binding protein, belongs to the lipocalin family). Concerning the 20 Lepidopteran proteins, 18 showed peptide coverages lower than 50% (table 2).

Protein expression in challenged larval hemolymph

The relationship between treatment ratios was compared to determine expression of each protein. Biased and unbiased statistical analyses were developed to establish the best differences. Table S4 displays the data ratios (non-biased) between the values of the different treatments and its p-values for the hypothesis test

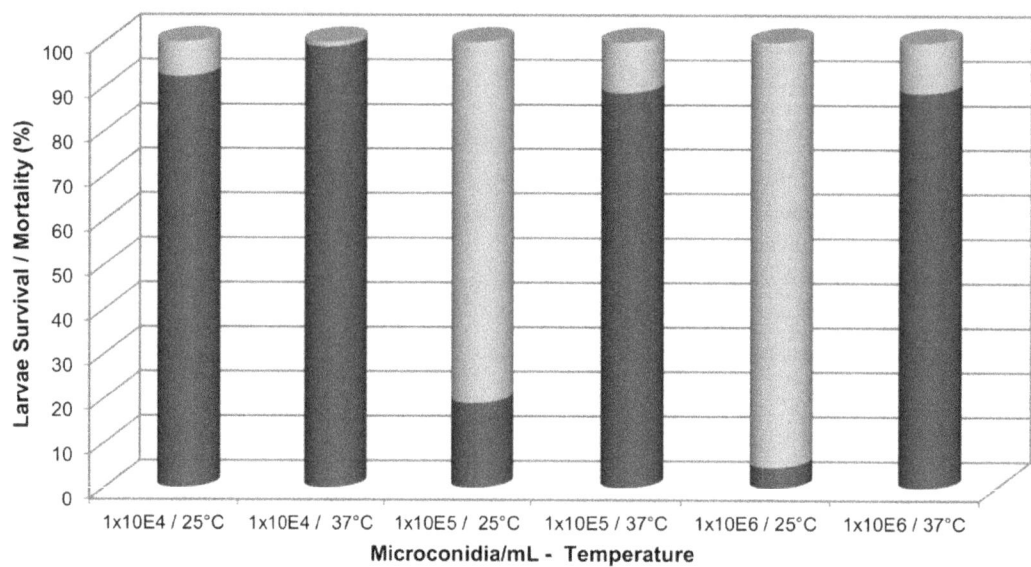

Figure 1. Survival/mortality percentage of larvae. Bars show *G. mellonella* larvae survivors in dark gray and dead larvae in light gray after being infected with 1×10^4, 1×10^5 and 1×10^6 microconidia/mL of *F. oxysporum* at 25 or 37°C. Comparing the results, larval survival was higher at 37 than at 25°C.

that the deviation from unity is by chance. The p-values lower than 0.05 indicate significant differences among the values of the compared treatments for 113 denominator (channel). From 17 proteins analyzed, 2 had only one peptide (non-significant, without p-value). For the other 15, 4 significant p-values for 115:113 ratio, 8 significant p-values for 116:113 ratio and 12 significant p-values for 117:113 ratio were observed (Data in table S5).

We analyzed 17 proteins inside each treatment that were affected (up- or down-regulated) by a pathogenic infection with *F. oxysporum* microconidia (10^4 or 10^6 microconidia/mL) at two temperatures (25 or 37°C). The relevant protein expression results from those proteins are summarized in figure 3, though only the three notable treatments (larvae with 10^4 microconidia/mL at 25 or 37°C and larvae with 10^6 microconidia/mL at 37°C) with controls (larvae without microconidia injection at 25°C) are shown in figure 3.

When the larvae were inoculated with 10^4 microconidia/mL at 25°C, some proteins lowered their concentrations compared with untreated larvae at 25°C; the proteins that were down-regulated included: arylphorin, hexamerin, larval hemolymph, apolipophorin, apolipophorin 3, juvenile hormone binding protein, transferrin precursor, cellular retinoic acid binding proteins and anionic antimicrobial peptides. Conversely, the following proteins were up-regulated under the same conditions compared with untreated larvae at 25°C: 27 kDa hemolymph, cationic 8 precursor, hemolin, transferrin precursor, prophenoloxidase subunit 2, lysozime and actin. The level of cecropin D-like peptide was evidently low and was down-regulated in 10^4 microconidia/mL at 25°C; this concentration was the lowest among the 17 proteins (figure 3).

Concerning protein expression with 10^4 microconidia/mL treatment at 37°C, compared with 10^4 microconidia/mL at 25°C or no treatment at 25°C, the results were quite different. With the exception of apolipophorin 3, all of the proteins were up-regulated with respect to the control (see untreated at 25°C, apolipophorin 3, figure 3) and showed higher protein concentrations. However, more interestingly, in each case of treatment with 10^4 microconidia/mL at 37°C, the protein concentration was

higher than 10^4 microconidia/mL at 25°C, including that of apolipophorin 3 (figure 3).

Comparing the larvae at 10^4 or 10^6 microconidia/mL treatments, regardless of temperature, we observed two protein expression groups: one with high concentration levels (to the left of the graph) and the other with low concentration levels (to the right of the graph; figure 4). The first group included arylphorin, hexamerin, 27 kDa hemolymph, cationic 8 precursor, apolipophorin, juvenile hormone binding, apolipophorin 3. The second group included hemolin, transferrin precursor, prophenoloxidase subunit 2, lysozyme, actin, cellular retinoic acid binding, anionic antimicrobial peptide and cecropine D-like peptide.

Concerning the 10^6 microconidia/mL treatment at 37°C, some of the proteins were slightly over-expressed compared with the 10^4 treatment, such as the following: apolipophorin, juvenile hormone binding, apolipophorin 3, hemolin and transferrin precursor (these proteins are in the middle of the plot, figure 3). This pattern differs from the remaining proteins that never surpassed the 10^4 microconidia/mL at the 37°C treatment.

Finally, we validated the results of expression on iTRAQ using RT-PCR, statistic correlations were found between the proteins analyzed for both techniques. Effectively, the genes selected: serpin, cationic protein 8p and 26 kDa ferritin subunit showed the same pattern of expression between the analyzed microconidia concentrations and temperatures. The p-values for the three proteins were greater than 0.05 indicating that both sets of data match, see data set in tables S6A, S6B, S6C, S6D, S6E, S6F and figures S1, S2.

Bioinformatics analysis and gene ontology of lepidoteran proteins

The iTRAQ approach was useful for identifying, not only protein expression levels at different microconidia concentrations, but also new homologous proteins in *G. mellonella* due to this insect's lack of whole genome annotation. This approach was interesting because the ProteinPilot and Mascot Distiller are primarily designed to identify proteins in the own genome of the

Table 2. List of proteins identified by iTRAQ, from the highest to the lowest protein score.

Accession	Protein name from *G. mellonella*	Peptides	Accession	Protein name from other Lepidotera species	Peptides
AAT76806	Apolipophorin	89	EHJ68005	apolipophorins, *Danaus plexippus*	2
AAA74229	Arylphorin	48	AGR44824	actin, *Bombyx mori*	7
AAQ63970	transferrin precursor	31	ADA84299	hexamerin receptor, *Corcyra cephalonica*	6
AAA19801	Hexamerin	22	Q6VU70	apolipophorin-3 precursor, *Hyphantria cunea*	1
P80703	apolipophorin-3; precursor	20	ADR64702	antennal esterase CXE5, *Spodoptera exigua*	1
AAQ75026	prophenoloxidase subunit 2	11	BAK82317	apolipophorin precursor, *Bombyx mori*	2
P83632	27 kDa hemolymph protein; precursor	12	EHJ79039	cellular retinoic acid binding protein, *Danaus plexippus*	3
ACU09501	Hemolin	21	BAM18997	imaginal disc growth factor 4, *Papilio polytes*	4
AAK64363	prophenoloxidase	14	Q25490	apolipophorin-2; precursor, *Manduca sexta*	3
AAN06604	juvenile hormone binding protein	8	NP_001037386	glyceraldehyde-3-phosphate dehydrogenase, *Bombix mori*	1
P82174	lysozyme; 1,4-beta-N-acetylmuramidase	6	AAG44959	hexamerin 2, *Corcyra cephalonica*	1
CAK22401	beta-1,3-glucan recognition protein precursor	1	AAY26453	moderately methionine rich storage protein, *Sesamia nonagrioides*	2
ADI87454	cationic protein 8 precursor	5	ACZ68116	masquerade-like serine proteinase, *Pieris rapae*	3
ABG91580	larval hemolymph protein	5	EHJ65451	moderately methionine rich storage protein, *Danaus plexippus*	2
AAL47694	32 kDa ferritin subunit	4	EHJ75277	serpin 1, *Danaus plexippus*	1
ACQ99193	proline-rich protein	2	ADX62478	isocitrate dehydrogenase, *Arcte modesta*	1
AAG41120	26kDa ferritin subunit	3	AEW46856	seminal fluid protein, *Chilo suppressalis*	1
P85210	cecropin-D-like peptide	1	BAM19609	peptidoglycan recognition protein, *Papilio xuthus*	2
P85216	anionic antimicrobial peptide 2	1	BAF42698	hemolymph storage protein 1, *Samia ricini*	1
ADK26057	kunitz-type protease inhibitor precursor	1	AAC35429	diapause associated protein 2, *Choristoneura fumiferana*	2

Left Column shows 20 proteins from *G. mellonella*. Right Column displays 20 proteins from other Lepidotera species, which are not identified with some sequence of *G. mellonella* protein reported in database. They matched other butterfly species.

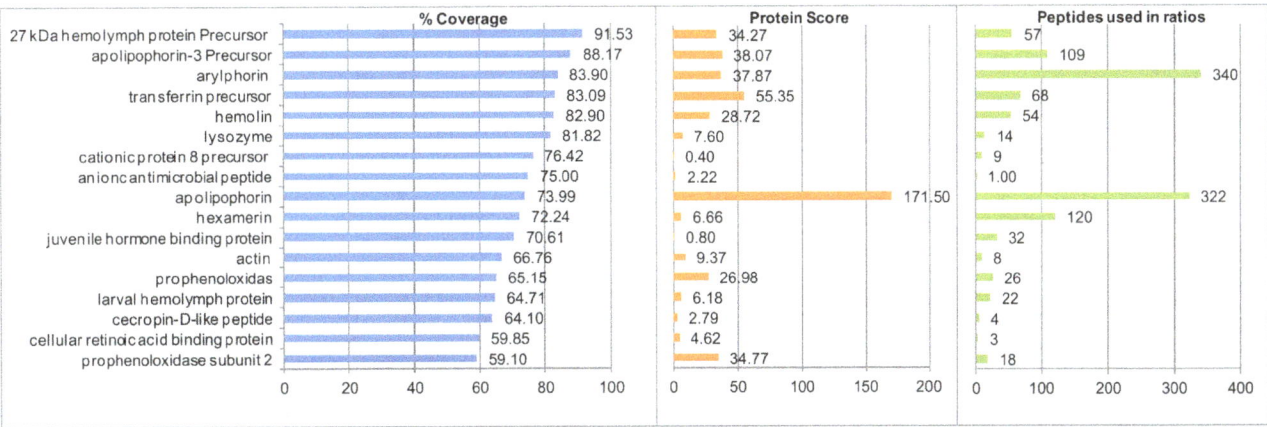

Figure 2. Data analysis of peptides from 17 proteins selected. Peptide percentage coverage is represented by the blue bar plot at the left; the protein score by the orange bar plot in the middle; and the peptides used in the ratios by the green bar plot to the right. The peptide ratios and protein scores were calculated using the ProteinPilot software. Selected proteins had more than 58% coverage.

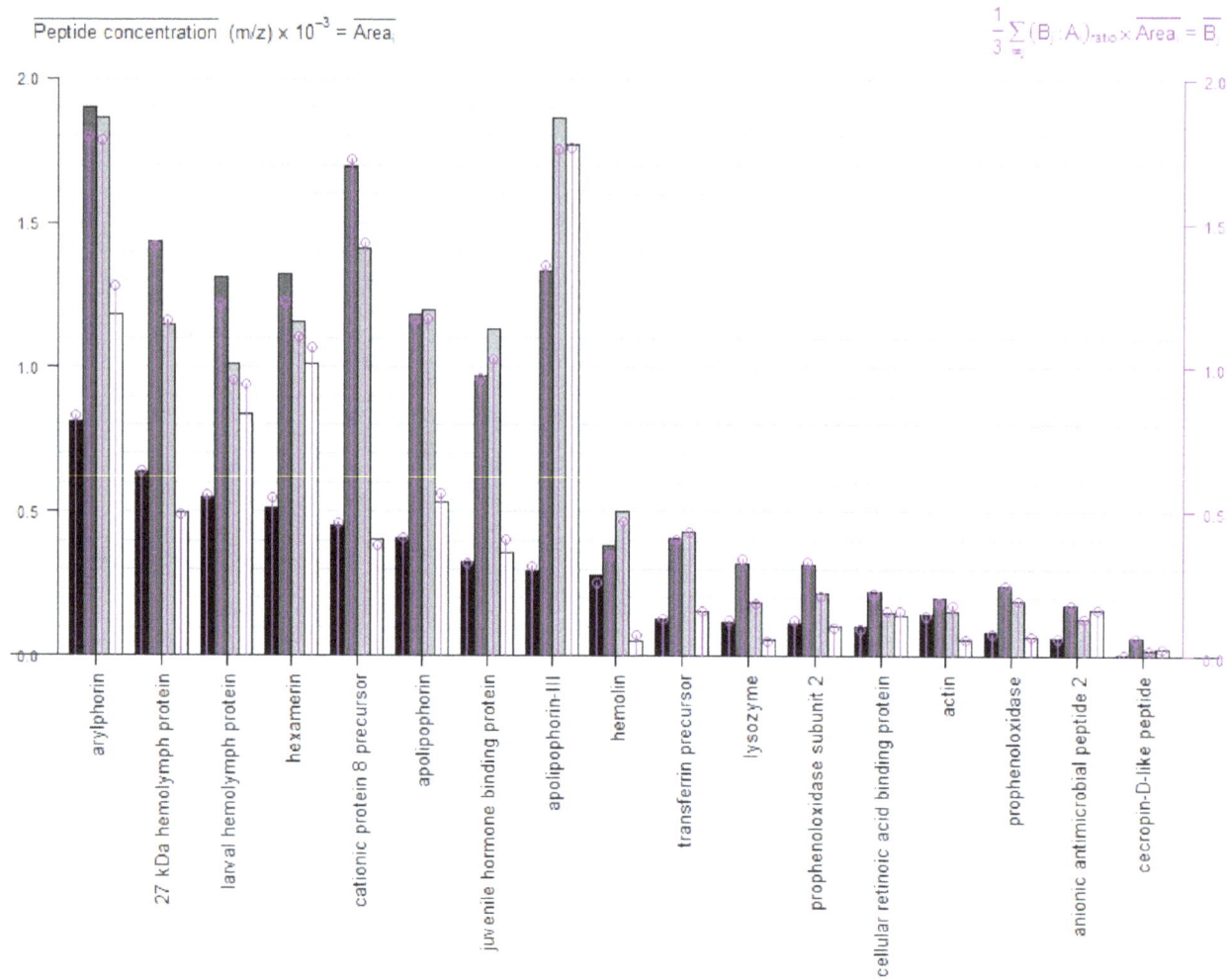

$\overline{\text{Peptide concentration}}\ (m/z) \times 10^{-3} = \overline{\text{Area}_i}$

$\frac{1}{3}\sum_{m_i}(B_i:A_i)_{\text{ratio}} \times \overline{\text{Area}_i} = \overline{B_i}$

Figure 3. Results of protein expression of 17 proteins. The plot represents a comparative data expression level of the quantitative peptide concentration from 0 to 2000 units from the selected main results of 17 proteins identified by iTRAQ using ProteinPilot. Black bars correspond to 10^4 microconidia/mL at 25°C, dark gray bars correspond to 10^6 microconidia/mL at 37°C, light gray bars correspond to 10^4 microconidia/mL at 37°C and white bars correspond to controls at 25°C untreated. Left y-axis axis is the average of the sum of the peptide areas, m/z (Area), as represented in black, gray and white bars. Right y-axis (purple lines) represent the average of the sum of the iTRAQ ratios per the mean of the sum from the peptide areas, m/z or Vp.

species considered. For example, of the selected 17 proteins, two were not annotated for *G. mellonella*: actin and cellular retinoic binding protein. Among the 40 lepidopteran proteins, we found 20 new ones that had not yet been annotated for *G. mellonella* (listed in table 2).

Our search in Uniprot, Pfam and KEGG allowed us to correlate the information with Gene Ontology and to associate the biological process with its immunological function. Figure 4 shows a plot (A) with the results of this search. We defined five biological functions for 17 proteins, supported by bioinformatics information metadata on the Gene Ontology data-base: transport, immune response, storage, oxide-reduction and catabolism. Of those 17 proteins, 6 are specialized in transport; 6 carry out immunological responses; 4 work in storage; 2 operate in oxide-reduction; and 2 more serve in catabolism. One protein appeared to possess an unknown function: the 27 kDa hemolymph protein precursor (figure 4A and 4B). Apolipophorin, arylphorin, lysozyme and actin are all essential proteins that have been quite well studied. Furthermore, some of these and other proteins were previously reported to play major roles in immunological processes, such as

apolipophorin, cationic 8 precursor, hemolin, anionic antimicrobial peptide 2, lysozyme, and cecropin D-like peptide. The role of prophenolxydase in the oxide-reduction process and the function of lysozyme and actin in the catabolic process are also well known. As discussed below, our results identified unknown proteins from *G. mellonella* that were most likely associated with transport, storage or immunological response challenges and affected by fungal microconidia invasion.

Further biological functions are also likely associated with immunological responses. However, the plot in figure 4A represents either one or two main functions of the 17 proteins studied, as related to the immunological defense in Lepidoptera larvae oriented towards fungal invasion. Figure 4B shows the biological process for every protein. Apolipophorin 3, arylphorin, apolipophorin, actin and lysozymes perform two functions that are likely associated with immunological defenses, as shown in our results. Conversely, hexamerin, larval hemolymph, cationic 8 precursor, juvenile hormone binding, transferrin precursor, cellular retinoic acid binding, prophenoloxydase subunit 3, hemolin, prophenoloxydase, and two peptides (anionic antimicrobial 2 and cecropin

A. Plot

Biological process

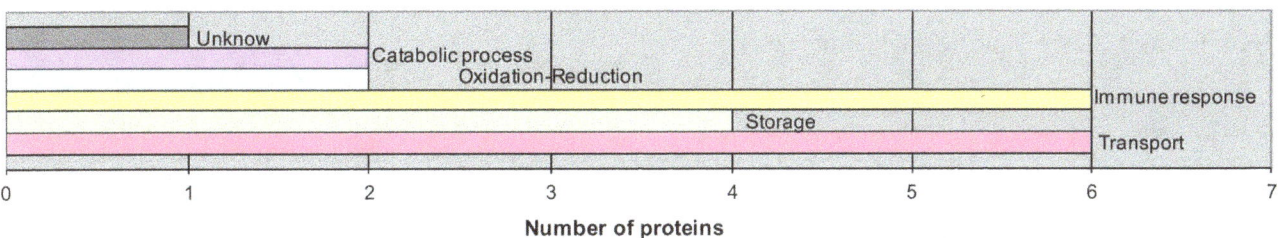

	Number of proteins

B. Grid

Hemolymph untreated at 25°C		Hemolymph with 10⁴ at 25°C		Hemolymph with 10⁴ at 37°C		Hemolymph with 10⁶ at 37°C	
apolipophorin-3	1773	arylphorin	812	arylphorin	1902	arylphorin	1864
arylphorin	1180	27 kDa hemolymph	636	apolipophorin-3	1696	apolipophorin-3	1863
hexamerin	1011	larval hemolymph	545	cationic 8 precursor	1435	cationic 8 precursor	1407
larval hemolymph	838	hexamerin	512	apolipophorin	1329	apolipophorin	1198
apolipophorin	531	cationic 8 precursor	449	hexamerin	1321	hexamerin	1154
27 kDa hemolymph	495	apolipophorin	408	27 kDa hemolymph	1310	27 kDa hemolymph	1145
cationic 8 precursor	400	juvenile hormone binding	325	juvenile hormone binding	1184	juvenile hormone binding	1132
juvenile hormone binding	356	apolipophorin-3	294	larval hemolymph	968	larval hemolymph	1011
anionic antimicrobial peptide 2	158	hemolin	280	hemolin	408	hemolin	502
transferrin precursor	153	actin	147	transferrin precursor	380	transferrin precursor	429
cellular retinoic acid binding	140	transferrin precursor	129	prophenoloxidase	321	prophenoloxidase subunit 2	219
prophenoloxidase subunit 2	102	lysozyme	119	prophenoloxidase	314	prophenoloxidase	189
hemolin	52	prophenoloxidase subunit 2	114	lysozyme	244	lysozyme	186
prophenoloxidase	66	cellular retinoic acid binding	103	actin	223	actin	152
actin	58	prophenoloxidase	80	cellular retinoic acid binding	201	cellular retinoic acid binding	152
lysozyme	55	anionic antimicrobial peptide 2	63	anionic antimicrobial peptide 2	176	anionic antimicrobial peptide 2	126
cecropin-D-like peptide	25	cecropin-D-like peptide	2	cecropin-D-like peptide	64	cecropin-D-like peptide	22

Figure 4. Biological process from 17 proteins. Colors in graph A and table B are related. A. Plot shows the number of proteins reported in databases in relationship with immune response. B. Grid represents either controls or microconidia concentrations in each treatment column (microconidia/mL). Column at the left of the protein name provides the color of the biological function. Some proteins have two or more functions, but for immune activity we selected two, including immune response. The column at the right of the protein name shows the concentration level of peptides. Gray represents an unknown function, but the protein domains were identified as follows: purple, catabolic process; blue, oxidation-reduction; yellow, immune response; green, storage; and red, transport.

D-like) contribute to one known function related to immunological response. There is no information concerning the 27 kDa hemolymph in the existing databases (figure 4B).

Hypothetical structure of unidentified proteins of G. mellonella

Two proteins have not been annotated in any protein data bank for *G. mellonella*: actin (actin 4) and cellular retinoic acid binding protein (lipocalin family). Those important orthologous proteins were inferred based on sequences of close Lepidoptera species, such as *B. mori* for actin 4 (GenBank: AGR44824.1 or AGR44827.1) and *D. plexippus* for cellular retinoic acid binding protein (GenBank: EHJ79039.1).

The eight peptides of actin 4 identified by iTRAQ 8 plex from *G. mellonella* exactly match 100% of the *B. mori* actin 4 (GenBank: AGR44824.1, 376 AA), the *D. melanogaster* actin (GenBank: AAA28314.1 or NP_511052.1, 376 AA), the *D. grimshaw* (GenBank: XP_001986647, 375 AA) actin, and the *Antherae pernyi* actin (GenBank: ADJ67594, 376 AA). Those peptides also match 98.91% with the next Lepidoptera species due to their amino acid mutation in S233A and with *D. plexippus* (GenBank: EHJ70060), *S. exigua* (ADJ67594.1, 376 AA) and *P. xuthus* (GenBank: BAG30799) actin (figure 5A). The sequence homology between 376 AA from *B. mori* and *D. melanogaster* is 100%. The alignment with these actin sequences has shown that the conserved domains are the nucleotide-binding domain of the sugar kinase/HSP70/actin superfamily and that the catalytic site residues are exactly the same as the *D. plexippus* sequence (K18, D157, K213, E214, T303, M305, Y306, K336, and the highly conserved secondary and tertiary structure; figure 5A and 5B in blue). The consensus sequence takes the secondary structure hypothesized by the Phyre and i-Tasser servers. We also hypothesized a tertiary structure using the 3MN6 actin structure (Protein Data Bank) from *D. melanogaster* as a pattern, comparing it with AGR44824.1 actin 4 from *B. mori*. The probable substitutions are shown in yellow and green (figure 5B), with the following 18 possible amino acid substitutions in *G. mellonella*: T149A, T163A, S233A, A272S, T279V, Q361E, S369G, (in yellow) E4D, E5D, V77I, T130S, I166V, S235T L300M, M326I, I331V (in green), H89del, and F276del (figure 5B).

The three peptides of the cellular retinoic acid binding protein of *G. mellonella*, identified by iTRAQ 8 plex, match exactly 100% of the *D. plexippus* cellular retinoic acid binding protein (GenBank: EHJ79039.1, 132 AA). A lower percentage match was found for *M. sexta* cellular retinoic acid binding protein (GenBank: AAC24317.1, 132 AA), *B. mori* cellular retinoic acid binding protein (GenBank: NP_001037364.1, 132 AA), *Plutella xylostella* cellular retinoic acid binding protein (GenBank: BAD26694.1, 132 AA) and *A. yamamai* cellular retinoic acid binding protein (GenBank: AGG56524.1, 132 AA). The alignment with those cellular retinoic acid binding protein sequences showed the conserved domains between them (lipocalin superfamily) and their consistent conserved secondary structure (figure 6A). The consensus sequence takes the secondary structure hypothesized by Phyre and i-Tasser servers. The probable substitutions showed the 20 most likely amino acids to be replaced in *G. mellonella*: E2D, V11I/V11T, T22A/T22V, I23L, E28D, T37N, R43K, Q44K/Q44R, D45E, N48G/N48E/N48D, F49Y, V50N, K65S, E68Q, T90I, A98P/A98Q/A98D, L101S, V106I, A120T, and V131A. We also hypothesized a tertiary structure using the EHJ79039.1 cellular retinoic acid binding protein from *D. plexippus* on Phyre server as a pattern. The 3D alignment between the hypothetical structure (orange) and 2CBR (Protein Data Bank, in blue with ATP) cellular retinoic acid binding protein (136 AA) from *Bos*

taurus was made using the Phyre server and visualized with UCSF Chimera version 1.8. This alignment is showing a structural homology with slightly structural differences, despite the low sequence homology with the *B. taurus* sequence (figure 6B).

Discussion

Immunological model and iTRAQ performance

G. mellonella larva is an excellent model to study biological process such as immunological responses and virulence mechanisms and the innate immune system. In our study, we infected the caterpillars with a *F. oxysporum* microconidia suspension. Their survival and mortality were statistically consistent. The hemolymph proteins were in high concentration, as noted by a Bradford assay.

The mold *F. oxysporum* is an inter-kingdom pathogen system and an emergent human, mammalian and vertebrate pathogen [37]. The very same mold strain might equally affect animals and plants [38]. We noted that the larvae of *G. mellonella* incubated at $37°C$ showed an enhanced humoral immune response after fungal infection compared with infected larvae incubated at $25°C$. This enhanced response was manifested by an increased expression of proteins, as other authors have noted with yeast or molds [4], [39], [40], [41]. The larvae hemolymph is able to inhibit *Fusarium* filamentation and after 48 h can still resist from 10^4 to 10^6 microconidia/mL. The quantitation of iTRAQ is achieved by the comparison of peak areas and resultant peak ratios for eight LC/MS/MS reporter ions, ranging from 113–119 to 121 Da. In our case, 119 and 121 were not considered. The ratio of the signal intensities from these tags acts as an indicator of the relative proportion of that peptide between the differently labeled samples [42]. The minimal amount of protein identification achieved after iTRAQ 8 plex analyses was first attributed to the lack of genome information for *G. mellonella*. However, the strategy of matching peptides with closely related species allowed us to resolve and identify *G. mellonella* proteins in the immune response. Thus, it should be feasible to expect a large number of proteins to be identified by their similarity with well characterized insects, such as *Anopheles gambiae*, *Aedes aegypti*, *Apis mellifera* and *D. melanogaster*. These results seem to indicate that lepidopteran species could have evolved independently, a long time ago, from other insects or, at least, that the *G. mellonella* hemolymph proteins, related to the innate immune system, may be absent in other insect species.

Protein expression induced by the fungal attack

Butterflies are holometabolous, meaning that pupae do not feed during metamorphosis; instead, they depend on nutrients that were previously accumulated in the larval period. Proteins, lipids and carbohydrates that have been selectively taken up by the larval fat body from the hemolymph before pupation [43] provide energy and amino acid building blocks for the imaginal tissues.

Banville *et al.* [44], found a down-regulated expression of a range of proteins associated with immune responses in a proteome analysis of 7-day starved larvae. In our case, we found the same result after a 2-day infection with *F. oxysporum* at $25°C$, in contrast with a remarkable proteome over-expression at $37°C$. The 48 hours of starvation with infection challenge triggered the immune response rapidly, increasing temperature from 25 to $37°C$, enhancing protein expression linked to the immune response. Unlike the larvae in the starved stage, which reduced their expression in a range of antimicrobial and immune proteins, including apolipophorin and arylphorin [44], these proteins had a remarkably increased expression at $37°C$.

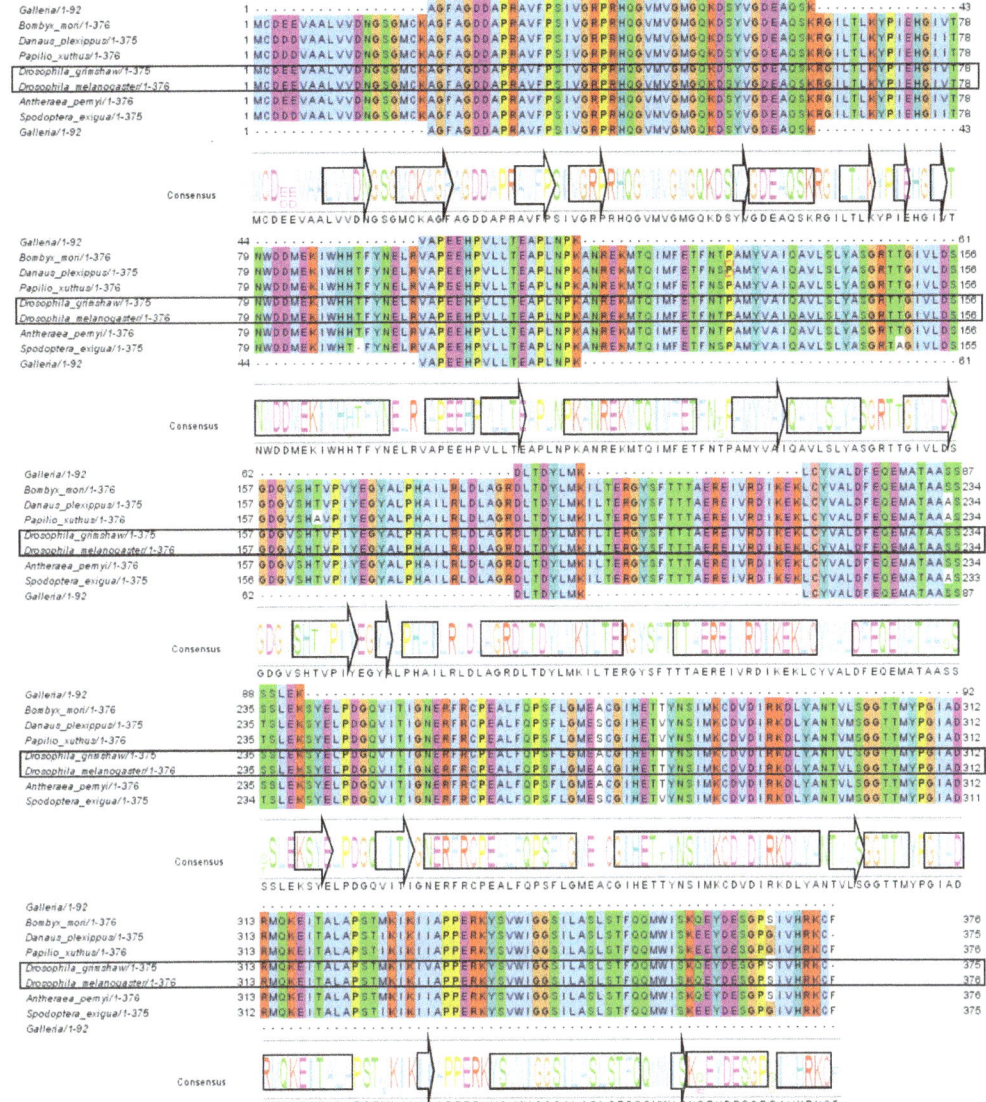

Figure 5. Actin 4 alignment and hypothetical 3D structure. A. The actin alignment of Lepidoptera protein sequences matches that of *G. mellonella*'s actin peptides (first discontinued sequences and dotted) and aligns well (100%) with the *B. mori* protein. Related species and *Drosophila* species were also included. A consensus was obtained on Jalview using a Clustal algorithm. The consensus sequence includes arrows (α helix) and cylinders (β-sheet) obtained by structural analysis. The alignment shows insertions, deletions or substitutions. B. Original tertiary 3MN6 (Protein Data Bank) actin pattern and probable amino acid substitutions are marked in yellow and green. The catalytic site residues can be found in blue.

In our study, arylphorin (AAA74229), which is possibly a hexamerin, had the best peptide coverage and was likely the most abundant protein in the hemolymph, together with apolipophorins. These data were found because arylphorin was down-regulated at 25°C, but markedly up-regulated at 37°C at the same microconidial concentration. Apolipophorin (ALP, AAT76806) is a lipoprotein that plays a role in lipid transport and is likely to be associated with antimicrobial activity because

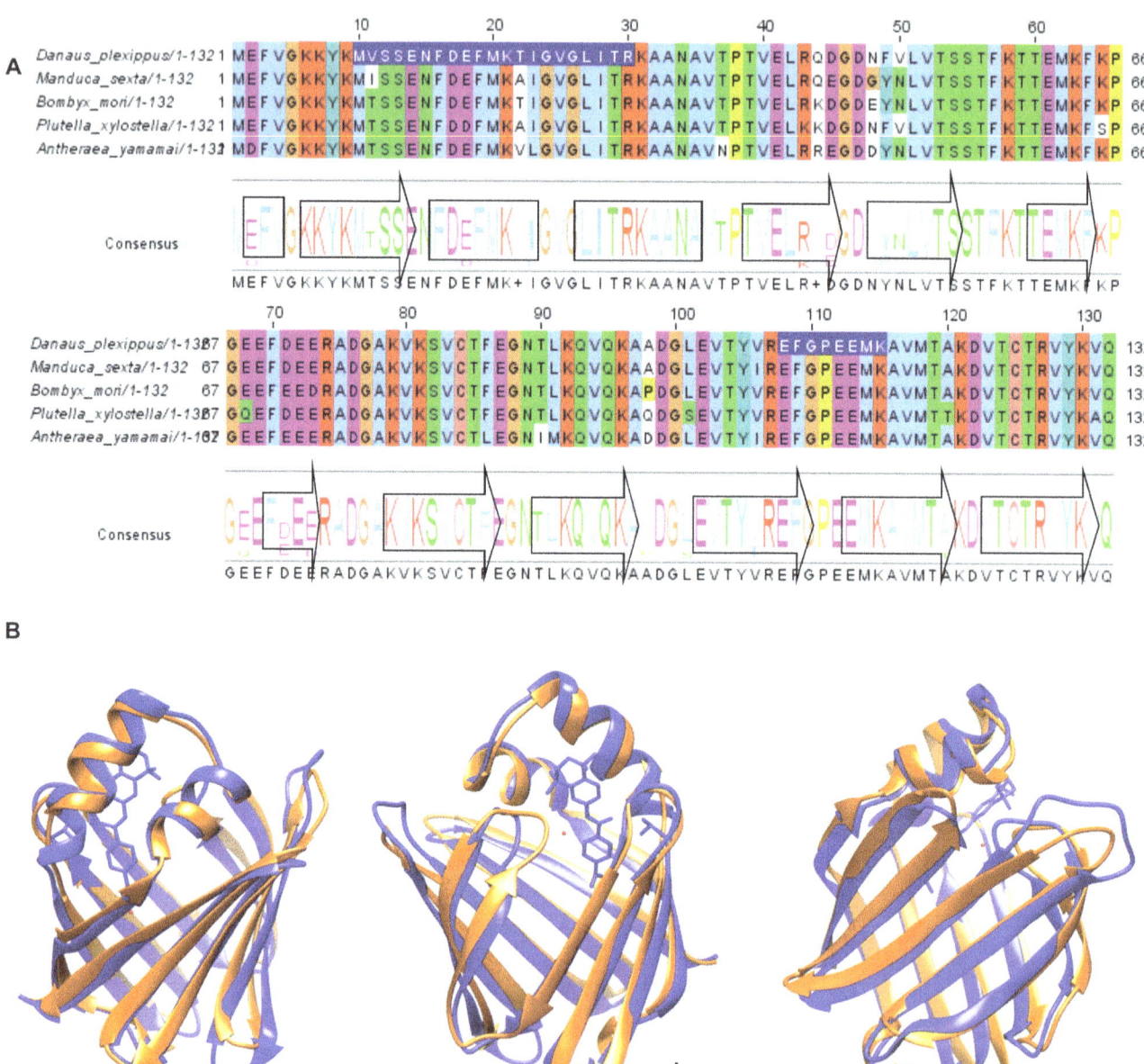

Figure 6. Cellular retinoic acid binding protein or lipocalin A. A. Alignment of different cellular retinoic acid binding proteins from five different Lepidoptera sequences. In the blue box, peptides identified by iTRAQ over the *D. plexippus* sequence can be found. A consensus was obtained on Jalview using Clustal algorithm, including the arrows (α- helix) and cylinders (β-sheet) obtained through a structural analysis. B. Three different side angles amid tertiary 2CBR (Protein Data Bank) cellular retinoic acid binding patterns with our hypothetical lipocalin structure. The alignment of the 3D structures (a,b,c) between the hypothetical *G. mellonella* CRABP (orange) and 2CBR (Protein Data Bank, in blue with ATP) cellular retinoic acid binding protein (136 AA) from *B. taurus*.

lipids (LPs) are secreted to prevent microbial invasions. Furthermore, the functional properties of lipophorin as a lipid carrier are well-characterized [45]. Recently, the study by Banville et al. [44] demonstrated the reduced expression of a range of antimicrobial peptides that are down-regulated in association with the immune response.

Apolipophorin III (apoLPIII, P80703) was the most down-regulated protein in this study, especially after 10^4 microconidia/ mL at 25°C, but it was slightly up-regulated at 10^6 microconidia/ mL (not significant). This protein seems to be unmodified under physiological conditions and all functions and effects observed must be inherent to the protein's structure [46]. Mammalian apolipoproteins, such as apolipoprotein E (apoE), are involved in lipopolisacharide (LPS) detoxification, phagocytosis, and, potentially, pattern recognition. The apoLp-III insect protein is homologous to apoE [47]. ApoLpIII stimulates increases in the hemolymph antibacterial activity [48] and super-oxide production by the blood cells (hemocytes) [49]. The protein was found to participate in immune reactions as an LPS-binding protein (LBP) or as a potentiator of bacteriolytic activity in the hemolymph, but was not involved in lipid transport functions [50], [51]. We believe it probable that ApoLpIII, stimulates antimycotic activity in the hemolymph.

The 27 kDa hemolymph protein precursor (P27k, P83632) has been poorly studied, and very little is known about it in Lepidoptera. Its sister protein in the hemolymph, Spz C-106, controls the expression of the antifungal peptide Drosomycin (Drs) by acting as a ligand of Told1 and inducing an intracellular signaling pathway [52]. In our study, this protein is evidently present in the hemolymph before infection and is up-regulated even at 25°C. However, its expression was remarkably increased at 37°C. These results are striking because, based on our previous meta-analysis, this remarkable expression was unexpected. Richards and Edwards [53], detected the 27 kDa hemolymph protein after 3 days of parasitization. Our results indicate that microconidial invasion stimulates this expression after 2 days.

Cationic protein 8 precursor (CP8, ADI87454) was an abundant protein that was markedly up-regulated at 25°C or 37°C in response to mycotic infection. This protein is a multiligand receptor that can recognize three microorganisms selected in other studies: Escherichia coli, Micrococcus luteus, and C. albicans [54]. The authors attempted to ascertain which immune responses were caused by the binding of the protein to these microorganisms. According to their results, CP8 in G. mellonella presented opsonin activity for the phagocytosis of the three microorganisms by hemocytes [54]. The iTRAQ data provided by this protein showed an extraordinarily up-regulated expression with microconidia invasion, regardless of the temperature.

Larval hemolymph protein (LHP, ABG91580) detected by iTRAQ showed that this glycoprotein is down-regulated at 25°C, but up-regulated at 37°C in response to the infection. This protein is considered another storage glycoprotein that can have different degrees of glycosylation, including different patterns and molecular sizes (LHP74, 76, 76) 81, 82 and 84 [55]. The expression of the two Lhp genes of known sequences (Lhp76 and Lhp82) were monitored in both diapausing and non-diapausing G. mellonella individuals [56], [57]. However, those exceptional studies were not specific for immune responses, and at present, this protein is poorly investigated in G. mellonella, as opposed to past decades during which the protein was profusely studied [55], [58], [59].

Hexamerin (AAA19801) was another abundant protein in the G. mellonella results for this microconidia infection, as the protein is down-regulated at 25°C and up-regulated at 37°C in response to

the infection. Storage hexamerins are important members of the hemocyanin superfamily. Although insects have storage hexamerins, very little is known about their characters or specific functions in G. mellonella and in general in insects [60]. Hexamerins are storage proteins with primordial functions in insect metamorphosis, but they may also have functions outside their role in storage, potentially even in immune responses [61].

Juvenile hormone binding protein (JHBP, AAN06604) is one of the main regulators of insect development and reproduction. It is crucial for proper insect development, acting as a transporter, protector and reservoir of the highly hydrophobic and chemically labile juvenile hormone, JH [62]. The expression was demonstrated to be maintained at a constant level when G. mellonella larvae were challenged with β-glucan [63]. JHBP has been found to down-regulate transferrin [64] and to bind to apolipophorin, arylphorin and hexamerin in the hemolymph, each of which are involved in JHBP molecular traffic [65]. In our study, JHBP maintained almost the same concentration at 25°C after microconidia infection, however, it was over-expressed or up-regulated at 37°C. Those results are in concordance with our findings, as JHBP was likely interacting with arylphorin, apolipophorin, hexamerin (high concentration) and transferrin precursors.

Transferrin has the ability to remove essential iron ions to withstand pathogen invasion, thereby retarding microbial colonization of the insect's hemocoel [66]. Banville et al. [44], found that ferritin and transferrin precursor were reduced in expression by approximately 50% in food-deprived larvae. In our study, the transferrin precursor (AAQ63970) was down-regulated at 25°C, but up-regulated at 37°C. Seitz, et al. [66] found genes that showed homologies with molecules, such as transferrin, which is known to be involved in immunomodulation after bacterial infection in G. mellonella. Transferrin, which is well known for its role in removing iron ions in vertebrates and thereby creating unfavorable environments for bacteria [67], might also be involved in an antibacterial iron-withholding strategies in insects [68]. Transferrin is likely involved in immunomodulation after mycotic infection to create an unfavorable environment in the hemolymph for fungal invasion because the removal of iron is very important for pathogens. Transferrin acts as an antimycotic mediator, preventing oxidative stress and enhancing survival against infections.

Hemolin (ACU09501), a member of the immunoglobulin protein superfamily, interacts in Lepidoptera as an opsonin, defending against potential pathogens and potentially playing a role in tissue morphogenesis. Hemolin also seems to be involved in insect defense against viral infections [69], but does not exhibit direct antibacterial activity. Instead, because it could be up-regulated 18-fold following bacterial injection in Hyalophora cecropia pupae and 30- to 40-fold in M. sexta larvae [70], this protein is thought to be the major inducible protein against fungi in the hemolymph. In our case, we found similar results because the protein was up-regulated 5- to 10-fold with F. oxysporum from 10^4 to 10^6 microconidia/mL at 37°C. The mechanism includes viral interaction with the hemolin anchored to the cellular membrane of the hemocytes. Because hemolin is associated with the hemocytes [71], [72], its function as an opsonin facilitates pathogen recognition and mediates hemocytic immune responses [73], [74]. We consider hemolin to have shown important activity in antimycotic immune responses, as its expression was always up-regulated when temperature or microconidia concentration were elevated.

We also found that prophenoloxidase (POO, AAK64363) and prophenoloxidase subunit 2 (PPOs2, AAQ75026) are important proteins in immune responses. These proteins make up the other coupled group found in this study, similar to apolipophorins and hexamerins. In general, phenoloxidase occurs in arthropod's hemolymph or hemocytes in its inactive form as a proenzyme called prophenoloxidase (PPO) [75], [76]. The activation of the prophenoloxidase triggers the enzymatic reaction that synthetize cytotoxic quinones. The latter protein can be toxic to the host tissue if it is not regulated to the site of infection [5]. The protein has been studied after conidia invasion. The sera from those larvae that received an injection of 10^5 conidia of *A. fumigatus* showed the greater binding of PPO to the conidia compared with the sera from control larvae; however, all pre-treatments resulted in higher levels of this protein's binding [77]. There was evidence of an interaction between the coagulation system and the prophenoloxidase activating cascade [78].

A major defense in the innate immune system in invertebrates is melanization in response to pathogens or damaged tissues. This process constitutes the first humoral defense and is controlled by the enzyme phenoloxidase (PO) [79]. The proPO-activating system is triggered by the presence of minute amounts of compounds of microbial origins, such as β-1,3-glucans, lipopolysaccharides, and peptidoglycans, ensuring that the system will become active in the presence of potential pathogens [79]. We found that PPO and PPO subunit 2 proteins had the same patterns of expression in different treatments and remained at the same concentration at 25°C with or without microconidia. However, these proteins were significantly up-regulated with *F. oxysporum* from 10^4 to 10^6 microconidia/mL at 37°C.

Lysozymes (P82174) are most likely the best studied antibacterial enzymes and act as a model for structure-function studies [80]. However, these proteases are poorly studied in insects' innate immune systems. The insect's immune response is mediated by pattern recognition proteins that signal to expression effectors, including the antibacterial lysozyme protein [81], [82]. We found that lysozymes comprise important antimycotic enzymatic activities, as noted by its antiviral or antibacterial activity in many *in vitro* studies. We observed the remarkable over-expression of lysozymes in all microconidia treatments in comparison to the controls. It appears that lysozymes take on other roles besides degrading the bacterial cell wall, and it has been shown that this protease inhibits the enzymatic activity of mosquito phenoloxidase, thereby modulating melanization [82], [83].

Recently, the study by Banville *et al.* [44] demonstrated the reduced expression of a range of antimicrobial peptides that are down-regulated in the expression of a range of proteins associated with the immune response. We found it an interesting coincidence that cecropin-D-like peptide (P85210) and anionic antimicrobial peptide 2 (P85216) were both present in low concentrations. iTRAQ is likely the most accurate technique for studying peptides. Cecropins might normally offer protection against bacteria, but we found a marked down-regulation of Cecropin-D-like and Anionic antimicrobial peptide 2 peptides. In larvae challenged with fungi, these peptides were slightly up-regulated at 37°C compared with controls and other proteins. It is possible that these peptides are important among insects' immune responses to bacteria, according to Mak *et al.* [84], but are irrelevant for immune responses to fungi. Other studies should be undertaken to better clarify their final role *in vivo*, as these peptides have been well studied *in vitro* [28], [85], [86]. An interesting transcriptome study by Vogel *et al.* [87] catalogued the immune responses of a large number of peptides and proteins to lipopolysaccharides. In our study with fungi, we found a low expression of the serine protease inhibitor; however, we believe that it is still difficult to correlate immune responses between the transcriptome and proteome with different antigens.

Expression in the immune response and structure of two renowned orthologous proteins

Unlike the most recent proteins discussed, actin (AGR44824 from *B. mori*) and cellular retinoic acid binding protein (CRABP, lipocallin family) are quite well studied in eukaryotic organisms. Although, they still have not been annotated in *G. mellonella* and are associated with many biological processes including immune responses. In this study, the actin expression was up-regulated continuously at any microconidia concentration or any set temperature. This result is in accordance with those of other authors because actin-5C, one of two cytoskeletal proteins, is up-regulated after short H_2O_2 treatments or other stress situations, such as anoxia or ethanol treatment in *D. melanogaster* [88]. The functional genomic analysis of phagocytosis also established the participation of actin cytoskeleton regulation proteins in innate immunity of *D. melanogaster* [89]. During the course of an infection, immune cells migrate toward the site of microbial entry with the goal of eliminating pathogens. This process requires not only cell migration but also changes in cell adhesion and phagocytosis, which universally depend on the dynamics of the actin network (i.e., the polymerization or depolymerization of actin filaments) [90].

Cellular retinoic acid binding protein (CRABP, EHJ79039 *D. plexippus*) is an important carrier of retinoic acid (RA or vitamin A) and, in mammals, is involved in inflammation and detoxification. RA is a potential chemotherapeutic agent in the treatment of cancers [91], [92] and is critical as a regulator of proper skin function [93]. There are few studies on CRABP and the innate immune system in insects; however, various works have characterized the protein in other organisms [94], [95]. This is an important matter in mammals because RA is a key modulator of the innate immune system [96]. As in other proteins in our study, CRABP was down-regulated with 10^4 microconidia/mL at 25°C, but up-regulated at 37°C (non-significant). CRABP either was not greatly affected or produced similar results to the controls at 10^6 microconidia/mL at 37°C, showing that CRABP most likely transports RA and lipids during stress conditions to stimulate the immune system.

The iTRAQ approach was not only useful to detect and quantify peptides; it was also useful to determine protein structures. It was interesting to address a 3D structural analysis using sequence peptides because the homologous protein information in closed species has been well supported, and the protein structure has been thoroughly investigated. Actin has been studied in insects [97] and three-dimensional structures have been characterized in *D. melanogaster*, such as the 3MN6 actin structure at the Protein Data Bank [98]. Compared with other lepidopteran or insect actins, our peptides are significantly conserved, allowing the inference of the actin's secondary or tertiary structure. Even if the *G. mellonella* actin structure is slightly different from *D. mellanogaster* actin, the homology of 7 peptides with *D. plexippous* actin allows the study of such structures in Lepidoptera. Non-synonymous substitutions do not seem to change the basic structure of the insect, as determined with *D. mellanogaster* 3MN6. One deletion was present only in the last AA C-terminus, and the number of hypothetical α-helices and β-sheets did not change domains.

The structure of CRABP has been well-studied in humans [99], and three-dimensional structures have been examined in 2CBR CRABP cow structure (Protein Data Bank). The structural homology of CRABPs between insects and mammals is considerable, reflecting their conserved functions among animals [99]. In *M. sexta*, the protein structure noticeably compared with those of mice and cows, as in our case [94]. Non-synonymous substitutions did not change the basic structure or domains. Finally, the number of hypothetical α-helices and β-sheets was a little shorter or longer than that in the 2CBR CRABP cow protein.

Conclusions

As in other works, our study verified that the use of isobaric tags for the relative and absolute quantification (iTRAQ) of protein expression is a powerful proteomic tool that can be combined with the accuracy of Mass Spectrometry for protein identification. This tool constitutes a consistent method to detect proteins associated with the innate immune system of *G. mellonella* in response to infection caused by *F. oxysporum*.

In addition, iTRAQ was a reliable quantitative proteomic approach to detect and quantify the expression levels of immune system proteins and peptides, showing differential expressions in varying temperatures and microconidia concentrations, especially when the larvae were resistant to microconidia treatment. Our proteomic study allowed us to monitor the improvement of the immune response of *G. mellonella* against challenges with *F. oxysporum* at 37°C. In particular, it was found that 10^4 microconidia/mL at 37°C over expressed many more proteins than other treatments. It is likely that most of the 17 main proteins contributed to caterpillar survival by protecting against the fungal infection and taking advantage of the heat shock, generating a real benefit in terms of evolutionary fitness.

The statistical validation of protein expression by RT-PCR allowed us to infer that iTRAQ expression data was consistent. It is likely that all proteins expressed in stress conditions (microconidia infection and high temperature) have not only been due to the exclusive effect of immunological response because some proteins were stimulated by temperature or metamorphic grown. In our case, infection and temperature increase were the main causes to over-express many of the 17 proteins studied, as supported by bibliographic metadata, but it is necessary to deepen in the *G. mellonella-F. oxysporum* model to better understand the innate immune system of Lepidoptera.

Supporting Information

Figure S1   Data from table S6D. Sets of validation iTRAQ results. Validation of iTRAQ results with q-PCR. The S6 sets of tables (A, B, C, D, E and F) and Figures S1 and S2 results.

Figure S2   Data from table S6E. Sets of validation iTRAQ results. Validation of iTRAQ results with q-PCR. The S6 sets of tables (A, B, C, D, E and F) and Figures S1 and S2 results.

Table S1   Bradford assay. Bradford quantification of proteins in hemolymph after caterpillar sacrifice.

Table S2   F and t tests for 10^4, 10^5, 10^6 microconidia/mL at 25 and 37°C. First statistical assessment for microconidia concentration at 25 and 37°C. The concentrations 10^7 and 10^8 displayed sn (no survivals) and nt (not tested).

Table S3   F and t tests for 10^4, 10^6 microconidia/mL at 25 and 37°C. Second statistical assessment for 10^4 and 10^6 microconidia/mL concentration at 25 and 37°C.

Table S4   The 59 proteins detected by iTRAQ. ITRAQ Result: S1–S4, C1–C4 bias ProGroup NCBInr. Denominator 113, from YEPD repository platform.

Table S5   The 17 proteins detected by iTRAQ.

Table S6   Sets of validation iTRAQ results. Validation of iTRAQ results with q-PCR. The S6 sets of tables (A, B, C, D, E and F) and Figures S1 and S2 results. **Table S6A**, Pair of channels selected randomly. Next Data (tables and plots) are the statistic assessment to evaluate the relationship between iTRAQ and q-PCR. **Table S6B**, iTRAQ protein ratio relationship. **Table S6C**, Q-PCR from mRNA. **Table S6D**, Comparison iTRAQ-Q-PCR. **Table S6E**, Comparison of Q-PCR-iTRAQ. **Table S6F**, Chi-square for iTRAQ-Q-PCR.

Acknowledgments

We thank Mrs. Nora Restrepo, Dean of The Faculty of Natural and Exact Sciences, Universidad de Antioquia, for her support and Drs. Shiftman and Abbott for the MS/MS core facilities at Yale University and their important advices.

Author Contributions

Conceived and designed the experiments: AMG MC ABP CP. Performed the experiments: AMG. Analyzed the data: AMG MC ABP CP. Contributed reagents/materials/analysis tools: AMG MC ABP CP. Wrote the paper: AMG MC ABP CP.

References

1. Bender JK, Wille T, Blank K, Lange A, Gerlach RG (2013) LPS structure and PhoQ activity are important for Salmonella Typhimurium virulence in the *Galleria mellonella* infection model. PLoS One 8: e73287.
2. Chamilos G, Lionakis MS, Lewis RE, Kontoyiannis DP (2007) Role of mini-host models in the study of medically important fungi. Lancet Infect Dis 7: 42–55.
3. Champion OL, Cooper IAM, James SL, Ford D, Karlyshev A, et al. (2009) *Galleria mellonella* as an alternative infection model for Yersinia pseudotuberculosis. Microbiology 155: 1516–1522. doi:10.1099/mic.0.026823-0
4. Jackson JC, Higgins L, Lin X (2009) Conidiation color mutants of *Aspergillus fumigatus* are highly pathogenic to the heterologous insect host *Galleria mellonella*. PLoS One 4: e4224.
5. Mylonakis E, Moreno R, El Khoury J, Idnurm A, Heitman J, et al. (2005) *Galleria mellonella* as a Model System To Study *Cryptococcus neoformans* Pathogenesis. Infect Immun 73: 3842–3850. doi:10.1128/IAI.73.7.3842
6. Fuchs BB, O'Brien E, El Khoury JB, Mylonakis E (2010) Methods for using *Galleria mellonella* as a model host to study fungal pathogenesis. Virulence 1: 475–482.
7. Bergin D, Reeves EP, Renwick J, Wientjes FB, Kavanagh K (2005) Superoxide Production in *Galleria mellonella* Hemocytes: Identification of Proteins Homologous to the NADPH Oxidase Complex of Human Neutrophils. Infect Immun 73: 4161–4170. doi:10.1128/IAI.73.7.4161
8. Jander G, Rahme LG, Ausubel FM (2000) Positive correlation between virulence of *Pseudomonas aeruginosa* mutants in mice and insects. J Bacteriol 182: 3843–3845.
9. Seed KD, Dennis JJ (2008) Development of *Galleria mellonella* as an alternative infection model for the *Burkholderia cepacia* complex. Infect Immun 76: 1267–1275.
10. Lebreton F, Le Bras F, Reffuveille F, Ladjouzi R, Giard JC, et al. (2011) *Galleria mellonella* as a model for studying *Enterococcus faecium* host persistence. J Mol Microbiol Biotechnol 21: 191–196. doi:10.1159/000332737
11. Aperis G, Fuchs BB, Anderson CA, Warner JE, Calderwood SB, et al. (2007) *Galleria mellonella* as a model host to study infection by the *Francisella tularensis* live vaccine strain. Microbes Infect 9: 729–734.
12. Harding CR, Schroeder GN, Reynolds S, Kosta A, Collins JW, et al. (2012) *Legionella pneumophila* pathogenesis in the *Galleria mellonella* infection model. Infect Immun 80: 2780–2790.

13. Mukherjee K, Altincicek B, Hain T, Domann E, Vilcinskas A, et al. (2010) *Galleria mellonella* as a model system for studying *Listeria* pathogenesis. Appl Environ Microbiol 76: 310–317.

14. Miyata S, Casey M, Frank DW, Frederick M, Ausubel FM (2003) Use of the *Galleria mellonella* Caterpillar as a Model Host To Study the Role of the Type III Secretion System in *Pseudomonas aeruginosa* Pathogenesis. Infect Immun 71: 2404–2413. doi:10.1128/IAI.71.5.2404

15. Ott L, McKenzie A, Baltazar MT, Britting S, Bischof A, et al. (2012) Evaluation of invertebrate infection models for pathogenic corynebacteria. FEMS Immunol Med Microbiol 65: 413–421.

16. Slater JL, Gregson L, Denning DW, Warn PA (2011) Pathogenicity of *Aspergillus fumigatus* mutants assessed in *Galleria mellonella* matches that in mice. Med Mycol 49 Suppl 1: S107–S113. doi: 10.3109/13693786.2010.494255

17. Brennan M, Thomas DY, Whiteway M, Kavanagh K (2002) Correlation between virulence of *Candida albicans* mutants in mice and *Galleria mellonella* larvae. FEMS Immunol Med Microbiol 34: 153–157.

18. Fraser MJ, Smith GE, Summers MD (1983) Acquisition of Host Cell DNA Sequences by Baculoviruses: Relationship Between Host DNA Insertions and FP Mutants of Autographa californica and *Galleria mellonella* Nuclear Polyhedrosis Viruses. J Virol 47: 287–300.

19. Simpson AA, Chipman PR, Baker TS, Tijssen P, Rossmann MG (1998) The structure of an insect parvovirus (*Galleria mellonella* densovirus) at 3.7 A resolution. Structure 6: 1355–1367.

20. Navarro-Velasco GY, Prados-Rosales RC, Ortíz-Urquiza A, Quesada-Moraga E, Di Pietro A (2011) *Galleria mellonella* as model host for the trans-kingdom pathogen *Fusarium oxysporum*. Fungal Genet Biol 48: 1124–1129. doi.org/10.1016/j.fgb.2011.08.004

21. Coleman JJ, Muhammed M, Kasperkovitz PV, Vyas JM, Mylonakis E (2012) *Fusarium* pathogenesis investigated using *Galleria mellonella* as a heterologous host. Fungal Biol 115: 1279–1289. doi101016/j.funbio201109005

22. Lehane MJ (2005) The biology of blood-sucking in insects. Cambridge: Cambridge University Press. 337p.

23. Lesch C, Theopold U (2007) Insect Immunology: Methods to study hemolymph clotting in insects. London: Academic Press. 360p.

24. Schmidt O, Theopold U, Beckage NE (2008) Insect Immunology: Insect and vertebrate inmunity: key similarities versus differences. London: Academic Press. 347 p.

25. Kingsmore SF (2006) Multiplexed protein measurement: technologies and applications of protein and antibody arrays. Nat Rev Drug Discov 5: 310–320.

26. Thongboonkerd V (2007) Practical points in urinary proteomics. J Proteome Res 6: 3881–3890.

27. Renwick J, Daly P, Reeves EP, Kavanagh K (2006) Susceptibility of larvae of *Galleria mellonella* to infection by *Aspergillus fumigatus* is dependent upon stage of conidial germination. Mycopathologia 161: 377–384. doi:10.1007/s11046-006-0021-1

28. Cytryńska M, Mak P, Zdybicka-Barabas A, Suder P, Jakubowicz T (2007) Purification and characterization of eight peptides from *Galleria mellonella* immune hemolymph. Peptides 28: 533–546.

29. Andrejko M, Mizerska-Dudka M, Jakubowicz T (2009) Antibacterial activity in vivo and in vitro in the hemolymph of *Galleria mellonella* infected with *Pseudomonas aeruginosa*. Comp Biochem Physiol B Biochem Mol Biol 152: 118–123.

30. Bradford MM (1976) A rapid and sensitive method for the quantitation of microgram quantities of protein utilizing the principle of protein-dye binding. Anal Biochem 72: 248–254.

31. Applied Biosystems AB Sciex (2010) Protocol. iTRAQ Reagents - 8 plex. Amine-Modifying Labeling Reagents for Multiplexed Relative and Absolute Protein Quantitation. AB Sciex Pte Ltd.

32. Altschul SF, Gish W, Miller W, Myers EW, Lipman DJ (1990) Basic local alignment search tool. J Mol Biol 215: 403–410.

33. Edgar RC (2004) MUSCLE: multiple sequence alignment with high accuracy and high throughput. Nucleic Acids Res 32: 1792–1797.

34. Larkin M, Blackshields G, Brown NP, Chenna R, McGettigan P, et al. (2007) Clustal W and Clustal X version 2.0. Bioinforma Appl Note 23: 2947–2948.

35. Kelley L, Sternberg MJE (2009) Protein structure prediction on the Web: a case study using the Phyre server. Nat Protoc 4: 363–371.

36. Zhang Y (2008) I-TASSER server for protein 3D structure prediction. BMC Bioinformatics 9: 1–8.

37. Assaf C, Goerdt S, Seibold M, Orfanos C (2000) Clinical picture Cutaneous hyalohyphomycosis. Lancet 356: 1185.

38. Ortoneda M, Guarro J, Madrid MP, Roncero MIG, Mayayo E, et al. (2004) *Fusarium oxysporum* as a Multihost Model for the Genetic Dissection of Fungal Virulence in Plants and. Infect Immun 72: 1760–1766. doi:10.1128/IAI.72.3.1760

39. Wojda I, Jakubowicz T (2007) Humoral immune response upon mild heat-shock conditions in *Galleria mellonella* larvae. J Insect Physiol 53: 1134–1144.

40. Mowlds P, Kavanagh K (2008) Effect of pre-incubation temperature on susceptibility of *Galleria mellonella* to infection by *Candida albicans*. Mycopathologia 165: 5–12.

41. Mylonakis E (2008) *Galleria mellonella* and the study of fungal pathogenesis: making the case for another genetically tractable model host. Mycopathologia 165: 1–3. doi:10.1007/s11046-007-9082-z

42. Unwin RD (2010) Quantification of proteins by iTRAQ. In: Cutillas PR, Timms JF, editors. LC-MS/MS in Proteomics. Methods in Molecular Biology.

43. Dordrecht: Springer Science+Bussiness media LLC. pp. 658: 205–215. doi: 10.1007/978-1-60761-780-8_12

43. Hansen IA, Gutsmann V, Meyer SR, Scheller K (2003) Functional dissection of the hexamerin receptor and its ligand arylphorin in the blowfly Calliphora vicina. Insect Mol Biol 12: 427–432.

44. Banville N, Browne N, Kavanagh K (2012) Effect of nutrient deprivation on the susceptibility of *Galleria mellonella* larvae to infection. Virulence 3: 497–503.

45. Ryan RO, van der Horst DJ (2000) Lipid transport biochemistry and its role in energy production. Ann Rev Entomol 45: 233–60.

46. Weise C, Franke P, Kopáček P, Wiesner A (1998) Primary structure of apolipophorin-III from the greater wax moth, *Galleria mellonella*. J Protein Chem 17: 633–641.

47. Whitten MM, Tew IF, Lee BL, Ratcliffe N (2004) A novel role for an insect apolipoprotein (apolipophorin III) in beta-1,3-glucan pattern recognition and cellular encapsulation reactions. J Immunol 172: 2177–2185.

48. Park SY, Kim CH, Jeong WH, Lee JH, Seo SJ, et al. (2005) Effects of two hemolymph proteins on humoral defense reactions in the wax moth, *Galleria mellonella*. Dev Comp Immunol 29: 43–51.

49. Dettloff M, Wiesner A (1999) Immune stimulation by lipid-bound apolipophorin III. In: Wiesner A, Dunphy GB, Marmaras VJ, Morishima I, Sugumaran M, Yamakawa M, editors. Techniques in Insect Immunology. Fair Haven: SOS Publications. pp. 243–251.

50. Bulet P, Hetru C, Dimarcq JL, Hoffmann D (1999) Antimicrobial peptides in insects; structure and function. Dev Comp Immunol 23: 329–344.

51. Casteels P (1998) Immune response in Hymenoptera. In: Brey PT, Hultmark D, editors. Molecular mechanisms of immune responses in insects. London: Chapman & Hall pp. 92–110.

52. Weber ANR, Tauszig-Delamasure S, Hoffmann J, Lelièvre E, Gascan H, et al. (2003) Binding of the *Drosophila* cytokine Spätzle to Toll is direct and establishes signaling. Nat Immunol 4: 794–800.

53. Richards EH, Edwards JP (1999) Parasitism of *Lacanobia oleracea* (Lepidoptera, noctuidae) by the ectoparasitic wasp Eulophus pennicornis, results in the appearance of a 27 kDa parasitism-specific protein in host plasma. Insect Biochem Mol Biol 29: 557–569.

54. Kim CH, Shin YP, Noh MY, Jo YH, Han YS, et al. (2010) An insect multiligand recognition protein functions as an opsonin for the phagocytosis of microorganisms. J Biol Chem 285: 25243–25250.

55. Miller SG, Silhacek DL (1982) Identification and purification of storage proteins in tissues of the greater wax moth *Galleria mellonella*. Insect Biochem 12: 277–292.

56. Godlewski J, Kludkiewicz B, Grzelak K, Cymborowski BX (2001) Expression of larval hemolymph proteins (Lhp) genes and protein synthesis in the fat body of greater wax moth (*Galleria mellonella*) larvae during diapause. J Insect Physiol 47: 759–766.

57. Godlewski J, Khudkiewicz B, Grzelak K, Beręsewicz M, Cymborowski B (2003) Hormonal regulation of the expression of two storage proteins in the larval fat body of the greater wax moth (*Galleria mellonella*). J Insect Physiol 49: 551–559.

58. Levenbook L (1985) Insect storage proteins. In: Kerkut GA, Gilbert LI, editors. Comprehensive Insect Physiology, Biochemistry and Pharmacology. Oxford: Pergamon Press. pp. 307–346.

59. Ray A, Memmel NA, Kumaran AK (1987) Developmental regulation of the larval hemolymph protein genes in *Galleria mellonella*. Roux Arch Dev Biol 196: 414–420.

60. Tang B, Wang S, Zhang F (2010) Two storage hexamerins from the beet armyworm Spodoptera exigua: cloning, characterization and the effect of gene silencing on survival. BMC Mol Biol 11: 1–13.

61. Ramos Martins J, Bitondi M (2012) Nuclear Immunolocalization of Hexamerins in the Fat Body of Metamorphosing Honey Bees. Insects 3: 1039–1055.

62. de Kort CAD, Granger NA (1996) Regulation of JH titers: the relevance of degradative enzymes and binding proteins. Arch. Insect Biochem Physiol 33: 1–26.

63. Mowlds P, Coates C, Renwick J, Kavanagh K (2010) Dose-dependent cellular and humoral responses in *Galleria mellonella* larvae following beta-glucan inoculation. Microbes Infect 12: 146–153.

64. Geiser DL, Winzerling JJ (2012) Insect transferrins: multifunctional proteins. Biochim Biophys Acta 1820: 437–451.

65. Zalewska M, Kochman A, Estève J-P, Lopez F, Chaoui K, et al. (2009) Juvenile hormone binding protein traffic – Interaction with ATP synthase and lipid transfer proteins. Biochim Biophys Acta 1788: 1695–1705.

66. Seitz V, Clermont A, Wedde M, Hummel M, Vilcinskas A, et al. (2003) Identification of immunorelevant genes from greater wax moth (*Galleria mellonella*) by a subtractive hybridization approach. Dev Comp Immunol 27: 207–215.

67. Kumagai T, Awai M, Okada S (1992) Mobilization of iron and iron-related proteins in rat spleen after intravenous injection of lipopolysaccharides (LPS). Pathol Res Pract 188: 931–41.

68. Ciencialova T, Neubauerova M, Sanda R, Sindelka J, Cvacka Z, et al. (2008) Mapping the peptide and protein immune response in the larvae of the fleshfly Sarcophaga bullata. J Pept Sci 14: 670–682.

69. Terenius O (2008) Hemolin-A lepidopteran anti-viral defense factor? Dev Comp Immunol 32: 311–316.

70. Bettencourt R, Lanz-Mendoza H, Lindquist KR, Faye I (1997) Cell adhesion properties of hemolin, an insect immune protein in the Ig superfamily. Eur J Biochem 250: 630–637.

71. Ladendorff NE, Kanost MR (1991) Bacteria-induced protein-P4 (hemolin) from Manduca sexta—a member of the immunoglobulin superfamily which can inhibit hemocyte aggregation. Arch Insect Biochem Physiol 18: 285–300.

72. Zhao L, Kanost MR (1996) In search of a function for hemolin, a hemolymph protein from the immunoglobulin superfamily. J Insect Physiol 42: 73–79.

73. Lanz-Mendoza H, Bettencourt R, Fabbri M, Faye I (1996) Regulation of the insect immune response: the effect of hemolin on cellular immune mechanisms. Cell Immunol 169: 47–54.

74. Yu X-Q, Zhu Y-F, Ma C, Fabrick J, Kanost MR (2002) Pattern recognition proteins in Manduca sexta plasma. Insect Biochem Mol Biol 32: 1287–1293.

75. Ashida M, Yamazaki HI (1990) Biochemistry of the phenoloxidase system in insects: With special respect to its activation, in Molting, and Metamorphosis. In: Ohnishi E, Ishizaki H, editors. Tokyo: Japan Sci. Soc. Press. pp. 239–256.

76. Brunet PCJ (1980) The metabolism of the aromatic amino acids concerned in the cross-linking of insect cuticle. Insect Biochem 10: 467–500.

77. Fallon JP, Troy N, Kavanagh K (2011) Pre-exposure of Galleria mellonella larvae to different doses of Aspergillus fumigatus conidia causes differential activation of cellular and humoral immune responses. Virulence 2: 413–421.

78. Li D, Scherfer C, Korayem AM, Zhao Z, Schmidt O, et al. (2002) Insect hemolymph clotting: evidence for interaction between the coagulation system and the prophenoloxidase activating cascade. Insect Biochem Mol Biol 32: 919–928.

79. Cerenius L, Söderhäll K (2004) The prophenoloxidase-activating system in invertebrates. Immunol Rev 198: 116–126.

80. Jolles P, Jolles J (1984) What's new in lysozyme research? Always a model system, today as yesterday. J Mol Cell Biochem 1984: 63: 165–89.

81. Kanost MR, Jiang H, Yu XQ (2004) Innate immune responses of a lepidopteran insect, Manduca sexta. Immunol Rev 198: 97–105

82. Sotelo-Mundo RR, López-Zavala AA, García-Orozco KD, Arvizu-Flores AA, Velázquez-Contreras EF, et al. (2007) The lysozyme from insect (Manduca sexta) is a cold-adapted enzyme. Protein Pept Lett 14: 774–778.

83. Li B, Paskewitz SM (2006) A role for lysozyme in melanization of Sephadex beads in Anopheles gambiae. J Insect Physiol 52: 936–942.

84. Mak P, Chmiel D, Gacek GJ (2001) Antibacterial peptides of the moth Galleria mellonella. Acta Biochim Pol 48: 1191–1195.

85. Hultmark D, Engstrom A, Bennich H, Kapur R, Boman HG (1982) Insect Immunity: Isolation and Structure of Cecropin D and Four Minor Antibacterial Components from Cecropia Pupae. 217: 207–217.

86. Brogden KA, Ackermann M, McCray Jr PB, Tack BF (2003) Antimicrobial peptides in animals and their role in host defences. Int J Antimicrob Agents 22: 465–478.

87. Vogel H, Altincicek B, Glöckner G, Vilcinskas A (2011) A comprehensive transcriptome and immune- gene repertoire of the lepidopteran model host Galleria mellonella. BMC Genomics 12: 308.

88. Courgeon AM, Maingourd M, Maisonhaute C, Montmory C, Rollet E, et al. (1993) Effect of hydrogen peroxide on cytoskeletal proteins of Drosophila cells: comparison with heat shock and other stresses. Exp Cell Res 204: 30–7.

89. Rämet M, Manfruelli P, Pearson A, Mathey-Prevot B, Ezekowitz RA (2002) Functional genomic analysis of phagocytosis and identification of a Drosophila receptor for E. coli. Nature 416: 644–8.

90. Fauvarque MO, Williams MJ (2011) Drosophila cellular immunity: a story of migration and adhesion. J Cell Sci 1: 1373–82. doi: 10.1242/jcs.064592

91. Lotan R (1996) Retinoids and their receptors in modulation of differentiation, development, and prevention of head and neck cancers. Anticancer Res 16: 2415–2420.

92. Chomienne C, Fenaux P, Degos L (1996) Retinoid diferentiation therapy in promyelocytic leukemia. FASEB J 10: 1025–1030.

93. Fisher GJ, Voorhees JJ (1996) Molecular mechanisms of retinoid actions in skin. FASEB J 10: 1002–1013.

94. Mansfield SG, Cammer S, Alexander SC, Muehleisen DP, Gray RS, et al. (1998) Molecular cloning and characterization of an invertebrate cellular retinoic acid binding protein. Proc Natl Acad Sci U S A 95: 6825–6830.

95. Evans JD, Wheeler DE (1999) Differential gene expression between developing queens and workers in the honey bee, Apis mellifera. Proc Natl Acad Sci U S A 96: 5575–5580.

96. Wojtal KA, Wolfram L, Frey-Wagner I, Lang S, Scharl M, et al. (2013) The effects of vitamin A on cells of innate immunity in vitro. Toxicol Vitr 27: 1525–1532.

97. Manseau LJ, Ganetzky B, Craigt EA (1988) Molecular and genetic Characerization of the Drosophila melanogaster 87E Actin Gene Region. Genetics 119: 407–420.

98. Ducka AM, Joel P, Popowicz GM, Trybus KM, Schleicher M, et al. (2010) Structures of actin-bound Wiskott-Aldrich syndrome protein homology 2 (WH2) domains of Spire and the implication for filament nucleation. Proc Natl Acad Sci U S A 107: 11757–11762.

99. Chaudhuri BN, Kleywegt GJ, Broutin-L'Hermite I, Bergfors T, Senn H, et al. (1999) Structures of cellular retinoic acid binding proteins I and II in complex with synthetic retinoids. Acta Crystallogr 55: 1850–1857.

11

Comparative Studies on Detergent-Assisted Apocytochrome b₆ Reconstitution into Liposomal Bilayers Monitored by Zetasizer Instruments

Michał A. Surma[9][¤], Andrzej Szczepaniak, Jarosław Króliczewski*[9]

Faculty of Biotechnology, University of Wroclaw, Wroclaw, Poland

Abstract

The present paper is a systematic, comparative study on the reconstitution of an apocytochrome b₆ purified from a heterologous system using a detergent-free method and reconstitution into liposomes performed using three different detergents: SDS, Triton X-100 and DM, and two methods of detergent removal by dialysis and using Bio-Beads. The product size, its distribution and zeta potential, and other parameters were monitored throughout the process. We found that zeta potential of proteoliposomes is correlated with reconstitution efficiency and, as such, can serve as a quick and convenient quality control for reconstitution experiments. We also advocate using detergent-free protein purification methods as they allow for an unfettered choice of detergent for reconstitution, which is the most crucial factor influencing the final product parameters.

Editor: Monica M. Jablonski, The University of Tennessee Health Science Center, United States of America

Funding: This research was supported by the University of Wrocław, Poland to AS. Publication fee was partially paid by the Wrocław Center for Biotechnology program KNOW (National Scientific Leadership Center) 2014-2018.

Competing Interests: The authors have read the journal's policy and have the following conflicts: Michal A. Surma has paid employment at Lipotype, GmbH. The following authors have no competing interests: Jarosław Kroeliczewski, Andrzej Szczepaniak.

* Email: jarekk@ibmb.uni.wroc.pl

¤ Current address: Lipotype GmbH, Dresden, Germany

9 These authors contributed equally to this work.

Introduction

Membrane proteins account for 20–25% of all open reading frames in sequenced genomes, and fulfill a wide range of functions in cells [1]. Our knowledge of the assembly of this class of proteins is still poor. The study of membrane proteins *in situ* is difficult for several reasons: proteins assemble in larger complexes; interact with other proteins, and with other cell components. Therefore model bilayer *in vitro* systems are often used for studying membrane proteins as they negate many of the problems associated with the use of membrane vesicles [2]. Performing *in vitro* studies requires solubilization of a membrane protein with a suitable detergent, followed by a functional reconstitution in a well-characterized bilayer system. However such studies are greatly hampered due to the difficulties in purifying membrane proteins in large amounts, which is the first limitation in this approach.

Two types of reconstitution system are routinely used: planar lipid bilayers and liposomes. The planar lipid bilayer system consists of a phospholipid bilayer supported within an approximately 1 mm diameter pore in a hydrophobic partition separating two aqueous chambers [3]. The liposome system consists of spherical lipid bilayers enclosing an aqueous compartment and separating it from the aqueous milieu [4]. Planar lipid bilayers have the advantage of a well-defined, though enforced, geometry and simple to assay transport between the aqueous compartments on either side of a membrane. In comparison to liposomes, the major disadvantages of planar bilayers include difficulties in controlling the extent of reconstitution of the protein into the bilayer and measuring the exact amount that has been reconstituted [5]. Planar lipid bilayers are also inherently unstable and very sensitive to the presence of free detergent [6]. Moreover, liposomal lipid composition is easy to manipulate, which taken together with the ease of controlling reconstituted protein amounts made proteoliposomes the leading tool for characterizing transmembrane transport by different membrane proteins [7–10].

Liposomes can be described by a number of properties. Two of the most important are particle size and zeta potential. Almost all particulate or macroscopic materials in contact with a liquid acquire an electronic charge on their surfaces. Zeta potential is an indicator of this charge that can be used to predict and control the stability of colloidal suspensions or emulsions and also is important for interaction of liposomes with biological systems *in vivo* [11].

In the present work, we have applied different detergents and methods of their removal, followed by measurements of zeta potential and other properties to investigate the process of reconstitution of apocytochrome b₆ into liposomes. Cytochrome b₆ is present in chloroplasts and functions in the b₆f complex that is localized between green plants photosystems I and II. It is an integral membrane protein with a mass of about 24 kDa that contains three hemes as cofactors. It is the most hydrophobic part of the b_6f complex and is comprised of four transmembrane helices [12]. Like many other integral proteins, cytochrome b_6 operates with an unknown uncleaved signal for membrane insertion and integration [13]. Previously it was shown that apocytochrome b_6 from spinach can be heterologously expressed as a fusion protein – maltose binding protein-apocytochrome b_6 (MBP-b_6) in *E. coli* – and that the expressed fusion protein integrates into the *E. coli* inner membrane [14–16]. In other reports it was reported that transmembrane cytochrome b_6 assembles spontaneously *in vitro* in the *E. coli* membrane [17]. The maltose binding protein is part of the maltodextrin transport system in E. coli that belongs to the periplasmic permease family. It is synthesized as a precursor in the cytoplasm and must be exported to the periplasm where it is folded and becomes functional. [18]. MBP serves as an initial high-affinity binding component of the active-transport system of maltooligosaccharides in bacteria and also participates in chemotaxis towards maltooligosaccharides. It is a monomeric protein with a mass of about 40 kDa and contains two globular domains separated by a deep groove with the oligosaccharide-binding site. Both domains exhibit similar packing of the secondary structure elements; they are composed of a core of β sheet flanked on both sides by helices. The maltose-binding protein consists of 40% α helix and 30% β sheet [19,20].

Materials and Methods

Egg phosphatidyl choline (EPC) and cholesterol were purchased from Lipid Products (England). Triton X-100, *n*-dodecyl-β-D-maltoside (DM) and sodium dodecyl sulfate (SDS) detergents were obtained from Sigma-Aldrich (USA). SM2 Bio-Beads (20–50 mesh; Bio-Rad) were prepared for application by extensive rinsing with methanol and 20 mM HEPES/HCl, pH 7.4 buffer [21]. 100 nm polycarbonate filters were purchased from Nucleopore Corporation. All other reagents were of analytical grade. MBP-apocytochrome b_6 fusion protein (MBP-b_6) was isolated using the *E. coli* overexpression system according to Króliczewski and Szczepaniak [14]. MBP-b_6 in detergent solutions (Triton X-100, DM or SDS) was prepared by exchange dialysis to 20 mM HEPES/HCl, pH 7.4 buffer containing respectively 0.1% Triton X-100, 0.025% DM or 0.3% SDS.

Circular dichroism (CD) measurements

Protein renaturation in detergent solutions was characterized by obtaining circular dichroism spectra recorded at RT in a quartz cuvette using a cell with a 1-mm light path from 190 to 260 nm, averaging 10 scans per sample with 1 nm step, with Jasco J-715 apparatus (JASCO Europe S.R.L, Italy). The spectra were noise reduced using the instrument algorithm and corrected for the buffer contributions without peak shape distortion and without altering peak maxima or intensities. Secondary structure content was estimated using the CDPro software (http://lamar.colostate.edu/~sreeram/CDPro), which permits analysis of a secondary structure by CONTIN, SELCON, and CDSSTR [22,23].

Liposome preparation

7:3 molar mixture of chloroform solutions of EPC and cholesterol was dried in a low-pressure rotary evaporator to obtain a thin film of lipids on the wall of round-bottomed flask. Multilamellar vesicles were prepared by rehydrating the dry lipid film with 20 mM HEPES/HCl, pH 7.4 buffer; the rehydration was assisted by agitation with glass beads. Unilamellar liposomes

of defined size were prepared by ten times extrusion of multilamellar vesicles through 100 nm Nucleopore filters using LIPEX extruder (Lipex Biomembranes, Canada) [24,25].

Liposome solubilization

Solubilization of liposomes was carried out by adding an increasing amount (1 to 200 μL) of a concentrated stock of detergents (the same volume of buffer was used as a control) to 1 mL liposome suspensions with a lipid concentration of 2.5 mg/mL with constant stirring at 20°C and monitored by the light scattering at a 90° angle (nephelometry) at 600 nm with Kontron SFM 25 (Kontron AG, Switzerland) spectrofluorimeter in 1 cm × 1 cm acrylic cuvettes (Sarstedt, Germany). The onset solubilization detergent concentration value was set as an end point of a stage where the light scattering shows a decrease followed by a slight increase (the partitioning of detergent monomers between the aqueous medium and the lipid bilayer); and the total solubilization point was set at the inflection point after which a large and significant decrease in scattered light (gradual liposome solubilization and lipid-detergent micelles forming) slowed down to a stage where the addition of detergent stopped decreasing the light scattering rapidly (e.g. the complete solubilization of lipid into small mixed micelles) [26].

Size distribution and zeta-potential of liposomes

The size and size distribution [27] and zeta-potential [28] of vesicles were measured using a Malvern Zetasizer 2000 instrument (Malvern Instruments, Malvern, UK). Samples were diluted to a total lipid concentration of approximately 0.25 mg/ml in 20 mM HEPES/HCl, pH 7.4 prior to measurements. Zetasizer measures distribution of size in a population and the sizes provided in Tables 1–3 reflect peaks of this distribution in nm. Each peak is characterized by a width at its base (in nm), which reflects homogeneity (or polydispersity) of the subpopulation under this peak. Each sample was measured five times and the result is expressed as an average value ± standard deviation.

MBP-b_6 protein reconstitution

Liposomes with a lipid concentration of 2.5 mg/mL in 1 mL of total volume were first treated with detergents (105 μL of of DM and Triton X-100 and 2 μL a concentrated stock solution of SDS was added) to reach the total solubilization concentration indicated by a decrease and subsequent stabilization of light scattering. Then, detergent solution of protein (15 μL of 800 μg/mL solution) was added to the equilibrated detergent-lipid mixtures to obtain 1:200 protein to lipid weight ratio. Finally, proteoliposomes were reconstituted by two different methods of detergent removal:

A. Detergent removal by dialysis. Detergent-lipid-protein mixtures were transferred to Roth QuickStep Micro Dialyzer System and dialysed at 4°C through 30 kDa Sigma-Aldrich dialysis membrane toward 20 mM HEPES/HCl, pH 7.4 buffer, four times for 1 h to 100 volumes and overnight to 150 volumes of buffer [29].

B. Detergent removal by Bio-Beads [21,26,30]. 300 mg/mL of wet Bio-Beads was added to equilibrated detergent-lipid-protein mixtures and everything was incubated for 6 h at 20°C with constant gentle stirring.

The resulting proteoliposomes was recovered from the ~1 mL volume of post-reconstitution mixture by centrifugation for 30 min at 65000 x g [31–33], washed three times and finally resuspended in 1 mL of 20 mM HEPES/HCl, pH 7.4 buffer.

Table 1. Reconstitution of protein using SDS.

	Diameter (peaks [nm])		Homogeneity (width of a peak [nm])	
	Control	Reconst.	Control	Reconst.
Dialysis				
LUV	113.7±11.2		69.9±15.0	
Post. mix.	94.4±8.8	93.5±8.6	56.4±13.6	55.3±9.6
Proteolip.	107.5±14.5	106.0±10.7	37.2±7.8	32.9±7.5
Bio-Beads				
LUV	116.9±10,2		75.4±14.8	
Post. mix.	85.3±9.3	83.3±9.4	54.1±14.6	57.2±18.8
Proteolip.	101.6±12.5	96.9±8.4	59.4±14.2	59.4±11.0

The size and the size distribution of reconstituted vesicles.
LUV – initial liposome solutions (the same initial batch was measured at this point, which later was divided in two and processed in parallel for control and reconstitution experiments); Post. mix. – proteoliposomes in post-reconstitution mixture (directly after detergent removal); Proteolip. – final proteoliposomes (after centrifugation and washing). Control experiments underwent the same routine only the detergent solution without protein was added to them. Given are the mean values of n = 5 independent experiments with standard deviations.

Efficiency of the MBP-b_6 protein reconstitution

To calculate the efficiency of the protein reconstitution, the protein concentration in proteoliposomes was measured using Bicinchoninic Acid Kit Sigma-Aldrich (St Louis, MO, USA) according to the producer protocol with control samples (no protein reconstituted) as blanks. Each sample was measured three times and the result is expressed as an average value ± standard deviation.

Determination of protein orientation in bilayer

To the proteoliposome fraction suspended in 100 µl of 20 mM HEPES/HCl, pH 7.4 buffer with 7.5 mM $CaCl_2$, proteinase K was added to a final concentration of 0.33 mg/mL and the mixture was incubated at 30°C for 1 h. After incubation time PMSF inhibitor was added to a final concentration of 0.66 mg/mL to stop proteolytic activity. Proteoliposomes were pelleted by ultracentrifugation as described above and the supernatant was discarded. Pellets were dissolved in 100 µL of 10% SDS and 33 µL of 36% trichloroacetic acid was added and mixture was incubated for 1 h on ice in order to precipitate protein. After centrifugation at 16000 x g the supernatant was discarded and the pellet was washed once with cold acetone. Samples were air-dried and prepared for SDS-PAGE. Densitometric analysis was performed using the Fiji distribution (http://fiji.sc) of ImageJ (http://http://imagej.nih.gov/ij) software ver. 1.49b and Java

1.6.0_65, using 600 dpi, 32-bit deep, black and white scanned image.

Western blot analysis

SDS-PAGE was performed using standard techniques and was followed by a semidry blotting. Antibodies raised against MBP were prepared in rabbits [34]. The protein was visualized using secondary peroxidase-conjugated goat anti-rabbit antibodies (Sigma-Aldrich, St Louis, MO, USA).

Statistical analysis

The two-tailed, unpaired t-test using the Holm-Sidak method, with alpha = 0.05 was used for determining the statistical significance of differences between the means for reconstituted and control, using Prism 6.0e (GraphPad Software Inc., La Jolla, CA, USA).

Results

Secondary protein structure in detergent solutions

To test whether MBP-b_6, isolated from overexpression bacterial system, maintains its native helical-rich structure after solubilization in a given detergent, circular dichroism spectrometry in the ultraviolet wavelength region (UV-CD) was performed followed by secondary structure analysis to estimate structure content and thus

Table 2. Reconstitution of protein using Triton X-100 removed with Bio-Beads.

	Diameter (peaks [nm])		Homogeneity (width of a peak [nm])	
	Control	Reconst.	Control	Reconst.
LUV	126.2±14,0		75.2±20.3	
Post. mix.	96.2±10.9	97.8±8.1	49.3±10.5	54±8.9
	386.6±55.6	494.3±78.5	165.4±45.8	377.1±98.0
Proteolip.	101.7±15.4	108.9±15.2	50.3±10.2	56.2±14.0
	400.3±89.5	498.7±57.4	167.5±28.5	172.2±39.0

The size and the size distribution of reconstituted vesicles.
Description the same as for the Table 1.

Table 3. Reconstitution of protein using DM.

	Diameter (peaks [nm])		Homogeneity (width of a peak [nm])	
	Control	Reconst.	Control	Reconst.
Dialysis				
LUV	122.5±15.2		65.6±18.9	
Post. mix.	198.2±20.1	208.5±19.8	72.7±15.1	102.7±22.8
	421.0±68.3	454.7±69.3	73.3±12.2	130.7±26.0
	755.7±105.3	828.4±105.6	110.4±26.5	146.5±35.5
	1200.8±110.8		146.8±35.0	
Proteolip.	163.6±17.2	443.4±58.0	82.1±19.9	89.1±19.9
	652.8±98.7	829.4±79.6	551.4±155.3	571±128.8
Bio-Beads				
LUV	108.1±14,5		62.6±10.3	
Post. mix.	138.8±18.0	140.3±16.7	93.1±18.4	85.9±17.4
Proteolip.	146.9±16.0	149.4±19.5	104.1±25.0	95.2±14.4

The size and the size distribution of reconstituted vesicles.
Description the same as for the Table 1.

the protein folding state. Secondary structure analysis (Table 4) shows that the protein in DM, as well as in Triton X-100 solution, recovered its structure. However the spectra obtained from SDS solution shows decreased ellipticity in the range between 200–210 nm, which indicates incomplete secondary structure recovery (Figure 1 and Table 4).

Solubilization of liposomes by DM, Triton X-100 and SDS

In order to reconstitute our protein from an isotropic solution of lipid-protein-detergent and lipid-detergent micelles, we first had to determine the detergent concentration that was necessary to fully solubilize the liposome suspension [35]. Previous studies have demonstrated that measuring changes in the optical properties of liposome suspensions upon treatment with detergents can serve as a convenient technique for tracing bilayer solubilization [35]. The solubilization perturbations occurring by gradual addition of detergents to liposomes, yielding a suspension turbidity decrease, which results in lower light scattering, were measured in nephelometrical manner.

The solubilization curves of 1 mL liposomal suspensions with a lipid concentration of 2.5 mg/ml upon adding increasing amounts of the detergents used in this study were plotted as described previously [26]. The onset and total solubilization detergent

concentration values were determined for DM $\approx$1.0 and $\approx$4.8 mM, Triton X-100 $\approx$1.3, $\approx$4.8 and SDS $\approx$0.05 and $\approx$0.9 mM respectively (Figure 2A–C). DM and Triton X-100 5 mM and SDS 1 mM concentrations were used in reconstitution experiments as completely solubilizing ones. However, because detergents were added externally, the time needed for full equilibration of the solution had to be determined. For the above lipid concentrations, experiments indicated that around 30 min to 1 h was needed to get a constant value of light scattering.

Liposomes and proteoliposome reconstitution

The overexpression system combined with a detergent independent method of protein purification allowed us to test the application of three different detergents in proteoliposome reconstitution. To follow the process of reconstitution, several parameters were measured during the procedure at several "checkpoints".

The following procedure was used. Particles size and size distribution, zeta-potential values and light scattering were measured for initial liposome solutions (LUV on Figure 3 and Tables 1–3). Then a given detergent was added to reach the total solubilizing concentration and after a 1 h incubation light scattering was measured to ensure total solubilization. Next, the proper detergent solution of protein was added and once again light scattering was measured. With every detergent used, liposomes were shown to be solubilized completely and no further significant changes in light scattering were observed after protein addition. After an additional 1 h incubation to equilibrate the mixtures, detergent was removed using a chosen method. Proteoliposomes in the post-reconstitution mixture (Post. mix. on Figure 3 and Tables 1–3) were then characterized by size and size distribution and zeta-potential was determined. Finally, to separate proteoliposomes from the remaining mixture, centrifugation was applied and again size and size distribution and zeta-potential measurements were carried out, as well as protein concentration for these final proteoliposomes (Proteolip. on Figure 3 and Tables 1–3). The difference in size and size distribution between the control experiment (liposomes reconstituted with the same detergent and by the same manner but in the

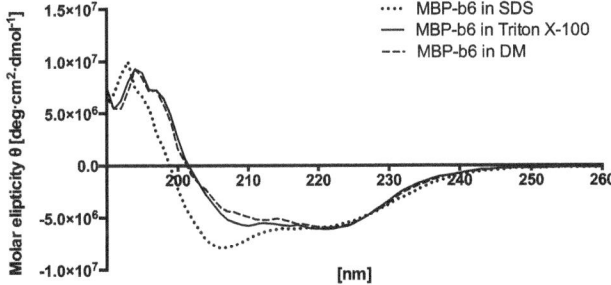

Figure 1. Circular dichroism spectra of MBP-cytochrome b₆ fusion protein in different detergent solutions.

Table 4. Predicted per cent secondary structure content of the MBP-b_6 in detergent solutions (Triton X-100, DM or SDS) from CD analysis.

	Secondary structure type			
	Helix	Sheet	Turn	Unordered
Calculated from the crystal structure of cytochrome b_6 (PDB code 1Q90) and MBP (PDB code 1PEB)	54,7	20,7	3,8	20,8
MBP-b_6 in DM	44,7	19,1	3,1	33,1
MBP-b_6 in Triton X-100	41,9	18,7	3,4	36,0
MBP-b_6 in in SDS	27,3	25,4	5,1	42,2

The percent secondary structure of the CD spectra was predicted using the CDSSTR algorithm found in the CDPro software suite.

absence of protein) and normal proteoliposomes were insignificant. This proves that reconstituted protein does not affect or only minimally affects these parameters, however significant changes in zeta-potential were observed.

Reconstitution using SDS

In this study, we used SDS mainly as a negative control in reconstitution. Using dialysis to remove detergent from solution one is able to reform liposomes after SDS solubilization. The obtained vesicles are characterized by a relatively small size and a narrow size distribution (Table 1). However reconstitution efficiency, indicated by a protein concentration in proteoliposomes, is almost close to zero ($1.1 \pm 0.9\%$; n = 3). Additionally, zeta-potential seems to be strongly influenced by the presence of this anionic detergent and after detergent dialysis it maintains highly negative values (Figure 3A; n = 5, p = 0.0004). Quite similar results were obtained for SDS removed with Bio-Beads (Table 1). Again we were able to reconstitute vesicles of uniform size, but protein reconstitution efficiency was virtually 0%. Here, zeta-potential values were not negative (Figure 3B). These two experiments confirmed that SDS should not be used for the reconstitution of MBP-apocytochrome b_6 into liposomes and casting uncertainty for its application with other proteins.

Reconstitution using Triton X-100

Reconstitution of proteoliposomes using this detergent was also performed using two methods of removal: using dialysis and adsorptive Bio-Beads. According to expectations, it was impossible to decrease Triton X-100 concentration to levels sufficient for proteoliposomes reconstitution even after four days of dialysis. On the other hand, Triton X-100 removal with Bio-Beads led to a vesicle population with heterogeneous size distribution with two main size peaks at around 100 nm and 500 nm, however, vesicles of near micrometer size were also counted (Table 2).

Protein reconstitution efficiency of that method reached $15.5 \pm 5.8\%$ (n = 3). Also negative zeta-potential increased after protein incorporation in comparison to the control experiment, which led to the hypothesis that these changes are caused by the protein presence in bilayer (Figure 3C).

Reconstitution using DM

After liposome solubilization in DM, detergent was removed either using Bio-Beads adsorptive material or during a dialysis procedure. In the latter case, the recreated vesicles had sizes between 400–800 nm and part of the population was even larger than 1000 nm, suggesting the presence of large, multilamellar structures (Table 3). In this case protein reconstitution efficiency reached $29.1 \pm 4.2\%$ (n = 3). Together with high reconstitution efficiency, the higher negative zeta-potential for proteoliposomes, in comparison with the protein free control, was observed (Figure 3D; n = 5, p = 1.5×10-5). DM removal using Bio-Beads also led to proteoliposome formation. This method yielded a very uniform vesicle population; with an average size around 140 nm and a very narrow size distribution (Table 3). Reconstitution efficiency was very high and reached $46.5 \pm 5.9\%$ (n = 3) of the MBP-b_6 added for reconstitution. Zeta-potential decrease was observed for these proteoliposomes as well, and in comparison with control liposomes proved to be highly significant (Figure 3E; n = 5, p = 0.002).

For proteoliposomes produced from DM solutions and detergent removed by Bio-Beads the protein orientation in the bilayer was also checked. To this end, the proteoliposomal fraction was treated with proteinase K, in order to cleave off and digest outward-directed MPB domains of MBP-b_6 fusion protein. The proteinase untreated fraction incubated under the same conditions served as a control. After incubation long enough to ensure total digestion of available MBP domains, SDS-PAGE and Western-blot with antibodies directed against MBP epitope (Figure 4) of obtained samples were performed and band intensities of proteinase treated and untreated samples were compared densitometrically. The ratio of outward-directed, MBP outside, protein (proteinase treated) to total protein incorporated in liposomes (untreated) was ~53% (n = 1). This means almost equal ratio of inward and outward pointed protein, indicating symmetrically reconstituted proteoliposomes (1:1 N- and C-termini out vs. in).

Discussion

Bilayer restoration upon detergent removal from homogenous solution of mixed lipid-detergent micelles has been proved to be an opposite process of bilayer solubilization [36] [37]. However the influence of reconstituted proteins on the properties of the end product (proteoliposomes) has not been examined in detail. In addition the choice of detergent is dictated mostly by the procedure used in protein purification [38], which often limits the application of different detergents in reconstitution experiments. In this work we compared the application of three different detergents: SDS, Triton X-100 and DM, removed by either dialysis or adsorption by Bio-Beads, in reconstituting apocytochrome b_6 in liposomes.

When purifying proteins from natural membranes, the choice of detergent is dictated on the one hand by its ability to effectively solubilize membranes, allowing efficient purification, and on the other hand by its stabilizing properties and compatibility with

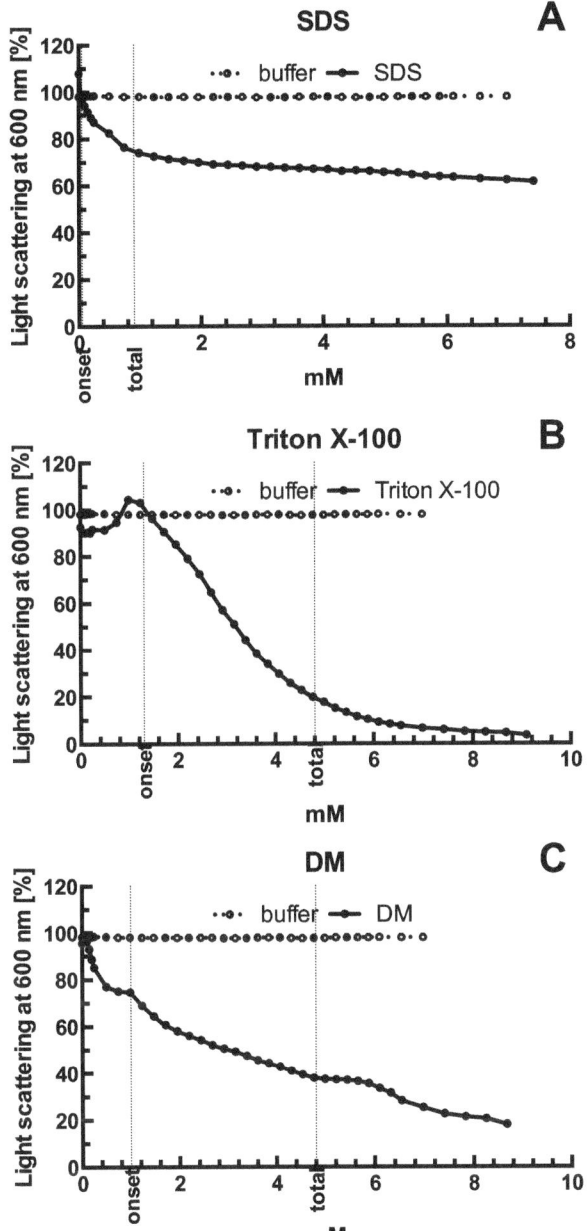

Figure 2. Solubilization of liposomes (black circles) by SDS (A), Triton X-100 (B) and DM (C) monitored by light scattering. The onset of solubilization concentration and the total lysis concentration are marked with dashed lines. As a control, a buffer without detergent (empty circles) was titrated into liposomes.

downstream purification procedures. However, purification yield from natural membranes is usually not high, making it difficult to obtain sufficient quantitates for reconstitution experiments. On the contrary, heterologous overexpression systems ensure high yields and purification procedures that do not rely on detergents (e.g. gel filtration of inclusion bodies solubilized in urea or GdCl), but a resultant protein must be refolded to its native structure. Here various detergents can be used, as long as they ensure proper folding. We used protein purified from an overexpression system; therefore we tested detergents for protein refolding first.

Folding was induced by dialysis of the urea-solubilized MBP–apocytochrome b_6 to a buffer containing an appropriate detergent. As shown above (Figure 1), the renaturation in Triton X-100 and DM was satisfactory. CD spectra and secondary structure analysis (Table 4) show that the protein recovered its native structure. As expected in SDS the decrease of ellipticity in the range of 200–210 nm was observed, indicating incomplete folding [14]. In previous studies it was shown that the disordered structure of the urea-solubilized heme-free form of cytochrome b_6 in the presence of SDS was converted into a partially folded structure with 37% α-helical content [14]. Incomplete restoration of the native structure could be a crucial cause of the reduced ability to reconstitute a protein into the bilayer.

Due to its strong surfactant properties, SDS is not so mild in its effect as other, non-ionic detergents [37]. However it is frequently used as a membrane mimetic environment in NMR studies of transmembrane peptides and the involvement of electrostatic interactions in the mechanism of peptide folding induced by SDS binding has been described [39]. However, despite the fact that its high critical micelle concentration allows for its effective removal from solutions using dialysis, this detergent is not commonly used in reconstitution experiments. Thus, it follows that, from the perspective of reconstitution, in this study SDS proved to be completely useless. Although vesicle restoration was observed in both experiments with SDS removed by dialysis or Bio-Beads, the protein incorporation rate into the bilayer (reconstitution efficiency) was virtually zero. It is worth mentioning that after dialysis a significantly lower negative zeta-potential of reformed vesicles was observed for a protein reconstituted sample than for a control sample (Figures 3A and B). In general a very low zeta potential suggests that negatively charged SDS was not completely removed from bilayers in this process. And the decreased zeta-potential of the protein-reconstituted sample indicates that SDS remained bound to trace amounts of bilayer-incorporated protein (the reconstitution efficiency was exactly 1.1% in this case). It is a well-established fact that SDS interacts with proteins, binds to them and grants them a high negative charge in an almost stoichiometric manner, which is then the basis for the standard SDS-PAGE [40]. This conclusion is further supported by the result of our next experiment where SDS was removed with Bio-Beads, which yielded 0% reconstitution efficiency and where zeta potential did not remain negative. It also provided initial evidence that zeta potential might be a valuable indicator of protein presence in a reconstituted bilayer. Here, SDS bound to protein enhanced this potential.

On the contrary, the non-ionic detergents, such as DM or Triton X-100 have long been known to be effective in isolation and reconstitution of a number of different membrane multi-spanning proteins, such as P-type ATPases [41] or bacteriorhodopsin [42]. Triton X-100 and DM are milder detergents that lyse liposomes at more than five times higher concentration than SDS.

Triton X-100's mild character ensures proper conditions to stabilize membrane protein structures and therefore it is also used for their purification from natural membranes [43]. However, its non-ionic character and low critical micelle concentration makes it difficult to remove by dialysis. But after discovering a quick and convenient method of Triton removal by Bio-Beads [21], which was well characterized for different conditions [30], this detergent became one of the most frequently used detergents in protein reconstitution. As expected, in our experiment we were unable to remove Triton X-100 and reform liposomes even after dialyzing extensively for a period of four days. However when Bio-Beads were used, 15% reconstitution efficiency was achieved. In comparison with SDS this is a high value, though still not

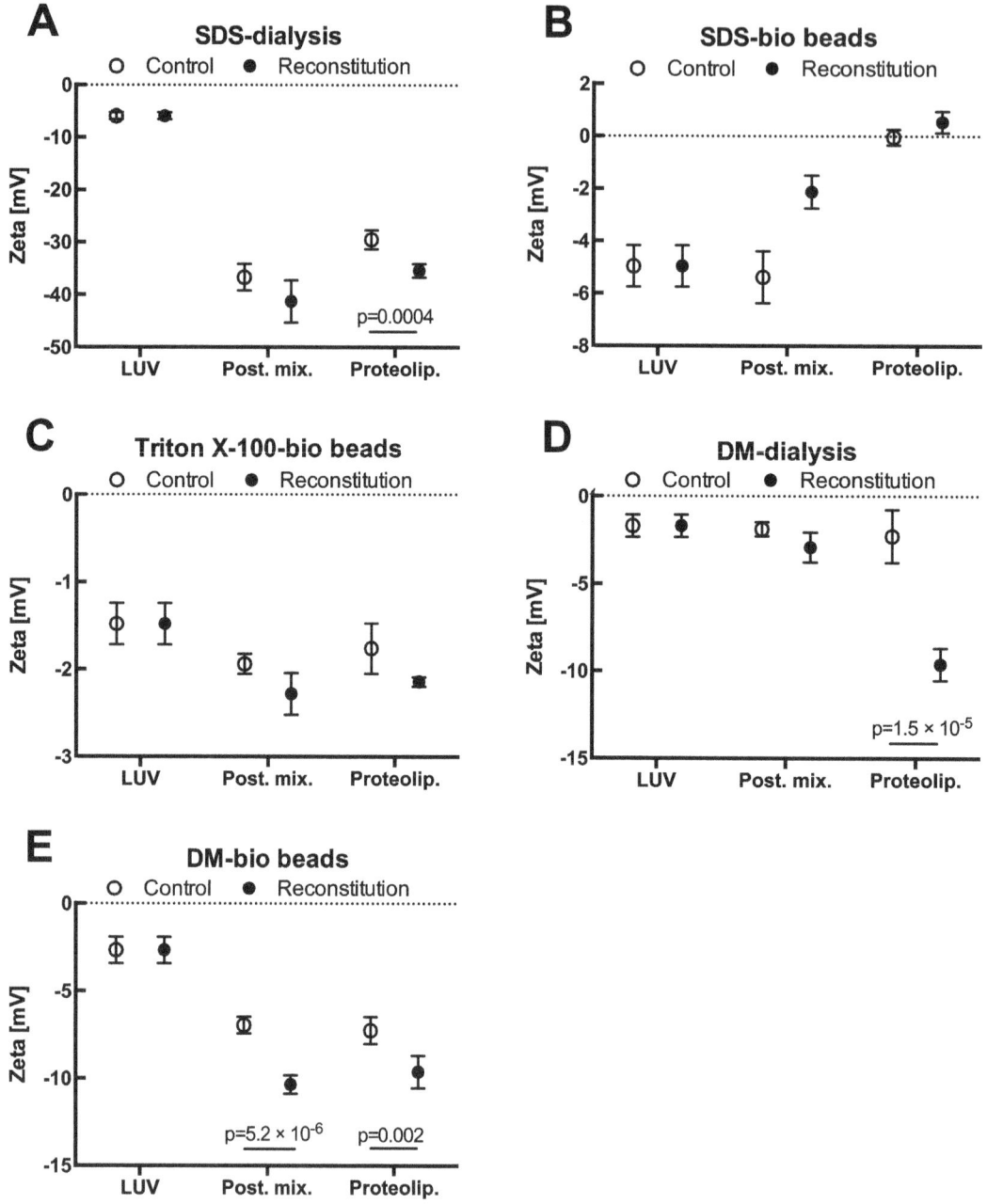

Figure 3. Zeta potential of liposomes during the reconstitution process for SDS removed by dialysis (A) or with Bio-Beads (B); Triton X-100 removed with Bio-Beads (C) and DM removed by dialysis (D) or with Bio-Beads (E). Plotted are the mean values of $n = 5$ independent experiments with error bars depicting standard deviations. Asterisks denote significant differences between control and reconstitution experiments. LUV – initial liposome solutions (the same initial batch was measured at this point, which later was divided in two and processed in parallel for control and reconstitution experiments); Post. mix. – proteoliposomes in post-reconstitution mixture (directly after detergent removal); Proteolip. – final proteoliposomes (after centrifugation and washing). Control experiments underwent the same routine only the detergent solution without protein was added to them.

satisfactory. Zeta potential was lower for proteoliposomes than for control experiments without protein, although not significantly (Figure 3C). With the reconstitution efficiency far above 0% we conclude that this difference is caused by the presence of protein, since Triton X-100 is a neutral charge detergent that does not influence this potential. The resultant population of proteoliposomes was not homogeneous. Apart from the vesicles approximately 100 nm in size, a large fraction with the size of 400–

500 nm was present. This implies two possibilities: we observed two populations of LUVs [36] or a larger subpopulation consisting of multilamellar vesicles (MLV). Additional data from previous studies suggests only LUVs are formed, but the authors did not observe such a large inhomogeneity [26,30]. The characteristics of the proteoliposome population play an important role in determining their suitability for particular applications. In most cases it is critical that the liposomes are homogenous in size, size

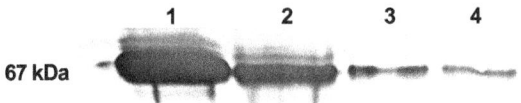

Figure 4. Orientation of MBP-apocytochrome b₆ in proteolipo-somes reconstituted with dodecyl maltoside removed with Bio-Beads was analyzed by Western blotting with anti-MBP antibodies. Purified, concentrated MBP-apocytochrome b₆ in DM solution (1); DM solution of MBP-apocytochrome b₆ added for reconstitution (2); untreated proteoliposomes (3); proteoliposomes treated with proteinase K (4). Lanes 3 and 4 contain the same amount of proteoliposomes.

distribution and lamellarity. Therefore despite the observation of non-zero reconstitution efficiency, Triton X-100 was demonstrated as an unsuitable detergent for reconstituting apocytochrome b₆.

The best results were obtained using dodecyl-β-D-maltoside. DM is a nonionic detergent with a low critical micellar concentration. Its hydrophobic moiety has an intermediate length, while a hydrophilic sugar headgroup is more bulky. Since the early 90's, DM has commonly been used in membrane solubilization and reconstitution of a wide range of functionally active membrane proteins [44–46]. Additionally DM solubilization and liposome formation from detergent-lipid micelles were examined in detail, which revealed a new "gel-like" phase in DM-lipid mixtures that strongly affects the properties of the final product [26,47]. DM removed by dialysis yielded reconstituted vesicles with protein reconstitution efficiency of approximately 29%, however great inhomogeneity of the resultant proteoliposomes was observed (Table 3). We link this observation directly with the slow rate of detergent removal achieved in dialysis. During lysis of liposomes by DM different structures and aggregates such as planar membrane; open vesicles; long, filamentous micelles etc. arise [26]. Liposome restoration is an opposing process, therefore when DM is removed slowly, all mentioned, intermediate structures appear, which then, after ongoing fusion and mixing events, will yield inhomogeneous population at the end of the process. On the contrary, removal of DM by Bio-Beads is a fast process, in which intermediate structures do not form and the final population is homogenous in size. In our experiments, cytochrome b₆ reconstituted with DM removed by Bio-Beads resulted in proteoliposomes with average size of around 150 nm and a narrow size distribution (Table 3). The differences in the final product properties with DM removed by dialysis and by adsorption on Bio-Beads correlates with previous observations of Lambert et.al. [26]. The reconstitution efficiency for this combination of detergent and its removal method was the highest observed and reached 46%. As for Triton X-100, for DM removed by both dialysis and Bio-Beads, a decrease in the zeta potential of the final proteoliposomes in comparison to the control was observed, and this difference proved to be statistically significant. We conclude that it reflects the highest reconstitution efficiency achieved and is in agreement with the protein net charge under the conditions used. The extramembranous fragments of protein, namely: MBP domain; N'-end; C'-end; loops between helices 1, 2 and 3 and the linker; have a net amino acid charge equal to -9.9 at the pH used in our reconstitution experiments (pH 7.4), which directly explains the negative zeta potential values, assuming the predicted reconstitution efficiency was reached. These results are in line with experiments, in which zeta potential was used to determine protein orientation in a bilayer, providing that in- and out-ends of it differs significantly in their net charge [48]. Considering these results, we advocate that zeta potential

measurements can be used to quickly and conveniently test whether reconstitution was successful; having the time- and simplicity-advantage over SDS-PAGE and Western blotting or a gradient ultracentrifugation. Zeta potential measurements are applicable when outer fragments of protein are charged and nonionic detergent (for ionic detergents zeta potential might be used to control detergent removal from a membrane) and uncharged lipids are used, as these can greatly influence zeta potential of proteoliposomes [49].

Another aspect of our study was the final orientation of protein in the bilayer. The orientation of incorporated proteins might have been influenced by the method of detergent removal, as again the rate of detergent removal can be crucial. When detergent is removed quickly, the protein incorporation occurs during formation of the vesicle favoring more symmetrical orientation [50]. Slow removal of detergent leads to preferential formation of liposomes and subsequent protein incorporation, which can favor orientation asymmetry of a reconstituted protein in a bilayer [51]. These facts are valid when reconstitution starts at lytic detergent concentrations; i.e. preformed liposomes are totally lysed with a detergent and after protein addition, an isotropic mixture consisting of three components (lipid-detergent-protein) and two component (lipid-detergent) micelles forms. When sublytic detergent concentrations are used; i.e. preformed liposomes are not lysed but their bilayer is only destabilized with a detergent, the protein orientation tends to be more asymmetrical [52]. Moreover, in such a process, charged lipids can also influence the protein orientation [48].

For the reconstitution using DM removed with Bio-Beads we determined the orientation of the incorporated protein by a proteolytic assay. As shown (Figure 4), incorporation of MBP-apocytochrome b₆ was almost ideally symmetrical (~53% asymmetry), which confirms the established model of reconstitution method by reverse solubilization [50,52–54]. On the basis of data on the reconstitution efficiency (46.5%), lipid concentration and protein content (2.5 mg/ml and 5.72 μg, respectively) in total volume (1 ml) of proteoliposomes of known size (150 nm), we calculated the number of protein molecules per one molecule of the liposome [55]. On average one liposome molecule contained 200 molecules of fusion protein incorporated into its bilayer.

From a practical point of view the reconstitution performed with Bio-Beads proved to be the most straightforward. In comparison to a dialysis it shortens the procedure from overnight to just six hours. Also handling is reduced, as Bio-Beads do not need to be changed during the reconstitution, as the dialysis buffer does; and also volumes are much lower (ca. 1 mL total volume with Bio-Beads in comparison to 100–150 mL during dialysis), which together easily allows for a parallel reconstitution of multiple samples. Also the efficiency of detergent removal (parameter not monitored in this work) is very good for Bio-Beads; reducing trace amounts of detergent and improving liposome stability and impermeability [50].

Conclusions

In this work, using apocytochrome b₆, we systematically compared its reconstitution into liposomes using three different detergents: SDS, Triton X-100 and DM, and two methods of their removal: slow – by dialysis and fast – using Bio-Beads. Despite the fact that two of the detergents used (Triton X-100 and DM) proved to successfully renature the protein to its native, helical rich structure after protein purification, their application in a reconstitution was dramatically different. Only DM removed with Bio-Beads gave satisfactory reconstitution efficiency – a crucial factor if downstream applications are considered such as activity measurements – and yielded proteoliposome populations of homogenous size and distribution. This points to the fact, that when protein reconstitution is considered, a detergent is the most crucial factor responsible for the end product parameters. Therefore, detergent-free methods of protein purification (e.g. from heterologous expression systems and not from native membranes) are advantageous here, as they do not predetermine the choice of a detergent to be used in a reconstitution.

Additionally we showed that zeta potential could be used as a convenient parameter for a quick assessment of reconstitution success and having an indisputable advantage in time and simplicity over e.g. SDS-PAGE followed by Western blotting or

gradient ultracentrifugation. Moreover the same parameter allows monitoring the removal of ionic detergents, if such were used.

Acknowledgments

We thank Dr. James Sáenz for the critical reading of this manuscript and Drs. Maria Stasiuk, Jerzy Gubernator and Daniel Krowarsch for useful hints and technical help.

Author Contributions

Conceived and designed the experiments: JK MAS AS. Performed the experiments: JK MAS. Analyzed the data: JK MAS. Contributed reagents/materials/analysis tools: JK MAS AS. Wrote the paper: JK MAS.

References

1. Wallin E, vonHeijne G (1998) Genome-wide analysis of integral membrane proteins from eubacterial, archaean, and eukaryotic organisms. Protein Sci 7: 1029 1038.
2. Hochstadt J, Quinlan D, Rader R, Li C-C, Dowd D (1975) Use of Isolated Membrane Vesicles in Transport Studies. In: Korn E, editor. Transport: Springer US. pp.117–162.
3. Miller C (1986) Ion Channel Reconstitution Why Bother? In: Latorre R, editor. Ionic Channels in Cells and Model Systems: Springer US. pp.257–271.
4. Szoka F, Papahadjopoulos D (1980) Comparative Properties and Methods of Preparation of Lipid Vesicles (Liposomes). Annu Rev Biophys Bio 9: 467–508.
5. Benes M, Billy D, Hermens WT, Hof M (2002) Muscovite (mica) allows the characterisation of supported bilayers by ellipsometry and confocal fluorescence correlation spectroscopy. Biol Chem 383: 337–341.
6. Hanke G, Bowsher C, Jones MN, Tetlow I, Emes M (1999) Proteoliposomes and plant transport proteins. J Exp Bot 50: 1715–1726.
7. Amalou Z, Gibrat R, Trouslot P, D'Auzac J (1994) Solubilization and Reconstitution of the Mg2+/2H+ Antiporter of the Lutoid Tonoplast from Hevea brasiliensis Latex. Plant Physiol 106: 79–85.
8. Ishijima S, Shigemi Z, Adachi H, Makinouchi N, Sagami I (2012) Functional reconstitution and characterization of the Arabidopsis Mg2+ transporter AtMRS2-10 in proteoliposomes. Biochim Biophys Acta 1818: 2202–2208.
9. Wallgren M, Lidman M, Pedersen A, Brännström K, Karlsson BG, et al. (2013) Reconstitution of the Anti-Apoptotic Bcl-2 Protein into Lipid Membranes and Biophysical Evidence for Its Detergent-Driven Association with the Pro-Apoptotic Bax Protein. PLoS ONE 8: e61452.
10. Pochini L, Scalise M, Galluccio M, Amelio L, Indiveri C (2011) Reconstitution in liposomes of the functionally active human OCTN1 (SLC22A4) transporter overexpressed in Escherichia coli. Biochem J 439: 227–233.
11. Scherphof GL, Velinova M, Kamps J, Donga J, van der Want H, et al. (1997) Modulation of pharmacokinetic behavior of liposomes. Adv Drug Del Rev 24: 179–191.
12. Cramer WA, Martinez SE, Furbacher PN, Huang D, Smith JL (1994) The cytochrome b(6)f complex. Curr Opin Struct Biol 4: 536–544.
13. Zak E, Sokolenko A, Unterholzner G, Altschmied L, Herrmann RG (1997) On the mode of integration of plastid-encoded components of the cytochrome bf complex into thylakoid membranes. Planta 201: 334–341.
14. Kroliczewski J, Szczepaniak A (2002) In vitro reconstitution of the spinach chloroplast cytochrome b6 protein from a fusion protein expressed in Escherichia coli. Biochim Biophys Acta 1598: 177–184.
15. Kroliczewski J, Hombek-Urban K, Szczepaniak A (2005) Integration of the thylakoid membrane protein cytochrome b6 in the cytoplasmic membrane of Escherichia coli. Biochemistry 44: 7570–7576.
16. Kroliczewski J, Gubernator B, Rogner M, Szczepaniak A (2011) On the mode of integration of the thylakoid membrane protein cytochrome b(6) into cytoplasmic membrane of Escherichia coli. Acta Biochim Pol 58: 335–343.
17. Dreher C, Prodohl A, Weber M, Schneider D (2007) Heme binding properties of heterologously expressed spinach cytochrome b(6): implications for trans-membrane b-type cytochrome formation. FEBS Lett 581: 2647–2651.
18. Ehrmann M, Ehrle R, Hofmann E, Boos W, Schlösser A (1998) The ABC maltose transporter. Mol Microbiol 29: 685–694.
19. Spurlino JC, Lu GY, Quiocho FA (1991) The 2.3-A resolution structure of the maltose- or maltodextrin-binding protein, a primary receptor of bacterial active transport and chemotaxis. J Biol Chem 266: 5202–5219.
20. Quiocho FA, Spurlino JC, Rodseth LE (1997) Extensive features of tight oligosaccharide binding revealed in high-resolution structures of the maltodextrin transport/chemosensory receptor. Structure 5: 997–1015.
21. Holloway PW (1973) A simple procedure for removal of triton X-100 from protein samples. Anal Biochem 53: 304–308.
22. Sreerama N, Woody RW (1994) Protein Secondary Structure from Circular Dichroism Spectroscopy: Combining Variable Selection Principle and Cluster Analysis with Neural Network, Ridge Regression and Self-consistent Methods. J Mol Biol 242: 497–507.

23. Sreerama N, Woody RW (2000) Estimation of Protein Secondary Structure from Circular Dichroism Spectra: Comparison of CONTIN, SELCON, and CDSSTR Methods with an Expanded Reference Set. Anal Biochem 287: 252–260.
24. Bangham AD, Standish MM, Watkins JC (1965) Diffusion of univalent ions across the lamellae of swollen phospholipids. J Mol Biol 13: 238–252.
25. Olson F, Hunt CA, Szoka FC, Vail WJ, Papahadjopoulos D (1979) Preparation of liposomes of defined size distribution by extrusion through polycarbonate membranes. Biochim Biophys Acta 557: 9–23.
26. Lambert O, Levy D, Ranck JL, Leblanc G, Rigaud JL (1998) A new "gel-like" phase in dodecyl maltoside-lipid mixtures: implications in solubilization and reconstitution studies. Biophys J 74: 918–930.
27. Zimm BH (1948) The Scattering of Light and the Radial Distribution Function of High Polymer Solutions. J Chem Phys 16: 1093–1099.
28. Weiser HB, Merrifield P (1950) The concept of critical zeta potential for hydrophobic sols. I. J Phys Colloid Chem 54: 990–998.
29. Milsmann MH, Schwendener RA, Weder HG (1978) The preparation of large single bilayer liposomes by a fast and controlled dialysis. Biochim Biophys Acta 512: 147–155.
30. Levy D, Bluzat A, Seigneuret M, Rigaud JL (1990) A systematic study of liposome and proteoliposome reconstitution involving Bio-Bead-mediated Triton X-100 removal. Biochim Biophys Acta 1025: 179–190.
31. John K, Kubelt J, Müller P, Wüstner D, Herrmann A (2002) Rapid Transbilayer Movement of the Fluorescent Sterol Dehydroergosterol in Lipid Membranes. Biophys J 83: 1525–1534.
32. Kubelt J, Menon AK, Muller P, Herrmann A (2002) Transbilayer movement of fluorescent phospholipid analogues in the cytoplasmic membrane of Escherichia coli. Biochemistry 41: 5605–5612.
33. John K, Schreiber S, Kubelt J, Herrmann A, Muller P (2002) Transbilayer movement of phospholipids at the main phase transition of lipid membranes: implications for rapid flip-flop in biological membranes. Biophys J 83: 3315–3323.
34. Kroliczewski J (1999) Structure of cytochrome b6f complex. In vitro reconstitution of the spinach chloroplast cytochrome b6 protein from a fusion protein expressed in Escherichia coli, interaction with the membrane. [Dissertation]: University of Wrocław. 179 p.
35. Goni FM, Alonso A (2000) Spectroscopic techniques in the study of membrane solubilization, reconstitution and permeabilization by detergents. Biochim Biophys Acta 1508: 51–68.
36. Ollivon M, Lesieur S, Grabielle-Madelmont C, Paternostre M (2000) Vesicle reconstitution from lipid-detergent mixed micelles. Biochim Biophys Acta 1508: 34–50.
37. Helenius A, Simons K (1975) Solubilization of membranes by detergents. Biochim Biophys Acta 415: 29–79.
38. Rigaud J-L, Chami M, Lambert O, Levy D, Ranck J-L (2000) Use of detergents in two-dimensional crystallization of membrane proteins. Biochim Biophys Acta 1508: 112–128.
39. Montserret R, McLeish MJ, Bockmann A, Geourjon C, Penin F (2000) Involvement of electrostatic interactions in the mechanism of peptide folding induced by sodium dodecyl sulfate binding. Biochemistry 39: 8362–8373.
40. Jonas A, Weber G (1971) Strong binding of hydrophobic anions by bovine serum albumin peptides covalently linked to lysozyme. Biochemistry 10: 4492–4496.
41. Levy D, Gulik A, Bluzat A, Rigaud JL (1992) Reconstitution of the sarcoplasmic reticulum Ca(2+)-ATPase: mechanisms of membrane protein insertion into liposomes during reconstitution procedures involving the use of detergents. Biochim Biophys Acta 1107: 283–298.
42. Freisleben HJ, Zwicker K, Jezek P, John G, Bettin-Bogutzki A, et al. (1995) Reconstitution of bacteriorhodopsin and ATP synthase from Micrococcus luteus into liposomes of the purified main tetraether lipid from Thermoplasma acidophilum: proton conductance and light-driven ATP synthesis. Chem Phys Lipids 78: 137–147.

43. Arnold T, Linke D (2001) The Use of Detergents to Purify Membrane Proteins. Current Protocols in Protein Science: John Wiley & Sons, Inc.

44. Boulter JM, Wang DN (2001) Purification and characterization of human erythrocyte glucose transporter in decylmaltoside detergent solution. Protein Expr Purif 22: 337–348.

45. Kragh-Hansen U, le Maire M, Noel JP, Gulik-Krzywicki T, Moller JV (1993) Transitional steps in the solubilization of protein-containing membranes and liposomes by nonionic detergent. Biochemistry 32: 1648–1656.

46. Ravaud S, Do Cao MA, Jidenko M, Ebel C, Le Maire M, et al. (2006) The ABC transporter BmrA from Bacillus subtilis is a functional dimer when in a detergent-solubilized state. Biochem J 395: 345–353.

47. Knol J, Sjollema K, Poolman B (1998) Detergent-mediated reconstitution of membrane proteins. Biochemistry 37: 16410–16415.

48. Tunuguntla R, Bangar M, Kim K, Stroeve P, Ajo-Franklin Caroline M, et al. (2013) Lipid Bilayer Composition Can Influence the Orientation of Proteorhodopsin in Artificial Membranes. Biophys J 105: 1388–1396.

49. Pérez JL, Acevedo R, Callicó A, Fernández Y, Cedré B, et al. (2009) A proteoliposome based formulation administered by the nasal route produces vibriocidal antibodies against El Tor Ogawa Vibrio cholerae O1 in BALB/c mice. Vaccine 27: 205–212.

50. Rigaud JL, Levy D, Mosser G, Lambert O (1998) Detergent removal by nonpolar polystyrene beads. Eur Biophys J 27: 305–319.

51. Ueno M, Tanford C, Reynolds JA (1984) Phospholipid vesicle formation using nonionic detergents with low monomer solubility. Kinetic factors determine vesicle size and permeability. Biochemistry 23: 3070–3076.

52. Helenius A, Sarvas M, Simons K (1981) Asymmetric and symmetric membrane reconstitution by detergent elimination. Studies with Semliki-Forest-virus spike glycoprotein and penicillinase from the membrane of Bacillus licheniformis. Eur J Biochem 116: 27–35.

53. Rigaud J-L, Pitard B, Levy D (1995) Reconstitution of membrane proteins into liposomes: application to energy-transducing membrane proteins. Biochim Biophys Acta 1231: 223–246.

54. Eytan GD (1982) Use of liposomes for reconstitution of biological functions. Biochimica et Biophysica Acta (BBA) - Reviews on Biomembranes 694: 185–202.

55. Torchilin V, Weissing V (1990) Liposomes: A Practical Approach; New RRC, editor: IRL Press/Oxford University Press.

12

Changing Folding and Binding Stability in a Viral Coat Protein: A Comparison between Substitutions Accessible through Mutation and Those Fixed by Natural Selection

Craig R. Miller[1,2,3◆], **Kuo Hao Lee**[3,4◆], **Holly A. Wichman**[1,3], **F. Marty Ytreberg**[3,5*]

1 Department of Biological Sciences, University of Idaho, Moscow, Idaho, 2 Department of Mathematics, University of Idaho, Moscow, Idaho, 3 Institute for Bioinformatics and Evolutionary Studies, University of Idaho, Moscow, Idaho, 4 Department of Biochemistry and Molecular Biophysics, Kansas State University, Manhattan, Kansas, 5 Department of Physics, University of Idaho, Moscow, Idaho

Abstract

Previous studies have shown that most random amino acid substitutions destabilize protein folding (i.e. increase the folding free energy). No analogous studies have been carried out for protein-protein binding. Here we use a structure-based model of the major coat protein in a simple virus, bacteriophage φX174, to estimate the free energy of folding of a single coat protein and binding of five coat proteins within a pentameric unit. We confirm and extend previous work in finding that most accessible substitutions destabilize both protein folding and protein-protein binding. We compare the pool of accessible substitutions with those observed among the φX174-like wild phage and in experimental evolution with φX174. We find that observed substitutions have smaller effects on stability than expected by chance. An analysis of adaptations at high temperatures suggests that selection favors either substitutions with no effect on stability or those that simultaneously stabilize protein folding and slightly destabilize protein binding. We speculate that these mutations might involve adjusting the rate of capsid assembly. At normal laboratory temperature there is little evidence of directional selection. Finally, we show that cumulative changes in stability are highly variable; sometimes they are well beyond the bounds of single substitution changes and sometimes they are not. The variation leads us to conclude that phenotype selection acts on more than just stability. Instances of larger cumulative stability change (never via a single substitution despite their availability) lead us to conclude that selection views stability at a local, not a global, level.

Editor: Eugene A. Permyakov, Russian Academy of Sciences, Institute for Biological Instrumentation, Russian Federation

Funding: KHL and FMY were supported by grant P20-RR016448-07S1 from the National Institutes of Health. CRM and HAW were supported by grant number R01-GM076040 from the National Institutes of Health. Computational resources were provided in part by grant R21-GM083827 from the National Institutes of Health. The content is solely the responsibility of the authors and does not necessarily represent the official views of the National Institutes of Health. The funders had no role in study design, data collection and analysis, decision to publish, or preparation of the manuscript.

Competing Interests: The authors have declared that no competing interests exist.

* Email: ytreberg@uidaho.edu

◆ These authors contributed equally to this work.

Introduction

Biological systems require proteins, and to function structured proteins require a minimum level of thermodynamic folding stability [1,2]. Most functioning proteins are marginally stable, with a folding thermodynamic stability between −5 and −15 kcal/mol [3–7]. The thermodynamic folding stability is an equilibrium measure of the fraction of folded to unfolded proteins given by the Gibbs free energy difference of folding, ΔG_{fold}, and can be experimentally determined by measuring the equilibrium constant [8–11]. Under equilibrium conditions, an increase in the thermodynamic folding stability of a protein corresponds to an increase in the fraction of time a protein is folded.

Protein folding stability can be broken down into several molecular interactions that depend on protein structure and environmental conditions [12–14]. Similarly, protein-protein binding stability, the equilibrium measure of the fraction of bound to unbound proteins, is also a function of these interactions. Hydrophobic interactions contribute to stability in proportion to the size of the protein and primarily tend to stabilize the globular

conformation [3,15,16]. Increased temperature can reduce the hydrophobic effect and the tendency for protein association reactions become enthalpy dominated [1,4,17–20]. Burying polar residues contributes to folding stability since the intramolecular hydrogen bonding and van der Waals interactions of polar groups in folded proteins are more favorable than similar interactions with water in unfolded proteins [21,22]. Changes in ion concentration or pH also alters the thermodynamic stability [23,24].

There is often a tradeoff between protein stability and protein function because proteins that are too stable can be less functional [2,19,25,26]. For example, a study of β-lactamase TEM-1 by Wang and collaborators showed that mutant enzymes with increased activity against antibiotics were less stable [27,28]. Similarly, five key active-site residues of AmpC β-lactamase have been characterized as decreasing the activity and increasing the stability of the enzyme [20,23,29]. These studies illustrate how changes in protein stability can result in changes of functional enzymatic activity.

Random substitutions of globular proteins tend to destabilize folding by decreasing the thermodynamic folding stability. Bloom and collaborators presented a thermodynamic framework to predict the probability that a protein retains its structure after one or more random amino acid substitutions, and highly simplified models of proteins were used to support their prediction that the substitutions tend to be destabilizing [4,7,8,15,22,23]. A study by Tawfik and collaborators showed that about 70% of random substitutions of globular proteins are destabilizing ($\Delta\Delta G >$ 0 kcal/mol), and that about 20% are highly destabilizing ($\Delta\Delta G >$ 2 kcal/mol) [15,17,24,25]. In another study they found that substitutions associated with new enzymatic functions are mostly destabilizing [1–3,5–7,17,19,26,27]. One reason that these findings are important is because it is thought that many monogenic diseases are caused, in part, by decreased protein thermodynamic stability [4,8,23,30–32]. A typical disease-causing mutation destabilizes protein folding by increasing the folding free energy by 2–3 kcal/mol [9–12,32,33].

Understanding the effect of random amino acid substitutions on protein-protein binding is critical to understanding protein evolution as well as potentially elucidating the biophysical mechanisms for some diseases. Since proteins frequently bind to other proteins to function, we hypothesize that either over-stabilizing or destabilizing protein-protein binding may cause loss of biological function (consistent with the ideas in [2,13–15,34–37]). For example, it has been shown that mis-assembly of homomers (self-interacting copies of a protein unit) is implicated in diseases [1,4,7,16,17,19,20,38]. One such disease is Parkinson's where the mis-assembly of protein complex I in brain mitochondria reduces the function of the complex [3,5–7,21,39]. The effect of amino acid substitutions on the aggregation rates of unfolded polypeptides can be correlated to physicochemical properties, such as hydrophobicity, protein structure and electric charge distribution [23,40,41].

Studying how substitutions alter protein stability is also integral to understanding and even predicting how viral and bacterial infectious diseases or agricultural insect pathogens evolve in real time. We expect that a limited tolerance to changes in both binding and folding stability in turn constrain and influence the adaptive pathways available to these organisms. For example, substitutions that would be adaptive (e.g. by conferring a new function like metabolizing an antibiotic) may not be if they destabilize the protein too much. In such cases, otherwise neutral substitutions that happen to stabilize a protein may, by chance, preadapt it to tolerate this type of destabilizing gain-of-function mutation [20,22,23,42]. Thus adaptation may not just be in response to direct selective forces; it may also be influenced circuitously by conditions like temperature and acidity that may select for changes in stability.

In this study, we determined how amino acid substitutions, accessible through a single mutation within a codon, change protein folding stability and protein-protein binding stability in a bacteriophage virus system. FoldX was used to estimate the changes in folding stability ($\Delta\Delta G_{fold}$) and binding stability ($\Delta\Delta G_{bind}$) for the coat protein F in the bacteriophage virus φX174 [7,8,12,15,24,25]. Folding and binding stabilities were calculated for all accessible substitutions for each amino acid residue in the major capsid protein (F). We examined the distribution of all accessible effects. We then compared the accessible substitutions with those observed in real evolving phage: first among the wild φX174-like phage, and second in the context of laboratory adaptations of φX174 [2,8,15,19,26,27,41,43–50]. We find that there are unexpected differences between accessible and observed substitutions. Observed substitutions tend to have smaller effects on stability than expected by chance. Substitutions observed in high temperature adaptations tend to stabilize folding but slightly destabilize binding. Finally their cumulative stability effects in lab adaptations can be considerably greater than individual effects suggesting that selection is acting on local aspects of protein stability.

Results and Discussion

The purpose of this study is examine the link between protein stability and natural selection by asking if and how substitutions fixed by selection differ from all accessible substitutions in their effects on both folding and binding stability. To do this we used the coat protein (protein F) from the phage φX174 as a model system (Figure 1A). As a first step in capsid formation in φX174, sets of five F proteins bind to form pentameric subunits (Figure 1B); twelve of these pentameric subunits then assemble in conjunction with several other proteins to form the capsid. We modeled the folding stability of individual F proteins (Figure 1C, 1D) and the binding stability of five folded mature F proteins into a single pentameric subunit (Figure 1B, 1D). More specifically, we used FoldX [1–3,5–7,15] to determine the effect on folding and binding stability of each amino acid change accessible within one DNA change from our reference sequence at every amino acid residue in the protein (Figure 1D). We choose this one DNA change criteria because nearly all the observed substitutions (discussed next) were within one DNA change. Stability effects were based on differences from our laboratory strain of φX174 (GenBank accession number AF176034 [4,8]) at 37°C and expressed as $\Delta\Delta G$ in units of kcal/mol. Substitutions fixed by natural selection came from two sources: (1) differences observed among wild phage that are closely related to φX174 [9–12], and (2) substitutions observed among 26 laboratory adaptation experiments using φX174 [13–15].

The resolutions of the protein structure used for this study is 3.0 Å. It is known that the FoldX folding and binding stability results are more accurate for high resolution structures (<1.8 Å) [51]. There is, however, no evidence that FoldX shows systematic bias for low resolution structures. Statistical methods that have high variance have lower power, or a reduced probability of detecting effects that exists. But if they are unbiased, they do not suffer from an elevated risk of false discoveries (or type I errors). We believe the use of FoldX in the current study is analogous: using a low resolution structure may have reduced our predictive power but it should not have elevated our type I error rate. Thus the significant differences we uncover despite this reduced power would likely be even more strongly supported if structure resolutions were higher.

As a method of evaluating whether our FoldX calculations are behaving as expected, we calculated the median effect on $\Delta\Delta G_{fold}$ and $\Delta\Delta G_{bind}$ of accessible substitutions at each residue. We then created heatmaps of the pentamer showing large median effects in red and low effects in blue. Since substitutions in residues along protein-protein interfaces have the potential to dramatically alter binding stability whereas residues far from an interface do not, we expect interface sites to show much larger binding effects. This is exactly what we observe (Figure 2A–B). By contrast, residues within the protein have more opportunity to interact with other residues of the same protein, leading us to expect that large-effect folding sites should be concentrated in the protein's interior and to thus have a very different pattern than binding effects. Again, this is what we observe (Figure 2C–D).

Patterns Among Accessible Substitutions

When we examine the effect of all substitutions within one DNA change, our results indicate that most accessible substitutions

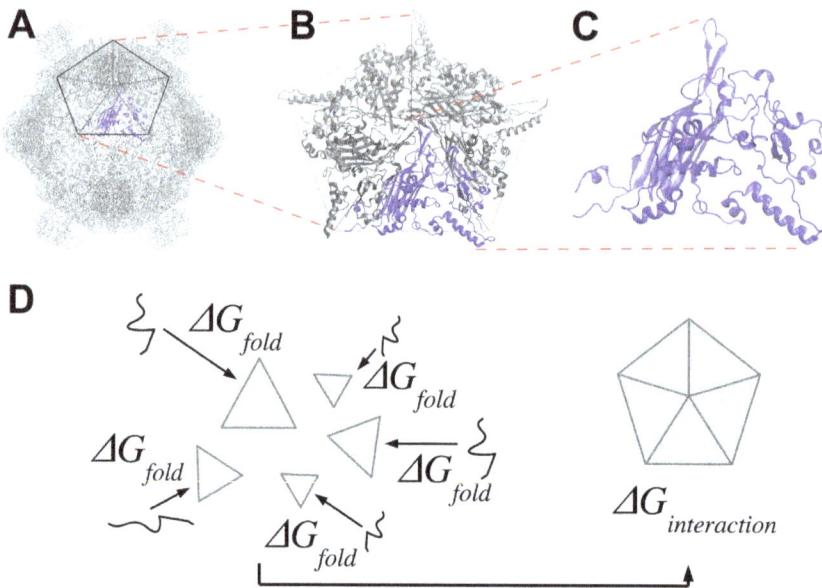

Figure 1. Model systems used in current study. (A) The capsid of φX174 consists of multiple copies of several kinds of proteins. The pentagon highlights a pentameric subunit that has five copies of coat protein F. (B) One pentameric subunit used in this study to estimate protein-protein binding stabilities, $\Delta\Delta G_{bind}$. (C) A single protein F used in this study to estimate protein folding stability, $\Delta\Delta G_{fold}$. (D) For each substitution within 1 DNA change of the reference sequence, we calculated $\Delta\Delta G_{fold}$ and $\Delta\Delta G_{bind}$ using FoldX and the conceptual model shown. For a given amino acid sequence of the F protein, we have $\Delta G_{bind} = \Delta G_{interaction} - 5\Delta G_{fold}$. Letting the subscripts *sub* and *ref* refer to the protein with and without a given substitution, the relative binding stability is then calculated as $\Delta\Delta G_{bind} = \Delta G_{bind,sub} - \Delta G_{bind,ref}$ and the relative folding stability is calculated as $\Delta\Delta G_{fold} = \Delta\Delta G_{fold,sub} - \Delta\Delta G_{fold,ref}$.

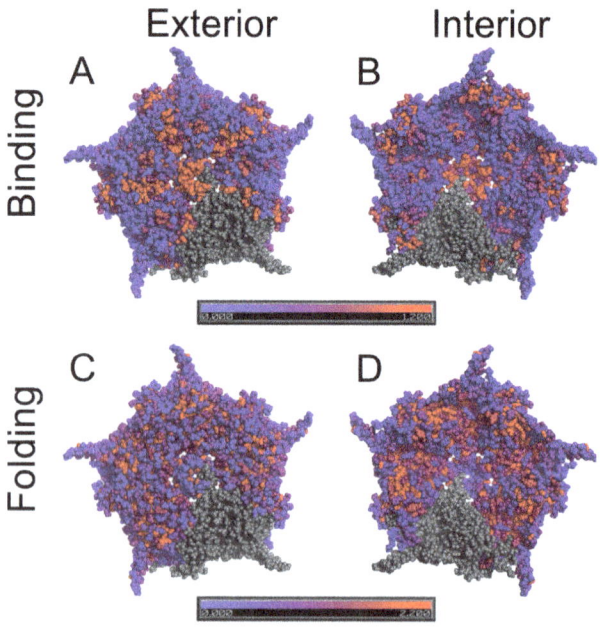

Figure 2. Heatmap of one pentamer showing median absolute effect size (i.e. |$\Delta\Delta G$|) at each residue among accessible substitutions. The figure illustrates that for binding stability, high effect residues are found along protein-protein interfaces while for folding stability, high effect residues are concentrated in the interior of the protein. Residues in red have large median effects; those in blue have small effects. Top panels (A and B) show effects on binding stability while lower two panels (C and D) show effects on folding stability. Left panels (A and C) show the exterior surface; right panels (B and D) show interior surface.

destabilize both folding and binding. For folding, 72.9% of the accessible substitutions have $\Delta\Delta G_{fold} > 0$. This agrees with previous studies that have shown random substitutions tend to be destabilizing [1,4,16,17,19,20]. We also find that a majority of accessible substitutions destabilize binding since 70.0% of the accessible substitutions have $\Delta\Delta G_{bind} > 0$. Note that 70% reflects destabilization of a single pentamer; in an expanded model that included multiple pentamers and interactions of the coat protein with other capsid proteins, we would expect this value would be higher. This prediction is supported by the graphic representation shown in Figure 2 where substitutions with moderate to strong destabilizing effects on binding tend to reside along the pentameric protein-protein interfaces (red sites in Figure 2 A–B) and not along the edges that would form the between-pentamer interfaces.

Examining the distribution of $\Delta\Delta G_{fold}$ and $\Delta\Delta G_{bind}$ of accessible substitutions shows that while most substitutions are destabilizing, they also tend to have small effects on stability (the white histogram bars in Figure 3A and C show accessible substitutions). For folding stability, 72.6% of the substitutions are between -2 and $+2$; for binding 91.1% are in this zone. If we had we included between pentamer-pentamer interactions, we expect that some of the substitutions along these interfaces would have been destabilizing and the distribution of $\Delta\Delta G_{bind}$ would be more spread out, like that of $\Delta\Delta G_{fold}$. Finally, the scatterplot of in Figure 3B shows that there is no correlation between $\Delta\Delta G_{fold}$ and $\Delta\Delta G_{bind}$ ($r^2 = 0.0003$, p = 0.39). This is not surprising given that substitutions having moderate to strong effects on binding stability occur at different residues than those having significant effects on binding stability (Figure 2).

Patterns among Observed Substitutions

We next characterized changes in stability for substitutions that have been observed in real evolving populations: either substitu-

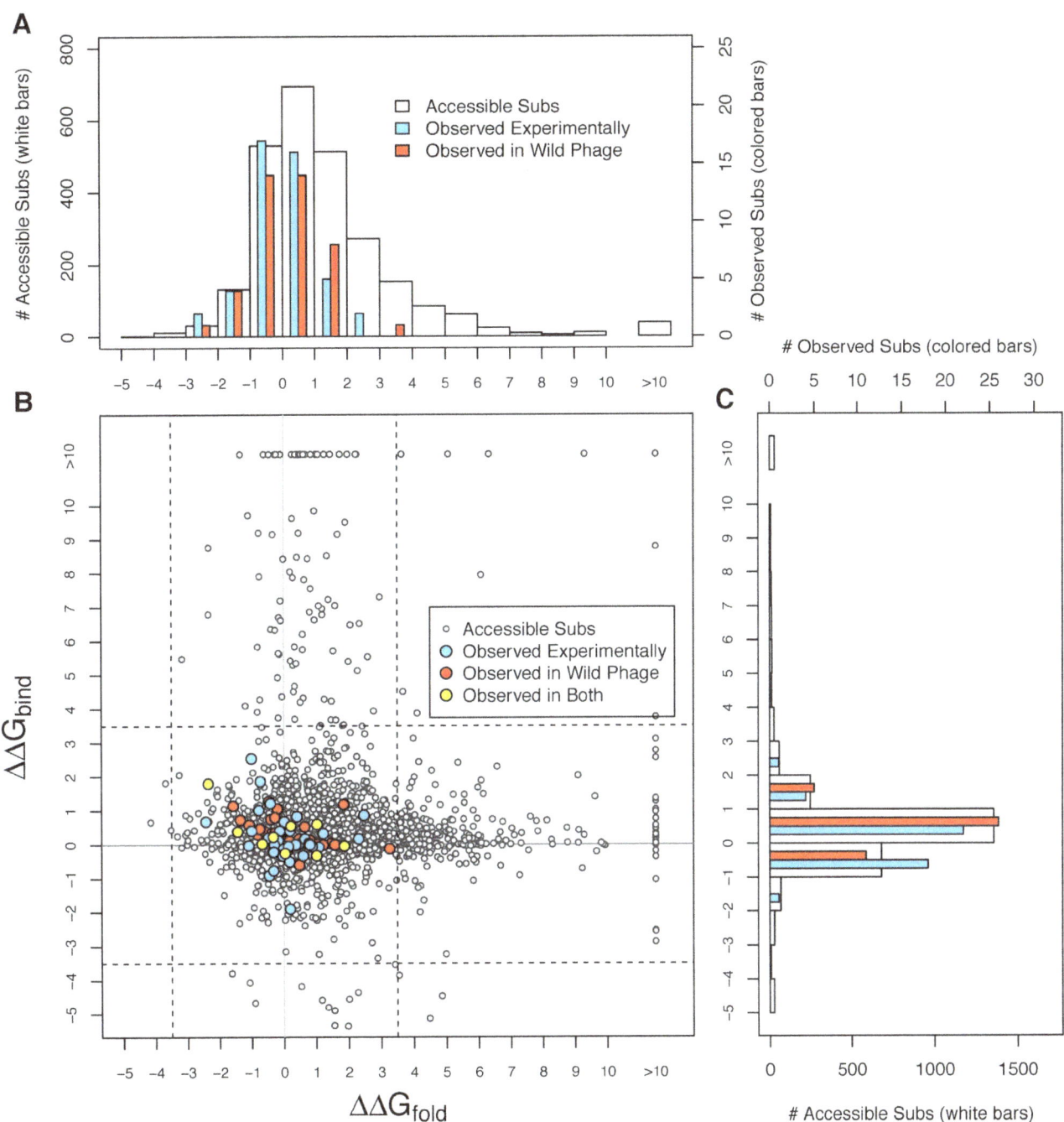

Figure 3. Comparison of stability effects between those accessible and those observed in the experimental and wild phage data.
The figure shows that all observed substitutions have small effects on both folding and binding stability. (A) Histogram of $\Delta\Delta G_{fold}$. (B) Scatterplot of $\Delta\Delta G_{fold}$ vs $\Delta\Delta G_{bind}$. (C) Histogram of $\Delta\Delta G_{bind}$. The dotted lines in (B) highlights the zone within which all observed substitutions fall. Note differences in scale between substitutions accessible (white bars) and those observed (red and blue bars) in the two histogram panels.

tions implicated by a comparison of the φX174-like wild phage, or substitutions observed during laboratory adaptations of φX174. We find that while observed substitutions can be stabilizing or destabilizing, none of them have large effects on stability (in Figure 3, colored histogram bars and points are observed substitutions). When the two datasets are combined, 79 unique substitutions are observed. Of these, 74 (93.7%) have $\Delta\Delta G_{fold}$ between −2 and +2, and 78 (98.7%) have $\Delta\Delta G_{bind}$ between −2 and +2 (Figure 3; Table 1). The six substitutions that fall outside

this zone are not far outside it, with the largest deviation being + 3.26 kcal/mol. The values for the two datasets viewed separately are quite similar but with smaller sample sizes (Table 1).

We conducted a randomization test to assess whether the observed substitutions differ significantly from the accessible substitutions. The answer is yes, observed substitutions are more concentrated near $\Delta\Delta G_{fold} = 0$ and $\Delta\Delta G_{bind} = 0$ than expected by chance. To perform the test, we took sets of 10,000 random samples from the accessible substitutions at the sample size of each

Table 1. The proportion of substitutions with $\Delta\Delta G$ within a stability zone around zero and the associated p-values.

Stability Zone	Set of Substitutions	n	$\Delta\Delta G_{fold}$ Prop (n)	p-value	$\Delta\Delta G_{bind}$ Prop (n)	p-value	$\Delta\Delta G_{fold}$ & $\Delta\Delta G_{bind}$ Prop (n)	p-value
−2 to +2	Accessible	2570	0.726 (1866)	–	0.911 (2340)	–	0.660 (1696)	–
	Experimental	46	0.913 (42)	0.0038	0.978 (45)	0.1456	0.891 (41)	0.0003
	Wild phage	42	0.952 (40)	0.0002	1.0 (42)	0.0364	0.952 (40)	<0.0001
	Experimental + Wild	79	0.937 (74)	0.0002	0.987 (78)	0.0114	0.924 (73)	<0.0001
−3.5 to +3.5	Accessible	2570	0.879 (2260)	–	0.950 (2441)	–	0.835 (2146)	–
	Experimental	46	1 (46)	0.0048	1 (46)	0.1876	1 (46)	0.0001
	Wild phage	42	1 (42)	0.0068	1 (42)	0.2292	1 (42)	0.0006
	Experimental + Wild	79	1 (79)	<0.0001	1 (79)	0.0310	1 (79)	<0.0001

Stability zone defined in the left column. The top row within each stability zone shows the accessible substitutions against which the other sets are compared. For $\Delta\Delta G_{fold}$, $\Delta\Delta G_{bind}$ and $\Delta\Delta G_{fold}$ & $\Delta\Delta G_{bind}$ together, the left column gives the proportion of substitutions in the stability zone with the actual number in parentheses. The right column gives the p-value associated with the null hypothesis that the observed counts (experimental, wild, and experimental + wild rows) are random samples from the accessible set and fall in the stability zone by chance. By $\Delta\Delta G_{fold}$ & $\Delta\Delta G_{bind}$ we mean the substitutions observed jointly within the zone by both measures of stability. Test are all two-sided and based on 10,000 random samples of accessible set.

observed set and asked how often the random sample has as many or more substitutions in the −2 to +2 stability zone as were actually observed. The test was done for folding stability alone, binding stability alone, or both folding and binding jointly. For the experimental and wild phage combined dataset, the two-sided p-values for folding, binding and the two jointly are 0.0002, 0.0114, and <0.0001 respectively (upper half of Table 1). For the two datasets individually, the smaller sample sizes lead to larger p-values, but except for binding in the experimental set, they remain significant. To check for robustness, we reran this test with the stability zone expanded to −3.5 to +3.5 and the results are very similar (Table 1).

The finding that observed substitutions differ from those accessible implies that selection acts on stability, either because stability or a trait highly correlated with it effects fitness or because the substitutions available to selection are constrained by their stability effects. We were interested in what selection surface could account for the differences between accessible and observed substitutions. To answer this, we assumed a simple model where that the probability of observing a substitution with a particular $\Delta\Delta G_{fold}$, $\Delta\Delta G_{bind}$ value in the data is proportional to the density of accessible substitutions in this stability region multiplied by the density of a selection function at this point. We assumed the selection function was a bivariate normal truncated below −3 and above +3 in both stability dimensions. We then determined what parameter values would make the observed data most probable. Before examining the results, it is helpful to consider interpretation of several of the most extreme possible selection functions. A very flat, plateau-like, selection function corresponds to stability acting purely as a filter, indifferent to the stability effects except whether they fall within the truncated zone or not. By contrast, a tight and perfectly symmetrical peak at zero would indicate selection strongly favors substitutions that change neither folding nor bindings stability. A long narrow ridge running along one axis indicates selection is indifferent to the stability the ridge is along but very sensitive to the other type of stability.

The best-fit selection functions are shown in Figure 4 with separate panels for the entire dataset combined, for the wild phage dataset, and the experimental datasets at high and normal temperatures. Averaging over the many conditions represented by our entire dataset (panel A), the selection function is centered on the origin indicating that selection favors substitutions that alter stability very little. The wild phage (panel B) are similar. The most interesting comparison is between the selection surfaces at high vs. normal temperatures (panels C and D). At high temperatures, the surface is a slightly elongated ridge running from the upper left quadrant down to the origin. In other words, selection favors substitutions with either little effect on stability or on those stabilize folding of the F protein and simultaneously destabilize binding of the pentamer (negative $\Delta\Delta G_{fold}$ and positive $\Delta\Delta G_{bind}$). At normal temperature, we see a selection surface that is roughly circular with a peak very near the origin.

A possible interpretation of these results is that the F protein is either at or is close to its optimal stability. This view asserts that at normal laboratory temperature substitutions conferring small changes to stability may be neutral or beneficial, but those that result in large changes are deleterious. The same is true at high temperatures except that the optimum stability appears to be slightly shifted from the ancestor. At both temperatures, all the changes we observe in stability across temperatures are small (< 2.5 kcal/mol). If this assertion that the protein is near or at the stability optimum is correct, we expect that the cumulative $\Delta\Delta G_{fold}$ and $\Delta\Delta G_{bind}$ over the course experiments (i.e. the sum $\Delta\Delta G_{fold}$ and $\Delta\Delta G_{bind}$ for all substitutions found in an experiment)

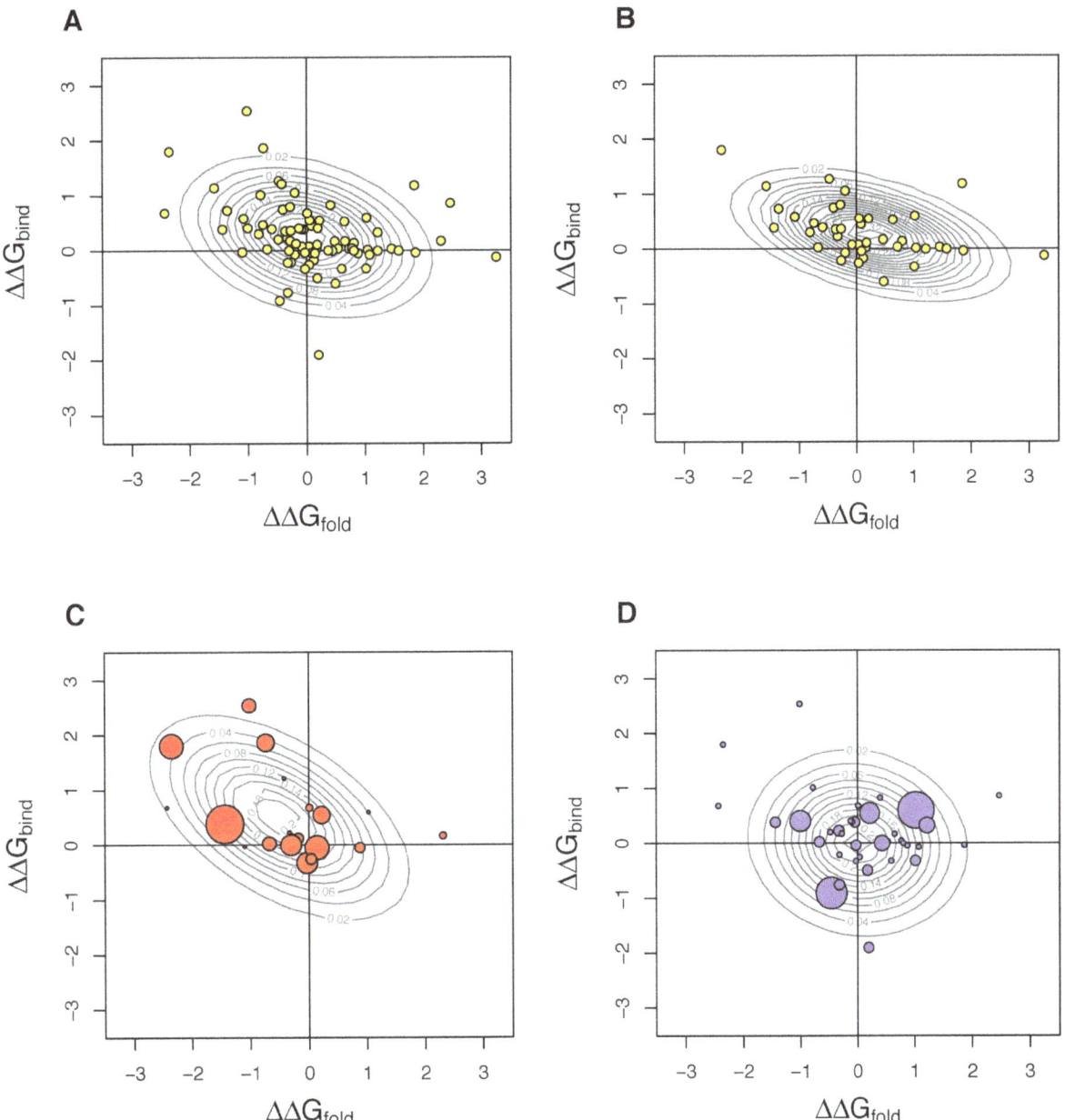

Figure 4. Estimated selection functions explaining the difference between accessible and observed substitutions. The figure shows that selection generally favors substitutions that have little effect on stability (peaks near the origin), but at high temperatures (in red), selection also favors substitutions that simultaneously stabilize folding and destabilize binding. The selection function is assumed to be a bivariate normal, the contour lines of equal probability of which are shown. Observed substitutions are colored circles. A) All 79 substitutions from both datasets combined. B) Wild phage dataset. C) Experimental data at high temperatures. D) Experimental data at normal temperature weighted by number of experiments observed in. In C and D substitutions are weighted by the number of experiments they appear in; size of symbols are scaled to show weighting. Density of accessible substitutions is shown in Figure 3B.

should also remain in the same zone as individual substitutions. By contrast, if cumulative $\Delta\Delta G_{fold}$ and $\Delta\Delta G_{bind}$ depart from this region, then we know selection is limiting the size of individual stability changes while still allowing larger shifts the protein's stability.

We tested these competing possibilities by looking at cumulative $\Delta\Delta G_{bind}$ and $\Delta\Delta G_{fold}$ in laboratory adaptation as a function of temperature. Temperature is a good candidate for examining this question for several reasons. First, it has a profound effect on fitness, so selection is strong. Second, certain substitutions are

observed repeatedly at high temperatures (e.g. L242F in Bull et al. 2000 [21]) indicating that they are adaptations to high temperature per se. Third, it is logical that protein stability links temperature to fitness since temperature affects stability, stability dictates the proportion of time the protein is folded and bound (as compared to unfolded and unbound), and we expect these proportions to affect viral assembly rate and therefore fitness.

The results, presented in Figure 5, show that the cumulative effects on stability often take the protein well outside the region where individual changes are found. If we look at adaptations that

A

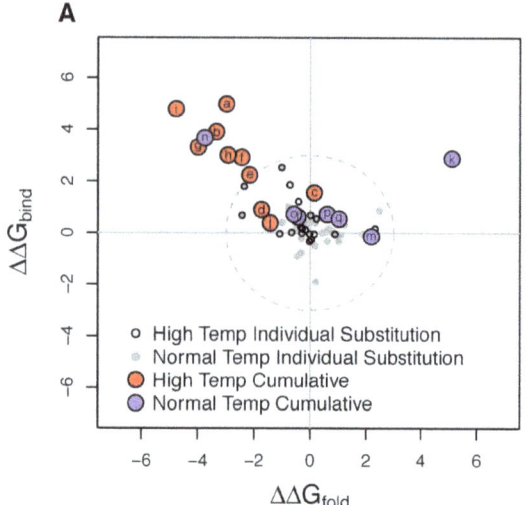

stability toward negative $\Delta\Delta G_{fold}$ and positive $\Delta\Delta G_{bind}$. A) Colored symbols show cumulative stability changes for all experiments beginning with ancestral φX174 and remaining at either high (≥42°C) or low (≤37°C) temperatures. Small open and grey points show the stability changes for individual substitutions. The dashed circle demarks the range within which all single substitutions fall. Letters within colored symbols indicate from where the experiment data is obtained (see end of legend). B) Cumulative stability changes in the Rain experiment [41]. The experiment had a branching design where temperature differed between each of the two branches as indicated. Number of substitutions on each branch indicated by + symbols. C) Cumulative stability changes in two unpublished 50-day chemostat experiments that were sampled every 10 days where temperature began at 37°C, was elevated to 42°C for part of period of time, and then returned to 37°C. The letters in panel A indicate the study where each dataset comes from: a–b [43], c–d [49], e–l [8], j [45], k [46], l [49], n–o [48], and q [47]. Experiments m and p are unpublished.

B

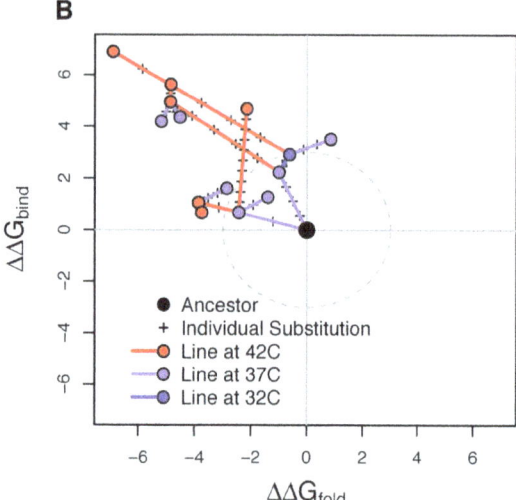

began with our ancestor (panel A), 7 of the 10 high temperature adaptations have cumulative effects outside the region of individual effects (denoted by the dashed circle). The most extreme case has $\Delta\Delta G_{fold} \approx -5$ and $\Delta\Delta G_{bind} \approx 5$, roughly twice the magnitude of departure from ancestor observed among the largest individual changes. At normal laboratory temperature, two of the seven experiments depart from the region of individual effects, but each in a different manner. In panel B we present the results from an experiment where adapting lines were split repeatedly, with each branch subjected to different hosts and/or temperatures [41]. Similar to panel A, we observe high temperatures tending to shift stability up and to the left. Here, the most extreme endpoint falls at $\Delta\Delta G_{fold} \approx -7$ and $\Delta\Delta G_{bind} \approx 7$, nearly three times the deviation found among individual changes. In panel C we show the results from two unpublished 50-day chemostat adaptations where temperature was initially normal (37°C), then high (42°C), and then returned to normal; populations were sampled every 10 days. For both populations we see only small cumulative changes, well within the range of individual effects.

Taken together, the cumulative $\Delta\Delta G$ results point to a few general conclusions. First, there is a lot of variation in the trajectory that stability takes under either temperature. This implies that selection must be acting on phenotypes beyond protein stability in these experiments. Second, cumulative changes can be much larger than individual changes. Because larger stability-changing substitutions are accessible, this suggests that selection favors several local modifications to stability over one large change that accomplishes the same thing at a global scale. Third, at high temperatures the stability trajectory tends to be toward negative $\Delta\Delta G_{fold}$ and positive $\Delta\Delta G_{bind}$. The negative change in $\Delta\Delta G_{fold}$ may be a way of counterbalancing the destabilizing effects of elevated temperature and leaving the protein highly functional.

The trend toward positive changes in $\Delta\Delta G_{bind}$ are, however, quite unexpected. In previous work on the related bacteriophage ID11 [22,23], we found the opposite patterns: a set of first-step substitutions that were highly beneficial at 37°C tended to stabilize binding (i.e. have negative $\Delta\Delta G_{bind}$ values). There are several differences between the ID11 study and the φX174 experiments reported here. Most importantly, while 37°C is near the optima for φX174, the optima for ID11 is around 32°C [24,25]; thus 37°C is a high temperature for ID11. Secondly, all of the changes reported for ID11 were first-step changes while each φX174 experiment reported accumulated many changes. Finally, those ID11 substitutions arose in flask adaptations where accessible hosts greatly outnumbered phage. Nearly all of the φX174 adaptations

C

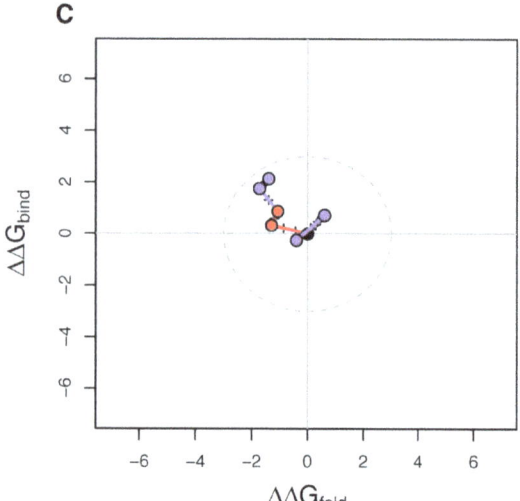

Figure 5. Cumulative changes in $\Delta\Delta G_{fold}$ and $\Delta\Delta G_{bind}$ across many lab adaptation experiments. The figure shows that cumulative stability changes frequently extend beyond individual changes and that high temperature changes are frequently beyond the range of individual changes and high temperatures (red) often push

occurred in chemostats where hosts greatly outnumbered by phage.

In flasks, logic dictates that a good strategy is to minimize the time to burst (and thereby allow subsequent infections and rapid exponential growth) while in chemostats it should pay to maximize the number of progeny in the current infection. Indeed, chemostat adaptations of φX174 commonly have mutations is in the D-promoter that serve to delay the time to burst [2,19,26,27]. One possible explanation for the tendency to destabilize binding at high temperatures is that this may slow capsid assembly. At high temperatures, cell growth is slowed and other aspects of phage reproduction like genome replication, translation and scaffolding construction are probably slowed as well. Slowing pentamer binding might bring the rate of capsid assembly into closer balance with other processes and ultimately increase burst size.

Summary

We have shown that in major capsid protein of φX174, the majority of accessible substitutions destabilize both protein folding and pentamer binding. The substitutions that are observed in the wild phage and in laboratory adaptations of φX174 have significantly smaller effects on stability than expected. However, in adaptations to temperatures above 42°C, there is tendency for substitutions to accumulate that confer stabilizing effects on folding, but destabilizing effects on binding. One possibility is that these changes leave F still functional, but slow the rate of pentamer and thereby capsid assembly in a way that increases burst size. Finally, the cumulative stability effects over the course of an adaptation are often greater than the range of individual changes suggesting that there are local as well as global constraints on protein stability.

Materials and Methods

Phage System

The organism used for this study is phage φX174, a virus that infects *Escherichia coli* and other bacteria [23]. Phage φX174 has 11 genes and is composed of several proteins depending on the stage of the assembly cycle [33]. The φX174 mature capsid (Figure 1A) is composed of 12 pentameric units containing proteins F, G, and J, plus 12 copies of H asymmetrically arranged inside the capsid [34–37]. The model system for the current study is the coat protein F which must both fold and then bind to form pentameric subunits in the early stage of the procapsid formation (Figure 1B; Figure 1C).

Stability Estimation

Changes in protein folding stabilities and protein-protein binding stabilities due to amino acid substitutions were estimated using FoldX [7]. FoldX was chosen for this study to balance accuracy and speed [3,5–7]. Given the large number of mutations studied here, it is not possible to use accurate statistical mechanical approaches such as all atom molecular dynamics simulation as we did in a previous study [23]. A total of 2570 substitutions (all substitutions at the 426 residues of protein F accessible with one DNA mutation) were estimated for each protein structure in unbound and pentameric system (Figure 1B, 1C). Initially, protein structures were equilibrated 15 times in succession using the "repairPDB" command in the FoldX software to obtain a fully minimalized conformation. Once the minimized conformation was obtained for each of the four model systems, then the binding and folding stabilities were estimated using the "BuildModel" command in FoldX (also see Figure 3). The estimated folding and

binding stability changes for all possible single substitutions from the reference sequence are available in the supplemental materials.

Observed Substitutions

Observed substitutions came from two different datasets: wild and experimental. The wild phage substitutions were based on the collecting, sequencing and phylogenetic work of Rokyta et al. [12] We obtained the F-protein amino acid sequences for 19 phage in the φX174-like clade, including φX174 itself. We used the consensus sequence of these to generate a putative ancestral sequence. Comparison of the 19 phage with this ancestral sequence yielded 42 unique substitutions among the wild phage. For the experimental set we constructed a database of many published [8,41,43–50] and two unpublished laboratory adaptations involving φX174. The dataset includes a total of approximately 29 different experiments (the count is complicated by the fact that some experiments involved branching lines). All but five of the experiments were conducted in chemostats (the others were in flasks); 17 of them began with our ancestor φX174 (the others used φX174 with substitutions already in the genome); 12 of them were at high temperatures (42–43.5°C), 13 at normal laboratory temperature (37°C), while 4 of them involved variable temperatures. Normal laboratory temperature is close to the optimal for φX174, while these high temperatures constitutes strong selection on this phage [15].

Statistical Analysis

To determine whether the observed substitutions were more narrowly clustered around $\Delta\Delta G$ of zero than expected, we did a set of randomization tests. We fist defined a zone around zeros as -2 to $+2$. We defined $n_{fold(real)}$, $n_{bind(real)}$ and $n_{fold+bind(real)}$ as, respectively, the number of real observed substitutions with $\Delta\Delta G_{fold}$ individually $\Delta\Delta G_{bind}$ individually, and $\Delta\Delta G_{fold}$ and $\Delta\Delta G_{bind}$ simultaneously inside this zone. For the wild phage, we drew samples of size 42 (the number of observed substitutions) without replacement from the pool of accessible substitutions and, each time, determined the number of substitutions within the zone by each criteria: $n_{fold(sim)}$, $n_{bind(sim)}$ and $n_{fold+bind(sim)}$. We did this 10,000 times and approximated p-values as twice the proportion of times the $n_{fold(sim)} \geq n_{fold(real)}$, $n_{bind(sim)} \geq n_{bind(real)}$, and $n_{fold+bind(sim)} \geq n_{fold+bind(real)}$. We then repeated this for the set of 46 experimentally observed substitutions, and the combined set of 79 substitutions. Finally, we redefined the zone as -3.5 to $+3.5$ and reran the analyses.

We estimated selection functions that could explain the disparities between accessible and observed substitutions. To do this we assumed that the approximate probability of observing a substitution in the data with a particular joint $\Delta\Delta G_{fold}$ and $\Delta\Delta G_{bind}$ value was proportional to the product of the density of accessible substitutions in this stability region and the density of the selection function at this point. The accessible densities were obtained by gridding the region between -3 and $+3$ at 0.25 increments and calculating the proportion of accessible substitutions within each square. We considered candidate bivariate normal distributions across a range of parameter values: μ_{fold} and μ_{fold} from -1 to $+1$ at 0.1 increments, σ_{fold} and σ_{bind} from 0.25 to 1.5 at 0.0625 increments, and ρ from -1 to $+1$ at 0.1 increments. For each we obtained the density at that $\Delta\Delta G_{fold}$, $\Delta\Delta G_{bind}$ value, multiplied by accessible density in that region, took the log, and summed over all substitutions in the dataset. The combination of parameter values that made this sum largest served as our estimated of the selection function. We did this for wild dataset alone, for the combined wild plus experimental dataset, for the experimental data at 37°C, and the experimental data at 42–43.5°C. In the last two cases we restricted ourselves to experiments that began with ancestral φX174 (excluding those that had previous adaptive changes). For these, we have ran the analysis both with each substitution represented once (unweighted) and with each substitution weighted by the number of different experiments it appeared in. We present the results from the weighted analysis, but the unweighted results were qualitatively the same.

Accession Numbers

The ancestral φX174 sequence is available at GenBank accession number AF176034. The model structure is based on Protein Data Bank accession number 2BPA.

Supporting Information

Table S1   FoldX estimates of $\Delta\Delta G_{fold}$ and $\Delta\Delta G_{bind}$ for all 8094 possible single substitutions in the φX174 F protein relative to the reference sequence. *site* is the residue number. Note in protein F the first amino acid, methionine, is

removed after translation. Numbering begins after its removal. *aa.from* and *aa.to* are the amino acids in the reference and the mutant respectively. *within.1.DNA.change* indicates substitutions that can be accessed by a single DNA change from the reference sequence (1 = yes, 0 = no). *wild.phg.sub* indicates substitutions we infer occurred in the evolution of the φX174-like wild phage by comparison of them with their consensus sequence (1 = yes, 0 = no). *lab.exp.sub* indicates substitutions found in a lab adaptation experiment (see paper for source of experiments; 1 = yes, 0 = no). *ddG.fold* and *ddG.bind* give changes in folding and binding stability, $\Delta\Delta G_{fold}$ and $\Delta\Delta G_{bind}$, respectively.

Author Contributions

Conceived and designed the experiments: KHL CRM HAW FMY. Performed the experiments: KHL. Analyzed the data: KHL CRM. Contributed reagents/materials/analysis tools: HAW FMY. Wrote the paper: KHL CRM HAW FMY.

References

1. Bloom JD, Raval A, Wilke CO (2007) Thermodynamics of neutral protein evolution. Genetics 175: 255–266. doi:10.1534/genetics.106.061754.
2. DePristo MA, Weinreich DM, Hartl DL (2005) Missense meanderings in sequence space: a biophysical view of protein evolution. Nat Rev Genet 6: 678–687. doi:10.1038/nrg1672.
3. Gromiha MM (2007) Prediction of protein stability upon point mutations. Biochem Soc Trans 35: 1569–1573. doi:10.1042/BST0351569.
4. Bloom JD, Silberg JJ, Wilke CO, Drummond DA, Adami C, et al. (2005) Thermodynamic prediction of protein neutrality. Proc Nat Acad Sci USA 102: 606–611. doi:10.1073/pnas.0406744102.
5. Capriotti E, Fariselli P, Casadio R (2005) I-Mutant2.0: predicting stability changes upon mutation from the protein sequence or structure. Nucleic Acids Res 33: W306–W310. doi:10.1093/nar/gki375.
6. Schymkowitz J, Borg J, Stricher F, Nys R, Rousseau F, et al. (2005) The FoldX web server: an online force field. Nucleic Acids Res 33: W382–W388. doi:10.1093/nar/gki387.
7. Guerois R, Nielsen JE, Serrano L (2002) Predicting changes in the stability of proteins and protein complexes: a study of more than 1000 mutations. J Mol Biol 320: 369–387. doi:10.1016/S0022-2836(02)00442-4.
8. Bull JJ, Badgett MR, Wichman HA, Huehenbeck JP, Hillis DM, et al. (1997) Exceptional convergent evolution in a virus. Genetics 147: 1497–1507.
9. Becktel WJ, Schellman JA (1987) Protein stability curves. Biopolymers 26: 1859–1877.
10. Robertson A, Murphy KP (1997) Protein structure and the energetics of protein stability. Chem Rev 97: 1251–1267.
11. Sanchez-Ruiz JM (2010) Protein kinetic stability. Biophys Chem 148: 1–15. doi:10.1016/j.bpc.2010.02.004.
12. Rokyta DR, Burch CL, Caudle SB, Wichman HA (2006) Horizontal gene transfer and the evolution of microvirid coliphage genomes. J Bacteriol 188: 1134–1142. doi:10.1128/JB.188.3.1134–1142.2006.
13. Khan S, Vihinen M (2010) Performance of protein stability predictors. Hum Mutat 31: 675–684. doi:10.1002/humu.21242.
14. Pace CN, Hebert EJ, Shaw KL, Schell D, Both V, et al. (1998) Conformational stability and thermodynamics of folding of ribonucleases Sa, Sa2 and Sa3. J Mol Biol 279: 271–286.
15. Wichman HA, Brown CJ (2010) Experimental evolution of viruses: Microviridae as a model system. Phil Trans R Soc B 365: 2495–2501. doi:10.1098/rstb.2010.0053.
16. Pace CN, Fu H, Fryar KL, Landua J, Trevino SR, et al. (2011) Contribution of Hydrophobic Interactions to Protein Stability. J Mol Biol 408: 514–528. doi:10.1016/j.jmb.2011.02.053.
17. Tokuriki N, Stricher F, Schymkowitz J, Serrano L, Tawfik DS (2007) The stability effects of protein mutations appear to be universally distributed. J Mol Biol 369: 1318–1332. doi:10.1016/j.jmb.2007.03.069.
18. Ross PD, Subramanian S (1981) Thermodynamics of Protein Association Reactions: Forces Contributing to Stability? Biochemistry 20: 3096–3102.
19. Tokuriki N, Stricher F, Serrano L, Tawfik DS (2008) How protein stability and new functions trade off. PLoS Comput Biol 4: e1000002.
20. Tokuriki N, Tawfik DS (2009) Stability effects of mutations and protein evolvability. Curr Opin Struct Biol 19: 596–604. doi:10.1016/j.sbi.2009.08.003.
21. Bull JJ, Badgett MR, Wichman HA (2000) Big-benefit mutations in a bacteriophage inhibited with heat. Mol Biol Evol 17: 942–950.
22. Pace CN, Horn G, Hebert EJ, Bechert J, Shaw K, et al. (2001) Tyrosine hydrogen bonds make a large contribution to protein stability. J Mol Biol 312: 393–404. doi:10.1006/jmbi.2001.4956.
23. Lee KH, Miller CR, Nagel AC, Wichman HA, Joyce P, et al. (2011) First-Step Mutations for Adaptation at Elevated Temperature Increase Capsid Stability in a Virus. PLoS ONE 6: e25640. doi:10.1371/journal.pone.0025640.t001.
24. Jha BK, Mitra N, Rana R, Surolia A, Salunke DM, et al. (2004) pH and cation-induced thermodynamic stability of human hyaluronan binding protein 1 regulates its hyaluronan affinity. J Biol Chem 279: 23061–23072. doi:10.1074/jbc.M310676200.
25. Knies JL, Kingsolver JG, Burch CL (2009) Hotter is better and broader: thermal sensitivity of fitness in a population of bacteriophages. Am Nat 173: 419–430.
26. Godoy-Ruiz R, Ariza F, Rodriguez-Larrea D, Perez-Jimenez R, Ibarra-Molero B, et al. (2006) Natural selection for kinetic stability is a likely origin of correlations between mutational effects on protein energetics and frequencies of amino acid occurrences in sequence alignments. J Mol Biol 362: 966–978. doi:10.1016/j.jmb.2006.07.065.
27. Brown CJ, Stancik AD, Roychoudhury P, Krone SM (2013) Adaptive regulatory substitutions affect multiple stages in the life cycle of the bacteriophage φX174. BMC Evol Biol 13: 66. doi:10.1146/annurev.mi.03.100149.002103.
28. Wang X, Minasov G, Shoichet BK (2002) Evolution of an antibiotic resistance enzyme constrained by stability and activity trade-offs. J Mol Biol 320: 85–95. doi:10.1016/S0022-2836(02)00400-X.
29. Beadle BM, Shoichet BK (2002) Structural bases of stability–function tradeoffs in enzymes. J Mol Biol 321: 285–296. doi:10.1016/S0022-2836(02)00599-5.
30. Wang Q, Johnson JL, Agar NYR, Agar JN (2008) Protein aggregation and protein instability govern familial amyotrophic lateral sclerosis patient survival. PLoS Biol 6: e170. doi:10.1371/journal.pbio.0060170.
31. Dobson CM (2003) Protein folding and misfolding. Nature 426: 884–890. doi:10.1038/nature02261.
32. Yue P, Li Z, Moult J (2005) Loss of protein structure stability as a major causative factor in monogenic disease. J Mol Biol 353: 459–473. doi:10.1016/j.jmb.2005.08.020.
33. Cherwa JE, Organtini LJ, Ashley RE, Hafenstein SL, Fane BA (2011) In Vitro Assembly of the φX174 Procapsid from External Scaffolding Protein Oligomers and Early Pentameric Assembly Intermediates. J Mol Biol 412: 387–396. doi:10.1016/j.jmb.2011.07.070.
34. McKenna R, Xia D, Willingmann P, Ilag LL, Krishnaswamy S, et al. (1992) Atomic structure of single-stranded DNA bacteriophage φX174 and its functional implications. Nature 355: 137–143. doi:10.1038/355137a0.
35. McKenna R, Xia D, Willingmann P, Ilag LL, Rossmann MG (1992) Structure determination of the bacteriophage ΦX174. Acta Crystallogr Sect B 48: 499–511. doi:10.1107/S0108768192001344.
36. Dokland T, McKenna R, Sherman DM, Bowman BR, Bean WF, et al. (1998) Structure Determination of the ΦX174 Closed Procapsid. Acta Crystallogr Sect D 54: 878–890.
37. Dokland T, Bernal RA, Burch AD, Pletnev S, Fane BA, et al. (1999) The role of scaffolding proteins in the assembly of the small, single-stranded DNA virus ΦX174. J Mol Biol 288: 595–608. doi:10.1006/jmbi.1999.2699.
38. Levy ED, Erba EB, Robinson CV, Teichmann SA (2008) Assembly reflects evolution of protein complexes. Nature 453: 1262–1265. doi:10.1038/nature06942.
39. Keeney PM, Xie J, Capaldi RA, Bennett JP Jr (2006) Parkinson's Disease Brain Mitochondrial Complex I Has Oxidatively Damaged Subunits and Is Functionally Impaired and Misassembled. J Neurosci 26: 5256–5264. doi:10.1523/JNEUROSCI.0984-06.2006.
40. Chiti F, Stefani M, Taddei N, Ramponi G, Dobson CM (2003) Rationalization of the effects of mutations on peptide and protein aggregation rates. Nature 424: 805–808. doi:10.1038/nature01891.

41. Rain MW (2001) Molecular evolution in two viruses: Using the ΦX174 virus to study phylogenetics, and phylogenetics to study the human immunodeficiency virus University of Idaho.

42. Suhre K, Sanejouand Y-H (2004) ElNemo: a normal mode web server for protein movement analysis and the generation of templates for molecular replacement. Nucleic Acids Res 32: W610–W614. doi:10.1093/nar/gkh368.

43. Wichman HA, Badgett MR, Scott L, Boulianne CM, Bull JJ (1999) Different trajectories of parallel evolution during viral adaptation. Science 285: 422–424.

44. Crill W, Wichman HA, Bull JJ (2000) Evolutionary reversals during viral adaptation to alternating hosts. Genetics 154: 27–37.

45. Wichman HA, Scott LA, Yarber CD, Bull JJ (2000) Experimental evolution recapitulates natural evolution. Phil Trans R Soc B 355: 1677–1684. doi:10.1098/rstb.2000.0731.

46. Wichman HA, Millstein J, Bull JJ (2005) Adaptive molecular evolution for 13,000 phage generations: a possible arms race. Genetics 170: 19–31. doi:10.1534/genetics.104.034488.

47. Pepin KM, Domsic J, McKenna R (2008) Genomic evolution in a virus under specific selection for host recognition. Infection, Genetics and Evolution 8: 825–834. doi:10.1016/j.meegid.2008.08.008.

48. Kronenberg ZN (2010) Asymmetrical adaption in a two-host viral chemostat University of Idaho.

49. Brown CJ, Millstein J, Williams CJ, Wichman HA (2013) Selection affects genes involved in replication during long-term evolution in experimental populations of the bacteriophage φX174. PLoS ONE 8: e60401. doi:10.1371/journal.pone.0060401.

50. Pepin KM, Wichman HA (2008) Experimental evolution and genome sequencing reveal variation in levels of clonal interference in large populations of bacteriophage φX 174. BMC Evol Biol 8: 85. doi:10.1186/1471-2148-8-85.

51. Schymkowitz JWH, Rousseau F, Martins IC, Ferkinghoff-Borg J, Stricher F, et al. (2005) Prediction of water and metal binding sites and their affinities by using the Fold-X force field. Proc Nat Acad Sci USA 102: 10147–10152.

13

Analysis of Nidogen-1/Laminin γ1 Interaction by Cross-Linking, Mass Spectrometry, and Computational Modeling Reveals Multiple Binding Modes

Philip Lössl[1¤], **Knut Kölbel**[1], **Dirk Tänzler**[1], **David Nannemann**[2], **Christian H. Ihling**[1], **Manuel V. Keller**[3], **Marian Schneider**[4], **Frank Zaucke**[3], **Jens Meiler**[2], **Andrea Sinz**[1]*

1 Department of Pharmaceutical Chemistry and Bioanalytics, Institute of Pharmacy, Martin Luther University Halle-Wittenberg, Halle (Saale), Germany, 2 Department of Chemistry and Center for Structural Biology, Vanderbilt University, Nashville, TN, United States of America, 3 Center for Biochemistry, Medical Faculty, University of Cologne, Cologne, Germany, 4 Research Group Artificial Binding Proteins, Institute of Biochemistry and Biotechnology, Martin Luther University Halle-Wittenberg, Halle (Saale), Germany

Abstract

We describe the detailed structural investigation of nidogen-1/laminin γ1 complexes using full-length nidogen-1 and a number of laminin γ1 variants. The interactions of nidogen-1 with laminin variants γ1 LEb2–4, γ1 LEb2–4 N836D, γ1 short arm, and γ1 short arm N836D were investigated by applying a combination of (photo-)chemical cross-linking, high-resolution mass spectrometry, and computational modeling. In addition, surface plasmon resonance and ELISA studies were used to determine kinetic constants of the nidogen-1/laminin γ1 interaction. Two complementary cross-linking strategies were pursued to analyze solution structures of laminin γ1 variants and nidogen-1. The majority of distance information was obtained with the homobifunctional amine-reactive cross-linker bis(sulfosuccinimidyl)glutarate. In a second approach, UV-induced cross-linking was performed after incorporation of the diazirine-containing unnatural amino acids photo-leucine and photo-methionine into laminin γ1 LEb2–4, laminin γ1 short arm, and nidogen-1. Our results indicate that Asn-836 within laminin γ1 LEb3 domain is not essential for complex formation. Cross-links between laminin γ1 short arm and nidogen-1 were found in all protein regions, evidencing several additional contact regions apart from the known interaction site. Computational modeling based on the cross-linking constraints indicates the existence of a conformational ensemble of both the individual proteins and the nidogen-1/laminin γ1 complex. This finding implies different modes of interaction resulting in several distinct protein-protein interfaces.

Editor: Bostjan Kobe, University of Queensland, Australia

Funding: KK is funded by the BMBF project. AS acknowledges financial support from the DFG (projects Si 867/13-1, 15-1, and 16-1), the BMBF (Pro-Net-T3), and the region of Saxony-Anhalt. JM is funded through NIH (R01 GM080403, R01 MH090192, R01 GM099842, and R01 DK097376) and NSF (CHE 1305874). The funders had no role in study design, data collection and analysis, decision to publish, or preparation of the manuscript.

Competing Interests: The authors have declared that no competing interests exist.

* Email: andrea.sinz@pharmazie.uni-halle.de

¤ Current address: Biomolecular Mass Spectrometry and Proteomics, Bijvoet Center for Biomolecular Research and Utrecht Institute for Pharmaceutical Sciences, Utrecht University, Padualaan 8, CH Utrecht, The Netherlands

Introduction

Laminins are the major non-collagenous proteins of basement membranes that are known to form networks through crucial non-covalent self-interactions [1,2]. Each member of the laminin protein family consists of three polypeptide chains with one copy of the α, β, and γ chain. Since the discovery of laminin [3], several nomenclatures have been developed, which are, however, not always completely systematic [4]. In this work, we apply the laminin nomenclature introduced by Aumailley et al. [5]. Electron microscopic studies of laminin-111 (α1-, β1-, and γ1-subunits) reveal a cross-shaped protein structure [6] with three subunits being connected within the central part according to a coiled-coil arrangement ('long arm'). The N-terminal regions of the laminin subunits are free ('short arms') (Figure 1). The globular domains at the N-termini of all three chains (LN domains) are required for

efficient polymerization as deletion mutants with two or fewer LN domains fail to form networks [2]. Furthermore, laminin-111 harbors more centrally located globular domains (α1 L4a, α1 L4b, β1 LF, and γ1 L4) as well as several 'laminin-type epidermal growth factor-like' (LE) modules. Three-dimensional structures of certain laminin domains are available, such as X-ray structures of the C-terminal LG domains (PDB entries 2JD4, 2WJS, 1QU0, 1DYK), the nidogen-binding region γ1 LEb2–4 (PDB entries 1KLO, 1NPE), and the α5, β1, and γ1 LN domains (PDB entries 2Y38, 4AQT, 4AQS), the latter of which were found to be in good agreement with previously reported computational models [7].

Nidogens (entactins) are sulfated monomeric glycoproteins that are ubiquitously present in basement membranes of higher organisms. While invertebrates possess only one nidogen, two nidogen isoforms, namely nidogen-1 (ca. 135 kDa) and nidogen-2

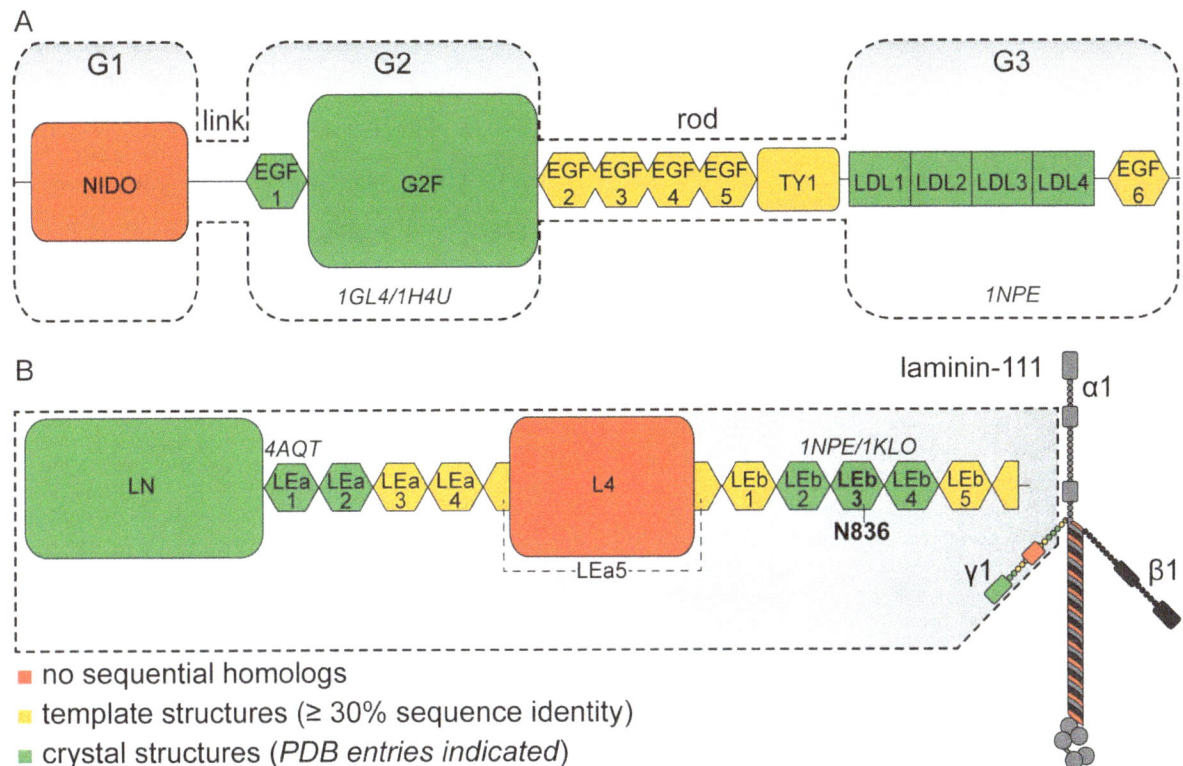

Figure 1. Arrangement of (A) nidogen-1 and (B) laminin γ1 short arm. The domains are color-coded with respect to the availability of crystal structures (green), template structures for comparative modeling (yellow) or neither of both (red). PDB IDs of the respective crystal structures are indicated in italics. (A) Nidogen-1 domain assignments are according to the UniProt KB entry P10493. Additionally, the historic domain names (G1–link–G2–rod–G3) are given. (B) Laminin domain designations follow the nomenclature of Aumailley et al. [5]. Laminin γ1 short arm is part of the heterotrimeric protein laminin-111, the overall structure of which is schematically depicted.

(ca. 150 kDa), have been identified in vertebrates [8–10]. Both isoforms exhibit a similar repertoire of binding partners [11], resulting in basement formation even after knocking out one nidogen isoform [12]. Electron microscopic studies have revealed nearly identical arrangements of nidogen-1 and -2 [9,13]. Both comprise three globular domains (G1–G3), which are connected by two rod-shaped domains ('link' and 'rod' regions) (Figure 1). Sequence analyses have confirmed the modular structures of nidogens. Identified motifs, such as epidermal growth factor (EGF)-like sequences, are also present in a number of other proteins of the extracellular matrix [14]. X-ray structures are available for G2 and G3 domains of nidogen-1 (PDB entries 1GL4, 1H4U, 1NPE). Nidogen is considered to be a stabilizer and adaptor protein within the basement membrane. In addition to isoform-specific interactions of nidogen-1 with fibulin-1 and -2 as well as of nidogen-2 with tropoelastin and type XVIII collagen, both nidogens bind to the essential basement membrane proteins perlecan, type IV collagen, and laminin [14].

Initial studies of nidogen-1 revealed an extraordinarily strong interaction with laminin [15] and led to the elucidation of a 1:1 stoichiometry between nidogen-1 and laminin in the complex [16]. The C-terminal G3 domain of nidogen-1 [16,17] as well as the EGF-like motif LEb3 in laminin γ1 [18–21] were identified as binding regions. A weaker nidogen-1 binding has been observed in laminin γ2 and γ3 [14]. Using radioligand binding assays with variants of this domain, amino acids Asp-834, Asn-836, and Val-838 were identified to be essential for the interaction of the laminin γ1 LEb3 domain with nidogen-1. The exchange of Asn-836 against aspartic acid resulted in a drastic decrease of nidogen-1

binding with a 25,000-fold loss in affinity [22]. The sequence numbering used here follows the amino acid sequences shown in Figure S1 with Asp-834, Asn-836, and Val-838 corresponding to Asp-800, Asn-802, and Val-804 in the classical laminin γ1 numbering [23].

In 2003, the three-dimensional structure of the complex between the G3 domain of nidogen-1 and LEb domains 2–4 of laminin γ1 was elucidated [24]. The G3 domain exhibits a ß-propeller composed of six LDL receptor YWTD modules, creating a concave interface for the amino acids Asp-834, Asn-836, and Val-838 within the loop of laminin γ1 LEb3. Complex formation is enhanced by additional interactions of laminin γ1 LEb2 with the ß-propeller.

Recently, the interaction of full-length nidogen-1 with laminin γ1 short arm has been investigated by size-exclusion chromatography, dynamic light scattering, and small-angle X-ray scattering [25]. These studies indicate that the interaction is mediated solely by the C-terminal domains, while the remaining regions of both proteins do not participate in complex formation.

For our studies investigating the interaction between nidogen-1 and laminin γ1 short arm, we chose an alternative approach providing 3D-structural insights into proteins. This strategy relies on chemical cross-linking and a subsequent mass spectrometry (MS)-based analysis of the created products [26,27]. Structural information can be obtained by the insertion of a chemical cross-linker between two functional groups within a protein. The cross-linker has a defined length and is connected via covalent bonds to functional groups of amino acid side chains, allowing the cross-linked amino acids to be identified after enzymatic digestion. This chemical cross-linking approach is also applied to study protein-protein interfaces. The sequence separation of cross-linked amino acids, combined with the cross-linker length, impose a distance constraint on the 3D structure of the protein complex [7,28–31]. Analysis of cross-linked peptides by MS makes use of several advantages: First, the mass of the protein or the protein complex under investigation is theoretically unlimited as the proteolytic peptides of the cross-linked proteins are analyzed – in case a "bottom-up" strategy is employed for MS protein analysis. Second, the experiment is rapid and requires very low (10^{-14}–10^{-15} mol) amounts of protein. Finally, as the cross-linking reaction can be executed in a native-like environment, protein structure and flexibility are accurately reflected. It is possible to study membrane proteins, post-translational modifications as well as splice variants. The broad range of cross-linking reagents with different specificities (primary amines, sulfhydryls, or carboxylic acids) and the wide range of distances (0 Å up to 20 Å) allow a setup of fine-tuned experimental strategies.

However, despite the straightforwardness of the cross-linking approach, the identification of the cross-linked products can be cumbersome due to the complexity of the reaction mixtures. Several strategies have been employed to enrich cross-linker-containing species by affinity chromatography or to facilitate the identification of the cross-linked products, e.g. by using MS/MS cleavable cross-linkers or isotope-labeled cross-linkers or proteins [26,27,32].

By combining sparse distance constraints from disulfide bonds and cross-links imposed by bis(sulfosuccinimidyl)glutarate (BS^2G), MS identification of the cross-linked products, and computational modeling we predicted a galactose-binding domain-like fold for laminin β1 and γ1 LN domains [7]. This fold was later confirmed by crystal structures of the α5 LN-LE1–2 [33], the β1 LN-LEa1–4, and the γ1 LN-LEa1–2 fragments [34].

In this work, we extend the cross-linking tool box from the exclusive use of amine-reactive cross-linkers towards the incorporation of photo-reactive amino acids that can deliver valuable short-distance information [35]. Combined with a mass spectrometric analysis of the created cross-links and computational modeling, we were able to gain detailed insights into the interaction mechanisms between full-length nidogen-1 and the laminin variants γ1 LEb2–4, γ1 LEb2–4 N836D, γ1 short arm, and γ1 short arm N836D. Our results suggest the existence of multiple nidogen-1/laminin γ1 interfaces in addition to the known interaction site.

Materials and Methods

Materials

The cross-linking reagent BS^2G and the photo-amino acids (photo-methionine and photo-leucine) were obtained from Thermo Fisher Scientific. The proteases trypsin (porcine), chymotrypsin, and GluC were obtained from Promega, all other chemicals were purchased from Sigma. Solvents used for nano-high performance liquid chromatography (HPLC) were spectroscopic grade (Uvasol, VWR). Milli-Q water was produced by a TKA Pacific system with X-CAD dispenser from Thermo Electron LED GmbH (part of Thermo Fisher Scientific).

Expression and Purification of Nidogen-1 and Laminin γ1 Variants

Genes encoding all proteins (murine amino acid sequences, see Figure S1) were expressed in human embryonic kidney (HEK) 293 EBNA cells with N-terminal (His)$_6$-tag (nidogen-1) or with N- or C-terminal double Strep tag II (laminin γ1 variants γ1 LEb2–4, γ1

LEb2–4 N836D, γ1 short arm, and γ1 short arm N836D). Incorporation of photo-reactive amino acids was achieved by growing the cells in a Leu- and Met-depleted medium (DMEM-LM, Thermo Fisher Scientific) to which photo-Met and photo-Leu were added. Strep II-tagged proteins were purified using Strept-Actin sepharose matrix (IBA), (His)$_6$-tagged nidogen-1 was purified with Nickel-NTA Superflow matrix (GE Healthcare) using an ÄKTA FPLC system (GE Healthcare). Amino acid sequences were confirmed by peptide fragment fingerprint analysis.

Surface Plasmon Resonance Spectroscopy (SPR)

Experiments were conducted at 25°C using a Biacore T100 instrument (GE Healthcare). Nidogen-1 was immobilized by covalent coupling on a Series S Sensor Chip CM 5 (GE Healthcare) using amine-coupling chemistry. A 1:1 mixture of 100 mM N-hydroxysulfosuccinimide and 400 mM 1-ethyl-3-(3-dimethylaminopropyl)carbodiimide hydrochloride (Pierce) was passaged over the sensor chip surface for 7 min at a flow rate of 10 μl/min to activate the matrix. For protein immobilization, a solution of 300 nM nidogen-1 in 10 mM sodium acetate (pH 5.5) was injected (contact time 35 min, flow rate 10 μl/min). Remaining activated groups were blocked by injecting 1 M ethanolamine (pH 8.5) for 7 min at a flow rate of 10 μl/min. As reference, a flow cell was prepared in the same manner omitting the injection of protein solution.

Binding assays using all laminin variants (wild type and N836D) as mobile analytes were performed as single-cycle kinetic experiments [36] at a constant flow rate of 30 μl/min. The analyte was diluted in running buffer (20 mM HEPES, 100 mM NaCl, 0.05% (v/v) Tween 20, pH 7.5) and 1 min injections of these dilutions were applied with increasing concentrations. Individual analyte injections were followed by a 40.6 min flow of running buffer to allow for partial dissociation. Blank cycles, where running buffer was injected instead of analyte solution, were performed prior to each analyte cycle to facilitate double referencing. After completion of each analysis cycle, the sensor chip surface was extensively washed with running buffer to achieve complete dissociation of the analyte. Response data were doubly referenced and kinetic parameters were determined as described previously [36] using BIAevaluation software 4.1.1 (GE Healthcare) to fit a 1:1 binding model to the data.

Enyzme-Linked Immunosorbent Assays (ELISA)

ELISA-based binding assays were performed in 96-well plates (Nunc Maxisorp, Thermo Fisher Scientific). Laminin fragments were coated overnight at 4°C with a concentration of 10 μg/ml in 50 μl Tris-buffered saline (TBS) per well (50 mM Tris-HCl, 150 mM NaCl, pH 7.4). The supernatant was discarded after coating and the plates were washed once with a TBS-T solution (TBS with 0,05% (v/v) Tween-20) with a volume of 400 μl/well. Unspecific binding sites were blocked with 50 μl/well of 1% (w/v) bovine serum albumin (BSA) in TBS-T solution for 1 h at room temperature. Increasing concentrations of nidogen-1 ranging from 0.03 nM to 234 nM in TBS-T/1% BSA were added to the wells until saturation was reached. After ligand incubation for 1 h at room temperature plates were washed three times with TBS-T solution. Nidogen-1 was detected with mouse monoclonal antibodies directed against the (His)$_6$-tag (Qiagen, dilution 1:2000). Antibodies were added in 50 μl TBS-T/1% BSA per well and incubated for 1 h at room temperature. Plates were then washed three times with TBS-T solution and the secondary horseradish peroxidase-conjugated α-mouse IgG antibodies raised in rabbit (Dako, 1:2000 diluted in TBS-T/1% BSA) were added. Plates were washed three times with TBS-T solution and one time with water. For signal detection, 50 μl/well tetramethylbenzidine solution (1-Step ultra TMB-ELISA solution, Thermo Fisher Scientific) was used. The color reaction was stopped with 50 μl/well of 10% H_2SO_4 solution and absorption at 450 nm was determined in an ELISA reader (Tecan), subtracting the background (0 nM ligand). As a second control, increasing ligand concentrations were added to uncoated wells. All measurements were performed in triplicates. K_d values were calculated using Origin v6.0.

Cross-Linking Reactions

Cross-linking reactions were conducted with 1 to 10 μM protein solutions in 20 mM HEPES, 100 mM NaCl, pH 7.5. Freshly prepared stock solutions of the homobifunctional amine-reactive N-hydroxysuccinimide ester BS^2G (in dimethylsulfoxide) were added in 200-fold molar excess to the protein solution. The reactions were conducted at room temperature and were quenched after 30 min by adding NH_4HCO_3 to a final concentration of 20 mM.

For photo-cross-linking, nidogen-1 (0.5 μM) was incubated with laminin γ1 short arm (1 μM) or γ1 LEb2–4 (10 μM) for 20 min at room temperature. Then, the samples were irradiated with UV-A light (max. 360 nm) at 8000 mJ/cm^2 in a home-built device [37].

In Gel Digestion

After SDS-PAGE of the cross-linking reaction mixtures, bands of interest were excised, reduced with dithiothreitol, alkylated with iodoacetamide, and digested. For *in situ* digestion, either GluC or Chymotrypsin was added and gel pieces were incubated at 4°C for 1 h before trypsin was added (enzyme:substrate ratio 1:100). The digestion was performed overnight at 37°C. Peptides were extracted and samples were concentrated in a vacuum concentrator to volumes between 60 to 120 μl before LC/MS/MS analysis.

In Solution Digestion

For *in solution* digestion, proteins were incubated with acetone (–20°C, 1 h) to precipitate them from solution. The pellet was dried, solubilized with 1.6 M urea, reduced, alkylated, and digested with a mixture of GluC and trypsin (enzyme:substrate ratio 1:50).

Nano-HPLC/Nano-ESI-LTQ-Orbitrap-MS/MS

Fractionation of peptide mixtures was carried out on an Ultimate nano-HPLC system (Thermo Fisher Scientific) using reversed phase C18 columns (precolumn: Acclaim PepMap, 300 μm • 5 mm, 5 μm, 100 Å, separation column: Acclaim PepMap, 75 μm • 250 mm, 3 μm, 100 Å, Thermo Fisher Scientific). After washing the peptides on the precolumn for 15 min with water containing 0.1% trifluoroacetic acid, peptides were eluted and separated using gradients from 0 to 40% B (gradient times varying between 30 to 90 min), 40 to 100% B (1 min), and 100% B (gradient times varying between 11 to 30 min), with solvent A: 5% ACN containing 0.1% formic acid, and solvent B: 80% ACN containing 0.08% formic acid. The nano-HPLC system was directly coupled to the nano-ESI source (Proxeon) of an LTQ-Orbitrap XL hybrid mass spectrometer (Thermo Fisher Scientific). Data were acquired in data-dependent MS/MS mode: Each high-resolution full scan (m/z 350 to 2000, R = 60,000) in the orbitrap was followed by five product ion scans in the LTQ (collision-induced dissociation with 35% normalized collision energy) of the five most intense signals in the full-scan mass spectrum (isolation window 1.5 u). Dynamic exclusion

(exclusion duration 120 sec, exclusion window −1 to 2 u) was enabled to allow detection of less abundant ions. Data acquisition was controlled via XCalibur 2.0.7 (Thermo Fisher Scientific) in combination with DCMS link 2.0.

Identification of Cross-Linked Products

Cross-linked products were analyzed with StavroX v2.0.6 [38]. MS and MS/MS data were automatically analyzed and annotated. All cross-links were manually validated. A maximum mass deviation of 3 ppm between calculated and experimental precursor masses was applied as well as a signal-to-noise ratio of ≥2. Primary amino groups (Lys side chains and N-termini) were considered as cross-linking sites for BS^2G; all amino acid residues were regarded as potential sites for UV-A-induced cross-linking of photo-Met and photo-Leu. Oxidation of Met was set as variable modification for all cross-linked proteins. Additionally, carbamidomethylation was included as fixed modification for Cys. Two missed cleavage sites were considered for each amino acid (Lys and Arg for trypsin; Tyr, Trp, and Phe for chymptrypsin; Glu for GluC).

Identification of Templates for Computational Modeling

The sequences of nidogen-1 and laminin γ1 short arm were split into separate domains as defined by the UniProt Knowledgebase (www.uniprot.org, nidogen-1: entry P10493, laminin γ1: entry P02468) and modeled independently. The domains were subdivided in three classes according to the availability of structural models or templates (Figure 1). The first class comprises the nidogen-1 LDL-receptor class B repeats (G3 domain), EGF-like domain 1 and G2 β-barrel domain as well as the laminin γ1 LN, LEa1, LEa2, and LEb2–4 domains, all of which have been characterized by X-ray crystallography. The remaining EGF-like domains of nidogen-1 and laminin γ1 short arm as well as the nidogen-1 thyroglobulin type-1 (TY1) domain were assigned to the second class because a DELTA-BLAST [39] search run had led to the identification of homologous domains with existing crystal structures that could serve as templates for comparative modeling. Finally, the third class comprises the nidogen-1 G1 domain and the laminin γ1 L4 domain, for which we did not identify any sequential homologues. Both domains were searched against the threading servers PHYRE2 [40], HHPred [41], PSIPRED (pDomTHREADER and GenTHREADER) [42], and I-TASSER [43] for fold recognition.

All following modeling experiments were performed with Rosetta v3.4. The Rosetta total scores reported herein were calculated using the score12 full-atom scoring function. Full command lines for each step are included in Files S1 and S2.

Comparative Modeling Based on Sequential Homologous Templates

After comparing the DELTA-BLAST sequences of potential template structures with the actual sequences of the corresponding PDB entries, sequence alignments of all target and template sequences were performed with ClustalW 2.1 [44]. Only templates showing ≥30% sequence identity were used for comparative modeling. The corresponding alignments are shown in Figure S2.

The LE domain crystal structures (PDB entries 4AQS, 4AQT, 1NPE, 1KLO, 2Y38) were considered as templates for the remaining LE domains, namely LEa3–5, LEb1 and LEb5. LEa5 is split in two parts by the L4 domain (Figure 1). The N-terminal LEa5.1 was not modeled as an individual domain since it comprises only ten residues. Similarly, LEb6 was not modeled as an individual domain since only nine residues are contained in the laminin γ1 short arm fragment used here.

We identified EGF-like domains in 14 X-ray structures (PDB entries 1GL4, 1YO8, 1SZB, 1TOZ, 1UZJ, 2BO2, 3P5B, 1NFU, 3H5C, 3POY, 3QCW, 3S94, 3V64 and 2W86) as templates for nidogen-1 EGF2–6. The nidogen-1 TY1 domain was modeled using TY1 domains within the crystal structures 1ICF and 2DSR as templates. Threading of the primary sequences onto the 3D template structures, modeling of missing loop regions, and clustering of the created models were carried out as reported previously [45]. For each template/target sequence pair, 1000 models were constructed. All structural models for one target sequence were ranked according to their Rosetta total score and the best-scoring 10% were used for clustering with automated detection of the clustering radius by Rosetta. An overview of all clusters is given in Table S1. The best-scoring structures within the top three clusters were considered as final models.

Comparative Modeling of Laminin γ1 L4

Using the threading servers listed above, 21 potential structural homologues were identified for the L4 domain, 13 of which exhibit a ß-sandwich topology and carbohydrate-binding activity. Structural alignment with MUSTANG [46] revealed that these templates share a common topology that is in accordance with the PSIPRED [47] and JUFO9D [48] secondary structure prediction for L4. Hence, we hypothesize that the laminin γ1 L4 domain shares the galactose-binding domain-like fold and built the comparative model based on these templates. ClustalW 2.1 alignments of the laminin γ1 L4 domain with all 13 templates identified by fold recognition (PDB entries 1GU3, 1GUI, 1K42, 1CX1, 1WKY, 1WMX, 3OEA, 2ZEW, 2ZEZ, 1DYO, 3F95, 3ZXJ, 1D7B) revealed sequence identities of 3–14%. Hence, the alignments had to be adjusted manually to guarantee for correctly aligned secondary structure elements (Figure S3). Adjustments were based on PSIPRED secondary structure prediction for L4, DSSP secondary structure annotations of the template structures [49] and manual inspection of all templates in Pymol v1.5 (Schrödinger LLC). After optimization of the alignments, comparative modeling was performed as described above. For each template, 1000–2000 models were generated (25,296 models in total) and the top 10% were selected for clustering (Table S2). To obtain informative clustering results, long loop regions (>5 residues) were not considered for root-mean square deviation (RMSD) calculation. The clustering radius was set to 2 Å.

De Novo Folding of the Nidogen-1 NIDO Domain

No likely structural homologues of the nidogen-1 NIDO domain were identified by fold recognition. Hence, models were generated by de novo folding using Rosetta AbinitioRelax within the Rosetta Topology Broker framework [50]. Initially, Rosetta fragment picker was used to create a fragment library consisting of the primary sequence split into overlapping 3-mers and 9-mers, each of them represented by 200 peptide structures, mimicking the entire distribution of conformations these segments are likely to adopt in a protein structure [45,51]. Full atom refinement during de novo modeling resulted in partial unfolding of the created models. Therefore, we generated 11,902 centroid models, omitting full-atom refinement after running AbinitioRelax. To filter for structures that are likely to occur in nature we pursued two complementary strategies.

First, the 10% best-scoring models were compared to a precompiled PISCES library of structurally diverse PDB models (soluble proteins, sequence ID <25%, resolution <2.0 Å) [52] using MAMMOTH [53]. This served to evaluate whether the

domain adopts a known fold. Structural homologues for two of the generated models were identified (>50% of target sequence aligned, MAMMOTH Z-score >5).

Second, we generated 10,000 to 20,000 models for sequences of six homologous NIDO domains present in other organisms. Homologous domains were identified with DELTA-BLAST. Non-redundant sequences with an identity of >75% to the murine NIDO domain were selected and manually compared using Jalview [54]. The sequences of Sarcophilus harrisii, Rattus norvegicus, Homo sapiens, Bos taurus, Cricetulus griseus, and Felis catus were chosen, each of them showing differences in diverse regions that are otherwise conserved (Figure S4). To identify common topologies among the different NIDO domains, models were ranked based on their total centroid score as well as their strand pairing energy score and MAMMOTH was used to compare the best 1% models of each homologue with the best 10% models of the murine NIDO domain. Models of murine NIDO were regarded as representatives of a common topology when significant structural similarities (MAMMOTH Z-score >5) to at least one model of each homologue were found.

As a result, 109 candidate models were identified and selected for full-atom refinement. To prevent distortion of secondary structure element arrangement, side chains and peptide backbone were relaxed sequentially, keeping one of both fixed ('-relax:bb_move false' or '-relax:chi_move false') before a final round of refinement in thorough relax mode was performed on the complete structure. We generated 50 models per input model, which were inspected by visualization with Pymol, ranked according to their Rosetta total score, and clustered with a radius of 1.5 Å. The best-scoring output structure for each template and the best-scoring structures representing the top 20 clusters were kept for further analysis. The two candidate models, identified by comparison with the PISCES library, were processed similarly. For each of them, the five best-scoring models and the best-scoring models representing individual clusters were included in the final list of potential NIDO models, resulting in a total of 132 models. These models were combined and re-scored (Table S3).

Incorporation of Cross-Linking Constraints into the Nidogen-1 G3/Laminin γ1 LEb2–4 Experimental Structure

We identified eleven BS^2G- and two photo-cross-links within the experimental structure of the nidogen-1 G3/laminin γ1 LEb2–4 complex (PDB entry 1NPE), which were used as distance constraints for adapting the 3D structure. The cross-linked residues and their respective Cα–Cα distances are listed in Table 1. The maximum Cα–Cα distance for lysine-lysine cross-links was calculated to 26 Å by adding the lysine side chain length (2×6.3 Å), the distance spanned by BS^2G (7.5 Å [55]) and a tolerance of 5.9 Å to account for structural flexibility [56]. Acceptable Cα–Cα distances for the "zero-length" photo-cross-links varied depending on the side-chain length of the cross-linked residue. Fulfillment of cross-linking restraints was evaluated with a flat harmonic scoring function that renders an energy penalty when the Euclidean distance between cross-linked residues exceeds the allowed Cα–Cα distance [56]. A Rosetta constraint file containing all cross-links was created as described elsewhere [57]. The standard deviation granted for each cross-link was 1 Å. Additionally, distances of the hydrogen bonds formed by Asp-834, Asn-836, and Val-838 in the laminin γ1 LEb3 domain were restrained as they have been reported to be essential for high-affinity interaction [23,24]. Structural refinement was carried out by generating 800 models using the Rosetta Relax application with an atom pair constraint scoring weight of 1.0 ('-constraints:cst_fa_weight 1.0'). Structures ranking among the top 20 in terms of total

Table 1. Overview of cross-links within the nidogen-1 G3/laminin γ1 LEb2–4 complex.

| | Cα–Cα distances (Å) | | |
| | | model | model |
cross-linked lysines	1NPE	(best atom-pair constraint score)	(best total score)
K-948 × K-953	10.4	10.9	11.1
K-1128 × K-1165	13.3	12.3	16.0
K-1072 × K-1128	16.7	19.1	16.2
K-948 × K-1144	17.9	16.4	17.6
K-850 (laminin) × K-1072 (nidogen-1)	20.9	17.5	16.9
K-948 × K-1152	22.2	21.2	22.2
K-1032 × K-1072	27.1	27.1	27.0
K-961 × K-1072	28.7	28.0	28.2
K-864 (laminin) × K-1152 (nidogen-1)	32.2	22.4	27.1
K-850 (laminin) × K-953 (nidogen-1)	33.0	29.5	29.4
K-1032 × K-1152	35.8	35.4	35.4
Photo-L-990 × Arg-1038	24.7	23.4	23.5
Photo-L-844 (laminin) × K-1072 (nidogen-1)	33.8	19.4	20.8

Cα–Cα distances of cross-linked residues were determined for the unmodified crystal structure (PDB entry 1NPE) as well as for the Rosetta models with the best Rosetta total score and atom-pair constraint score, respectively (shown in Figure 6). For intermolecular contacts, residues are assigned to the respective protein. All other cross-links are located within nidogen-1.

score and atom-pair constraint score (reflecting the fulfillment of distance constraints) were selected as potential models of the nidogen-1 G3/laminin γ1 LEb2–4 core complex (Table S4).

Results

As there are only very limited structural data available for full-length nidogen-1/laminin γ1 complexes, we sought to investigate the complexes created between nidogen-1 and laminin variants γ1 LEb2–4, γ1 LEb2–4 N836D, γ1 short arm, and γ1 short arm N836D by applying a combination of chemical cross-linking and high-resolution nano-HPLC/nano-ESI-LTQ-Orbitrap mass spectrometry. In the laminin γ1 short arm and γ1 LEb2–4 N836D point mutants, the Asn residue, which is crucial for nidogen binding, was exchanged for an acidic Asp. The obtained distance constraints were then used to identify previously unknown nidogen/laminin interfaces, to generate 3D-structural models of all their individual domains, and to generate a refined model of the high-affinity nidogen-1/laminin γ1 binding site based on the known 3D structure. In addition, SPR and ELISA assays allowed us to derive kinetic constants of the nidogen-1/laminin γ1 interaction. Together, the data presented herein shed new light on the mechanisms underlying nidogen/laminin interaction.

Deriving Kinetic Constants of the Nidogen/Laminin Interaction

The binding affinities between nidogen-1 and laminin γ1 short arm and γ1 LEb2–4 were investigated with surface plasmon resonance spectroscopy by measuring single-cycle kinetics [36]. Representative examples of the obtained sensorgrams are depicted in Figure S5A. Apparent K_d values were determined to be in the (sub-)nanomolar range, confirming high binding affinities between nidogen-1 and laminins (Table S5). However, the nidogen-1 binding activity of laminin γ1 short arm variant was found to be one order of magnitude lower than that of laminin γ1 LEb2–4, which is almost exclusively caused by differences in the association

phase. Nidogen-1 binding was also investigated for N836D variants of laminin γ1 short arm and γ1 LEb2–4. When analyte concentrations were increased ca. 100-fold compared to the laminin γ1 'wild-type' variants, signals were detected for nidogen-1/laminin γ1 LEb2–4 N836D binding (Figure S5A), which, however, did not allow to derive kinetic constants. For nidogen-1 and laminin γ1 short arm N836D, no interaction was detected by SPR.

In addition, the binding affinities between nidogen-1 and laminin γ1 short arm, γ1 LEb2–4 and their respective N836D variants were investigated by ELISA-based binding assays (Figure S5B). These assays revealed no differences in binding affinities of the laminin γ1 short arm and the LEb2–4 fragment to nidogen-1. Both interactions showed strong binding with apparent K_d values of 1.1 and 1.4 nM (Table S5). Binding of the respective N836D variants to nidogen-1 showed a considerable loss of binding affinity with apparent K_d-values of 34 and 45 nM (γ1 LE2–4 N836D and γ1 short arm N836D, respectively).

Cross-Linking of Nidogen-1/Laminin γ1 Complexes

For gaining insights into the interaction between nidogen-1 and laminin γ1 variants on the molecular level, the proteins were cross-linked using the homobifunctional cross-linker BS^{2}G. Additionally, we pursued a complementary approach by incorporating the unnatural diazirine-containing amino acids photo-Met and photo-Leu instead of methionine and leucine [58] into nidogen-1, laminin γ1 LEb2–4, and laminin γ1 short arm. MS/MS analysis of the non-cross-linked proteins revealed 13–25% of all leucines and methionines to be partially replaced by their photo-reactive counterparts. After UV-A-induced or BS^{2}G-mediated cross-linking, the reaction mixtures were separated by SDS-PAGE and analyzed by LC/MS/MS. Experiments were conducted in the presence of varying laminin concentrations to optimize the efficiency of heterodimer formation between nidogen-1 and laminin. The verified cross-links are summarized in Figure 2 and in Tables S6 and S7.

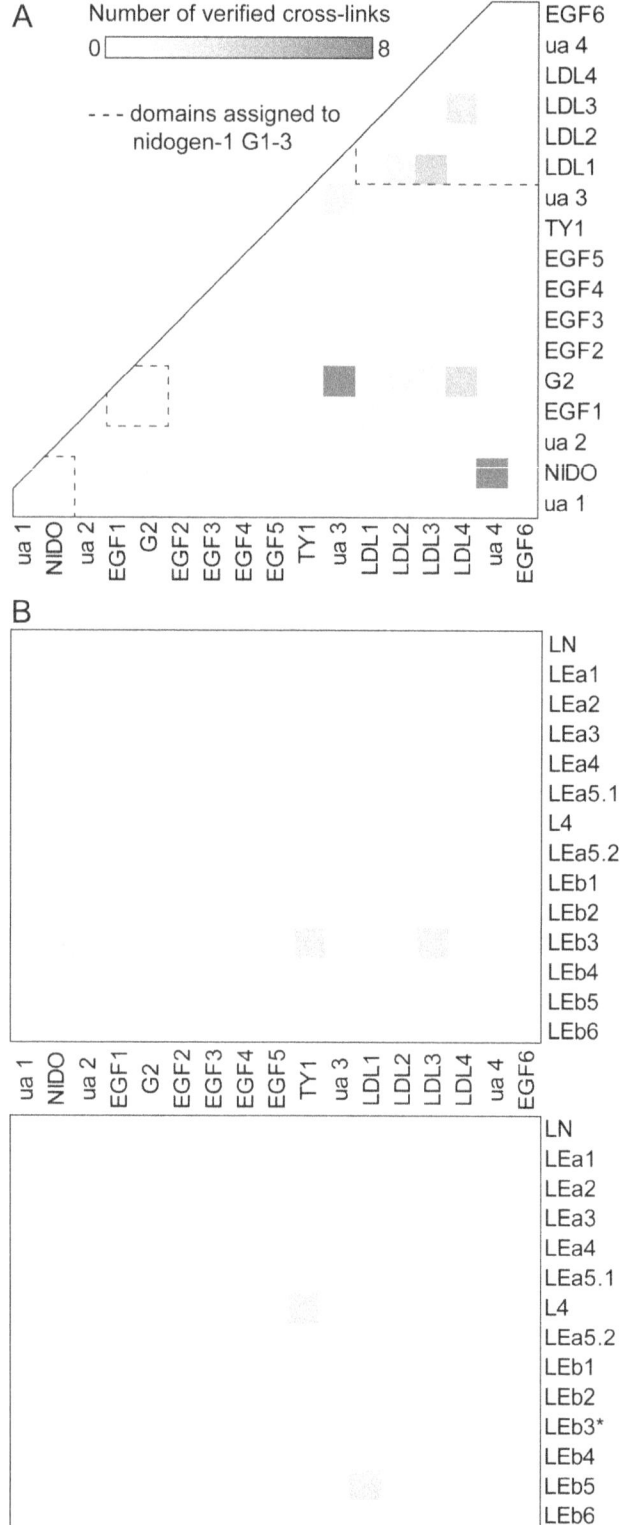

border represent contacts between domains being close to each other in the protein sequence. (B) Cross-links between nidogen-1 and laminin γ1 wild type (upper panel) as well as N836D variants (lower panel). The LEb3 domain, bearing the N836D mutation, is marked with an asterisk.

We confirmed 47 intramolecular BS^2G cross-links within nidogen-1, delivering 26 non-redundant distance constraints. Although the efficiency of photo-amino acid incorporation at the partially modified Leu and Met sites was moderate (~35% for photo-Met, ~3% for photo-Leu, see Figure S6), we identified two additional non-redundant photo-cross-links within nidogen-1, one connecting the link region with the G3 domain and one within the G3 domain, highlighting the sensitivity of our cross-linking/MS approach.

More than half of the intermolecular cross-links were inter-domain contacts, in which all globular nidogen-1 domains were connected with each other (Figure 2A). The majority of distance constraints were found in the G3 domain as well as between G2 and G3 domains, while the 'link' region (ua 2) and EGF domains 2–4 within the 'rod' region were not involved in any cross-link. As an example, the fragment ion mass spectrum representing a cross-link between the G2 and the G3 domain of nidogen-1 is shown in Figure 3.

Ten out of 19 non-redundant BS^2G cross-links between nidogen-1 and laminin γ1 were exclusively found in experiments with laminin γ1 wild type, six only in experiments with N836D variants, and three for both proteins. Additionally, one intermolecular contact between nidogen-1 G3 and laminin γ1 LEb3 was reproducibly identified in two consecutive photo-cross-linking experiments. With the exception of the link region, all nidogen-1 domains were involved in intermolecular cross-links with the nidogen binding motif of laminin γ1 (LEb2–4) as well as with additional regions within laminin γ1 short arm (Figure 2B). However, when laminin γ1 LEb2–4 N836D variants were used as interaction partners, the distribution of cross-links changed considerably with only two nidogen-1/laminin γ1 LEb2–4 contacts being identified (Figure 2C). Instead, L4 and LEb5 domains were repeatedly found to be cross-linked to nidogen-1.

Examining the intramolecular laminin γ1 cross-links underpins the complementarity of the two cross-linking strategies applied. Whereas only one BS^2G cross-link within laminin γ1 variants was identified (LEb2 with LEb5), UV-A irradiation revealed connections of the L4 domain to LEa1 and LEa2 domains as well as one contact between the LEa4 and LEb6 domains. Notably, we exclusively identified cross-links between sequentially non-adjacent laminin domains.

Structural Characterization of Laminin γ1 and Nidogen-1 Domains by Comparative Modeling

Structures of the laminin γ1 LEa3–5.2, LEb1, and LEb5 domains as well as for the nidogen-1 domains EGF2–6 and TY1 were derived by comparative modeling (Figure S7). Template structures were identified by sequence homology search using DELTA-BLAST. Since we did not observe cross-links within these domains only known disulfide linkages served as distance constraints. 'Laminin-type EGF-like' (LE) domains exhibit a characteristic disulfide linkage pattern that is different from classical EGF-like domains [21,24,33,34]. Hence, only crystal structures of LE domains were used as templates for the remaining LE modules and nidogen-1 EGF domain models were based on the structures of classical EGF-like repeats.

Structural templates for the laminin γ1 L4 domain were identified by fold recognition as described in 'Materials & Methods'. A topology dominated by ß-sheets was predicted by

Figure 2. Diagonal plots of all cross-links identified. Cross-links are assigned to domains based on the UniProt KB entry P10493 (nidogen-1) and the laminin nomenclature of Aumailley et al. [5]. Unannotated areas within the sequences are named 'ua'. Corresponding to the number of inter-domain contacts, areas of intersection are color-coded from white (none) to dark grey (maximum). (A) Intramolecular cross-links within nidogen-1. The globular domains G1, G2, and G3 are denoted by dotted lines. Cross-links located nearby the diagonal

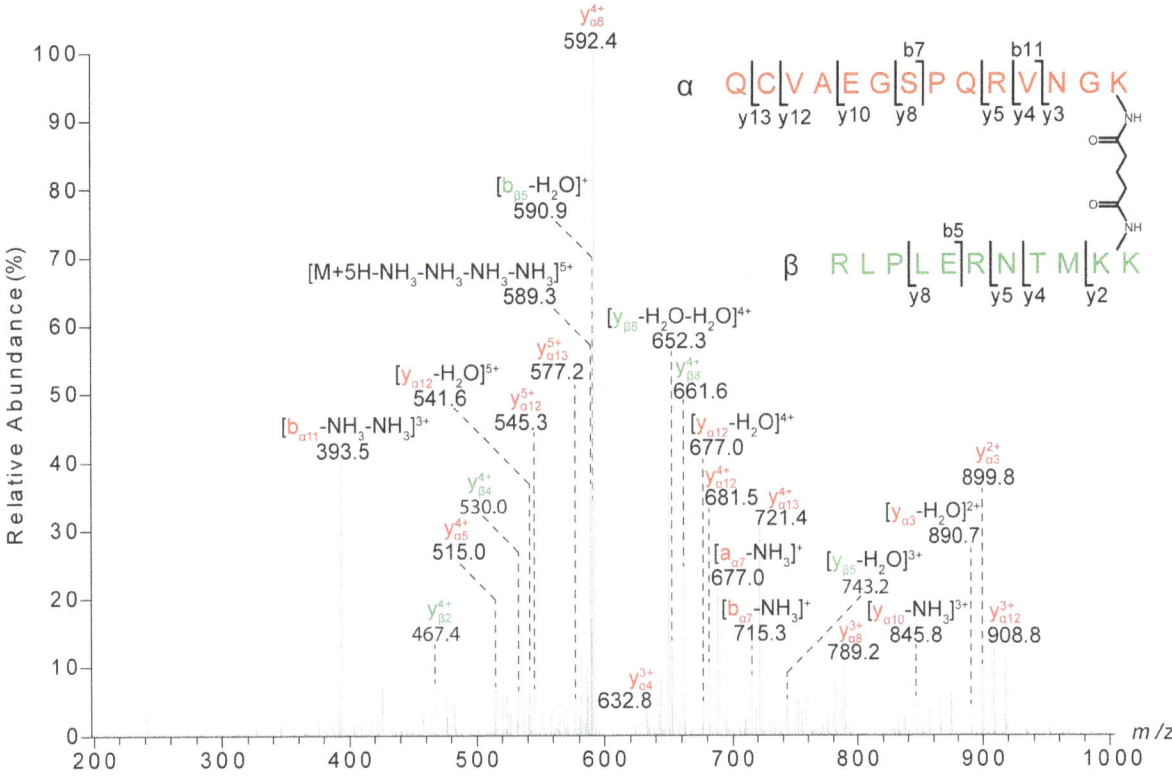

Figure 3. Nano-HPLC/nano-ESI-LTQ-Orbitrap-MS/MS analysis of cross-linked peptides derived from nidogen-1 G2 and G3 domain. The cross-linked product comprises amino acids 407–420 of the G2 domain (α-peptide, red) and 939–949 of the G3 domain (β-peptide, green), in which K-407 is connected to K-948/949.

PSIPRED [47] (Figure S8A) and JUFO9D [48]. In compliance with this prediction, 13 potential templates with a common ß-sandwich topology were identified by the applied threading servers (Figure S8B). Comparative modeling of laminin γ1 L4 was thus based on these structural homologues. The best 10% of all generated models were clustered and further validated by generating score-vs-RMSD plots, assuming each of the best-scoring models of the top five clusters as the native structure (RMSD = 0). Convergence of the models towards a minimum Rosetta total score and RMSD underpins the validity of the structures with the lowest score (Figure S8C). The two final models, both adopting a galactose-binding domain-like fold, match the secondary structure predictions for the L4 domain and exhibit energetically favorable residue conformations throughout the structure (Figure 4).

Generating an *Ab Initio* Model of the Nidogen-1 G1 Domain

G1, the *N*-terminal globular region of nidogen-1, essentially consists of a NIDO domain (Figure 1). As neither sequential nor structural homologues were identified for nidogen-1 NIDO we decided for a *de novo* folding strategy. This approach is exceptionally challenging because of NIDO's size (156 amino acids). Thus, we could not only rely on the Rosetta scoring function, but had to pursue alternative validation strategies [59]. First, we sought to identify known protein folds among the generated models by comparing them to a PISCES library of diverse PDB structures using MAMMOTH structural alignment (see 'Materials & Methods'). Second, we reasoned that the topologies being sampled during *de novo* folding can be substan-

tially influenced by subtle changes in the protein sequence. However, folding of highly similar sequences should result in identical tertiary structures. Similar topologies, sampled for several closely homologous sequences, are thus more likely to resemble the native structure. Therefore, we performed *de novo* folding not only for the murine NIDO domain, but also for homologues from six additional organisms with sequence identities larger than 75%. Taken together, these strategies resulted in the identification of 132 candidate structures after full-atom refinement and clustering (Table S3) and the ten top-scoring models were found to originate from four initial NIDO centroid models. Three of these models were validated by one BS²G cross-link, which had been identified within the NIDO domain, but was not used as a distance constraint during the modeling process. The determined Cα–Cα distances were within the range of the cross-linker, as shown in Figure 5. Additionally, a coarse clustering (radius = 10 Å) was performed to check whether further final candidate structures can be traced back to common centroid models. In view of the variety of conformations being sampled during the generation of centroid models, clusters were merged, when their best-scoring member structures originated from the same centroid model. We identified six additional centroid models that are represented by more than three full-atom refined models among the 132 final candidate structures (Table S8). For two of these centroid models, structural homologues in the PISCES PDB library have been identified. The best-scoring full-atom structural models for the six centroid models are shown in Figure S9. In all these NIDO models the cross-linking distance constraint is fulfilled. Taken together, our *de novo* approach suggests that NIDO adopts a compact topology containing a ß-sheet with at least four strands and two α-helices.

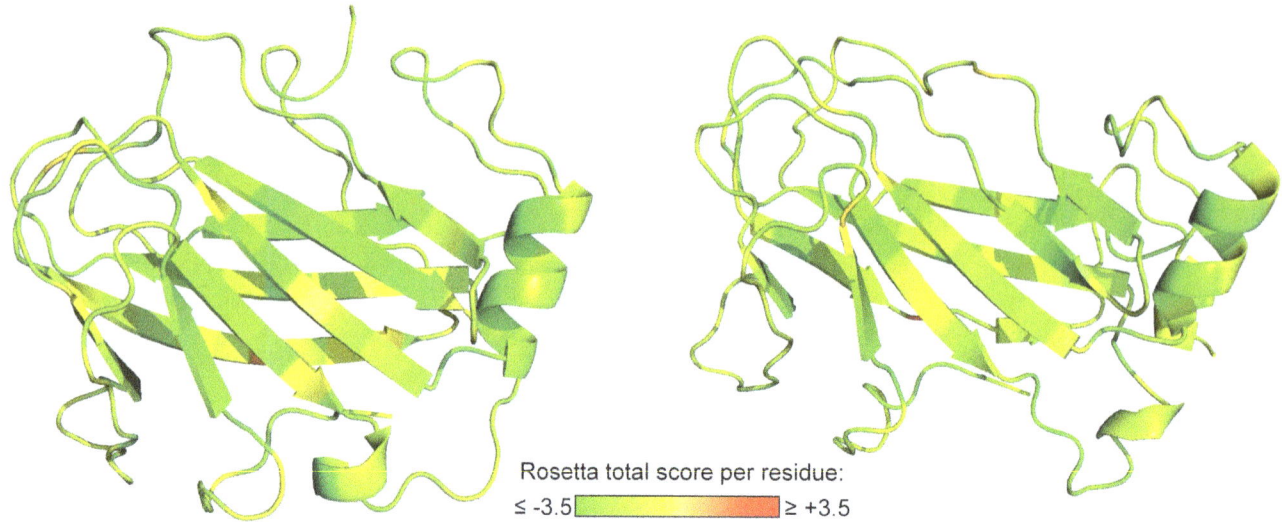

Rosetta total score per residue:

≤ -3.5 $\geq +3.5$

Figure 4. Structural models for laminin γ1 L4. Shown are the two best-scoring models generated by comparative modeling based on 13 structural homologues that have been identified by fold recognition. The residues are colored according to their Rosetta total score. Scores below zero (yellow-green color) indicate energetically favorable conformations.

Incorporation of Cross-Linking Distance Constraints into the Nidogen-1 G3/Laminin γ1 LEb2–4 Core Structure

Eleven BS^2G cross-links originate from the known interaction region between the LEb2–4 domains of laminin γ1 and the G3 domain of nidogen-1. The distance constraints obtained by our cross-linking experiments should thus be in agreement with the known 3D structure of the nidogen-1 G3/laminin γ1 LEb2–4 complex (PDB entry 1NPE) [24]. According to the spacer arm length of BS^2G (7.5 Å [55]) and the length of the cross-linked lysine side chains (2×6.3 Å), the maximum Euclidian Cα–Cα distance between the cross-linked residues should be 20.1 Å. However, longer distances are frequently observed, when cross-links are mapped in structural models. Therefore, it is common practice to grant a distance tolerance of 6–7 Å to account for structural flexibility. Recently, a rationale for this approach was presented by studying the lysine–lysine distances within 807 proteins during molecular dynamics simulations [60]. This analysis gave evidence that the ε-amino groups of lysines, with initial Cα–Cα distances of up to 38 Å, will move inside the range of the amine-reactive cross-linker DSS (spacer arm length 11.4 Å) during a 50 ns-simulation. Considering the shorter spacer arm length of BS^2G, the maximum Euclidean Cα–Cα distance that is likely to allow cross-linking can, therefore, be estimated to 34 Å. In an earlier study, similar results were obtained by cross-linking seven model proteins and comparing the identified cross-links with their respective crystal structures [61]. Mapping our cross-links into the X-ray structure of the nidogen-1 G3/laminin γ1 LEb2–4 complex resulted in Cα–Cα distances between 10.4 and 35.8 Å. By integrating the crystal structure and our cross-linking data in a Rosetta-based modeling approach, we aimed to derive structural models that better reflect plausible in-solution conformations of the nidogen-1 G3/laminin γ1 LEb2–4 complex, which would be signified by a decrease in observed Cα–Cα distances. The best-scoring models of the complex are depicted in Figure 6. The laminin γ1 fragment is considerably bent compared to the X-ray structure (Figure 6C), which is conceivable as the three LE repeats do not form a compact tertiary structure around a defined hydrophobic core. Both the disulfide bond pattern of the LE domains and the β-propeller structure of the G3 domain are well maintained indicating that the structural rearrangements in our models are reasonable. Notably, the spanned Cα–Cα distances of all except one BS^2G cross-link are significantly reduced compared to the X-ray structure, suggesting that the models give a better picture of the conformations the nidogen-1 G3/laminin γ1 LEb2–4 complex can adopt in solution (Table 1). However, one has to be aware that the Rosetta models, just as the crystal structure, represent conformational samples and do not reflect the entire conformational space of the protein complex.

In this context, it has to be pointed out that the single cross-link that exceeds the 34 Å distance limit of the X-ray structure is in conflict with the models. Visual inspection of the structural models shows that the respective lysines cannot be cross-linked, since the spacer arm would have to traverse through the center of the β-propeller, which is blocked by residues of the G3 domain. We interpret this cross-link as an intermolecular contact between two nidogen-1 molecules. Forcing Rosetta to fulfill this distance constraint resulted in unfolding of the G3 domain, confirming that a sound model cannot be forced to match the experimental data, but complies only with sterically feasible cross-links. In addition to the BS^2G cross-links we obtained two distance constraints by photo-cross-linking. The respective Cα–Cα distances of the cross-links, in which photo-Leu is connected to Arg and Lys, were 24.7 and 33.7 Å within the X-ray structure as well as 23.4–23.5 Å and 19.4–20.8 Å within the models (Table 1). This is well above the expected maximum values of 10.4 Å for a photo-Leu/Lys cross-link and 11.4 Å for a photo-Leu/Arg cross-link, which were determined from the side chain lengths within representative amino acid crystal structures deposited in the Cambridge Structural Database [62]. However, in both cases, one of the cross-linked amino acids is located in a loop region. It is conceivable that the obtained Cα–Cα distances are longer than the maximum expected distances as the loop regions are flexible. Granting a similar distance tolerance as for BS^2G results in maximum allowed Cα–Cα distances of 23.4 Å and 24.4 Å, respectively, both of which are met by our models. Intriguingly, the cross-link between photo-Leu-844 (laminin) and Lys-1072 (nidogen-1) concurs with a BS^2G cross-link pointing to the same region (laminin Lys-850 with nidogen-1 Lys-1072).

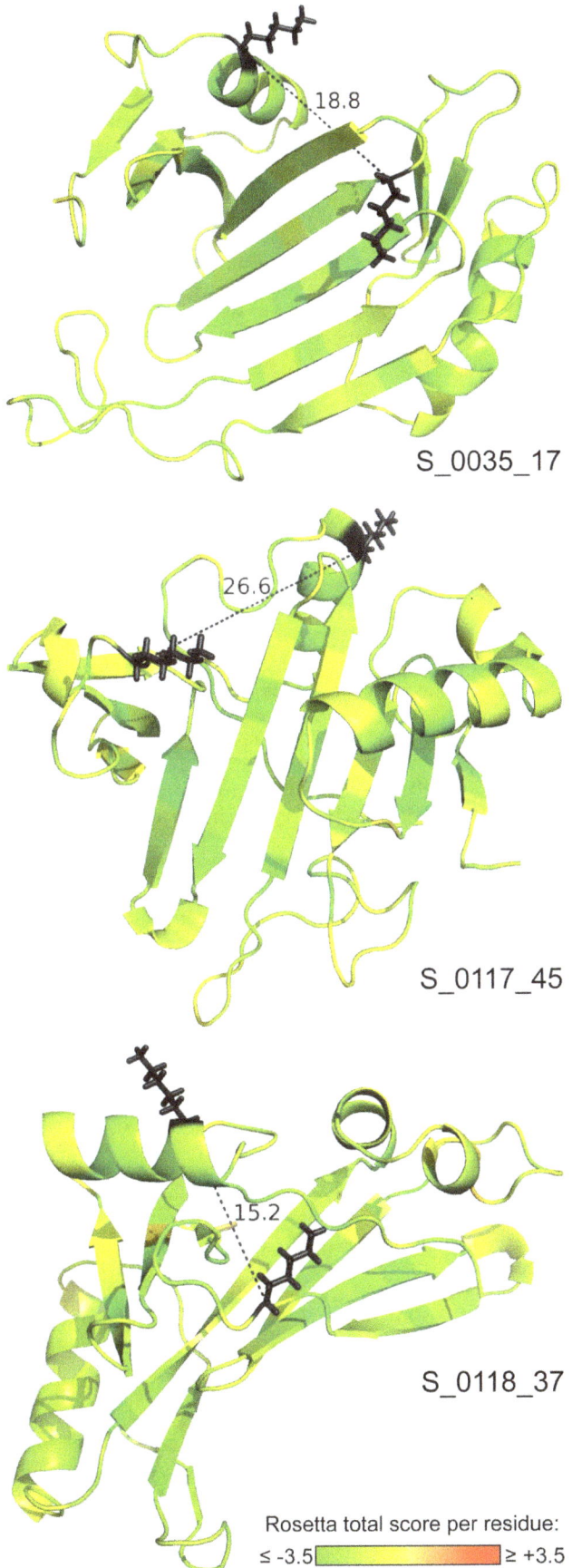

S_0035_17

S_0117_45

S_0118_37

Rosetta total score per residue:

≤ -3.5 ▰▰▰▰ ≥ +3.5

Figure 5. *De novo* **folded models of nidogen-1 G1.** The ten best-scoring structures among the final models were all derived from four initial centroid models of the G1 domain. Structures representing three of these centroid models are shown here. These models comply with the single distance constraint in this region that was identified by cross-linking/MS. Cross-linked residues are displayed as black sticks. Cα–Cα distances are given in Å. The residues are colored according to their Rosetta total score. Scores below zero (yellow-green color) indicate energetically favorable conformations. The identifiers of the underlying centroid models are indicated.

Discussion

The primary goal of this work was to gain novel insights into the nidogen-1/laminin γ1 interaction in solution by combining amine-reactive and photo-chemical cross-linking with high-resolution mass spectrometry and computational modeling. Additionally, we probed the affinity of nidogen-1 to different laminin γ1 variants by means of SPR and ELISA-based binding assays.

So far, quantitative analyses of the nidogen-1/laminin affinity have only been performed using the laminin P1 fragment, which is produced by limited pepsin proteolysis and comprises parts of all three chains (α1, β1, γ1) of laminin-111 [9,11,13,63]. Apparent K_d values determined in those studies ranged between 0.5 nM and 1 nM, complying with the dissociation constants derived from the ELISA assays (Table S5). SPR analysis yielded similar results for the nidogen-1/laminin γ1 LEb2–4 interaction, but a lower apparent K_d value (~12 nM) for laminin γ1 short arm. In the SPR experiments, nidogen-1 was immobilized in random orientations by covalently linking lysine residues to the sensor chip surface, while the laminin variants were used as mobile analytes. Notably, the lower affinity of laminin γ1 short arm is solely caused by a slower association – the dissociation rate constant is similar for both laminin variants (Table S5). Therefore, we conclude that the deviations in the SPR experiments are caused by different *in solution* properties of laminin γ1 short arm compared to laminin γ1 LEb2–4. This hypothesis is supported by the ELISA experiments where nidogen-1 was used as ligand, while the laminin variants were immobilized through passive adsorption in a 96-well plate [64], resulting in similar nidogen-1-binding affinities. These different experimental setups may introduce different degrees of steric hindrance that probably do not affect the binding of the relatively small laminin γ1 LEb2–4 fragment (25 kDa), but hamper the interaction with the laminin γ1 short arm variant, which is considerably bulkier (113 kDa). In other words, the laminin γ1 short arm molecules might not be equally binding-competent due to different *in solution* conformations. This results in a slower association and consequently, a lower apparent K_d value when using laminin γ1 short arm as mobile analyte.

Moreover, interactions between nidogen-1 and laminin γ1 N836D variants were verified by both affinity assays. The ELISA results suggest a ~30-fold loss in affinity upon mutating Asn-836 to aspartic acid. This finding is in contrast to a previous study by Pöschl *et al.* who found this mutation to cause a 25,000-fold loss in affinity, practically abolishing any interaction [23]. Our data suggest a much less dramatic influence of laminin γ1 Asn-836 on nidogen-1 binding.

The large number of inter-domain contacts verified for nidogen-1 implies a globular conformation rather than a linear domain arrangement. This is in agreement with a previous structural investigation of nidogen-1 by electron microscopy showing a wide range of conformations, including both linear as well as globular structures [13]. This variability within nidogen-1 might be caused by high flexibility of the elongated regions connecting the globular domains ('link' and 'rod'), which could likewise be an explanation

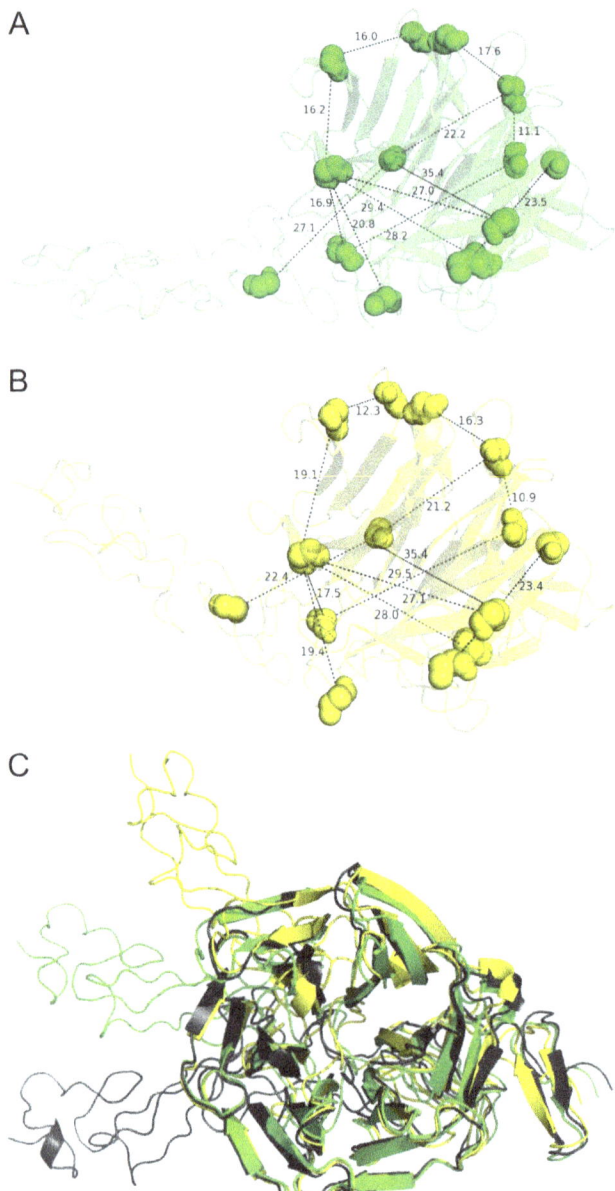

Figure 6. Refined models of the nidogen-1 G3/laminin γ1 LEb2–4 complex. Based on PDB entry 1NPE and the identified cross-links, modified structural models for the high-affinity interaction region of laminin γ1 and nidogen-1 were generated. Cross-linked residues are displayed as spheres. Cα–Cα distances are given in Å. (A) Model with the best Rosetta total score and a Rosetta atom-pair constraint score ranking among the top 2.5%. (B) Model with the best Rosetta atom-pair constraint score and a Rosetta total score ranking among the top 2.5%. (C) Alignment of both models and the unmodified crystal structure 1NPE (black). The orientation of LEb2–4 clearly has changed during structural refinement. The β-propeller fold of the G3 domain is still intact.

for the almost complete lack of cross-links in these regions (Figure 2A). Within laminin γ1 short arm variants, intramolecular contacts were exclusively found between non-adjacent domains supporting a compact globular protein architecture as well. The cross-links also give hints on the existence of an ensemble of defined conformations for both nidogen-1 and laminin γ1 short arm.

The intermolecular cross-links between nidogen-1 and the laminin variants strongly indicate additional interaction regions next to the known nidogen-1 G3/laminin γ1 LEb2–4 binding site. Even laminin γ1 LEb2–4 was found to form contacts to all globular nidogen-1 domains and the rod region (Figure 7). We conclude that our cross-linking experiments allowed us to pick up different nidogen-1/laminin γ1 complexes that are present in solution, while the published X-ray structure reflects only one interaction 'snapshot' – most likely the best crystallizable conformation.

The existence of additional interaction patterns was further substantiated by cross-linking experiments using nidogen-1 and laminin γ1 N836D variants, which were unanimously shown to interact regardless of the N836D substitution within the laminin γ1 LEb3 domain. Interestingly, contacts of nidogen-1 to LEb3 were almost completely abolished, while alternative laminin γ1 domains were found to be cross-linked to nidogen-1 (Figure 2). . Taken together, these results suggest alternative nidogen-1/ laminin γ1 binding modes, conceivably as a result of the conformational flexibility of both proteins, which had already been implied by electron microscopy [65]. When performing cross-linking experiments, one has to keep in mind that amine-reactive cross-linking depends on the reactivity of lysines, which is influenced by solvent exposure of their side chains and local pK_a values [66]. Considering the spacer length of BS^2G, the observation of an intermolecular cross-link is not *per se* equivalent to physical binding at exactly that site. While our cross-linking results indeed indicate additional binding sites, it is still imaginable that laminin γ1 LEb2–4 remains the primary anchoring site of nidogen-1. In fact, solid phase binding studies have proven that the nidogen-1 G3 domain is essential for nidogen-1/laminin γ1 interaction [63,67]. Therefore, nidogen-1 G3/laminin γ1 LEb2–4 binding is most likely crucial for a high-affinity interaction and binding at alternative sites proceeds with substantially lower affinity.

This finding seems reasonable also in view of the stabilizing function of nidogen-1 within basement membranes, connecting the laminin and the type IV collagen network [14]. Given the mechanical stress basement membranes have to withstand [2], secondary interactions at alternative binding sites may occur when the basement membrane is in a more relaxed state. In contrast, the high-affinity anchoring interaction between nidogen-1 G3 and laminin γ1 LEb2–4 is likely to be continuously present, thereby ensuring a high mechanical stability. Although this binding region represents one of the smallest high-affinity interfaces known so far [24], our SPR, ELISA, and cross-linking/MS data indicate a certain robustness against the single point mutation N836D, which is plausible in light of the physiological importance of the nidogen/ laminin interaction [68].

In line with these findings, we were not able to fit all cross-linking distance constraints for nidogen-1 and laminin γ1 into one single model of the protein complex. Assigning the cross-links to defined conformations within an ensemble of co-existing structural arrangements is currently beyond the bounds of the method. Consequently, a computational model of the entire nidogen-1/ laminin γ1 short arm complex based on cross-linking distance constraints would be overly speculative. Comparative modeling and *de novo* folding using the Rosetta modeling suite, however, enabled us to create models of all nidogen-1 and laminin γ1 short arm domains that have not been structurally characterized so far. The structural models of laminin γ1 L4 and nidogen-1 NIDO are of particular interest as there are not any sequential homologues with known structures. We were able to identify structural homologues of laminin γ1 L4 suggesting a galactose-binding

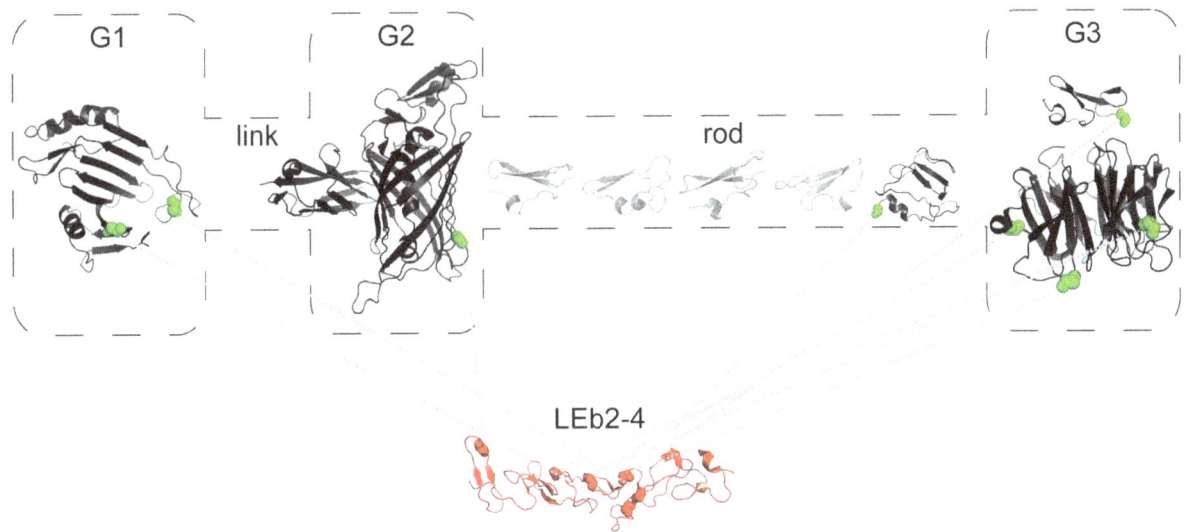

Figure 7. Contacts of laminin γ1 LEb2–4 wild type with nidogen-1. The LEb2–4 structure (red) is taken from PDB entry 1NPE. Nidogen-1 (grey) is schematically depicted as a combination of the crystal structures 1GL4 (G2 domain) and 1NPE (G3 domain) and representative models of the remaining domains. Residues involved in intermolecular contacts are shown as spheres. Gray dotted lines represent verified cross-links.

domain-like fold of this domain. Interestingly, this fold is also adopted by the N-terminal LN domains of laminin α5, β1, and γ1 [7,33,34]. To date, experimental evidence for carbohydrate-binding activity of laminins only exists for C-terminal laminin α LG domains [69].

Structures of the nidogen-1 NIDO domain were obtained by *de novo* folding. The proposed models either exhibit structural similarities to known PDB structures or were consistently modeled based on six NIDO sequences of evolutionary closely related organisms. All but one of the final models were further validated as they fulfill the only cross-linking constraint identified in this region. These models exhibit a compact topology with a β-sheet surrounded by two α-helices as central elements.

Finally, integrating the cross-links found within the nidogen-1 G3/laminin γ1 LEb2–4 complex and the known X-ray structure confirms the main structural features suggested by X-ray crystallography, yet indicating a more bent topology of the LEb2–4 domains (Figure 6C). Again, this finding depicts the flexibility of laminin γ1 resulting in overlying conformations of the nidogen-1/laminin γ1 complex, which can be captured by chemical cross-linking and thus reflect the whole picture of nidogen-1/laminin γ1 interaction in solution.

Conclusions

With our approach integrating chemical cross-linking, mass spectrometry, and computational modeling, we were able to structurally characterize conformations of nidogen-1/laminin γ1 complexes in solution. We applied two complementary cross-linking approaches, one using a classical homobifunctional amine-reactive cross-linker, the other one relying on the incorporation of unnatural photo-reactive amino acids.

Cross-links between laminin γ1 short arm and nidogen-1 were found in all protein regions. Therefore, it is likely that both proteins exhibit several additional contact regions apart from their known interaction site. In addition, different modes of interaction resulting in several distinct protein-protein interfaces can be imagined. Our results indicate that Asn-836 within laminin γ1 LEb3 domain is not essential for complex formation. Conclusively,

this work describes the first structural insights into the conformational dynamics of the nidogen-1/laminin γ1 complex and provides, for the first time, structural models of all nidogen-1 and laminin γ1 short arm domains. Chemical cross-linking, MS, and computational modeling allowed elucidating different conformations of the nidogen-1/laminin γ1 complex, which exist simultaneously in solution but are not reflected in the X-ray structure.

Supporting Information

Figure S1 Amino acid sequences of (A) nidogen-1 and (B) laminin γ1 short arm.

Figure S2 Clustal W2.1 sequence alignments for comparative modeling. Shown are pairwise sequence alignments of all nidogen-1 and laminin γ1 target sequences to template sequences with sequence identities ≥30%. The scheme for template sequences is termed 'PDB-entry_domain-name_chain-identifier'. Annotations comply with the Clustal nomenclature with identical (*), conserved (:) and semi-conserved (.) residues being denoted.

Figure S3 Manual sequence alignments for modeling of the laminin γ1 L4 domain. Shown are pairwise sequence alignments to all template sequences. Alignments are manually optimized to obtain maximum overlap of secondary structure elements. The scheme for template sequences is termed 'PDB-entry_chain-identifier'.

Figure S4 Jalview sequence alignment of nidogen-1 NIDO domains from different organisms. All NIDO domains share sequence identities >75% but exhibit short sequence stretches that are diverse.

Figure S5 Probing the nidogen-1/laminin γ1 interaction with SPR and ELISA assays. (A) Single-cycle kinetic experiments were performed by injecting mobile analyte (laminin) at increasing concentrations followed by partial dissociation. Initially, experiments were carried out with 6.25 nM, 12.5 nM, 25 nM, 50 nM and 100 nM laminin γ1 LEb2–4 wild type and N836D. Binding of laminin γ1 LEb2–4 N836D was additionally probed with 100-fold increased concentrations. System artefact signals (~30 min after each injection) were removed from the sensorgrams. (B) ELISA assays were performed in 96-well plates with immobilized laminin γ1 variants. Nidogen-1 was added in increasing concentrations (0.03–234 nM) until saturation was reached (incubation time: 1 h). Error bars represent standard deviations.

Figure S6   Incorporation efficiency of photo-amino acids into nidogen-1 and laminin γ1 short arm. (A) Met and Leu variants that were considered during MS analysis, including the reaction products of the photo-amino acids identified in [35] (1: photo-Leu, 5: photo-Met, 2 and 6: alkene; 3 and 7: alcohol; 4: unmodified Leu; 8 and 9: unmodified and oxidized Met). (B and C) MS-based label-free quantification of photo-amino acid incorporation. The pie charts show the number of leucines (blue) and methionines (red) within nidogen-1 (B) and laminin γ1 short arm (C) that remained unmodified (light shades) or were partially replaced by their photo-reactive counterparts (dark shades). The bars represent the relative abundance of partially modified peptides, containing the Leu (blue) and Met (red) variants listed in (A) [70].

Figure S7   Homology models of (A) nidogen-1 and (B) laminin γ1 short arm domains. Alignments of the best-scoring models representing the top three clusters are shown. Disulfide bridges are depicted as black sticks. All models were generated based on X-ray structures sharing more than 30% sequence identity with the respective domains.

Figure S8   Comparative modeling of laminin γ1 L4. (A) PSIPRED secondary structure prediction for the L4 domain. A β-sheet-rich fold and one long α-helix are predicted. (B) MUSTANG alignment of 13 potential structural homologs of L4 identified by fold recognition using several threading servers. All template candidates exhibit a β-sandwich topology. The number of β-strands is in line with the predicted secondary structure of L4. Instead of an α-helix, all structures contain a long loop region. (C) Rosetta total score of the top 10% of all generated models plotted against their RMSD from the best-scoring structure. Only α-helices, β-sheets and short loops (≤5 residues) were included in RMSD calculations. The models are converging to a minimum in score and RMSD indicating that the best-scoring models are valid. The two best-scoring models shown in Figure 4 are marked with red circles.

Figure S9   Best-scoring nidogen-1 NIDO models originating from common centroid models. The full-atom candidate structures of the NIDO domain were examined for common initial centroid models. Next to the centroid models underlying the structures depicted in Figure 5, we identified six centroid models that form the basis for more than three full-atom refined candidate structures. Shown are the best-scoring final candidate structures representing these initial centroid models. The Cα–Cα distances corresponding to the cross-link located within the models are given in Å. The residues are colored according to their Rosetta total score. Scores below zero (yellow-

green color) indicate energetically favorable conformations. The identifiers of the underlying centroid models are given.

Table S1   Results of Rosetta clustering of comparative nidogen-1 and laminin γ1 domain models. The best 10% of all generated models were clustered. The ideal clustering radius was automatically determined by the Rosetta algorithm. Shown are clusters with a size >1.

Table S2   Results of Rosetta clustering of the laminin γ1 L4 domain models. The best 10% of all generated models were clustered using a clustering radius of 1 Å. Shown are clusters with a size >1. The best-scoring models of clusters 1 and 2 were chosen as final models of the L4 domain.

Table S3   Scores of the final nidogen-1 NIDO domain models. Models sharing the 'centroid model identifier' originate from the same initial low-resolution centroid model. Models, for which structural homologues within the PDB have been identified, are listed in italics. The remaining models share a similar topology to models generated based on highly homologous sequences of NIDO domains derived from related organisms.

Table S4   Scores of the final models of the nidogen-1 G3/laminin γ1 LEb2–4 complex. The listed models rank among the top 20 of 800 generated models considering both Rosetta total score and atom pair constraint score, which reflects their compliance with the cross-linking distance constraints.

Table S5   Affinities and kinetic parameters of the nidogen-1/laminin γ1 interaction. For SPR measurements, given values for k_a and k_d are the weighted mean from two individual measurements and K_d was calculated from these values as $K_d = k_d/k_a$. All ELISA-based measurements were performed in triplicates and K_d values were determined by non-linear regression of the saturation binding curves. The values in parentheses represent standard deviations.

Table S6   Verified products of BS^2G-mediated cross-linking. Peptide sequences written in parentheses are part of the protein affinity tags and do thus not belong to the native amino acid sequences of the proteins studied. Oxidized methionines within the peptide sequences are denoted with 'm'. Loss of water or ammonia is indicated by addition of '$-H_2O$' or '$-NH_3$' to the fragment ion.

Table S7   Verified products of UV A-induced cross-linking. Peptide sequences written in parentheses are part of the protein affinity tags and do thus not belong to the native amino acid sequences of the proteins studied. For ambiguous cross-links, all potential cross-linked amino acids are listed. Within the peptide sequences, photo-leucine and photo-methionine are assigned with 'z' and 'o', respectively. Oxidized methionines are denoted with 'm'. Loss of water or ammonium is indicated by addition of '$-H_2O$' or '$-NH_3$' to the fragment ion.

Table S8   Rosetta clustering results of the final nidogen-1 NIDO domain models. The clustering radius was set to 10 Å. Clusters represented by models originating from the same initial low-resolution centroid model were merged. Shown are clusters with more than three member structures. Models, for which structural homologues within the PDB have been identified, are listed in italics. The remaining models share a similar topology to models generated based on highly homologous sequences of NIDO domains derived from related organisms.

File S1　Command line execution commands and flags used for computational modeling with Rosetta.

File S2　Rosetta loops files generated for comparative modeling. The position of the loops was determined based on the sequence alignments of the target sequences (listed in italics) to the respective templates. The scheme for template sequences is termed 'PDB-entry_domain-name_chain-identifier'.

References

1. Yurchenco PD, O'Rear JJ (1994) Basal lamina assembly. Curr Opin Cell Biol 6 (5): 674–681.
2. Yurchenco PD (2011) Basement membranes: cell scaffoldings and signaling platforms. Cold Spring Harb Perspect Biol 3 (2): a004911.
3. Timpl R, Rohde H, Robey PG, Rennard SI, Foidart JM, et al. (1979) Laminin - a glycoprotein from basement membranes. J Biol Chem 254 (19): 9933–9937.
4. Tunggal P, Smyth N, Paulsson M, Ott MC (2000) Laminins: structure and genetic regulation. Microsc Res Techniq 51 (3): 214–227.
5. Aumailley M, Bruckner-Tuderman L, Carter WG, Deutzmann R, Edgar D, et al. (2005) A simplified laminin nomenclature. Matrix Biol 24 (5): 326–332.
6. Beck K, Hunter I, Engel J (1990) Structure and function of laminin: anatomy of a multidomain glycoprotein. FASEB J 4 (2): 148–160.
7. Kalkhof S, Haehn S, Paulsson M, Smyth N, Meiler J, et al. (2010) Computational modeling of laminin N-terminal domains using sparse distance constraints from disulfide bonds and chemical cross-linking. Proteins 78 (16): 3409–3427.
8. Carlin B, Jaffe R, Bender B, Chung AE (1981) Entactin, a novel basal lamina-associated sulfated glycoprotein. J Biol Chem 256 (10): 5209–5214.
9. Kohfeldt E, Sasaki T, Göhring W, Timpl R (1998) Nidogen-2: a new basement membrane protein with diverse binding properties. J Mol Biol 282 (1): 99–109.
10. Timpl R, Dziadek M, Fujiwara S, Nowack H, Wick G (1983) Nidogen: a new, self-aggregating basement membrane protein. Eur J Biochem 137 (3): 455–465.
11. Salmivirta K (2002) Binding of mouse nidogen-2 to basement membrane components and cells and its expression in embryonic and adult tissues suggest complementary functions of the two nidogens. Exp Cell Res 279 (2): 188–201.
12. Miosge N, Sasaki T, Timpl R (2002) Evidence of nidogen-2 compensation for nidogen-1 deficiency in transgenic mice. Matrix Biol 21 (7): 611–621.
13. Fox JW, Mayer U, Nischt R, Aumailley M, Reinhardt D, et al. (1991) Recombinant nidogen consists of three globular domains and mediates binding of laminin to collagen type IV. EMBO J 10 (11): 3137–3146.
14. Ho MSP, Boese K, Mokkapati S, Nischt R, Smyth N (2008) Nidogens - extracellular matrix linker molecules. Microsc Res Techniq 71 (5): 387–395.
15. Dziadek M, Paulsson M, Timpl R (1985) Identification and interaction repertoire of large forms of the basement membrane protein nidogen. EMBO J 4 (10): 2513–2518.
16. Paulsson M, Aumailley M, Deutzmann R, Timpl R, Beck K, et al. (1987) Laminin-nidogen complex. Extraction with chelating agents and structural characterization. Eur J Biochem 166 (1): 11–19.
17. Mann K, Deutzmann R, Timpl R (1988) Characterization of proteolytic fragments of the laminin-nidogen complex and their activity in ligand-binding assays. Eur J Biochem 178 (1): 71–80.
18. Mayer U, Nischt R, Pöschl E, Mann K, Fukuda K, et al. (1993) A single EGF-like motif of laminin is responsible for high affinity nidogen binding. EMBO J 12 (5): 1879–1885.
19. Gerl M, Mann K, Aumailley M, Timpl R (1991) Localization of a major nidogen-binding site to domain III of laminin B2 chain. Eur J Biochem 202 (1): 167–174.
20. Baumgartner R, Czisch M, Mayer U, Pöschl E, Huber R, et al. (1996) Structure of the nidogen binding LE module of the laminin gamma 1 chain in solution. J Mol Biol 257 (3): 658–668.
21. Stetefeld J, Mayer U, Timpl R, Huber R (1996) Crystal structure of three consecutive laminin-type epidermal growth factor-like (LE) modules of laminin gamma 1 chain harboring the nidogen binding site. J Mol Biol 257 (3): 644–657.
22. Pöschl E, Mayer U, Stetefeld J, Baumgartner R, Holak TA, et al. (1996) Site-directed mutagenesis and structural interpretation of the nidogen binding site of the laminin gamma 1 chain. EMBO J 15 (19): 5154–5159.
23. Sasaki M, Yamada Y (1987) The laminin B2 chain has a multidomain structure homologous to the B1 chain. J Biol Chem 262 (35): 17111–17117.
24. Takagi J, Yang YT, Liu JH, Wang JH, Springer TA (2003) Complex between nidogen and laminin fragments reveals a paradigmatic beta-propeller interface. Nature 424 (6951): 969–974.

25. Patel TR, Bernards C, Meier M, McEleney K, Winzor DJ, et al. (2014) Structural elucidation of full-length nidogen and the laminin-nidogen complex in solution. Matrix Biol 33: 60–67.
26. Sinz A (2014) The advancement of chemical cross-linking and mass spectrometry for structural proteomics: from single proteins to protein interaction networks. Expert Rev Proteomics Sept. 16: 1–11, online available.
27. Sinz A (2006) Chemical cross-linking and mass spectrometry to map three-dimensional protein structures and protein–protein interactions. Mass Spectrom Rev 25 (4): 663–682.
28. Kalisman N, Adams CM, Levitt M (2012) Subunit order of eukaryotic TRiC/CCT chaperonin by cross-linking, mass spectrometry, and combinatorial homology modeling. Proc Natl Acad Sci USA 109 (8): 2884–2889.
29. Leitner A, Joachimiak LA, Bracher A, Mönkemeyer L, Walzthoeni T, et al. (2012) The molecular architecture of the eukaryotic chaperonin TRiC/CCT. Structure 20 (5): 814–825.
30. Rappsilber J (2011) The beginning of a beautiful friendship: cross-linking/mass spectrometry and modelling of proteins and multi-protein complexes. J Struct Biol 173 (3): 530–540.
31. Rinner O, Seebacher J, Walzthoeni T, Mueller LN, Beck M, et al. (2008) Identification of cross-linked peptides from large sequence databases. Nat Meth 5 (4): 315–318.
32. Müller MQ, Dreiocker F, Ihling CH, Schäfer M, Sinz A (2010) Cleavable cross-linker for protein structure analysis: reliable identification of cross-linking products by tandem MS. Anal Chem 82 (16): 6958–6968.
33. Hussain S, Carafoli F, Hohenester E (2011) Determinants of laminin polymerization revealed by the structure of the α5 chain amino-terminal region. EMBO Rep 12 (3): 276–282.
34. Carafoli F, Hussain S, Hohenester E (2012) Crystal structures of the network-forming short-arm tips of the laminin β1 and γ1 chains. PLoS ONE 7 (7): e42473.
35. Kölbel K, Ihling CH, Sinz A (2012) Analysis of peptide secondary structures by photoactivatable amino acid analogues. Angew Chem Int Ed Engl 51 (50): 12602–12605.
36. Karlsson R, Katsamba PS, Nordin H, Pol E, Myszka DG (2006) Analyzing a kinetic titration series using affinity biosensors. Anal Biochem 349 (1): 136–147.
37. Schaks S, Maucher D, Ihling CH, Sinz A (2012) Investigation of a calmodulin/peptide complex by chemical cross-linking and high-resolution mass spectrometry. In: König S, editor. Biomacromolecular mass spectrometry. Tips from the bench. Hauppauge: Nova Science Publishers. 1–18.
38. Götze M, Pettelkau J, Schaks S, Bosse K, Ihling CH, et al. (2012) StavroX - a software for analyzing crosslinked products in protein interaction studies. J Am Soc Mass Spectrom 23 (1): 76–87.
39. Boratyn GM, Schäffer AA, Agarwala R, Altschul SF, Lipman DJ, et al. (2012) Domain enhanced lookup time accelerated BLAST. Biol Direct 7: 12.
40. Kelley LA, Sternberg MJE (2009) Protein structure prediction on the Web: a case study using the Phyre server. Nat Protoc 4 (3): 363–371.
41. Söding J, Biegert A, Lupas AN (2005) The HHpred interactive server for protein homology detection and structure prediction. Nucleic Acids Res 33 (Web Server issue): W244–8.
42. Buchan DWA, Minneci F, Nugent TCO, Bryson K, Jones DT (2013) Scalable web services for the PSIPRED Protein Analysis Workbench. Nucleic Acids Res 41 (Web Server issue): W349–57.
43. Zhang Y (2008) I-TASSER server for protein 3D structure prediction. BMC Bioinformatics 9: 40.
44. Larkin MA, Blackshields G, Brown NP, Chenna R, McGettigan PA, et al. (2007) Clustal W and Clustal X version 2.0. Bioinformatics 23 (21): 2947–2948.
45. Combs SA, DeLuca SL, DeLuca SH, Lemmon GH, Nannemann DP, et al. (2013) Small-molecule ligand docking into comparative models with Rosetta. Nat Protoc 8 (7): 1277–1298.
46. Konagurthu AS, Whisstock JC, Stuckey PJ, Lesk AM (2006) MUSTANG: a multiple structural alignment algorithm. Proteins 64 (3): 559–574.

Acknowledgments

Prof. Gunter Fischer and Dr. Cornelia Schiene-Fischer are acknowledged for generously providing their cell culture facilities. The authors are indebted to Prof. Mats Paulsson for continuous support and valuable discussions.

Author Contributions

Conceived and designed the experiments: PL KK FZ JM AS. Performed the experiments: PL DT DN CHI MVK MS. Analyzed the data: PL KK CHI MS MVK AS. Contributed reagents/materials/analysis tools: DT CHI FZ JM AS. Contributed to the writing of the manuscript: PL DN MS FZ JM AS.

47. Jones DT (1999) Protein secondary structure prediction based on position-specific scoring matrices. J Mol Biol 292 (2): 195–202.

48. Leman JK, Mueller R, Karakas M, Woetzel N, Meiler J (2013) Simultaneous prediction of protein secondary structure and transmembrane spans. Proteins 81 (7): 1127–1140.

49. Joosten RP, te Beek TAH, Krieger E, Hekkelman ML, Hooft RWW, et al. (2010) A series of PDB related databases for everyday needs. Nucleic Acids Res 39 (Database): D411.

50. Lange OF, Baker D (2012) Resolution-adapted recombination of structural features significantly improves sampling in restraint-guided structure calculation. Proteins 80 (3): 884–895.

51. Gront D, Kulp DW, Vernon RM, Strauss CEM, Baker D, et al. (2011) Generalized Fragment Picking in Rosetta: Design, Protocols and Applications. PLoS ONE 6 (8): e23294.

52. Wang G, Dunbrack RL (2003) PISCES: a protein sequence culling server. Bioinformatics 19 (12): 1589–1591.

53. Ortiz AR, Strauss CE, Olmea O (2002) MAMMOTH (matching molecular models obtained from theory): An automated method for model comparison. Protein Sci 11 (11): 2606–2621.

54. Waterhouse AM, Procter JB, Martin DMA, Clamp M, Barton GJ (2009) Jalview Version 2–a multiple sequence alignment editor and analysis workbench. Bioinformatics 25 (9): 1189–1191.

55. Green NS, Reisler E, Houk KN (2001) Quantitative evaluation of the lengths of homobifunctional protein cross-linking reagents used as molecular rulers. Protein Sci 10 (7): 1293–1304.

56. Herzog F, Kahraman A, Boehringer D, Mak R, Bracher A, et al. (2012) Structural probing of a protein phosphatase 2A network by chemical cross-linking and mass spectrometry. Science 337 (6100): 1348–1352.

57. Kahraman A, Herzog F, Leitner A, Rosenberger G, Aebersold R, et al. (2013) Cross-link guided molecular modeling with ROSETTA. PLoS ONE 8 (9): e73411.

58. Suchanek M, Radzikowska A, Thiele C (2005) Photo-leucine and photo-methionine allow identification of protein-protein interactions in living cells. Nat Meth 2 (4): 261–267.

59. Borek F (1961) A new two-stage method for cross-linking proteins. Nature 191 (4795): 1293–1294.

60. Merkley ED, Rysavy S, Kahraman A, Hafen RP, Daggett V, et al. (2014) Distance restraints from crosslinking mass spectrometry: mining a molecular dynamics simulation database to evaluate lysine-lysine distances. Protein Sci. 23 (6): 747–759.

61. Leitner A, Walzthoeni T, Kahraman A, Herzog F, Rinner O, et al. (2010) Probing native protein structures by chemical cross-linking, mass spectrometry, and bioinformatics. Mol Cell Proteomics 9 (8): 1634–1649.

62. Allen FH (2002) The Cambridge Structural Database: a quarter of a million crystal structures and rising. Acta Crystallogr., B 58 (Pt 3 Pt 1): 380–388.

63. Ries A, Göhring W, Fox JW, Timpl R, Sasaki T (2001) Recombinant domains of mouse nidogen-1 and their binding to basement membrane proteins and monoclonal antibodies. Eur J Biochem 268 (19): 5119–5128.

64. Butler JE (2000) Solid supports in enzyme-linked immunosorbent assay and other solid-phase immunoassays. Methods 22 (1): 4–23.

65. Aumailley M, Wiedemann H, Mann K, Timpl R (1989) Binding of nidogen and the laminin-nidogen complex to basement membrane collagen type IV. Eur J Biochem 184 (1): 241–248.

66. Guo X, Bandyopadhyay P, Schilling B, Young MM, Fujii N, et al. (2008) Partial acetylation of lysine residues improves intraprotein cross-linking. Anal Chem 80 (4): 951–960.

67. Bechtel M, Keller MV, Bloch W, Sasaki T, Boukamp P, et al. (2012) Different domains in nidogen-1 and nidogen-2 drive basement membrane formation in skin organotypic cocultures. FASEB J 26 (9): 3637–3648.

68. Sasaki T (2004) Laminin: the crux of basement membrane assembly. J Cell Biol 164 (7): 959–963.

69. Hohenester E, Yurchenco PD (2013) Laminins in basement membrane assembly. Cell Adh Migr 7 (1): 56–63.

70. Lössl P, Sinz A (2014) Combining amine-reactive cross-linkers and photo-reactive amino acids for 3D-structure analysis of proteins and protein complexes. Meth Mol Biol (submitted).

14

Analysis of the Na$^+$/Ca^{2+} Exchanger Gene Family within the Phylum Nematoda

Chao He[1,2], Damien M. O'Halloran[1,2]*

1 Department of Biological Sciences, The George Washington University, Washington, D.C., United States of America, **2** Institute for Neuroscience, The George Washington University, Washington, D.C., United States of America

Abstract

Na$^+$/Ca^{2+} exchangers are low affinity, high capacity transporters that rapidly transport calcium at the plasma membrane, mitochondrion, endoplasmic (and sarcoplasmic) reticulum, and the nucleus. Na$^+$/Ca^{2+} exchangers are widely expressed in diverse cell types where they contribute homeostatic balance to calcium levels. In animals, Na$^+$/Ca^{2+} exchangers are divided into three groups based upon stoichiometry: Na$^+$/Ca^{2+} exchangers (NCX), Na$^+$/Ca^{2+}/K$^+$ exchangers (NCKX), and Ca^{2+}/Cation exchangers (CCX). In mammals there are three NCX genes, five NCKX genes and one CCX (NCLX) gene. The genome of the nematode *Caenorhabditis elegans* contains ten Na$^+$/Ca^{2+} exchanger genes: three NCX; five CCX; and two NCKX genes. Here we set out to characterize structural and taxonomic specializations within the family of Na$^+$/Ca^{2+} exchangers across the phylum Nematoda. In this analysis we identify Na$^+$/Ca^{2+} exchanger genes from twelve species of nematodes and reconstruct their phylogenetic and evolutionary relationships. The most notable feature of the resulting phylogenies was the heterogeneous evolution observed within exchanger subtypes. Specifically, in the case of the CCX exchangers we did not detect members of this class in three Clade III nematodes. Within the *Caenorhabditis* and *Pristionchus* lineages we identify between three and five CCX representatives, whereas in other Clade V and also Clade IV nematode taxa we only observed a single CCX gene in each species, and in the Clade III nematode taxa that we sampled we identify NCX and NCKX encoding genes but no evidence of CCX representatives using our mining approach. We also provided re-annotation for predicted CCX gene structures from *Heterorhabditis bacteriophora* and *Caenorhabditis japonica* by RT-PCR and sequencing. Together, these findings reveal a complex picture of Na$^+$/Ca^{2+} transporters in nematodes that suggest an incongruent evolutionary history of proteins that provide central control of calcium dynamics.

Editor: Christian Braendle, Centre National de la Recherche Scientifique & University of Nice Sophia-Antipolis, France

Funding: Funding was from The George Washington University (GW) Columbian College of Arts and Sciences, GW Office of the Vice-President for Research, and the GW Department of Biological Sciences. The funders had no role in study design, data collection and analysis, decision to publish, or preparation of the manuscript.

Competing Interests: The authors have declared that no competing interests exist.

* Email: damienoh@gwu.edu

Introduction

Na$^+$/Ca^{2+} exchangers are a family of proteins that provide homeostatic balance to the cell's calcium concentration. Na$^+$/Ca^{2+} exchangers are divided into three groups in animals based upon their stoichiometry: Na$^+$/Ca^{2+} exchangers (NCX) which exchange sodium for calcium, Na$^+$/Ca^{2+}/K$^+$ exchangers (NCKX) which exchange sodium for potassium and calcium, and Ca^{2+}/Cation exchangers (CCX; also called NCLX) which exchange sodium or lithium for calcium [1–3]. Na$^+$/Ca^{2+} exchangers have been shown to regulate calcium exchange at the cell membrane, endoplasmic reticulum, mitochondrion, and at the nucleus [4–6]. NCX, NCKX, and CCX exchangers are low affinity/high capacity ion transporters and can rapidly expel (forward mode) or introduce (reverse mode) calcium ions to the cell or organelle. All Na$^+$/Ca^{2+} exchangers contain a tandemly repeated protein motif, the alpha repeat, which invariably occurs in two blocks of five transmembrane domains separated by a cytoplasmic loop; with some variations, the residues of the alpha repeat are conserved among all three classes of Na$^+$/Ca^{2+} exchanger and in all organisms; and

where perturbed experimentally, these residues have been shown to be crucial for exchanger function [1]. NCX proteins are comprised of ten transmembrane domains [7,8], including an intracellular loop between TM5 and TM6 that contains the calcium binding domain 1 (CBD1) and calcium binding domain 2 (CBD2) that represent regulatory domains required for intracellular ion sensing [9–11]. At the primary sequence level, these tandem CBD1 and CBD2 domains both correspond with the CalX-beta motif, which is found tandemly repeated in essentially all NCX-class exchangers examined, and which is therefore a diagnostic marker distinguishing NCX-class from CCX-class exchangers [11]. The NCX and NCKX exchangers share sequence similarity in the transport α-repeat domains: G(S/G)SAPE within the α1 repeat, and GTS(I/V)PD within the α2 repeat. The CCX exchanger has a unique conserved sequence within the α-repeats: GNG(A/S)PD in α1 and (G/S)(N/D)SxGD in α2. Three NCX genes, five NCKX genes, and one CCX gene have been cloned and identified in mammals. Mammalian NCXs (NCX1-3) are highly expressed in cardiac muscle, skeletal muscle, and the central nervous system [5,12,13]. Mammalian NCKX1-5 are widely expressed in various cells including rod and cone photoreceptor cells, retinal ganglion cells, platelets, vascular smooth muscles, uterus, brain tissue, intestine, lungs, thymus, and epidermal cells [14–17]. The mammalian CCX exchanger NCLX (also termed NCKX6) is expressed in all tissues examined including the brain, thymus, heart, skeletal muscles, lungs, kidneys, intestines and testes and has been shown to localize to mitochondria [18–20]. Functionally, Na^+/Ca^{2+} exchangers contribute to the normal physiology of a wide variety of cells and tissues. Na^+/Ca^2 based exchange is considered the principal method of Ca^{2+} removal during heartbeat [21], and within the hippocampus NCX2 and NCX3 has been shown to contribute to plasticity and learning behavior [22,23].

In *Caenorhabditis elegans*, a total of ten Na^+/Ca^{2+} exchanger genes have been identified in the *C. elegans* genome (designated *ncx-1* to *ncx-10*) [1,24,25]. There are three NCX genes (*ncx-1*, *ncx-2* and *ncx-3*); *ncx-4* and *ncx-5* encode for proteins that belong to the NCKX branch; and *ncx-6* – *ncx-10* encode for CCX representatives in *C. elegans*. Here we set out to characterize structural and taxonomic specializations within the family of Na^+/Ca^{2+} exchangers across the phylum Nematoda. We sourced members of the Na^+/Ca^{2+} exchanger family in the following twelve nematode species (Clade designations described by Blaxter et al. [26]): Clade IV - *Strongyloides ratti*, Clade V - *Haemonchus contortus*, *Heterorhabditis bacteriophora*, *Caenorhabditis elegans*, *Caenorhabditis brenneri*, *Caenorhabditis japonica*, *Caenorhabditis briggsae*, *Caenorhabditis remanei*, *Pristionchus pacificus*, Clade III - *Brugia malayi*, *Loa loa*, and *Ascaris suum*. From these sequences we then reconstructed the phylogenetic relationship for NCX, NCKX, and NCLX across all twelve species and investigated rates of selection for each transporter type. Na^+/Ca^{2+} exchangers are highly conserved across mammalian taxa at the protein and syntenic levels [1,27], and from our analysis we observed an unexpected level of heterogeneity in copy number within nematodes, in particular within the CCX subtype where we detected several putative examples of gene gain and/or loss. We detected between three and five CCX members across *Caenorhabditis* and *P. pacificus* species, and single CCX proteins for *H. contortus*, *H. bacteriophora*, and *S. ratti*, and did not detect any CCX members within *B. malayi*, *L. loa*, or *A. suum*. We also provided re-annotation for gene structure predictions for CCX members within *C. japonica* and *H. bacteriophora* by RT-PCR and sequencing.

Materials and Methods

Sequences

The genomes of the nematodes sampled (Strongyloides ratti, Haemonchus contortus, Heterorhabditis bacteriophora, Caenorhabditis elegans, Caenorhabditis brenneri, Caenorhabditis japonica, Caenorhabditis briggsae, Caenorhabditis remanei, Pristionchus pacificus, Brugia malayi, Loa loa, and Ascaris suum) were searched for NCX, NCKX and CCX protein sequences with bidirectional BlastX and BlastP searches using WormBase ver.WS243 [28], Ensembl [29], Nematode.net [30], InParanoid [31]; and OrthoMCL [32] using curated NCX, NCKX, and CCX sequences from C. elegans, Drosophila melanogaster, and mammals as starting material. Specific structural signatures unique to each sodium calcium exchanger subtype were used to parse matches (Figure 1A), which were filtered through Interpro [33] and SMARTDB [34] to organize hits into either: Sodium Calcium Exchangers (NCX), Potassium dependent Sodium Calcium Exchangers (NCKX), or Sodium Calcium Lithium Exchanger (NCLX [aka CCX]) subtypes (Figure 1B).

Sequence Analysis

Proteins were aligned using the multiple sequence alignment software MUSCLE v3.8.31 [35], and gaps were systematically stripped after alignment. The appropriate model was selected using Prottest v3 [36,37] and found to be LG+I+G+F for NCX phylogeny with gamma distribution parameter = 1.06 and four substitution rate categories and proportion of invariant sites = 0.05, WAG+G+F for NCKX phylogeny with gamma distribution parameter = 0.8 and four substitution rate categories, and WAG+I+G+F for CCX phylogeny with gamma distribution parameter = 1.03 and four substitution rate categories and proportion of invariant sites = 0.04. Phylogenetic relationships were inferred by reconstructing trees by Maximum Likelihood using the PhyML command-line application [38] as described previously [39]. Signatures of selection were detected using the single-likelihood ancestor counting (SLAC), random effects likelihood (REL) [40], and mixed effects model of evolution (MEME) [41] methods implemented in the HyPhy package [42,43]. DNA sequences were tested for best fit models using jModelTest [44] and recoded into codon based alignments using Pal2Nal [45]. High resolution images of alignments were obtained using Geneious [46]. Pairwise patterns of molecular diversity (π) for NCX, NCKX, and CCX exchangers between *C. elegans* and *C. briggsae* were calculated using DnaSP ver.5 [47]. NCX protein structure was predicted using Phyre [48] which incorporated the resolved NCX structure from Liao et al (PDB ID 3V5U) [8], and visualized using RasMol [49]. The alignments from our sequence analyses were used to generate a position specific weight matrix (PSWM) based upon the divergent alpha repeat structures detected across the twelve nematode species we analyzed. Using this PSWM we developed a web-based tool called 'N(em)CX' that searches for divergent NCX-like proteins. The server side script was written in Perl using the CGI.pm Perl module to generate output html. N(em)CX is available here: http://ohalloranlab.net/NemCX.html

Strains and maintenance

Caenorhabditis japonica strain DF5081 was maintained at 20°C by mating males and females on NGM plates seeded with *E. coli* strain OP50 [50,51]. *Heterorhabditis bacteriophora* strain TTO1 animals (kindly provided by John Hawdon) were cultured *in vivo* at 25°C in *Galleria mellonella* (wax moth) larvae using standard protocols [52]. Parasitized larvae were placed on water traps [53] to collect the infective-stage nematodes. The water traps were constructed and nematodes harvested as described previously [54,55]. Parasitic juvenile stages of *H. bacteriophora* were harvested by obtaining *G. mellonella* cadavers 6 to 8 days post-infection and cutting them open in a Petri dish containing M9 buffer, and the emerging nematodes were washed twice with distilled water. Female adult nematodes were obtained by dissecting *G. mellonella* cadavers 5 to 7 days post-infection in a Petri dish containing M9 buffer and females were picked using an aspirator.

DNA, RNA isolation, and RT-PCR

Genomic DNA was isolated by harvesting animals and collecting in a 1.5 ml tube. 200 µl of Lysis buffer (60 g/ml proteinase K, 10 mM Tris-Cl, pH 8.3, 50 mM KCl, 2.5 mM $MgCl_2$, 0.45% IGEPAL, 0.45% Tween-20, 0.01% gelatin) was added to the tube and then frozen at −80°C for 10 mins followed by incubation at 60°C for 1 hr followed by 95°C for 15 mins. In the case of *H. bacteriophora*, animals were crushed into a fine powder using a pestle and mortar. Tubes were then centrifuged at 13,000 rpm for 1 min and ~50 µl gDNA supernatant isolated for PCR. For *H. bacteriophora*, an extra step of phenol-chloroform

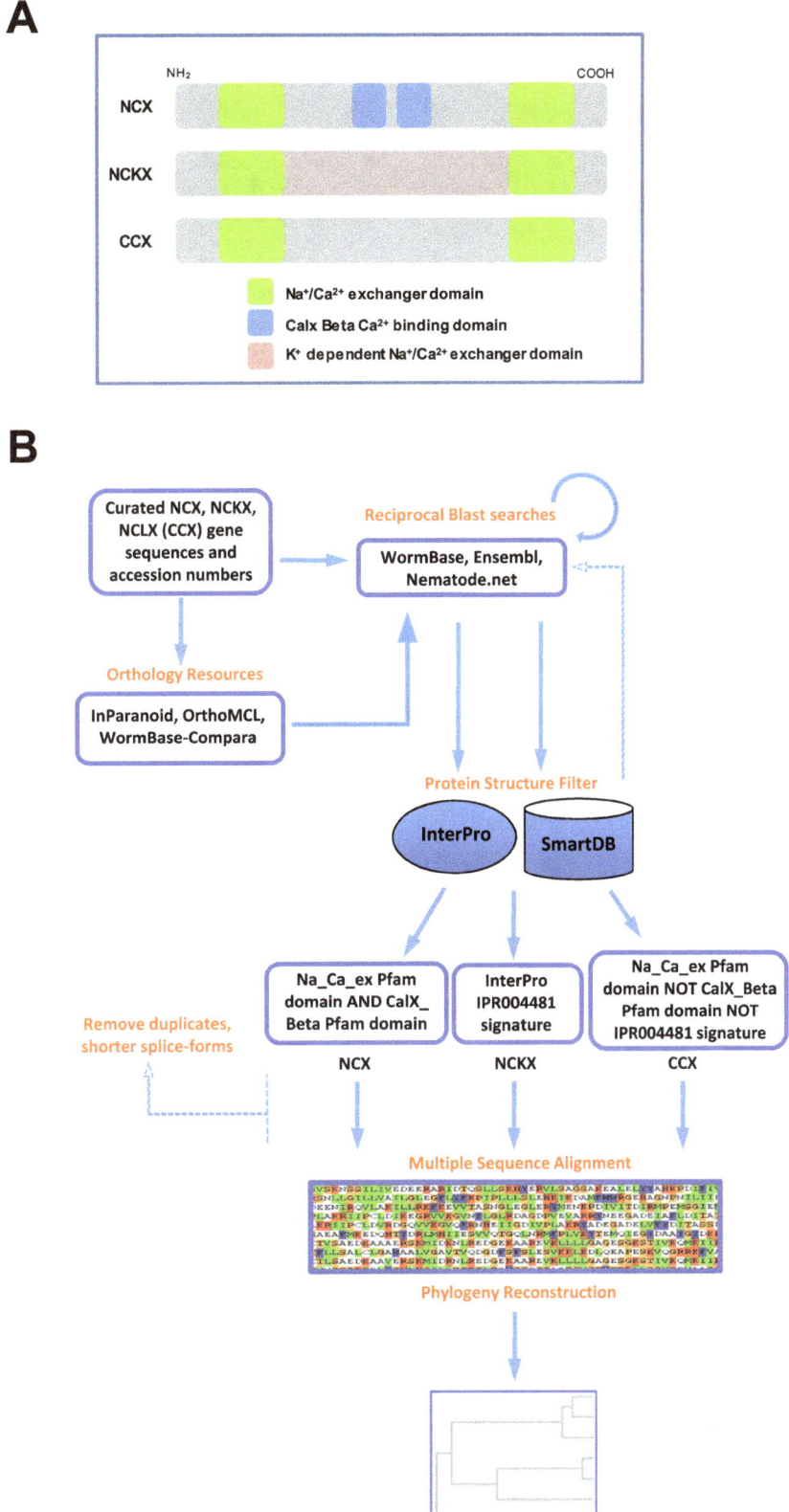

Figure 1. Structures within NCX, NCKX, and CCX proteins, and an overview of the pipeline used to detect orthologs of these proteins in twelve species of nematodes. (**A**) Cartoon depicting structures within the NCX, NCKX, and CCX (NCLX) proteins. (**B**) Overview of a pipeline used to detect orthologous sodium calcium exchanger genes in twelve different species of nematodes.

purification was performed by adding a 1:1 volume to gDNA supernatant. Total RNA was isolated from mixed stage animals using 1 ml Trizol Reagent (Invitrogen, Life Technologies, Carlsbad, CA). A 20 G syringe (Becton-Dickinson 3 ml syringe) was used to break down material and 200 µl of chloroform was added to isolate RNA from the sample. Tubes were vigorously shaken followed by centrifuging at 13,000 rpm for 10 min. The clear supernatant mixed with 1 volume of 70% EtOH was then cleaned using an RNA mini kit (Invitrogen, Life Technologies, Carlsbad, CA). The concentration and purity of the RNA samples were determined using spectrophotometry and ethidium bromide visualization of intact 18S and 28S RNA bands after agarose gel electrophoresis. Total RNA was treated with DNase (Thermo Scientific, Waltham, MA) by incubating at 37°C for 30 min followed by 65°C for 10 min, and reverse transcribed (RT) with MMLV reverse transcriptase (50 U, USB, Affymetrix, Santa Clara, CA) in 5× RT polymerase chain reaction (PCR) buffer (500 mM KCl and 100 mM Tris-HCl, pH 8.3, 7.5 mM $MgCl_2$), 4 U RNase inhibitor, 10 mM each of the dNTPs. The final PCR products were electrophoresed on 1.5% agarose gels. Primer pairs used for *Caenorhabditis japonica* were CJ-F TACGTGAGC-CATGGACATCACA and CJ-R TCGATACGTTGGATTGA-GAATC, and primer pairs used for *Heterorhabditis bacteriophora* were: Hba-F TGCTCTACTCCTTGCTTCGTGCCCCGTA, Hba-R GGGAGTTACATTCATGGCATTTGGCAATGG using the following cycling conditions for *C. japonica*: 95°C for 3 mins, 95°C for 30 sec, 56°C for 30 sec, and 72°C extension for 3 mins 30 sec for gDNA and 1 min for cDNA; and the following cycling conditions for *H. bacteriophora*: 95°C for 3 mins, 95°C for 30 sec, 56°C for 30 sec, and 72°C extension for 1 min 40 sec for gDNA and 1 min for cDNA.

Results

Detecting Orthologous Sodium Calcium Exchangers

Na^+/Ca^{2+} exchangers from *Caenorhabditis elegans* (NCX-1 to NCX-10) and mammals (NCX1-3, NCKX1-5, NCLX) were used to search various databases and resources: WormBase ver. WS243 [28], Nematode.net [30], Ensembl [29], InParanoid [31], and OrthoMCL [32] to obtain orthologs from the following nematodes: *Strongyloides ratti, Haemonchus contortus, Heterorhabditis bacteriophora, Caenorhabditis elegans, Caenorhabditis brenneri, Caenorhabditis japonica, Caenorhabditis briggsae, Caenorhabditis remanei, Pristionchus pacificus, Brugia malayi, Loa loa, and Ascaris suum*. All hits were filtered through InterPro [33] and SmartDB [34] to separate into NCX, NCKX or CCX proteins based upon unique structures (Figure 1A) within each subtype of exchanger (Figure 1B). Only NCKX members harbor a potassium dependent exchanger domain (InterPro IPR004481 signature), whereas NCX proteins contain exchanger domains (Na_Ca_ex Pfam domain) and CalX beta domains (CalX_ Beta Pfam domain), and finally CCX members do not contain the CalX beta domains but do contain the exchanger domains (Na_Ca_ex Pfam domain). In each case we used validated NCX, NCKX, and NCLX (CCX) proteins from mammals as positive controls to ensure the pipeline parsed the hits appropriately.

NCX Phylogeny

The NCX Na^+/Ca^{2+} exchangers were mostly conserved across all species examined (Figure 2A). NCX-1 and NCX-2 are most closely related in each case with NCX-3 representing a more diversified clade. The alpha repeat domains were highly conserved and adopted the consensus GSSAPE for the α1 repeat and GTS(I/V/L)PD for the α2 repeat. Together, these repeats comprise four transmembrane domains (TM2, TM3, TM7, TM8) that form a diamond shaped transport vestibule [8]. We identified three NCX genes for each *Caenorhabditis* species that we examined except in the case of *C. brenneri* where we only identified two NCX genes. For each of the NCX clusters, the *Caenorhabditis* orthologs grouped together (Figure 2A). For the NCX-1 and NCX-3 groups, the *H. bacteriophora* and *H. contortus* orthologs grouped together, and for NCX-3, the *B. malayi, L. loa*, and *A. suum* orthologs grouped close together (Figure 2A). Outside of the *Caenorhabditis* genus, we identified three NCX members for each species except *P. pacificus, B. malayi* and *L. loa* which each were assigned two NCX genes. In the case of *P. pacificus*, each NCX gene grouped into the NCX-1 cluster. Similarly, for *S. ratti*, two of its three NCX genes grouped into the NCX-2 cluster. In cases where multiple hits were detected within each cluster, we re-examined the alignment to ensure alternative splicing was not a misleading factor. We examined selection across the NCX *Caenorhabditis* taxa using MEME [41] and found a global dN/dS value = 0.130, we also used SLAC [40] and found a global dN/dS value = 0.138955 ($p<0.01$). We also conducted a sliding window analysis of nucleotide diversity between *C. elegans* and *C. briggsae* NCX genes using DnaSP ver. 5 [47], and observed similar polymorphic patterns across the *ncx-1* and *ncx-3* sequences (average π score = 0.2075 for *ncx-1*, and 0.1955 for *ncx-3*), and slightly elevated levels of diversity for *ncx-2* (average π score = 0.369) (Figure 3A–3B). However, elevated nucleotide diversity for *ncx-2* does not hold for other *Caenorhabditis* pairs: for example using *C. elegans* and *C. japonica* the average *ncx-2* π score = 0.173 (sampling variance = 0.007), and using *C. elegans* and *C. remanei* the average *ncx-2* π score = 0.11 (sampling variance = 0.003). Next, we tested specific sites for positive selection using REL [40] within the *Caenorhabditis* taxa, and from this analysis we did not detect any sites undergoing positive selection. We also tested for episodic diversifying selection using MEME and detected two significant ($p<0.01$) sites: codon 455, which is positioned close to the second calcium binding domain (CBD2) between TM5 and TM6, which detects local intracellular calcium levels, and codon 925 which is located after TM9 in the intracellular loop that connects with TM10 (see Figure 3C). The crystal structure for the NCX from *Methanococcus jannaschii* (NCX_Mj) has been resolved [8], and using this structure we examined structural diversity across NCX proteins in nematodes. The NCX-1 and NCX-3 clusters are the most divergent amongst the three NCX clusters (Figure 2A), and so to examine structural differences between diverse NCX proteins in nematodes we selected a representative from the NCX-1 cluster (*C. elegans* NCX-1) and a representative from the NCX-3 cluster (*A. suum* GS_14034) for *in silico* structural analysis. For NCX-1 from *C. elegans* 32% of the residues (285 residues in total) were modelled at 100% confidence and yielded 33% alpha helical and 20% beta strand structures. The *A. suum* GS_14034 NCX was modelled at 100% confidence for 36% of the sequence (289 residues) including 36% alpha helical and 22% beta strand structures. In each case the single highest scoring modelling template was the resolved NCX_Mj structure (PDB: 3V5U) from Liao et al. [8]. In the case of NCX-1 we observed a longer beta strand structure connecting TM8 and TM9 from residues 801–817 than is predicted for *A. suum* GS_14034 (see red arrowheads, Figure 3C), and similarly another lengthy beta structure is predicted for NCX-1 connecting TM4 to TM5, and also a shorter alpha helical structure within TM4 of NCX-1 (red arrowheads, Figure 3C), that is not predicted for *A. suum* GS_14034 (Figure 3C). In the case of *A. suum* GS_14034, TM4 is composed of consistent alpha helical structure from residues 165–185, and the linker connecting TM8 and TM9 is significantly shorter in *A. suum* GS_14034 from residues 742–748 (Figure 3C).

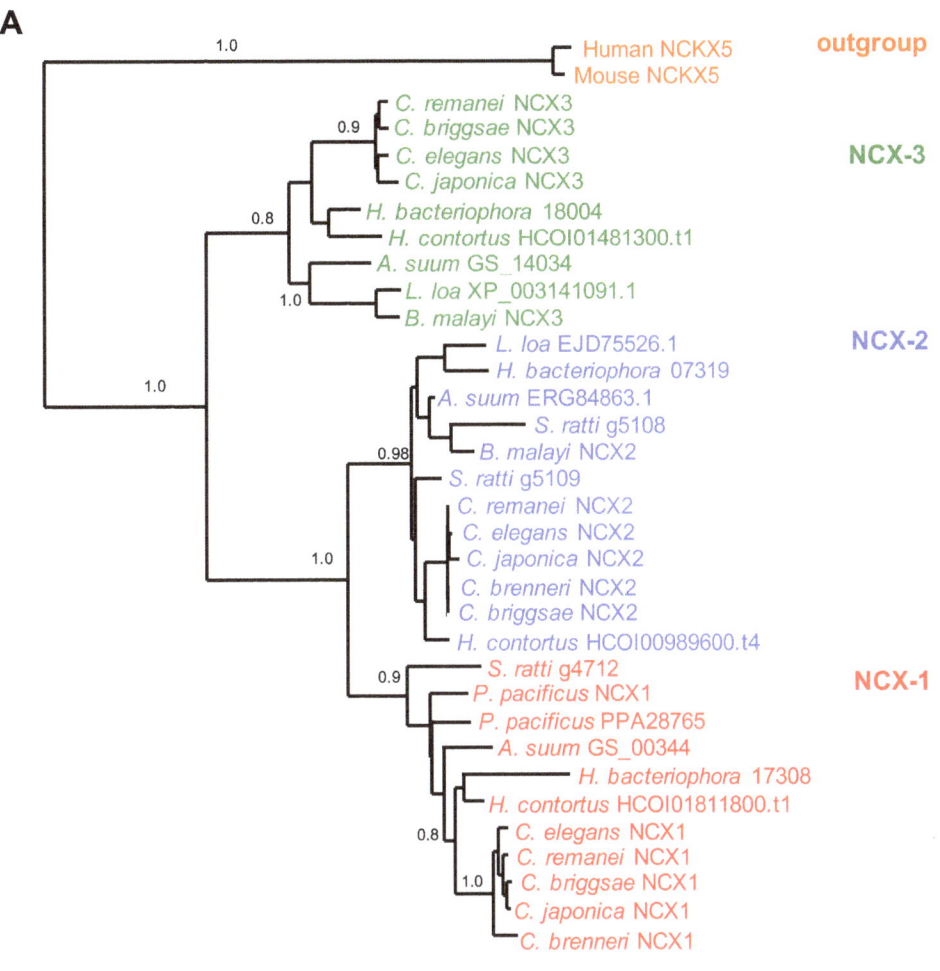

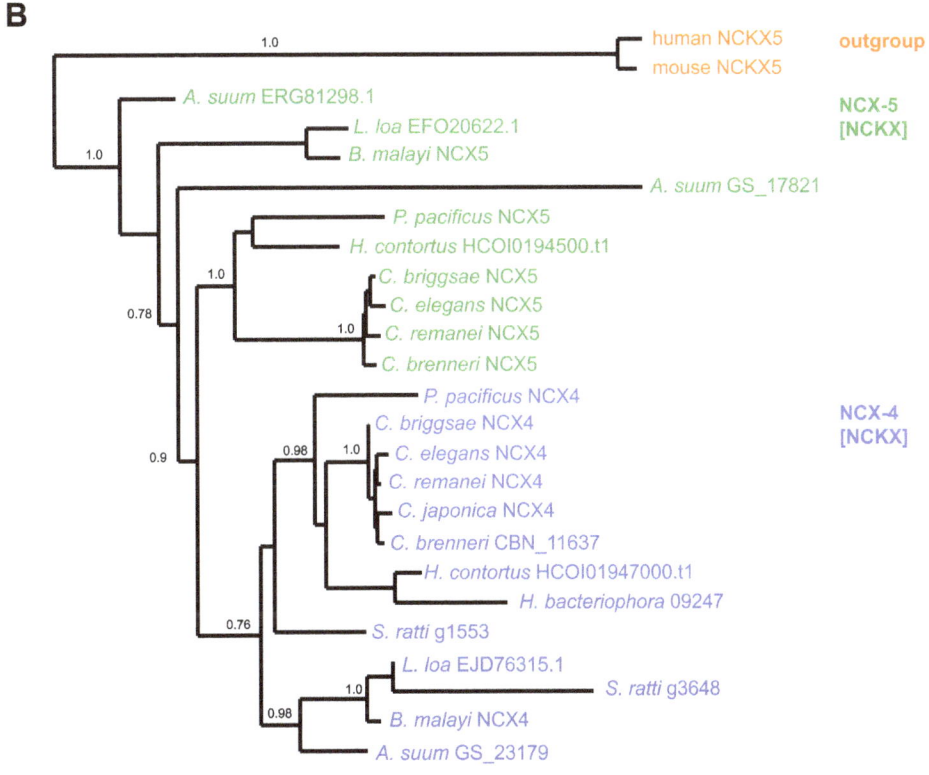

Figure 2. Phylogenetic analysis of NCX and NCKX exchangers in various nematodes. (**A**) Phylogenetic analysis of NCX type exchangers from *Strongyloides ratti, Haemonchus contortus, Heterorhabditis bacteriophora, Caenorhabditis elegans, Caenorhabditis brenneri, Caenorhabditis japonica, Caenorhabditis briggsae, Caenorhabditis remanei, Pristionchus pacificus, Brugia malayi, Loa loa*, and *Ascaris suum*. Inferred phylogeny was constructed using PhyML [38] and derived from amino acid alignments using MUSCLE [35]. The NCKX5 exchanger from human and mouse was used as an outgroup. (**B**) Phylogenetic analysis of NCKX type exchangers from *S. ratti, H. contortus, H. bacteriophora, C. elegans, C. brenneri, C. japonica, C. briggsae, C. remanei, P. pacificus, B. malayi, L. loa*, and *A. suum*. Inferred phylogeny was constructed using PhyML [38] using the model WAG+G+F determined from Prottest [36] and derived from amino acid alignments using MUSCLE [35]. The NCKX type exchangers from human and mouse served as an outgroup.

NCKX Phylogeny

Members of the NCKX family exhibited much diversity at the protein level and broadly assembled into representatives of NCX-4 and NCX-5 clusters. In each cluster, the *Caenorhabditis* species grouped together although we did not detect an NCX-5 member for *C. japonica* (Figure 2B). In all other species we detected two NCKX genes except for *H. bacteriophora* for which we only detected one NCKX gene and *A. suum* for which we detected three NCKX genes. In the NCX-4 cluster *H. contortus* and *H. bacteriophora* orthologs grouped together. Within the NCX-5 cluster we observed more diversity and longer branch lengths, especially true in the case of *A. suum* GS_17821 which only shares 34.9% percent identity with its nearest neighbor, *B. malayi* NCX-5. In almost all cases the highly conserved aspartic acid residue

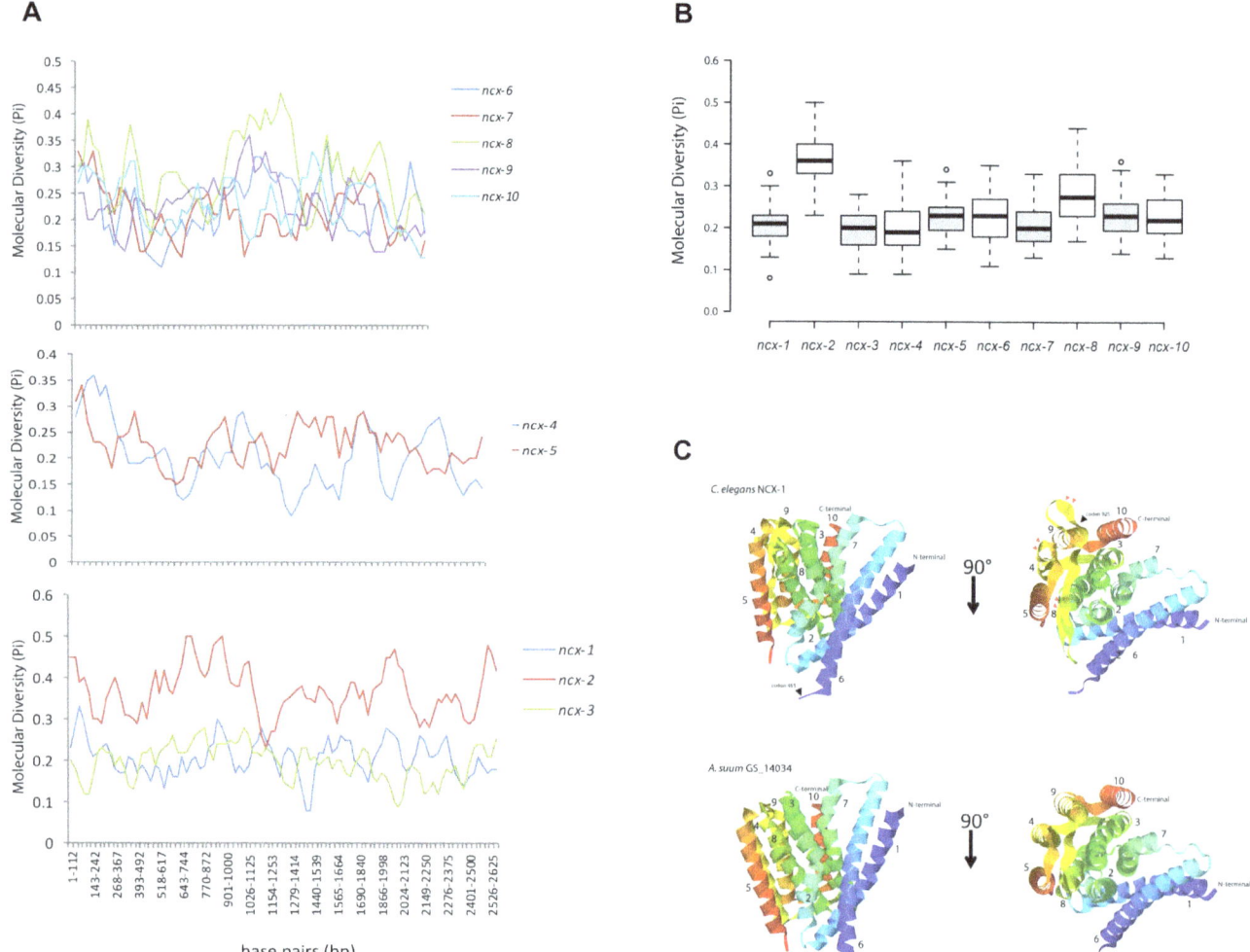

Figure 3. Sequence and structural analysis of nematode exchangers. (**A**) Sliding window analysis of nucleotide diversity using DnaSP version 5 [47] with 100 bp windows and 25 bp steps for CCX (upper graph), NCKX (middle graph), and NCX exchanger (lower graph) gene pairs between *C. elegans* and *C. briggsae*. (**B**) Box plot analysis of molecular diversity for each *ncx* gene (*ncx-1* to *ncx-10*) between *C. elegans* and *C. briggsae*. Whiskers extend to data points that are less than 1.5 × interquartile range away from 1st/3rd quartile; center lines show the medians and outliers are shown as dots. (**C**) Ribbon model of NCX-1 from *C. elegans* and an NCX-3 ortholog from *A. suum* (GS_14034). N and C termini are indicated, and red arrowheads refer to structural differences between each NCX. Numbers refer to the transmembrane (TM) domains. Extracellular side is up in left views. The position of two candidate sites in NCX-1 undergoing episodic diversifying selection are indicted - codon 455 and codon 925. In each case the right view is rotated by 90°. Structural predictions were made using Phyre [48] and visualized using RasMol [49]. In each case the single highest scoring modelling template was the resolved NCX (NCX_Mj) structure (PDB: 3V5U) from Liao et al. [8].

A

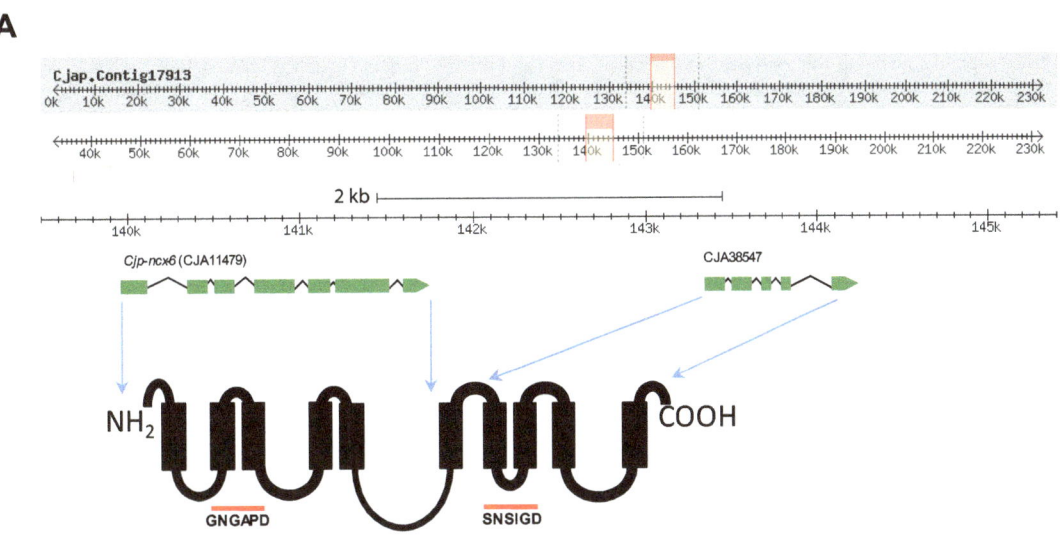

B

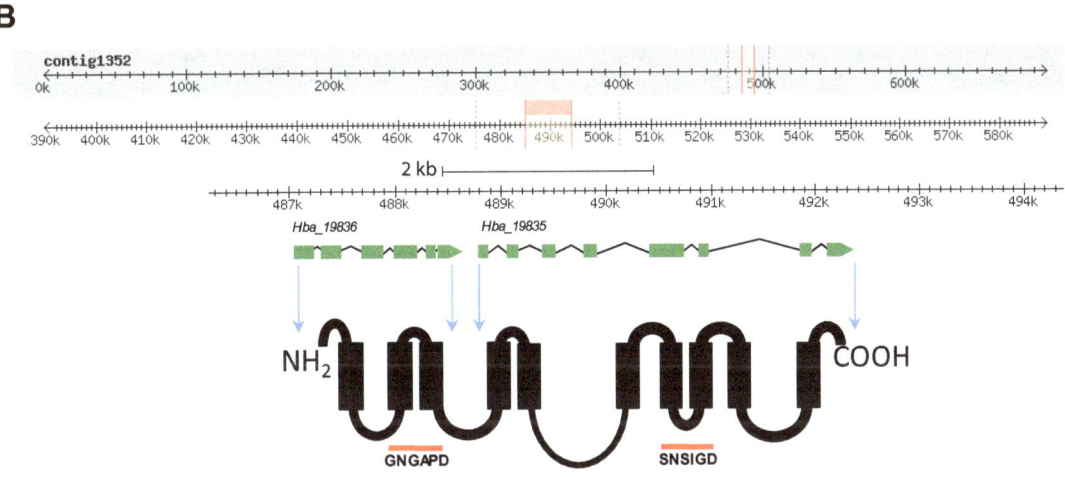

C

D

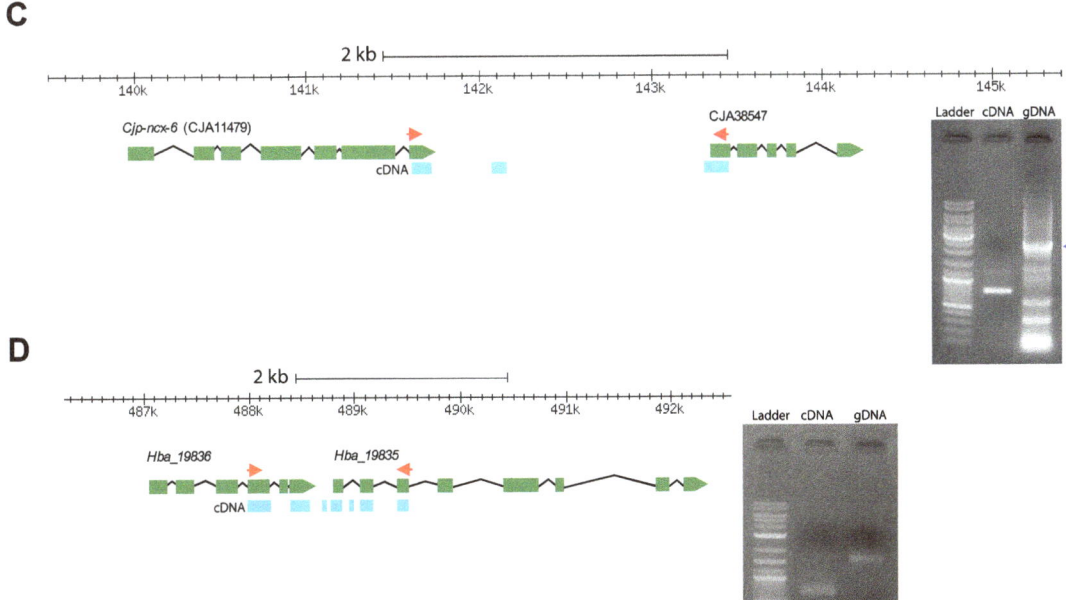

Figure 4. Predicted gene structure of CCX proteins from *Caenorhabditis japonica* and *Heterorhabditis bacteriophora*. (**A**) Predicted gene structures from *Cja11479* and *Cja38547* on contig 17913 from *Caenorhabditis japonica*. Translation of each predicted gene produces approximately half an NCLX-like protein containing the α1 repeat sequence GNGAPD and the α2 domain sequence SNSIGD. Blue arrows indicate the approximate position of the predicted coding sequence mapped to the translation. (**B**) Predicted gene structure from *Hba_19835* and *Hba_19836* on contig 1352 from *Heterorhabditis bacteriophora*. Translations generate approximately half an NCLX-like protein containing the α1 repeat sequence GNGAPD and the α2 domain sequence SNSIGD. Blue arrows indicate the approximate position of the predicted coding sequence mapped to the protein translation. (**C**) PCR Primers were designed that spanned the final exon of the upstream predicted gene *Cja11479* and the first exon of the downstream predicted gene *Cja38547*. Using genomic DNA as template for PCR we observed a band at 2443 bp (third lane in gel inset), and using reverse transcribed RNA as template for PCR we observed a band at 813 bp (second lane in gel inset). Primers are indicated by red arrowheads, and resulting cDNA sequence mapped to gene structure is indicated by the blue rectangles. First lane in the gel electrophoresis image is a GenRuler DNA ladder mix (Thermo Scientific - SM0334). Purple arrowhead at inset indicates the position of the gDNA band corresponding to the 2443 bp fragment. (**D**) Primers were designed that spanned the third to last exon of the upstream predicted gene *Hba_19836* and the third exon of the downstream predicted gene *Hba_19835*. Using genomic DNA as template for PCR we observed a band at 1623 bp (third lane in gel inset), and using reverse transcribed RNA as template for PCR we observed a band at 674 bp (second lane in gel inset). Primers are indicated by red arrowheads, and resulting cDNA sequence mapped to the gene structure is denoted by blue rectangles. First lane in the gel electrophoresis inset image is a GenRuler DNA ladder mix (Thermo Scientific - SM0334).

within the α2 repeat domain that confers potassium dependence [56] was present with the exception of the following proteins: *S.ratti*-g3648, *A.suum*-ERG81298.1, and *H.bacteriophora*_09247 - each of these proteins contained the conserved α1 repeat domain but were atypical for the α2 repeat domain. We also examined selection across the NCKX *Caenorhabditis* genus and found a global dN/dS value using SLAC of 0.118693 ($p<0.01$), and a global dN/dS value of 0.115 using MEME ($p<0.01$). A sliding window of nucleotide diversity was generated for each NCKX gene pair between *C. elegans* and *C. briggsae* and revealed similar patterns of DNA polymorphisms across each exchanger (average π value for *ncx-4* = 0.203, and average π value for *ncx-5* = 0.22) (Figure 3A–3B). Next, we tested for site specific evidence of positive selection using REL, and found evidence for one site undergoing positive selection: codon 9, which is located prior to the first predicted TM segment. Finally, we tested for specific

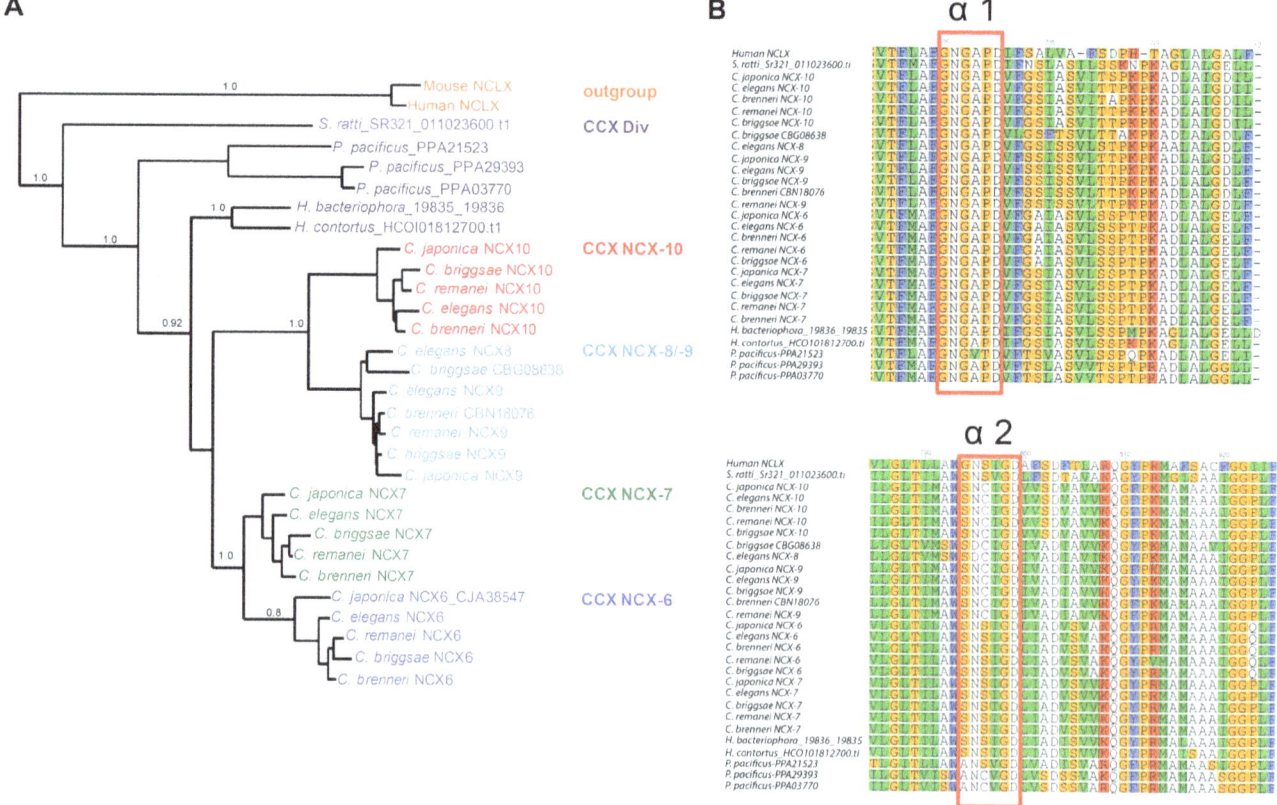

Figure 5. Phylogenetic analysis of NCLX (CCX) exchangers from various nematodes. (**A**) Phylogenetic analysis of NCLX type exchangers from *Strongyloides ratti*, *Haemonchus contortus*, *Heterorhabditis bacteriophora*, *Caenorhabditis elegans*, *Caenorhabditis brenneri*, *Caenorhabditis japonica*, *Caenorhabditis briggsae*, *Caenorhabditis remanei*, and *Pristionchus pacificus*. Inferred phylogeny was constructed using PhyML [38] using the model WAG+I+G+F determined from Prottest [36] and derived from amino acid alignments using MUSCLE [35]. 'CCX div' denotes a divergent CCX cluster that does not group with the *Caenorhabditis* CCX protein clusters (i.e. NCX-6 to NCX-10). The NCLX exchanger from human and mouse were used as an outgroup. (**B**) Alignment of α repeat domains within CCX (NCLX) proteins from various nematodes and also human NCLX. Alignments were generated using MUSCLE [35] of NCLX type exchangers from Human, *S. ratti*, *H. contortus*, *H. bacteriophora*, *C. elegans*, *C. brenneri*, *C. japonica*, *C. briggsae*, *C. remanei*, and *P. pacificus*.

branches undergoing episodic diversification using MEME but did not detect any evidence of episodic diversification ($p<0.01$).

CCX (NCLX) Phylogeny

In analyzing the CCX group, we noted that for *C. japonica* two tandem predicted genes, *Cja11479* and *Cja38547*, were annotated as separate protein-coding genes, and at the translated protein level we found that each gene was predicted to encode only one half of a single CCX protein (Figure 4A). We observed a similar scenario with the *H. bacteriophora* tandem predicted genes, *Hba_19835* and *Hba_19836*, which are annotated as two separate protein coding genes, and from our analysis we found these separate predicted genes would encode one single CCX protein (Figure 4B). Interestingly, each of these predicted genes are annotated as separate genes on account of the stop codon at the end of the final exon of the upstream gene prediction in each case. Na^{+}/Ca^{2+} exchangers have been shown to exhibit many alternatively spliced isoforms [57,58], however, in these two cases considering the size of the encoded protein predicted by the shorter isoform, which would only encode a single α-repeat domain and lack critical structures necessary for sodium calcium exchange, it seems unlikely that such an isoform would be generated. To investigate this further, we examined by RT-PCR whether a single mRNA could be detected bridging both predicted genes in each case. We designed primers that spanned the final exon of the upstream gene prediction and the first exon of the downstream predicted gene in the case of *C. japonica*, and the third to last exon of the upstream gene prediction and third exon of the downstream predicted gene in the case of *H. bacteriophora* (see red arrowheads in Figure 4C–4D). Using genomic DNA as template we observed a band at 2443 bp in the case of *C. japonica*, and using reverse transcribed RNA as template we observed a cDNA band at 813 bp (see inset of gel in Figure 4C). This suggests that together these predicted genes likely produce an individual mRNA. We sequenced the RT-PCR product and found that the cDNA sequence ended 5 bp prior to the currently annotated stop codon in the final exon of *Cja11479*, and then continued in what is currently annotated as noncoding sequence for 125 bp, and then once again continued 29 bp upstream of the currently annotated start codon of the *Cja38547* gene prediction (Figure 4C, blue rectangles indicate cDNA). BlastP interrogation of the *C. elegans* genome using the protein translation from this cDNA sequence provides a top match with the *C. elegans* NCX-6 predicted protein, which includes our cDNA sequence that covers what is currently annotated as non-coding sequence between each predicted gene. This suggests misannotation in the current gene structure prediction at this locus for *C. japonica*. We adopted the same approach to investigate the *H. bacteriophora* predicted genes, *Hba_19836* and *Hba_19835*, and similarly found that a single transcript could be detected that bridges both predicted genes, suggesting again the possibility of misannotation in the current gene structure prediction at this locus (Figure 4D). We observed a band at 1623 bp using gDNA as template and a band at 674 bp using cDNA as template. We sequenced this RT-PCR cDNA product and found that the cDNA sequence ended prior to the currently annotated stop codon in the final exon of *Hba_19836*, and then continued in what is currently annotated as non-coding sequence between both predicted genes, and then continued 5 bp upstream of the currently annotated start codon of the *Hba_19835* gene (Figure 4D, blue rectangles indicate cDNA). Taken together, this also suggests misannotation in the current gene structure predictions at this locus for *H. bacteriophora*. Therefore, in the analysis that follows on the CCX exchanger phylogeny we used the translation from concatenated *Cja11479*

and *Cja38547* gene predictions in the case of *C. japonica* and the translation from concatenated *H. bacteriophora* predicted genes *Hba_19835* and *Hba_19836* for our phylogenetic analyses. Our cDNA sequences for *C. japonica* and *H. bacteriophora* were deposited at NCBI's GenBank (accession number KJ873055 for *C. japonica* and KM009146 for *H. bacteriophora*).

The CCX nematode phylogeny revealed the most unexpected reconstruction, most notably is the gene expansion specific to the *Caenorhabditis* genus (Figure 5A). Within each *Caenorhabditis* species that we examined we detected between four and five CCX genes. We detected three CCX genes for *P. pacificus*, and all other species examined had either a single CCX member or no CCX representative. The CCX encoding genes from *H. contortus*, *H. bacteriophora*, *P. pacificus* and *S. ratti* were more divergent ('CCX div', Figure 5A) than those observed for *Caenorhabditis* species. Although the CCX group exhibited much diversity at the nucleotide level (Figure 3A and 3B), at the structural level this group is highly conserved as evident in the alignment of the α1 and α2 repeat domains alongside the human NCLX (Figure 5B), which comprise the GNGAPD motif for α1 and (A/S)N(S/C)(V/I)GD for the α2 repeat domain. We also examined selection across the CCX *Caenorhabditis* taxa and found a global dN/dS value = 0.1629 using SLAC ($p<0.01$), and a global dN/dS value = 0.1550 using MEME ($p<0.01$). We examined nucleotide diversity for each CCX gene and observed unique patterns of diversity (Figure 3A) but similar overall rates of variation (Figure 3B) for each CCX gene with the exception of *ncx-8*, which exhibited elevated levels of diversity compared with the other CCX exchangers, particularly within the region between each transport domain (Figure 3A). We identified two candidate sites undergoing positive selection: codon 17 which is positioned in the extracellular N terminal before the first TM segment, and codon 520 which is located in a large predicted intracellular loop prior to the second α repeat domain. We also implemented MEME to search for episodic diversification and found one significant ($p<0.01$) example at codon 754 of *C. briggsae* NCX-7 in close proximity to the second transport repeat domain.

Discussion

Within the *Caenorhabditis* genus we observed significant lineage specific expansions within the NCLX exchanger group, suggesting the possibility of relatively recent gene duplication events. Within the five *Caenorhabditis* species we examined, the NCLX-type genes *ncx-6* and *ncx-7* are positioned in tandem sequence within their respective physical maps; the *ncx-8* and *ncx-9* NCLX-type genes are also in tandem sequence in the *C. elegans* genome, and *ncx-10* is closely linked on the same arm of chromosome V in *C. elegans*; in *C. briggsae ncx-8, ncx-9*, and *ncx-10* genes are all located within 15 kb of each other on chromosome V; in *C. brenneri* and *C. remanei*, *ncx-9* and *ncx-10* are within 10 kb and 3 kb of each other respectively. This linkage organization for subsets of NCLX-type exchangers may lend support to the hypothesis that some of these genes have arose relatively recently within *Caenorhabditis* species. While these apparent serial and parallel gene duplication events may be relatively recent in terms of nematode evolution, data from our group suggests that at least in the case of *ncx-6* and *ncx-9*, these genes are contributing to important neuronal functions at the behavioral and developmental levels (Vishal Sharma, Katrin Bode, and D.O'H, unpublished data), suggesting that these duplicated genes have acquired neo-functionalized roles in the animal. It will be interesting moving forward to characterize mutants in exchangers of the other NCLX-type genes in an effort to understand how sequence specificity may lend itself to functional specializations within this

Na$^+$/Ca^{2+} exchanger subtype. Furthermore, searching more nematode genomes as they become more annotated will add more resolution to the timing of gene accretion within the NCLX subtype by also testing the alternative possibility that these NCLX-type duplicates may have been lost in other nematode lineages. It was also surprising that we did not detect NCLX-type orthologs within the genomes of the Clade III nematodes (*B. malayi*, *L. loa*, and *A. suum*) that we examined. Gene loss is one possibility to explain this observation, however, it is unexpected considering the central role NCLX proteins have been shown to play in mammalian systems [18,19,62]. Alternative hypotheses include, diversification or low sequence coverage, each of these scenarios may have precluded their detection using our approach, and further annotation and functional analysis will be required to resolve these questions. One place that we might find clues as to the function of Na$^+$/Ca^{2+} exchanger duplicates is in the case of NCX4 [59]: NCX4 has been found exclusively in teleost, amphibian, and reptilian genomes and is not present in mammalian genomes, and interestingly NCX4 is thought to have been lost from the mammalian genome [27,59]. NCX4 has been shown to function as an NCX-type exchanger, and in zebrafish is ubiquitously expressed with highest levels in the brain and eyes [60,61]. Reduction of NCX4 activity by morpholinos in zebrafish embryos has been shown to affect left-right patterning causing heterotaxia, situs inversus, as well as reversed cardiac looping [61]. These data demonstrate that functional specializations within the NCX family can vary significantly across species.

Na$^+$/Ca^{2+} exchangers are central regulators of calcium homeostasis in a wide variety of cell types. Not surprisingly, defects in Na$^+$/Ca^{2+} exchange have been implicated in numerous diseases and pathologies including epilepsy, multiple sclerosis, Parkinson's disease, Alzheimer's disease, as well as brain ischemia [21,63–66]. Understanding this family of proteins means also understanding differences within this family, and an entry point into this problem is using comparative genomics to resolve structural and taxonomic specializations. This is the approach we adopted here by identifying Na$^+$/Ca^{2+} exchanger genes from a broad spectrum of nematodes, and using this sequence data to reconstruct molecular phylogenetic relationships and tease apart the selective pressures shaping this family of proteins. From our analyses we uncover a pervasive theme of constraint across the Na$^+$/Ca^{2+} exchanger family and reveal a significant level of heterogeneity within subtypes of this family. Specifically, in the case of the NCLX subtype of Na$^+$/Ca^{2+} exchangers we observed lineage specific expansions as well as possible gene loss. Together, these findings reveal a complex picture of Na$^+$/Ca^{2+} transporters in nematodes that suggests an incongruent evolutionary history of an important family of proteins that provide central control of calcium dynamics.

Acknowledgments

We thank Theresa Stiernagle and Aric Daul for the Caenorhabditis Genetics Center strains used which is supported by the National Institutes of Health Office of Research Infrastructure Programs (P40 OD010440). We would like to thank The George Washington University Columbian College of Arts and Sciences, GW Office of the Vice-President for Research, and the Department of Biological Sciences for Funding to C.H and D.O'H.

Author Contributions

Conceived and designed the experiments: DO. Performed the experiments: CH DO. Analyzed the data: CH DO. Contributed reagents/materials/analysis tools: CH DO. Wrote the paper: DO.

References

1. Cai X, Lytton J (2004) The cation/ca(2+) exchanger superfamily: Phylogenetic analysis and structural implications. Mol Biol Evol 21: 1692–1703. 10.1093/molbev/msh177.

2. Baker PF, Blaustein MP, Hodgkin AL, Steinhardt RA (1969) The influence of calcium on sodium efflux in squid axons. J Physiol 200: 431–458.

3. Blaustein MP (1968) Barbiturates block sodium and potassium conductance increases in voltage-clamped lobster axons. J Gen Physiol 51: 293–307.

4. Cai X, Lytton J (2004) Molecular cloning of a sixth member of the K+-dependent na+/Ca2+ exchanger gene family, NCKX6. J Biol Chem 279: 5867–5876. 10.1074/jbc.M310908200.

5. Minelli A, Castaldo P, Gobbi P, Salucci S, Magi S, et al. (2007) Cellular and subcellular localization of na+-Ca2+ exchanger protein isoforms, NCX1, NCX2, and NCX3 in cerebral cortex and hippocampus of adult rat. Cell Calcium 41: 221–234. 10.1016/j.ceca.2006.06.004.

6. Gobbi P, Castaldo P, Minelli A, Salucci S, Magi S, et al. (2007) Mitochondrial localization of na+/Ca2+ exchangers NCX1-3 in neurons and astrocytes of adult rat brain in situ. Pharmacol Res 56: 556–565. 10.1016/j.phrs.2007.10.005.

7. Ren X, Philipson KD (2013) The topology of the cardiac na(+)/ca(2)(+) exchanger, NCX1. J Mol Cell Cardiol 57: 68–71. 10.1016/j.yjmcc.2013.01.010; 10.1016/j.yjmcc.2013.01.010.

8. Liao J, Li H, Zeng W, Sauer DB, Belmares R, et al. (2012) Structural insight into the ion-exchange mechanism of the sodium/calcium exchanger. Science 335: 686–690. 10.1126/science.1215759; 10.1126/science.1215759.

9. Giladi M, Boyman L, Mikhasenko H, Hiller R, Khananshvili D (2010) Essential role of the CBD1-CBD2 linker in slow dissociation of Ca2+ from the regulatory two-domain tandem of NCX1. J Biol Chem 285: 28117–28125. 10.1074/jbc.M110.127001 [doi].

10. Wu M, Wang M, Nix J, Hryshko LV, Zheng L (2009) Crystal structure of CBD2 from the drosophila na(+)/ca(2+) exchanger: Diversity of ca(2+) regulation and its alternative splicing modification. J Mol Biol 387: 104–112. 10.1016/j.jmb.2009.01.045 [doi].

11. Wu M, Le HD, Wang M, Yurkov V, Omelchenko A, et al. (2010) Crystal structures of progressive Ca2+ binding states of the Ca2+ sensor Ca2+ binding domain 1 (CBD1) from the CALX na+/Ca2+ exchanger reveal incremental conformational transitions. J Biol Chem 285: 2554–2561. 10.1074/jbc.M109.059162 [doi].

12. Nicoll DA, Quednau BD, Qui Z, Xia YR, Lusis AJ, et al. (1996) Cloning of a third mammalian na+-Ca2+ exchanger, NCX3. J Biol Chem 271: 24914–24921.

13. Li Z, Matsuoka S, Hryshko LV, Nicoll DA, Bersohn MM, et al. (1994) Cloning of the NCX2 isoform of the plasma membrane na(+)-Ca2+ exchanger. J Biol Chem 269: 17434–17439.

14. Altimimi HF, Szerencsei RT, Schnetkamp PP (2013) Functional and structural properties of the NCKX2 na(+)-ca (2+)/K (+) exchanger: A comparison with the NCX1 na (+)/ca (2+) exchanger. Adv Exp Med Biol 961: 81–94. 10.1007/978-1-4614-4756-6_8; 10.1007/978-1-4614-4756-6_8.

15. Lee SH, Kim MH, Park KH, Earm YE, Ho WK (2002) K+-dependent na+/Ca2+ exchange is a major Ca2+ clearance mechanism in axon terminals of rat neurohypophysis. J Neurosci 22: 6891–6899. 20026723.

16. Lytton J, Li XF, Dong H, Kraev A (2002) K+-dependent na+/Ca2+ exchangers in the brain. Ann N Y Acad Sci 976: 382–393.

17. Yang H, Yoo YM, Jung EM, Choi KC, Jeung EB (2010) Uterine expression of sodium/potassium/calcium exchanger 3 and its regulation by sex-steroid hormones during the estrous cycle of rats. Mol Reprod Dev 77: 971–977. 10.1002/mrd.21245; 10.1002/mrd.21245.

18. Palty R, Silverman WF, Hershfinkel M, Caporale T, Sensi SL, et al. (2010) NCLX is an essential component of mitochondrial na+/Ca2+ exchange. Proc Natl Acad Sci U S A 107: 436–441. 10.1073/pnas.0908099107; 10.1073/pnas.0908099107.

19. Cai X, Lytton J (2004) Molecular cloning of a sixth member of the K+-dependent na+/Ca2+ exchanger gene family, NCKX6. J Biol Chem 279: 5867–5876. 10.1074/jbc.M310908200.

20. Palty R, Hershfinkel M, Sekler I (2012) Molecular identity and functional properties of the mitochondrial na+/Ca2+ exchanger. J Biol Chem 287: 31650–31657. 10.1074/jbc.R112.355867; 10.1074/jbc.R112.355867.

21. Levesque PC, Leblanc N, Hume JR (1991) Role of reverse-mode na(+)-Ca2+ exchange in excitation-contraction coupling in the heart. Ann N Y Acad Sci 639: 386–397.

22. Jeon D, Yang YM, Jeong MJ, Philipson KD, Rhim H, et al. (2003) Enhanced learning and memory in mice lacking na+/Ca2+ exchanger 2. Neuron 38: 965–976.

23. Molinaro P, Cataldi M, Cuomo O, Viggiano D, Pignataro G, et al. (2013) Genetically modified mice as a strategy to unravel the role played by the na(+)/ca (2+) exchanger in brain ischemia and in spatial learning and memory deficits. Adv Exp Med Biol 961: 213–222. 10.1007/978-1-4614-4756-6_18; 10.1007/978-1-4614-4756-6_18.

24. Sharma V, O'Halloran D (2014) Recent structural and functional insights into the family of sodium calcium exchangers. genesis, The journal of Genetics and Development 52: 93.

25. Sharma V, He C, Sacca-Schaeffer J, Brzozowski E, Herranz DM, et al. (2013) Insight into the family of na+/Ca2+ exchangers of caenorhabditis elegans. Genetics. 10.1534/genetics.113.153106.

26. Blaxter ML, De Ley P, Garey JR, Liu LX, Scheldeman P, et al. (1998) A molecular evolutionary framework for the phylum nematoda. Nature 392: 71–75. 10.1038/32160 [doi].

27. On C, Marshall CR, Chen N, Moyes CD, Tibbits GF (2008) Gene structure evolution of the na+-Ca2+ exchanger (NCX) family. BMC Evol Biol 8: 127-2148-8-127. 10.1186/1471-2148-8-127 [doi].

28. Harris TW, Baran J, Bieri T, Cabunoc A, Chan J, et al. (2014) WormBase 2014: New views of curated biology. Nucleic Acids Res 42: D789–93. 10.1093/nar/gkt1063 [doi].

29. Flicek P, Amode MR, Barrell D, Beal K, Billis K, et al. (2014) Ensembl 2014. Nucleic Acids Res 42: D749–55. 10.1093/nar/gkt1196 [doi].

30. Martin J, Abubucker S, Heizer E, Taylor CM, Mitreva M (2012) Nematode.net update 2011: Addition of data sets and tools featuring next-generation sequencing data. Nucleic Acids Res 40: D720–8. 10.1093/nar/gkr1194 [doi].

31. Ostlund G, Schmitt T, Forslund K, Kostler T, Messina DN, et al. (2010) InParanoid 7: New algorithms and tools for eukaryotic orthology analysis. Nucleic Acids Res 38: D196–203. 10.1093/nar/gkp931 [doi].

32. Li L, Stoeckert CJ Jr, Roos DS (2003) OrthoMCL: Identification of ortholog groups for eukaryotic genomes. Genome Res 13: 2178–2189. 10.1101/gr.1224503 [doi].

33. Hunter S, Jones P, Mitchell A, Apweiler R, Attwood TK, et al. (2012) InterPro in 2011: New developments in the family and domain prediction database. Nucleic Acids Res 40: D306-12. 10.1093/nar/gkr948 [doi].

34. Liebich I, Bode J, Frisch M, Wingender E (2002) S/MARt DB: A database on scaffold/matrix attached regions. Nucleic Acids Res 30: 372–374.

35. Edgar RC (2004) MUSCLE: Multiple sequence alignment with high accuracy and high throughput. Nucleic Acids Res 32: 1792–1797. 10.1093/nar/gkh340.

36. Darriba D, Taboada GL, Doallo R, Posada D (2011) ProtTest 3: Fast selection of best-fit models of protein evolution. Bioinformatics 27: 1164–1165. 10.1093/bioinformatics/btr088; 10.1093/bioinformatics/btr088.

37. Abascal F, Zardoya R, Posada D (2005) ProtTest: Selection of best-fit models of protein evolution. Bioinformatics 21: 2104–2105. 10.1093/bioinformatics/bti263.

38. Guindon S, Gascuel O (2003) A simple, fast, and accurate algorithm to estimate large phylogenies by maximum likelihood. Syst Biol 52: 696–704.

39. O'Halloran D (2014) A practical guide to phylogenetics for nonexperts. J Vis Exp (84): e50975. doi: e50975. 10.3791/50975 [doi].

40. Kosakovsky Pond SL, Frost SD (2005) Not so different after all: A comparison of methods for detecting amino acid sites under selection. Mol Biol Evol 22: 1208–1222. msi105 [pii].

41. Murrell B, Wertheim JO, Moola S, Weighill T, Scheffler K, et al. (2012) Detecting individual sites subject to episodic diversifying selection. PLoS Genet 8: e1002764. 10.1371/journal.pgen.1002764 [doi].

42. Pond SL, Frost SD, Muse SV (2005) HyPhy: Hypothesis testing using phylogenies. Bioinformatics 21: 676–679. bti079 [pii].

43. Delport W, Poon AF, Frost SD, Kosakovsky Pond SL (2010) Datamonkey 2010: A suite of phylogenetic analysis tools for evolutionary biology. Bioinformatics 26: 2455–2457. 10.1093/bioinformatics/btq429 [doi].

44. Darriba D, Taboada GL, Doallo R, Posada D (2012) jModelTest 2: More models, new heuristics and parallel computing. Nat Methods 9: 772. 10.1038/nmeth.2109 [doi].

45. Suyama M, Torrents D, Bork P (2006) PAL2NAL: Robust conversion of protein sequence alignments into the corresponding codon alignments. Nucleic Acids Res 34: W609-12. 34/suppl_2/W609 [pii].

46. Kearse M, Moir R, Wilson A, Stones-Havas S, Cheung M, et al. (2012) Geneious basic: An integrated and extendable desktop software platform for the organization and analysis of sequence data. Bioinformatics 28: 1647–1649. 10.1093/bioinformatics/bts199 [doi].

47. Librado P, Rozas J (2009) DnaSP v5: A software for comprehensive analysis of DNA polymorphism data. Bioinformatics 25: 1451–1452. 10.1093/bioinformatics/btp187 [doi].

48. Kelley LA, Sternberg MJ (2009) Protein structure prediction on the web: A case study using the phyre server. Nat Protoc 4: 363–371. 10.1038/nprot.2009.2 [doi].

49. Sayle RA, Milner-White EJ (1995) RASMOL: Biomolecular graphics for all. Trends Biochem Sci 20: 374. S0968-0004(00)89080-5 [pii].

50. Brenner S (1974) The genetics of caenorhabditis elegans. Genetics 77: 71–94.

51. Kiontke K, Hironaka M, Sudhaus W (2002) Description of caenorhabditis japonica n. sp. (nematoda: Rhabditida) associated with the burrower bug parastrachia japonensis (heteroptera: Cydnidae) in japan. Nematology 4: 933.

52. Woodring JL, Kaya HK (1998) Steinernematid and heterorhabditid nematodes: A handbook of biology and techniques. Southern Cooperative Series Bulletin Arkansas Agricultural Experiment Station, Fayetteville, Arkansas 331.

53. White GF (1927) A method for obtaining infective nematode larvae from cultures. Science 66: 302–303. 66/1709/302-a [pii].

54. O'Halloran DM, Burnell AM (2003) An investigation of chemotaxis in the insect parasitic nematode heterorhabditis bacteriophora. Parasitology 127: 375–385.

55. O'Leary SA, Stack CM, Chubb MA, Burnell AM (1998) The effect of day of emergence from the insect cadaver on the behavior and environmental tolerances of infective juveniles of the entomopathogenic nematode heterorhabditis megidis (strain UK211). J Parasitol 84: 665–672.

56. Kang KJ, Shibukawa Y, Szerencsei RT, Schnetkamp PP (2005) Substitution of a single residue, Asp575, renders the NCKX2 K+-dependent na+/Ca2+ exchanger independent of K+. J Biol Chem 280: 6834–6839. 10.1074/jbc.M412933200.

57. Schulze DH, Polumuri SK, Gille T, Ruknudin A (2002) Functional regulation of alternatively spliced na+/Ca2+ exchanger (NCX1) isoforms. Ann N Y Acad Sci 976: 187–196.

58. Quednau BD, Nicoll DA, Philipson KD (1997) Tissue specificity and alternative splicing of the na+/Ca2+ exchanger isoforms NCX1, NCX2, and NCX3 in rat. Am J Physiol 272: C1250–61.

59. Marshall CR, Fox JA, Butland SL, Ouellette BF, Brinkman FS, et al. (2005) Phylogeny of na+/Ca2+ exchanger (NCX) genes from genomic data identifies new gene duplications and a new family member in fish species. Physiol Genomics 21: 161–173. 00286.2004 [pii].

60. On C, Marshall CR, Perry SF, Le HD, Yurkov V, et al. (2009) Characterization of zebrafish (danio rerio) NCX4: A novel NCX with distinct electrophysiological properties. Am J Physiol Cell Physiol 296: C173–81. 10.1152/ajpcell.00455.2008 [doi].

61. Shu X, Huang J, Dong Y, Choi J, Langenbacher A, et al. (2007) Na, K-ATPase alpha2 and Ncx4a regulate zebrafish left-right patterning. Development 134: 1921–1930. dev.02851 [pii].

62. Nita II, Hershfinkel M, Fishman D, Ozeri E, Rutter GA, et al. (2012) The mitochondrial Na+/Ca2+ exchanger upregulates glucose dependent Ca2+ signalling linked to insulin secretion. PLoS One 7: e46649. 10.1371/journal.pone.0046649; 10.1371/journal.pone.0046649.

63. Pannaccione A, Secondo A, Molinaro P, D'Avanzo C, Cantile M, et al. (2012) A new concept: Abeta1-42 generates a hyperfunctional proteolytic NCX3 fragment that delays caspase-12 activation and neuronal death. J Neurosci 32: 10609–10617. 10.1523/JNEUROSCI.6429-11.2012; 10.1523/JNEUROSCI.6429-11.2012.

64. Kovacs R, Kardos J, Heinemann U, Kann O (2005) Mitochondrial calcium ion and membrane potential transients follow the pattern of epileptiform discharges in hippocampal slice cultures. J Neurosci 25: 4260–4269. 10.1523/JNEUROSCI.4000-04.2005.

65. Wood-Kaczmar A, Deas E, Wood NW, Abramov AY (2013) The role of the mitochondrial NCX in the mechanism of neurodegeneration in parkinson's disease. Adv Exp Med Biol 961: 241–249. 10.1007/978-1-4614-4756-6_20 [doi].

66. Sisalli MJ, Secondo A, Esposito A, Valsecchi V, Savoia C, et al. (2014) Endoplasmic reticulum refilling and mitochondrial calcium extrusion promoted in neurons by NCX1 and NCX3 in ischemic preconditioning are determinant for neuroprotection. Cell Death Differ. 10.1038/cdd.2014.32 [doi].

15

Identification of Functionally Important Residues of the Rat P2X4 Receptor by Alanine Scanning Mutagenesis of the Dorsal Fin and Left Flipper Domains

Vendula Tvrdonova[1,2], Milos B. Rokic[1,3], Stanko S. Stojilkovic[3], Hana Zemkova[1]*

1 Department of Cellular and Molecular Neuroendocrinology, Institute of Physiology Academy of Sciences of the Czech Republic, Prague, Czech Republic, 2 Department of Physiology of Animals, Faculty of Science, Charles University, Prague, Czech Republic, 3 Section on Cellular Signaling, Program in Developmental Neuroscience, National Institute of Child Health and Human Development, National Institutes of Health, Bethesda, Maryland, United States of America

Abstract

Crystallization of the zebrafish P2X4 receptor in both open and closed states revealed conformational differences in the ectodomain structures, including the dorsal fin and left flipper domains. Here, we focused on the role of these domains in receptor activation, responsiveness to orthosteric ATP analogue agonists, and desensitization. Alanine scanning mutagenesis of the R203-L214 (dorsal fin) and the D280-N293 (left flipper) sequences of the rat P2X4 receptor showed that ATP potency/efficacy was reduced in 15 out of 26 alanine mutants. The R203A, N204A, and N293A mutants were essentially non-functional, but receptor function was restored by ivermectin, an allosteric modulator. The I205A, T210A, L214A, P290A, G291A, and Y292A mutants exhibited significant changes in the responsiveness to orthosteric analog agonists 2-(methylthio)adenosine 5'-triphosphate, adenosine 5'-(γ-thio)triphosphate, 2'(3'-O-(4-benzoylbenzoyl)adenosine 5'-triphosphate, and α,β-methyleneadenosine 5'-triphosphate. In contrast, the responsiveness of L206A, N208A, D280A, T281A, R282A, and H286A mutants to analog agonists was comparable to that of the wild type receptor. Among these mutants, D280A, T281A, R282A, H286A, G291A, and Y292A also exhibited increased time-constant of the desensitizing current response. These experiments, together with homology modeling, indicate that residues located in the upper part of the dorsal fin and left flipper domains, relative to distance from the channel pore, contribute to the organization of the ATP binding pocket and to the initiation of signal transmission towards residues in the lower part of both domains. The R203 and N204 residues, deeply buried in the protein, may integrate the output signal from these two domains towards the gate. In addition, the left flipper residues predominantly account for the control of transition of channels from an open to a desensitized state.

Editor: Jon Brown, University of Exeter, United Kingdom

Funding: This work was supported by the Grant Agency of the Czech Republic (P304/12/G069), the Centrum of Biomedicine Research (CZ.1.07/2.3.00/30.0025), the "BIOCEV" project with the Biotechnology and Biomedicine Centre of the Academy of Sciences and Charles University in Vestec (CZ.1.05/1.1.00/02.0109) from the European Regional Development Fund, the Grant Agency of Charles University in Prague (3446/2011), the Academy of Sciences of the Czech Republic (Research Project No. RVO 67985823, and the Intramural Research Program of the NICHD, NIH. The funders had no role in study design, data collection and analysis, decision to publish, or preparation of the manuscript.

Competing Interests: The authors declare that no competing interests exist.

* Email: zemkova@biomed.cas.cz

Introduction

The purinergic P2X receptors (P2XRs) are ATP-gated ion channels that are permeable to Na^+, K^+, Ca^{2+}, and small organic cations. Seven subunits of P2XRs have been identified in mammals [1], and functional receptors are composed of three homologous or heterologous subunits [2]. Each subunit consists of a large, glycosylated, and cystine-rich extracellular domain that contributes to the formation of the intersubunit ATP binding sites, two transmembrane domains that form the pore of the channel, and intracellular N- and C- termini that contribute to gating specificity [3]. Previous studies using single-point mutagenesis have identified most of conserved amino acid residues involved in ATP binding and have shown that ATP binding occurs at the interface between adjacent receptor subunits, assuming that ATP stabilizes the P2X trimer [4–10]. In contrast to the large number of studies using the native ligand, ATP, there are very few studies providing structural information derived from the use of orthosteric ATP analog agonists. Understanding receptor interactions with these analog agonists may provide significant insights aiding the design of drugs that compete with the native ligand.

The recent crystallization of the zebrafish P2X4R receptor (zfP2X4R) in the absence (closed state; PDB entry codes: 3H9V and 4DW0) and presence (open state; 4DW1) of ATP has confirmed the predicted topology and locations of the ATP binding sites in P2XRs. The authors suggest that the architecture of the P2XRs resembles a dolphin, with a rigid central extracellular body domain, a flexible head, a left flipper (LF), a right flipper, and a dorsal fin (DF). The crystal structure of zfP2XR in the apo-closed state and the ATP-bound open state has also provided structural insights into the mechanisms of ATP binding, the opening of ion channel pore, and a series of conformational changes associated with channel gating [11,12]. These insights have enabled a better understanding of precrystallization studies focused on the structural-functional characterization of P2XR transmembrane domains [13–21] and facilitated further studies focused on extracellular vestibule function [22–24] and molecular dynamics to model conformation transitions [25].

Following ATP binding, the head, upper body, and LF domains of one subunit and the lower body and DF domains of another subunit undergo marked movement that results in the closing of the ATP binding site jaw [11]. During this movement, the LF and DF domains remain in close proximity (Fig. 1B). This promotes expansion of the upper vestibule, leading to the activation of P2XRs [26]. The P2X6R receptor lacks most of the LF domain (Fig. 1A) and is incapable of forming functional homomeric channels [27]. However, it can form functional heteromeric channels with P2X2 and P2X4 subunits [28,29], which may indicate that one or two complete LF domains per receptor are needed to activate the channel after ATP binding.

Crystallographic data also indicate that the DF and LF domains are intrinsically unfolded and lack secondary structures. These regions have significant conformational flexibility due to higher Debye-Waller factors (B-factors; Fig. 1B). This reflects the thermal fluctuation of atoms in zfP2X4R crystals, as assessed by X-ray scattering techniques, around their average positions and provide important information about protein dynamics [30]. Most importantly, the specific role(s) of non-structural and low-conserved DF and LF regions (Fig. 1A) is not well understood. In particular, we do not know the physiological relevance of having these domains positioned between an ATP binding site and the downstream K313-I333 β-sheet that has been previously identified in rat P2X4R (rP2X4R) as important for transmission of signal from the binding site to the channel gate [31].

We examined the hypothesis that the DF and LF domains may influence the organization of the ATP binding pocket, transmission of ATP-induced signal from ATP binding pocket to the gate, and receptor desensitization. To do this, we used 26 mutants generated by alanine scanning mutagenesis of the R203-L214 (DF) and D280-N293 (LF) sequences. These regions are highly variable between P2XRs, and only few of these residue mutants have been previously characterized electrophysiologically (Table 1). We expressed the wild type (WT) rP2X4R and alanine mutants in HEK293 cells and studied the current responses induced by the application and withdrawal of ATP or its analog agonists 2-(Methylthio)adenosine 5′-triphosphate (2-MeS-ATP), Adenosine 5′-(γ-thio)triphosphate (ATPγS), 2′(3′-O-(4-Benzoylbenzoyl)adenosine 5′-triphosphate (BzATP), and α,β-methyleneadenosine 5′-triphosphate (α,β-meATP), both in the presence and absence of ivermectin (IVM), an allosteric regulator of P2X4R [32–34].

Methods

Cells culture and transfection

To express the recombinant channels, we used human embryonic kidney (HEK) 293T cells (American Type Culture Collection, Rockville, MD, USA) grown in Dulbecco modified Eagle's medium (Thermo Fisher Scientific, Waltham, MA) supplemented with 10% fetal bovine serum (Sigma-Aldrich, St Louis, MO), 50 U/ml penicillin and 50 μg/ml streptomycin (both Thermo Fisher Scientific, Waltham, MA) in a humidified 5% CO_2 and 95% air at 37°C. Cells were cultured in 75 cm^2 plastic culture flasks (NUNC, Rochester, NY) for 36–72 hours until reaching 80–95% confluence. Before the day of transfection, the cells were plated on 35 mm culture dishes (Sarstedt, Newton, NC) and

incubated at 37°C for at least 24 h. Transfection was done using 2 μg of either WT or mutant receptor DNA with 2 μl of JetPrime reagent in 2 ml of Dulbecco modified Eagle's medium, according to manufacturer's instructions (PolyPlus-transfection, Illkirch, France). Transfected cells were identified by the fluorescence signal of EGFP using the Olympus IX71 inverted research microscope with fluorescence illuminators (Model IX71; Olympus, Melville, NY).

DNA constructs

cDNAs encoding the sequences of the rP2X4 and mutated subunits were subcloned into the pIRES2-EGFP vector (Clontech, Mountain View, CA, USA). To generate the mutants, oligonucleotides (synthesized and provided by VBC-Genomics, Vienna, Austria or Sigma Aldrich) containing specific mutagenesis mismatches were introduced into the rP2X4/pIRES2-EGFP template using PfU Ultra DNA polymerase (Thermo Fisher Scientific). A High-Speed Plasmid Mini Kit (Geneaid, Taipei City, Taiwan) was used to isolate the plasmids for transfection. Dye terminator cycle sequencing (ABI PRISM 3100, Applied Biosystems, Foster City, CA) was used to identify and verify the presence of the mutations. The sequencing was performed by the DNA Sequencing Laboratory, Institute of Microbiology, ASCR, Prague.

Patch clamp recordings

ATP-induced currents were recorded from whole cells clamped to −60 mV using an Axopatch 200B patch-clamp amplifier (Axon Instruments, Union City, CA). The recordings were captured and stored using the Digidata 1322A and pClamp9 software package (Axon Instruments). During the experiments, the cell culture was perfused with a bath solution containing: 142 mM NaCl, 3 mM KCl, 2 mM CaCl$_2$, 1 mM MgCl$_2$, 10 mM 4-(2-Hydroxyethyl)piperazine-1-ethanesulfonic acid (HEPES) and 10 mM D-glucose, adjusted to pH 7.3 with 1 M NaOH. The patch electrodes were filled with a solution containing: 154 mM CsCl, 11 mM EGTA and 10 mM HEPES, adjusted to pH 7.2 with 1.6 M CsOH. The whole-cell configuration was used to abolish the influence of natively present metabotropic receptors for ATP and we used intracellular cesium to block any kind of possible background potassium conductance. Potency of ATP was measured based on the activation of naïve (not previously stimulated) receptors using a short (2–5 s) application of various concentrations of ATP. The results are expressed as molar concentration of ATP required to produce 50% of the maximal response (EC$_{50}$). One or two responses were recorded from one cell, if not otherwise stated, and responses from different cells were pooled. The maximum current amplitude (I$_{max}$) was measured in response to application of supramaximal concentrations (100–1000 μM) of ATP. Responsiveness to P2XR agonists 2-MeS-ATP, ATPγS, BzATP, and α,β-meATP, all applied in 100 μM concentration, was expressed as percentage of response in comparison to 100 μM ATP treatment for selected mutants. In all mutants, the whole cell currents were also measured in the presence of 3 μM IVM, which was dissolved in dimethyl sulfoxide, stored in stock solutions at 10 mM, and diluted to required concentrations in bath solution in the day of experiment. The control and drug containing solutions were applied via a rapid (exchange time 30–40 ms) perfusion system (RSC-200, BIOLOGIC, Claix, France). All other chemicals are from Sigma-Aldrich.

Calculations

The concentration-response data points were fitted with the equation $y = I_{max}/[1 + (EC_{50}/x)^h]$, where y is the amplitude of the current evoked by ATP, I_{max} is the maximum current amplitude

A

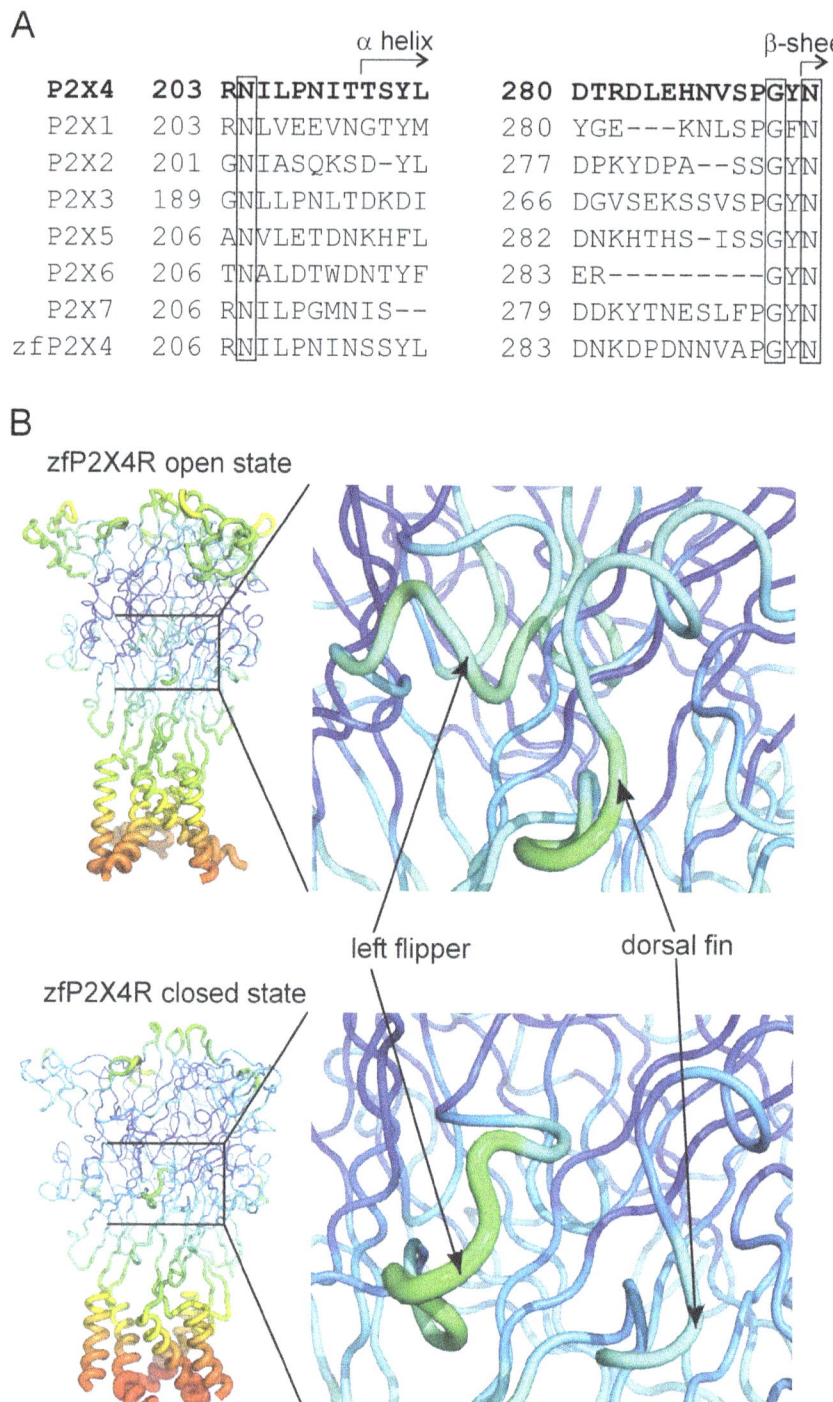

                                α helix   β-sheet

```
P2X4    203 RNILPNITTSYL      280 DTRDLEHNVSPGYN
P2X1    203 RNLVEEVNGTYM      280 YGE---KNLSPGFN
P2X2    201 GNIASQKSD-YL      277 DPKYDPA--SSGYN
P2X3    189 GNLLPNLTDKDI      266 DGVSEKSSVSPGYN
P2X5    206 ANVLETDNKHFL      282 DNKHTHS-ISSGYN
P2X6    206 TNALDTWDNTYF      283 ER--------GYN
P2X7    206 RNILPGMNIS--      279 DDKYTNESLFPGYN
zfP2X4  206 RNILPNINSSYL      283 DNKDPDNNVAPGYN
```

B

Figure 1. Structural and tridimensional organization of the DF and LF domains. (A) Alignment of amino acid sequences from R203-L214 (DF domain) and D280-N293 (LF domain) using seven rP2X and zfP2X subunits. Structurally, these regions are composed of random coils that terminate with short α-helix and β-sheet structures (indicated by arrows). Conserved amino acid residues are shown in boxes. (B) The models of the zfP2X4.1R are shown in the open (4DW1) or closed (4DW0) state, and the Debye-Waller factor (B-factor) indicates the degree to which the electron density is spread (miniatures). The model shows the elevated B-factor values within the region of intersubunit interaction (magnified segments). Higher B-factors are indicated with thicker cylinders and a red-shifted color, while the lowest B-factors are represented with the thinnest cylinders and a blue-shifted color.

induced by 100–1000 µM ATP, EC_{50} is the agonist concentration producing 50% of the maximal response, h is the Hill coefficient, and x is the concentration of ATP (SigmaPlot 2000 v9.01; SPSS Inc., Chicago, IL). Hill coefficient was fixed to 1.3 in all experiments, a value obtained for the WT receptor by fitting. The kinetics of deactivation (current decay evoked by washout of

Table 1. Summary of the changes in estimated EC_{50} values for ATP and changes in desensitization at the DF and LF alanine/cysteine mutants of P2X1-4R residues from published data.

P2X1R	P2X2R	P2X3R	P2X4R
DF			
N204A: ↑ 3,5x[45]	N202A: n.f.[6]	-	N204A: n.i.
-	-	-	I205A: ↑ 2,5x[49]
-	-	-	L206A: ≈[49]
-	-	T196A: ≈[54]	T210A: n.i.
-	-	E197A: D[53]	T211A: n.i.
-	-	M200C: ↑ 8x[58]	L214A: ↑ 8x[41], 3x[49]
LF			
-	D277A:: ≈[56]	D266A: D[53]	D280A ↑ >100x[10,52]
-	-	S267A: ≈[59]	T281A: n.i.
-	-	S269A: D[54]	D283A: n.i.
-	-	-	H286A: ↑ 2-4x[10,52,57,60]
-	-	-	N287A: ≈[10,52]
-	-	-	V288A: ↑ 2x[49]
-	-	S275A: D[55]	S289A: n.i.
P287C: ≈[9]	-	-	P290A: n.i.
G288A/C: ≈[47], ↑ 10x[9]	-	-	G291A: n.i.
N290A: ↑ >50x[45]	N288A: ↑ >100x[6]	N279A: ↑ 20x[4]	N293A: n.i.

n. f., non-functional mutants; ↑ mutant with significant increased EC_{50} in comparison to WT (values represent fold increase); ≈ close to WT receptor; D, affected time-constant of the desensitizing current response; -, non-investigated position; n.i., non-investigated P2X4R mutants that were analyzed in this study.

agonist) and desensitization (current decay in the continuous presence of agonist) were fitted by a single exponential function $(y = A_1 \exp(-t/\tau_1))$ or by the sum of two exponentials $(y = A_1 \exp(-t/\tau_2) + A_2 \exp(-t/\tau_2))$, respectively, using the program Clampfit 10 (Axon Instruments), where A_1 and A_2 are the relative amplitudes of the first and second exponentials, and τ_1 and τ_2 are the time constants. The derived time constants for deactivation and desensitization were labeled as τ_{off} and τ_{des}, respectively. Weight desensitization constant was calculated as $y = [(A_1\tau_{des1}) + (A_2\tau_{des2})]/(A_1 + A_2)$. Correlation coefficient was calculated using linear regression wizard (SigmaPlot 2000 v9.01). Data points are presented as mean ± SEM values. Significant differences (**p< 0.01 and *p<0.05) between means were determined using SigmaStat 2000 v9.01. The data for alanine mutants were analyzed by an ANOVA and Tukey's post hoc test.

Homology modeling

The rP2X4R (P51577) and zfP2X4.1R (Q6NYR1) share 61% identity at the amino-acid level, measured with the Basic Local Alignment Search Tool (BLAST; The UniProt Knowledgebase), a value sufficient to build a homology model of the rP2X4R using the automated mode of the SWISS-MODEL server [35]. We extracted a tertiary structure template from the Brookhaven Protein Data Bank under the accession number 4DW0 for the receptor in the apo-closed state and 4DW1 for the zfP2X4.1R in the ATP-bound open state. Model quality was estimated by a SWISSMODEL through a Qualitative Model Energy Analysis (QMEAN) score, which represents a composite scoring function describing the major geometrical aspects of protein structures by taking into consideration five different structural descriptors [36], and which was 0.593. The graphical representations of the protein

structure were prepared using PyMOL software (DeLano Scientific LLC, USA).

Results

1. Identification of DF and LF mutants with affected ATP potency and efficacy

To address the structure-function relationship between the LF and DF regions of rP2X4R, we performed single-point mutagenesis on sequences encompassing the LF and DF regions R203-L214 (DF) and D280-N293 (LF) (Fig. 1A). The crystal structure of the zfP2X4R showed elevated B-factor values in these regions, indicating conformational flexibility (Fig. 1B). For the initial electrophysiological characterization, we examined the EC_{50} and I_{max} values to determine ATP potency and efficacy, respectively. The results from experiments on both mutant and WT receptors are summarized in Figs. 2A and 3 and Table S1.

There were no significant effects on the EC_{50} or I_{max} values for P207A, I209A, T211A, S212A, Y213A, D283A, L284A, E285A, N287A, V288A, and S289A mutant receptors. The EC_{50} values for mutants R203A, N204A, and N293A could not be determined because they displayed a very low ATP-induced current ($I_{max} \leq$ 0.2 nA). Mutants I205A, L214A, D280A, R282A, and P290A showed a significant reduction (p<0.01) in I_{max}, and with exception of P290A, their EC_{50} values were approximately 10-fold rightward shifted when compared to the WT receptor. The EC_{50} values for T210A, H286A, and G291A mutants were 6- to 10-fold rightward shifted but these receptors showed no significant difference in I_{max} values when compared to the WT receptor (Fig. 3). Slightly less significant increases (p<0.05) in EC_{50} values were observed in the L206A, N208A, T281A, P290A, and Y292A mutants. With the exception of P290A, none of these mutants

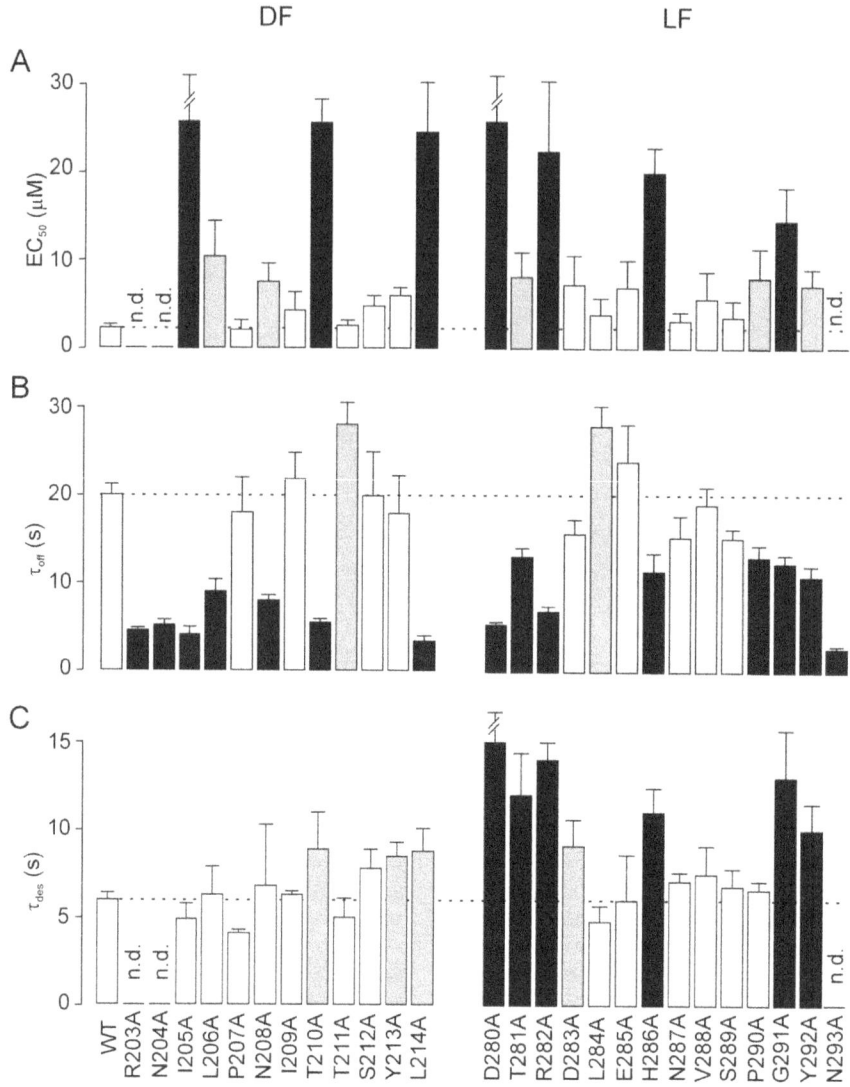

Figure 2. Characterization of rP2X4R–DF and -LF residue mutants. Effect of alanine substitutions on the potency of ATP (A), deactivation (B), and desensitization (C) kinetics. Summary histograms show the concentration of ATP producing a half-maximal current (EC$_{50}$), deactivation time constants (τ_{off}) were estimated by the monoexponential fit of the decay of current in response to 2 s of stimulation with 1–3 µM ATP after 4–6 min of preincubation with 3 µM IVM and desensitization time constants (τ_{des}) were derived from the biexponential fit of the response to 60 s of stimulation with 100 µM ATP for WT and alanine mutants of the dorsal fin (DF) and the left flipper (LF) domains. Values shown (and also given in Table S1) are the mean ± SEM of 21–63 measurements per mutant and 267 measurements for the WT. Significant differences between the WT and the mutant receptors are shown in gray (p<0.05) or black (p<0.01) columns. Horizontal dotted lines illustrate the values for WT receptor and n.d. indicates that the value could not be determined.

exhibited significant changes in their I$_{max}$ value. Thus, in 15 of 26 alanine mutants located at the interface of the LF and DF domains the potency and/or efficacy of ATP was significantly reduced, indicating the relevance of these residues in receptor functions.

2. IVM rescues the low-functioning DF- and LF-rP2X4R mutants

The effect of IVM on I$_{max}$ was tested initially during ongoing responses to 100 µM ATP to determine whether the low current amplitudes observed in R203A, N204A, I205A, L214A, D280A, R282A, P290A, and N293A mutants could be rescued. The application of 3 µM IVM increased immediately the amplitude of ATP-induced responses in all low-functioning mutants (Fig. 4A). Next, we performed quantitative analysis of I$_{max}$ in WT and all

alanine mutants before and after 4–6 min pretreatment with IVM (Table S1). The WT receptor was potentiated 1.5-fold by IVM, while the low-functioning mutants were potentiated 3.7- to 16-fold. In the presence of IVM, the I$_{max}$ values of all low-functioning mutants were comparable to those of WT receptors, except for N293A (Fig. 4B). These experiments indicate that R203, N204, I205, L214, D280, R282, R290, and N293 residues play a critical role in agonist binding and/or channel gating.

The IVM-induced rescue of I$_{max}$ values made it possible to examine the deactivation time constant (τ_{off}) for these mutants, which inversely correlates with EC$_{50}$ values [37,38]. As a result, we were able to more precisely characterize the potency of ATP under comparable conditions. A prolongation of current decay comparable to that observed in WT receptor would suggest that normal ATP potency has been maintained. Alternatively, a

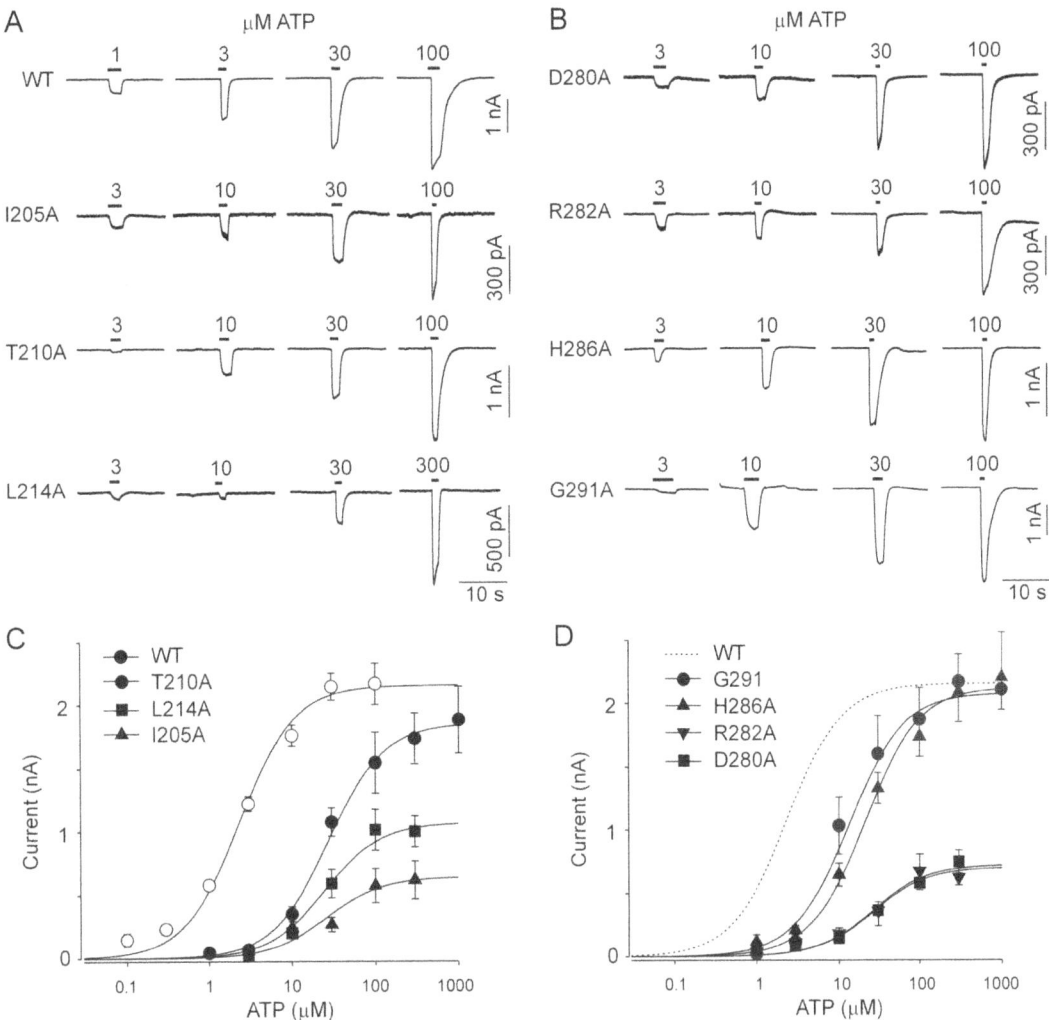

Figure 3. DF and LF mutants exhibit a rightward shift in EC$_{50}$. (A, B) Example records of ATP-induced currents from cells expressing the WT receptor and I205A, T210A, and L214A DF mutants (A) and D280A, R282A, H286A, and G291A LF mutants (B). Currents were stimulated by a short (2–5 s) application of different concentrations of ATP (1–1000 μM), indicated by horizontal bars above the traces. Experiments were performed on naïve receptors, and traces from different cells are shown. (C, D) Concentration response curves for WT, I205A, T210A, and L214A DF mutants (C) and D280A, R282A, H286A, and G291A LF mutants (D). Data points are presented as the mean ± SEM from 7–35 measurements per mutant, per concentration and 78 measurements for WT.

decrease or increase in the rate of decay would argue for reduced or enhanced ATP potency, respectively [10]. The deactivation time constant was examined in all alanine mutants by mono-exponential fitting of the decay of current after washout of a non-desensitizing concentration of agonist (1 or 3 μM ATP) in the presence of 3 μM IVM. Example traces from WT and mutant receptors with changed deactivations are shown in Fig. S1. The results of τ_{off} measurements are summarized in Fig. 2B and Table S1.

In parallel with the rightward shift changes in EC$_{50}$ values, we observed a significantly ($p<0.01$) accelerated rate of deactivation in non-responding mutants (R203A, N204A, and N293A) and all rightward shifted mutants (I205A, L206A, N208A, T210A, L214A, D280A, T281A, R282A, H286A, P290A, G291A, and Y292A). Less significant ($p<0.05$) prolonged deactivation times were found in the L284A and T211A mutants. There was a significant correlation, with highly comparable slopes, between the EC$_{50}$ vs. τ_{off} values for both DF and LF mutants (Fig. 5A). These data confirmed that residues in both domains contribute

significantly to receptor activation as well as to receptor deactivation, i.e., that deactivation is a reverse process occurring through the same signal transmission pathway.

3. Desensitization kinetics of LF- and DF-rP2X4R mutants

Next, we determined the desensitization kinetics of alanine mutants at the interface of the LF and DF domains. In the presence of 100 μM ATP for 60 s, the WT receptor current declined biexponentially with $\tau_{des1}=1.3\pm0.2$ s and $\tau_{des2}=9.0\pm0.7$ s; the slow component contributed to the decay with $63\pm3.0\%$ (Table S2; Fig. S1). The decay of current was also biexponential in all mutants, but in some cases monoexponential fit was the best, and we used a weighted desensitization time constant (τ_{des}) for comparison between the mutants and the P2X4R-WT (WT, $\tau_{des}=6.0\pm0.4$ s; Table S1 and Fig. 2C). The LF mutants D280A, T281A, R282A, D283A, H286A, G291A, and Y292A exhibited 1.5- to 2.6-fold slower desensitization kinetics when compared to the WT receptor. Less significantly (1.3- to 1.5-fold; $p<0.05$) prolonged τ_{des} were observed in DF

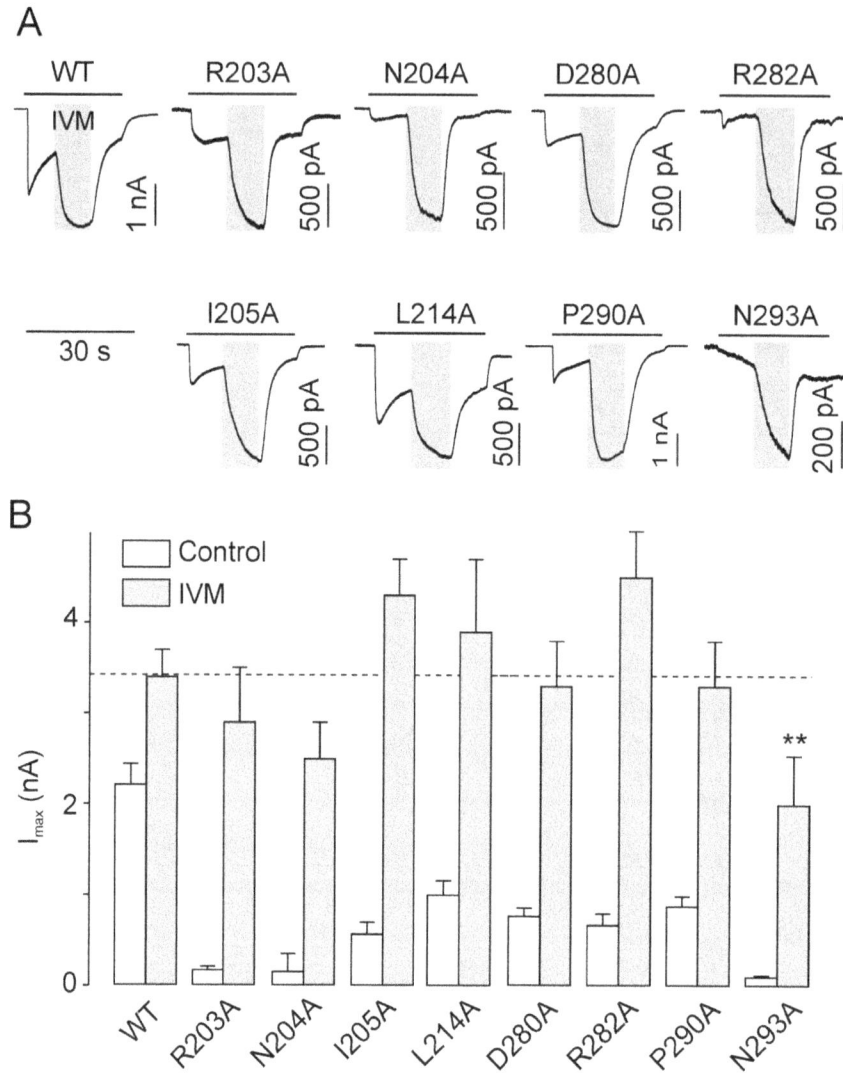

Figure 4. Ivermectin rescues the I_{max} of low-functioning mutants. (A) Acute effect of 3 µM ivermectin (IVM) applied for 10 s (gray areas) during ongoing stimulation with 100 µM ATP for 30 s (horizontal bars) in cells expressing the WT, the DF mutants (R203A, N204A, I205A, and L214A), or the LF mutants (D280A, R282A, P290A and N293A). Recordings are examples of traces similar to 3–5 traces per mutant and 30 per WT receptor. (B) Summary data showing the potentiating effect of IVM preapplication (for 4–6 min) on I_{max} in WT and alanine mutant receptors. The I_{max} values were derived from measurements taken in the absence (open bars) or in the presence (filled bars) of IVM. Values are presented as the mean ± SEM from 5–8 measurements per mutant and 15 measurements per WT. IVM treatment rescued the I_{max} of all low-functioning receptors, except in the case of N293A, which is an ATP binding mutant. The statistical significance was determined by an ANOVA comparing the WT I_{max} and the I_{max} of mutant receptors in the presence of IVM. **, $p<0.01$.

mutants T210A, Y213A, and L214A. The remaining mutants (I205A, L206A, P207A, N208A, I209A, T211A, S212A, L284A, E285A, N287A, V288A, S289A, and P290A) displayed no changes in the desensitization rate. Plotting τ_{des} versus EC_{50} (Fig. 5B) revealed a significant correlation for LF, but not for DF mutants. These results indicate that clusters of residues rather than individual amino acids, are responsible for the desensitization rate of P2X4R and that the LF domain plays a dominant role in this process.

4. The influence of the LF and DF domains of rP2X4R on agonist selectivity

To examine whether mutations in the LF and DF domains alter the responsiveness to orthosteric ligands, we compared the efficacy of ATP with four partial agonists for P2X4R. In WT receptor,

100 µM was maximal concentration for all analogue agonists, except α,β-meATP (2-MeS-ATP, $EC_{50} = 7.9\pm1.0$ µM; ATPγS, $EC_{50} = 8.4\pm1.8$ µM; BzATP, $EC_{50} = 11.1\pm2.9$ µM; α,β-meATP, $EC_{50} = 62\pm18$ µM; Fig. S2A, *upper panel*) and the agonist efficacy profile of the WT receptor was ATP (100%) >2-MeS-ATP (67%) > ATPγS (50%) > BzATP (38%) > α,β-meATP (32%). We examined all of the functional mutants that displayed significant changes in ATP potency and/or deactivation kinetics (I205A, L206A, N208A, T210A, T211A, L214A, D280A, T281A, R282A, L284A, H286A, P290A, G291A, and Y292A) and two substitution-insensitive mutants (S212A, Y213A) (Table 2).

The mutants could be divided into one of three groups. The *Group I* was composed of mutants with changes in ATP potency/efficacy and deactivation kinetics that also exhibited a significant decrease in the relative responsiveness to some or all of the

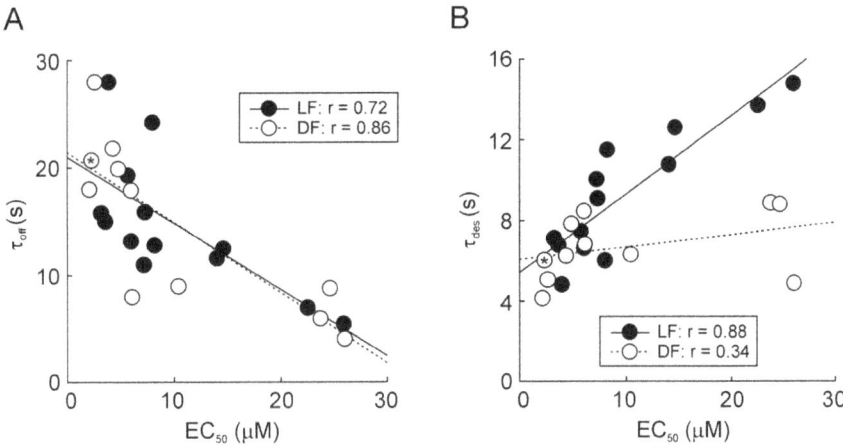

Figure 5. Deactivation and desensitization properties depend on the potency of ATP. (A, B) Correlation between EC_{50} and the deactivation time constant τ_{off} (A) and EC_{50} and desensitization time constant τ_{des} (B) for alanine DF and LF mutants. DF mutants are shown as open circles and LF mutants as closed circles. WT receptors are shown as an asterisk in an open circle. Values are derived from Table S1. Correlation analysis was performed as described in the Methods.

orthosteric agonists applied in 100 µM concentration (Table 2, Fig. S2B). This group includes I205A, T210A, L214A, P290A, G291A, and Y292A mutants. The I205A and L214A mutants displayed a significantly reduced responsiveness to all agonists when compared to the WT receptor, and the profile (2-MeS-ATP > ATPγS > BzATP ≥ α,β-meATP) was preserved (Table 2). This suggests that these residues also play a role in the mechanism by which an ATP-induced signal is coupled to the channel gating. In contrast, there were changes in the agonist profile for the T210A, P290A, G291A, and Y292A mutants. For example, the

Table 2. The relative responsiveness of the wild type (WT) and selected rP2X4R mutants to P2XR agonist analogs.

Receptor	Domain	2-MeS-ATP	ATPγS	BzATP	α,β-meATP
WT		67.3±4.6	49.6±3.5	38.2±3.4	31.8±3.3
Group I					
I205A	DF	35.6±1.9**	11.3±2.2**	4.4±2.6**	4.6±1.0**
T210A	DF	59.2±6.9	34.2±4.1*	23.9±3.2**	8.8±2.3**
L214A	DF	21.4±1.7**	7.9±2.0**	5.9±0.5**	6.5±1.5**
P290A	LF	50.7±3.7	37.0±11.0	16.9±3.2**	17.8±2.4**
G291A	LF	36.9±8.8**	31.6±6.1*	53.0±4.0**	2.8±0.5**
Y292A	LF	49.8±8.6	42.3±7.0	28.9±8.2	16.2±4.4**
Group II					
L206A	DF	68.3±6.5	39.6±8.4	33.7±8.0	27.5±2.5
N208A	DF	60.0±11.2	42.2±5.2	33.4±5.4	28.5±2.3
D280A	LF	56.3±7.8	55.7±3.8	25.4±1.5	33.1±6.0
T281A	LF	69.6±5.1	50.2±5.9	26.7±3.9	28.2±5.5
R282A	LF	55.1±12.9	45.5±5.2	25.7±3.7	27.3±7.9
H286A	LF	67.9±1.2	39.8±6.2	25.6±4.0	30.0±6.8
Group III					
S212A	DF	65.1±19.7	61.7±3.9	47.5±5.2	19.2±6.2
Y213A	DF	63.8±16.4	37.1±5.5	27.3±5.5	31.3±6.7
T211A	DF	79.3±6.7	61.8±6.1	23.4±4.7*	39.1±7.2
L284A	LF	69.0±3.0	32.4±8.2	20.6±0.2**	22.2±3.9

ATP and agonists were applied in 100 µM concentrations for 2 s with a washing interval of 60 s. The data are the mean ± SEM, relative to ATP efficacy (100%), from 26 to 37 measurements for the WT receptor and from 3 to 18 measurements per mutant. *Group I:* Mutants that exhibited changes in ATP potency/efficacy, deactivation kinetics, and changes in the relative responsiveness to orthosteric analog agonists. *Group II:* Mutants that exhibited changes in ATP potency/efficacy and deactivation, but no changes in the relative responsiveness to analog agonists. *Group III:* Mutants that showed no significant changes in ATP potency/efficacy. The statistical significance was determined by an ANOVA comparing the responsiveness to agonists between WT and mutant receptors: **, p<0.01, *, p<0.05. DF, Dorsal Fin; LF, Left Flipper.

profile for the G291A mutant was BzATP > 2-MeS-ATP > ATPγS > α,β-meATP (Fig. S2A, *lower panel*), indicating changes in the folding of the jaw for ATP.

The *Group II* of mutants showed changes in ATP potency/ efficacy and deactivation, but not in the relative responsiveness to orthosteric agonists to induce current. This includes the L206A, N208A, D280A, T281A, R282A, and H286A mutants (Table 2). The four members of *Group III*, T211A, S212A, Y213A, and L284A, showed no significant changes in ATP potency/efficacy or gating, and among them, only T211A and L284A displayed slightly (p<0.05) prolonged deactivation times (Fig. 2B). These two mutants also showed a significant decrease in the responsiveness to BzATP but no decrease in the responsiveness to 2-MeS-ATP, ATPγS, and α,β-meATP (Table 2). Therefore, residues T211 and L284 were not considered as residues of interest.

5. Model prediction for the positions of residues of interest in the DF and LF domains

We developed the rP2X4R homology model, as described in Materials and Methods, to identify the position of residues in the DF and LF domains in the ATP-bound open state. The data presented in Table 2 are also summarized in Fig. 6 as a receptor structure view. Native residues from the *Group I* mutants are located close to the ATP molecule, near the N293 residue. All of the residues from the *Group II* mutants are located downstream of the ATP binding domain and near the R203 and N204 residues that are burrowed in the protein. This topology suggests that the I205, T210, L214, P290, G291 and Y292 (green spheres) contribute to the organization of the structure of the ATP binding pocket and therefore dictate the specificity of responsiveness to the synthetic orthosteric ligands as well as the transmission of the conformational change induced by ATP binding. Residues L206, N208, D280, T281, R282, and H286 (red spheres) are important for transmitting the signal from the ATP binding cleft. Residues of mutants that have shown significant gating impairment (R203, N204 and N293) are shown as gray spheres.

The N293 residue is in close proximity (less than 5 angstroms) to ATP and, together with Y292, may directly interact with the β-

sheet segment of the K313-I333 sequence, which is responsible for the signal transmission from the ATP binding site to the pore [31] (Fig. S3A). The R203 and N204 residues are situated at the bottom of the ATP binding site (Fig. S3B) and play a crucial role in the transmission of signals towards the pore. The model also indicates that residues D283, H286, V288, and S289, from the adjacent subunit, are in the proximity of R203, while the K190, N191, N204, I205, and L206 residues are in close proximity within the same subunit. Close residues for N204 (I205, L206, and Y274) are also located within the same subunit (Fig. S3B). This suggests that the R203 and N204 residues may integrate the output signal from two neighboring subunits towards the gate and their mutants may display radical conformational misfolding of the DF and LF domains.

Discussion

The interface between the DF and LF domains, formed by sequences R203-L214 and D280-N293 in rP2X4R, is one of the most variable parts of the P2XRs [12]. Alignments of these regions for seven rat P2X subunits indicate that only three of 26 amino acids are fully conserved (N204, G291, and N293). Four hydrophobic residues, at positions I205, L206, L214, and V288, are partially conserved, and the residual amino acids of this interface are variable (Fig. 1A). Such variability in the structure of the DF and LF domains could indicate that they are not essential for receptor function or that they contribute to receptor subtype specificity in terms of agonist binding and/or gating, and desensitization.

In this study, the physiological relevance of the residues that comprise the DF and LF domains of rP2X4R was systematically analyzed for the first time by substituting each residue with an alanine. This approach enabled us to eliminate interactions between the side chains and to study the effects of that elimination on rP2X4R structure and activity. Alanine was also used as a substituent because its polarity is in the middle of the polarity scale [39] when compared to other residues. Moreover, alanine

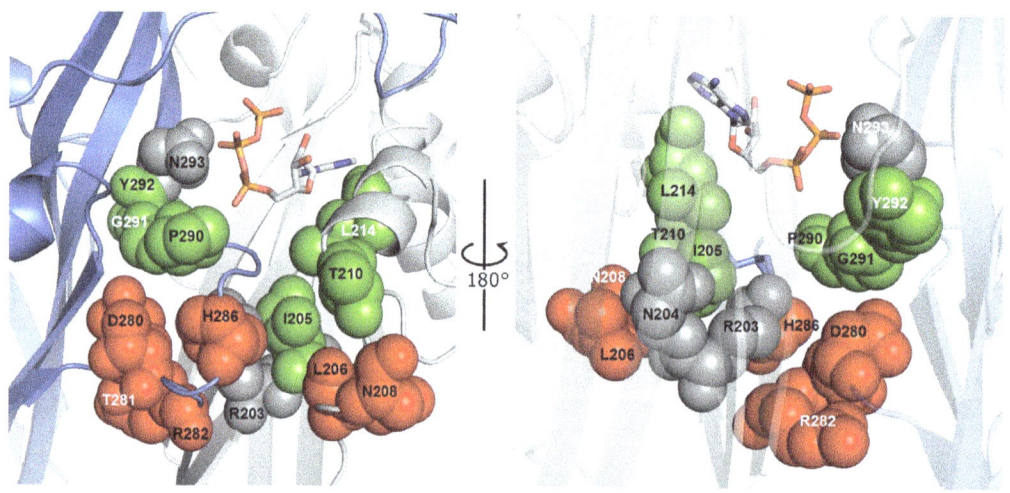

Figure 6. The structure of the ATP binding site in a rP2X4R homology model. Two panels show the position of affected residues (rotated 180°) at the interface between the LF and DF domains. The low-response residues without defined EC$_{50}$ values are grey spheres. The amino acid residues presented as green spheres demonstrate the topology of mutants with changes in ATP potency and/or efficacy (EC$_{50}$ and I$_{max}$), and agonist profile (*Group I* from Table 2). The amino acids presented in red spheres illustrate the position of residues whose mutation has affected ATP potency and/or efficacy without changing the action of ATP analogs (*Group II* from Table 2). In both panels, the ATP molecule is situated between two adjacent P2X4R subunits (blue and gray). The ATP molecule is shown in a wireframe model.

scanning mutagenesis has been widely used in research on P2XRs [6,18,23,40].

Electrophysiological and pharmacological characterization of the mutants revealed that substitution of 15 of 26 residues in the R203-L214 and D280-N293 sequences significantly attenuated the receptor function: R203A, N208A, T210A, T281A, R282A, P290A, G291A and Y292A, not previously studied, and N204A, I205A, L206A, L214A, D280A, H286A and N293A, previously studied across different P2XR subtypes (for overview see Table 1). However, the receptor function for all identified mutants, including almost non-functional mutants R203A, N204A and N293A, was rescued by the addition of IVM, an allosteric agonist of P2X4R [41].

In general, IVM allosterically potentiates the I_{max} of P2X4R, causes a leftward shift in the ATP concentration response curve and significantly prolongs deactivation [32,33]. Single channel analysis showed that IVM increases the probability of channel opening [34]. We have recently found that IVM induces dilation of the pore of the P2X4R ion channel and that the IVM-dependent transition from open to dilated state temporally coincides with receptor sensitization, which rescues the receptor from desensitization and subsequent internalization [42]. This suggests that the observed increase in the number of cell surface P2X4Rs after 2–30 min of preincubation with IVM [43] is not due to insertion of new receptors to the plasma membrane, but rather reflects its influence on channel pore dilation. The use of P2X4-pHluorin123 also revealed that IVM does not acutely increase the fraction of P2X4Rs in the plasma membrane [44]. Therefore, it is reasonable to conclude that the trafficking of mutant receptors is not affected, i.e., they are expressed at the plasma membrane, but ATP has reduced binding affinity and/or potency to activate them.

In two of three nonfunctional mutants, N204A and N293A, alanine substitutes conserved asparagine residue. The N293 amino acid was previously identified in the crystal structure of zfP2X4R as an ATP binding residue. This residue is a part of the NFR motif, which is important for the recognition of the triphosphate moiety of ATP [11]. In our experiments, the I_{max} of N293A could not be fully rescued by IVM, similar to previous observations of two other P2X4R ATP binding mutants, K67A and R295A [10]. A decrease in agonist potency was also observed in the corresponding N293A mutants of other receptors, including P2X1R-N290A [45], P2X2R-N288A [6], and P2X3R-N279A [4]. These studies further support the importance of N293 residue in the formation of the ATP binding pocket. In further agreement with our data, mutation of the conserved N204 residue is nonfunctional in P2X1R [6] and causes a 3-fold decrease in ATP potency in P2X2R [45]. An arginine in the position equivalent to 203 is present in P2X1R and P2X7R, and a lysine substitution of residue R206 enhances the sensitivity of P2X7R to activation by ATP [46].

The 12 mutants were functional but exhibited significant changes in the EC_{50}, I_{max}, τ_{off}, and/or τ_{des} values. These mutants were divided into two groups based on the relative responsiveness to stimulation with ATP and four P2XR agonist analogs. *Group I* is composed of mutants with altered relative response to the agonist analogs and includes the I205A, T210A, L214A, P290A, G291A, and Y292A mutants. In contrast, *Group II* mutants showed no change in the responsiveness to analogs and includes the L206A, N208A, D280A, T281A, R282A, and H286A mutants (Table 2). The model prediction for the positions of these residues supports the conclusions that *Group I* residues contribute to the formation of the large ATP binding pocket in addition to signal transmission, while *Group II* residues contribute to signal

transmission only. Therefore, both the DF and LF domain residues participate significantly in receptor function.

In our receptor model, in close proximity to the bound ATP molecule and asparagine 293, are the Y292, G291, and P290 residues. The substitution of these residues with alanine altered the ATP potency and/or efficacy, deactivation kinetics, and agonist profile. Among them, the most affected was the G291A mutant that exhibited an approximately 6-fold rightward shifted EC_{50} value and large changes in agonist selectivity profile. Glycine 291 is conserved in all rat P2XRs (Fig. 1A), and the corresponding cysteine mutant of P2X1R showed a 10-fold decrease in ATP potency [9], while the alanine mutant had little effect [47]. However, its role in ligand selectivity and ATP potency in other P2XR subtypes remains to be determined. These data, combined with the topology of residues in the receptor model, suggest that the N293-P290 sequence forms part of a wall in the ATP binding cleft and contributes to signal transmission through downstream LF domain residues (Fig. 6). The model predicts that this segment will also act on the transmission of signals to the gate, possibly by interactions with the K313-I333 β-sheet (Fig. S3A). Consistent with this hypothesis, mutagenesis of the Y315 and G316 residues significantly affects receptor function [31].

The partially conserved hydrophobic residue L214 has also been implicated in the recognition of the ATP ribose ring [11,48], which is fully consistent with our data. We observed that the L214A mutant displays full recovery of I_{max} in the presence of IVM, and has reduced responsiveness to all orthosteric agonists. However, the agonist profile of the WT receptor was preserved. The DF mutants I205A and T210A also exhibited a reduced potency/efficacy for ATP and its analogs. Topologically, these native residues may account for the bottom part of the ATP binding pocket (Fig. 6). A recent study on hydrophobic interactions between the LF and DF domains during receptor activation has identified several non-polar residues, including L214 and I205, that are important for the coordinated relative movements of these domains after ATP binding [49]. Therefore, we suggest that L214 residue plays a dual role in receptor functions: agonist binding and signal transmission.

The topology of T210 in zfP2X4R revealed that the residue is situated nearby the α-helix containing L214 involved in ATP recognition [11], (Fig. 1A). We observed changes in agonist profile for T210A mutant, suggesting that T210 could contribute to coordination of agonist position in the binding cleft. This explanation needs assumption that the T210 side chain position is variable and might be oriented towards the binding pocket, similarly as L214, and that orientation of ATP is different from that predicted by crystal, suggesting the existence of several ATP binding modes [50,51]. Further experiments are needed to explain the role of T210 in receptor function.

The homology model of rP2X4R predicts that the *Group II* amino acids are clustered into two subgroups: one composed of D280-H286 LF residues and the other composed of L206-N208 DF residues. The position of these residues is consistent with their roles in signal transmission. Fig. 6 suggests that the influence of ATP binding on gating is transmitted downstream through two signal transmission lines. The first is composed of N293, Y292, G291, and P290 towards D280, T281, R282, and H286 (from top to bottom) in the LF domain. The other unit appears to be composed of L214, T210, I205, N204, L206, and N208 (from top to bottom) in the DF domain. The model also suggests that R203 and N204 are positioned to accept the signal from the binding domain through both lines and from two neighboring subunits, and to integrate it towards the gate region.

Finally, seven out of the 13 LF mutants tested showed significantly slower rates of receptor desensitization and our correlation analysis of the relationship between EC_{50} vs. τ_{des} suggests that that the LF domain plays the major role in the transition from the open to the desensitized state, with signal transmission through the N293-D280 sequence (Fig. 2C and 5B). Alanine substitution of the corresponding positions D266A [53], S269A [54], but also S275A [55], prolongs desensitization of P2X3R, but the P2X2R-D277A mutant was normal [56]. These data indicate that this group of polar and charged residues might play a receptor-specific role in desensitization.

In conclusion, we have shown that the interface between DF and LF domains has dual roles in rP2X4R function. One role is the formation of the ligand-binding pocket and the other is for the transmission of signals from the pocket toward the gate. Both domains contribute to the specificity of binding sites for orthosteric agonists by residues in the upper part of interface, relative to distance from the channel pore, and to the transmission of signals towards the gate by residues in the lower part of the interface. The R203 and N204 may integrate the influence of both lines of transmission. The LF domain appears to have two additional roles: the transmission of signals towards the gate in the second transmembrane domain through the K313-I333 β-sheet and the control of desensitization of receptors.

Supporting Information

Figure S1   Deactivation and desensitization responses of WT and selected DF and LF mutants. (A) An example of the WT response and that of the L214A, D280A, and H286A mutant receptors when stimulated with 3 μM ATP for 2 s in the presence of IVM. Cells were preincubated with 3 μM IVM for 4–6 min, and the deactivation time constants (τ_{off}) were estimated by the monoexponential fit of decay of current after removal of the agonist. (B) The desensitization of WT, L214A, D280A, and H286A receptors when stimulated with 100 μM ATP for 60 s (gray traces), and the curves obtained by fitting (black). Weighted desensitization time constants (τ_{des}) were derived from mono-exponential (D280A) or biexponential fitting.

Figure S2   Responsiveness of the WT and mutant receptors to ATP analogue agonists. (A) Concentration response curves for WT and G291A receptors. Currents were stimulated by a short (2–5 s) application of different (1–300 μM) concentrations of ATP, 2-MeS-ATP (2MeS), ATPγS, BzATP (Bz) and α, βme-ATP (αβme). Even if the full dose response curve for analogue agonists could not be constructed for G291A, these experiments clearly show differences in agonist profile between the WT (*upper panel*) and G291A (*lower panel*) receptor. Experiments were performed on naïve receptors, and data points are presented as the mean ± SEM from 3–27 measurements per agonist, per concentration for both WT and G291A. (B) Example responses to 100 μM of ATP and several P2XR agonists, including 2-MeS-ATP (2MeS), ATPγS, BzATP (Bz) and α, βme-ATP (αβme), recorded from cells expressing the WT receptor and selected mutant receptors from from *Group I* (I205A, T210A, L214A, and G291A) and *Group II* (D280A and H286A). Each trace represents a continuous response from a single cell.

Figure S3   The structure of the ATP binding site in the rP2X4R homology model. (A) The possible interaction of N293 and Y292 residues with Y315 and G316 residues from the β-sheet segment from the K313-I333 sequence. (B) The multiple interactions of residues R203 (yellow spheres) and N204 (green spheres) with partners (all in cyan wireframes) from the same (K190, N191, N204, I205, L206, Y274) and adjacent (D283, H286, V288, S289) subunits. Two adjacent rP2X4R subunits are represented in blue and gray.

Table S1   Characterization of the DF and LF alanine mutants of rP2X4R.

Table S2   Desensitization parameters for the DF and LF mutants of the rP2X4R.

Author Contributions

Conceived and designed the experiments: VT SS HZ. Performed the experiments: VT MR HZ. Analyzed the data: VT MR SS HZ. Contributed reagents/materials/analysis tools: VT SS HZ. Contributed to the writing of the manuscript: VT MR SS HZ.

References

1. North RA (2002) Molecular physiology of P2X receptors. Physiol Rev. 82: 1013–67.
2. Nicke A, Baumert HG, Rettinger J, Eichele A, Lambrecht G, et al. (1998) P2X1 and P2X3 receptors form stable trimers: a novel structural motif of ligand-gated ion channels. Embo J. 17: 3016–28.
3. Coddou C, Yan Z, Obsil T, Huidobro-Toro JP, Stojilkovic SS (2011) Activation and regulation of purinergic P2X receptor channels. Pharmacol Rev. 63: 641–83.
4. Bodnar M, Wang H, Riedel T, Hintze S, Kato E, et al. (2011) Amino acid residues constituting the agonist binding site of the human P2X3 receptor. J Biol Chem. 286: 2739–49.
5. Ennion S, Hagan S, Evans RJ (2000) The role of positively charged amino acids in ATP recognition by human P2X(1) receptors. J Biol Chem. 275: 29361–7.
6. Jiang LH, Rassendren F, Surprenant A, North RA (2000) Identification of amino acid residues contributing to the ATP-binding site of a purinergic P2X receptor. J Biol Chem. 275: 34190–6.
7. Marquez-Klaka B, Rettinger J, Bhargava Y, Eisele T, Nicke A (2007) Identification of an intersubunit cross-link between substituted cysteine residues located in the putative ATP binding site of the P2X1 receptor. J Neurosci. 27: 1456–66.
8. Roberts JA, Digby HR, Kara M, El Ajouz S, Sutcliffe MJ, et al. (2008) Cysteine substitution mutagenesis and the effects of methanethiosulfonate reagents at P2X2 and P2X4 receptors support a core common mode of ATP action at P2X receptors. J Biol Chem. 283: 20126–36.
9. Roberts JA, Evans RJ (2007) Cysteine substitution mutants give structural insight and identify ATP binding and activation sites at P2X receptors. J Neurosci. 27: 4072–82.
10. Zemkova H, Yan Z, Liang Z, Jelinkova I, Tomic M, et al. (2007) Role of aromatic and charged ectodomain residues in the P2X4 receptor functions. J Neurochem. 102: 1139–50.
11. Hattori M, Gouaux E (2012) Molecular mechanism of ATP binding and ion channel activation in P2X receptors. Nature. 485: 207–12.
12. Kawate T, Michel JC, Birdsong WT, Gouaux E (2009) Crystal structure of the ATP-gated P2X(4) ion channel in the closed state. Nature. 460: 592–8.
13. Egan TM, Haines WR, Voigt MM (1998) A domain contributing to the ion channel of ATP-gated P2X2 receptors identified by the substituted cysteine accessibility method. J Neurosci. 18: 2350–9.
14. Haines WR, Voigt MM, Migita K, Torres GE, Egan TM (2001) On the contribution of the first transmembrane domain to whole-cell current through an ATP-gated ionotropic P2X receptor. J Neurosci. 21: 5885–92.
15. Jelinkova I, Vavra V, Jindrichova M, Obsil T, Zemkova HW, et al. (2008) Identification of P2X(4) receptor transmembrane residues contributing to channel gating and interaction with ivermectin. Pflugers Arch. 456: 939–50.
16. Jiang LH, Rassendren F, Spelta V, Surprenant A, North RA (2001) Amino acid residues involved in gating identified in the first membrane-spanning domain of the rat P2X(2) receptor. J Biol Chem. 276: 14902–8.
17. Khakh BS, Egan TM (2005) Contribution of transmembrane regions to ATP-gated P2X2 channel permeability dynamics. J Biol Chem. 280: 6118–29.

18. Li Z, Migita K, Samways DS, Voigt MM, Egan TM (2004) Gain and loss of channel function by alanine substitutions in the transmembrane segments of the rat ATP-gated P2X2 receptor. J Neurosci. 24: 7378–86.

19. Migita K, Haines WR, Voigt MM, Egan TM (2001) Polar residues of the second transmembrane domain influence cation permeability of the ATP-gated P2X(2) receptor. J Biol Chem. 276: 30934–41.

20. Rassendren F, Buell G, Newbolt A, North RA, Surprenant A (1997) Identification of amino acid residues contributing to the pore of a P2X receptor. Embo J. 16: 3446–54.

21. Silberberg SD, Chang TH, Swartz KJ (2005) Secondary structure and gating rearrangements of transmembrane segments in rat P2X4 receptor channels. J Gen Physiol. 125: 347–59.

22. Allsopp RC, El Ajouz S, Schmid R, Evans RJ (2011) Cysteine scanning mutagenesis (residues Glu52-Gly96) of the human P2X1 receptor for ATP: mapping agonist binding and channel gating. J Biol Chem. 286: 29207–17.

23. Rokic MB, Stojilkovic SS, Vavra V, Kuzyk P, Tvrdonova V, et al. (2013) Multiple roles of the extracellular vestibule amino acid residues in the function of the rat P2X4 receptor. PLoS One. 8: e59411.

24. Samways DS, Khakh BS, Dutertre S, Egan TM (2011) Preferential use of unobstructed lateral portals as the access route to the pore of human ATP-gated ion channels (P2X receptors). Proc Natl Acad Sci U S A. 108: 13800–5.

25. Du J, Dong H, Zhou HX (2012) Gating mechanism of a P2X4 receptor developed from normal mode analysis and molecular dynamics simulations. Proc Natl Acad Sci U S A. 109: 4140–5.

26. Lorinczi E, Bhargava Y, Marino SF, Taly A, Kaczmarek-Hajek K, et al. (2012) Involvement of the cysteine-rich head domain in activation and desensitization of the P2X1 receptor. Proc Natl Acad Sci U S A. 109: 11396–401.

27. Collo G, North RA, Kawashima E, Merlo-Pich E, Neidhart S, et al. (1996) Cloning of P2X5 and P2X6 receptors and the distribution and properties of an extended family of ATP-gated ion channels. J Neurosci. 16: 2495–507.

28. Le KT, Babinski K, Seguela P (1998) Central P2X4 and P2X6 channel subunits coassemble into a novel heteromeric ATP receptor. J Neurosci. 18: 7152–9.

29. Ormond SJ, Barrera NP, Qureshi OS, Henderson RM, Edwardson JM, et al. (2006) An uncharged region within the N terminus of the P2X6 receptor inhibits its assembly and exit from the endoplasmic reticulum. Mol Pharmacol. 69: 1692–700.

30. Carugo O (1999) Correlation between occupancy and B factor of water molecules in protein crystal structures. Protein Eng. 12: 1021–4.

31. Yan Z, Liang Z, Obsil T, Stojilkovic SS (2006) Participation of the Lys313-Ile333 sequence of the purinergic P2X4 receptor in agonist binding and transduction of signals to the channel gate. J Biol Chem. 281: 32649–59.

32. Jelinkova I, Yan Z, Liang Z, Moonat S, Teisinger J, et al. (2006) Identification of P2X(4) receptor-specific residues contributing to the ivermectin effects on channel deactivation. Biochem Biophys Res Commun. 349: 619–25.

33. Khakh BS, Proctor WR, Dunwiddie TV, Labarca C, Lester HA (1999) Allosteric control of gating and kinetics at P2X(4) receptor channels. J Neurosci. 19: 7289–99.

34. Priel A, Silberberg SD (2004) Mechanism of ivermectin facilitation of human P2X4 receptor channels. J Gen Physiol. 123: 281–93.

35. Schwede T, Kopp J, Guex N, Peitsch MC (2003) SWISS-MODEL: An automated protein homology-modeling server. Nucleic Acids Res. 31: 3381–5.

36. Benkert P, Tosatto SC, Schomburg D (2008) QMEAN: A comprehensive scoring function for model quality assessment. Proteins. 71: 261–77.

37. Rettinger J, Schmalzing G (2004) Desensitization masks nanomolar potency of ATP for the P2X1 receptor. J Biol Chem. 279: 6426–33.

38. Zemkova H, He ML, Koshimizu TA, Stojilkovic SS (2004) Identification of ectodomain regions contributing to gating, deactivation, and resensitization of purinergic P2X receptors. J Neurosci. 24: 6968–78.

39. Karplus PA (1997) Hydrophobicity regained. Protein Sci. 6: 1302–7.

40. Samways DS, Migita K, Li Z, Egan TM (2008) On the role of the first transmembrane domain in cation permeability and flux of the ATP-gated P2X2 receptor. J Biol Chem. 283: 5110–7.

41. Coddou C, Stojilkovic SS, Huidobro-Toro JP (2011) Allosteric modulation of ATP-gated P2X receptor channels. Rev Neurosci. 22: 335–54.

42. Zemkova H, Khadra A, Rokic MB, Tvrdonova V, Sherman A, et al. (2014) Allosteric regulation of the P2X4 receptor channel pore dilation. Pflugers Arch. 2014 Jun 11. [Epub ahead of print].

43. Toulme E, Soto F, Garret M, Boue-Grabot E (2006) Functional properties of internalization-deficient P2X4 receptors reveal a novel mechanism of ligand-gated channel facilitation by ivermectin. Mol Pharmacol. 69: 576–87.

44. Xu J, Chai H, Ehinger K, Egan TM, Srinivasan R, et al. (2014) Imaging P2X4 receptor subcellular distribution, trafficking, and regulation using P2X4-pHluorin. J Gen Physiol. 144: 81–104.

45. Roberts JA, Evans RJ (2006) Contribution of conserved polar glutamine, asparagine and threonine residues and glycosylation to agonist action at human P2X1 receptors for ATP. J Neurochem. 96: 843–52.

46. Adriouch S, Bannas P, Schwarz N, Fliegert R, Guse AH, et al. (2008) ADP-ribosylation at R125 gates the P2X7 ion channel by presenting a covalent ligand to its nucleotide binding site. Faseb J. 22: 861–9.

47. Digby HR, Roberts JA, Sutcliffe MJ, Evans RJ (2005) Contribution of conserved glycine residues to ATP action at human P2X1 receptors: mutagenesis indicates that the glycine at position 250 is important for channel function. J Neurochem. 95: 1746–54.

48. Zhang L, Xu H, Jie Y, Gao C, Chen W, et al. (2014) Involvement of ectodomain Leu 214 in ATP binding and channel desensitization of the P2X4 receptor. Biochemistry. 53: 3012–9.

49. Zhao WS, Wang J, Ma XJ, Yang Y, Liu Y, et al. (2014) Relative motions between left flipper and dorsal fin domains favour P2X4 receptor activation. Nat Commun. 5: 4189.

50. Jiang R, Lemoine D, Martz A, Taly A, Gonin S, et al. (2011) Agonist trapped in ATP-binding sites of the P2X2 receptor. Proc Natl Acad Sci U S A. 108: 9066–71.

51. Huang LD, Fan YZ, Tian Y, Yang Y, Liu Y, et al. (2014) Inherent Dynamics of Head Domain Correlates with ATP-Recognition of P2X4 Receptors: Insights Gained from Molecular Simulations. PLoS One. 9: e97528.

52. Yan Z, Liang Z, Tomic M, Obsil T, Stojilkovic SS (2005) Molecular Determinants of the Agonist Binding Domain of a P2X Receptor Channel. Mol Pharmacol. 67: 1078–88.

53. Fabbretti E, Sokolova E, Masten L, D'Arco M, Fabbro A, et al. (2004) Identification of negative residues in the P2X3 ATP receptor ectodomain as structural determinants for desensitization and the Ca2+ sensing modulatory sites. J Biol Chem. 279: 53109–115.

54. Stanchev D, Flehmig G, Gerevich Z, Norenberg W, Dihazi H, et al. (2006) Decrease of current responses at human recombinant P2X3 receptors after substitution by Asp of Ser/Thr residues in protein kinase C phosphorylation sites of their ecto-domains. Neurosci Lett. 393: 78–83.

55. Petrenko N, Khafizov K, Tvrdonova V, Skorinkin A, Giniatullin R (2011) Role of the ectodomain serine 275 in shaping the binding pocket of the ATP-gated P2X3 receptor. Biochemistry. 50: 8427–36.

56. Friday SC, Hume RI (2008) Contribution of extracellular negatively charged residues to ATP action and zinc modulation of rat P2X2 receptors. J Neurochem. 105: 1264–75.

57. Coddou C, Morales B, Gonzalez J, Grauso M, Gordillo F, et al. (2003) Histidine 140 plays a key role in the inhibitory modulation of the P2X4 nucleotide receptor by copper but not zinc. J Biol Chem. 278: 36777–85.

58. Kowalski M, Hausmann R, Dopychai A, Grohmann M, Franke H, et al. (2014) Conformational flexibility of the agonist binding jaw of the human P2X3 receptor is a prerequisite for channel opening. Br J Pharmacol. 2014 Jul 2. doi: 10.1111/bph.12830. [Epub ahead of print].

59. Wirkner K, Stanchev D, Koles L, Klebingat M, Dihazi H, et al. (2005) Regulation of human recombinant P2X3 receptors by ecto-protein kinase C. J Neurosci. 25: 7734–42.

60. Xiong K, Stewart RR, Hu XQ, Werby E, Peoples RW, et al. (2004) Role of extracellular histidines in agonist sensitivity of the rat P2X4 receptor. Neurosci Lett. 365: 195–9.

16

Missense Mutation Lys18Asn in Dystrophin that Triggers X-Linked Dilated Cardiomyopathy Decreases Protein Stability, Increases Protein Unfolding, and Perturbs Protein Structure, but Does Not Affect Protein Function

Surinder M. Singh[1], Swati Bandi[1], Dinen D. Shah[1], Geoffrey Armstrong[2], Krishna M. G. Mallela[1,3]*

1 Department of Pharmaceutical Sciences, Skaggs School of Pharmacy and Pharmaceutical Sciences, University of Colorado Anschutz Medical Campus, Aurora, Colorado, United States of America, 2 Department of Chemistry and Biochemistry, University of Colorado Boulder, Boulder, Colorado, United States of America, 3 Program in Structural Biology and Biochemistry, University of Colorado Anschutz Medical Campus, Aurora, Colorado, United States of America

Abstract

Genetic mutations in a vital muscle protein dystrophin trigger X-linked dilated cardiomyopathy (XLDCM). However, disease mechanisms at the fundamental protein level are not understood. Such molecular knowledge is essential for developing therapies for XLDCM. Our main objective is to understand the effect of disease-causing mutations on the structure and function of dystrophin. This study is on a missense mutation K18N. The K18N mutation occurs in the N-terminal actin binding domain (N-ABD). We created and expressed the wild-type (WT) N-ABD and its K18N mutant, and purified to homogeneity. Reversible folding experiments demonstrated that both mutant and WT did not aggregate upon refolding. Mutation did not affect the protein's overall secondary structure, as indicated by no changes in circular dichroism of the protein. However, the mutant is thermodynamically less stable than the WT (denaturant melts), and unfolds faster than the WT (stopped-flow kinetics). Despite having global secondary structure similar to that of the WT, mutant showed significant local structural changes at many amino acids when compared with the WT (heteronuclear NMR experiments). These structural changes indicate that the effect of mutation is propagated over long distances in the protein structure. Contrary to these structural and stability changes, the mutant had no significant effect on the actin-binding function as evident from co-sedimentation and depolymerization assays. These results summarize that the K18N mutation decreases thermodynamic stability, accelerates unfolding, perturbs protein structure, but does not affect the function. Therefore, K18N is a stability defect rather than a functional defect. Decrease in stability and increase in unfolding decrease the net population of dystrophin molecules available for function, which might trigger XLDCM. Consistently, XLDCM patients have decreased levels of dystrophin in cardiac muscle.

Editor: Sukesh R. Bhaumik, Southern Illinois University School of Medicine, United States of America

Funding: This work was funded by the American Heart Association grant 11SDG4880046, Jane and Charlie Butcher grants in Genomics and Biotechnology, and the ALSAM Foundation through the Skaggs Scholars Program. The funders had no role in study design, data collection and analysis, decision to publish, or preparation of the manuscript.

Competing Interests: The authors have declared that no competing interests exist.

* Email: krishna.mallela@ucdenver.edu

Introduction

X-linked dilated cardiomyopathy (XLDCM) involves progressive heart muscle degeneration and is a lethal disorder leading to death at an early age in male patients [1]. No cure is currently available for XLDCM. The only viable option for treating XLDCM is heart transplant. Mutations in the gene coding for a vital muscle protein dystrophin trigger XLDCM [1–12]. Although dystrophin mutations are in general linked to Duchenne/Becker muscular dystrophy (DMD/BMD) [13], no apparent sign of skeletal muscle degeneration were observed in XLDCM patients [1], which distinguishes them from DMD/BMD patients. The major function of dystrophin is to link intracellular cytoskeleton with the extracellular matrix [14]. This tethering helps in maintaining the integrity of the myocardial membrane against

continuous cycles of stretches and contractions, and plays an important role in cell adhesion [15,16]. Disease-causing mutations in dystrophin include premature stop codons, frameshift mutations, and missense mutations. Premature stop codons and frameshift mutations express partial or incorrect protein, and hence trigger the disease because of the lack of a functional protein. However, how missense mutations trigger the disease is not clear, in which the expressed protein is of similar length to that of the wild-type protein. Our goal is to understand the disease mechanisms at the fundamental protein level at an atomic resolution. Such molecular knowledge will help in developing effective therapies to treat XLDCM, for example, small molecule therapeutics and compensatory gene constructs. For this purpose, we need to first understand the effect of disease-causing mutations on the structure and function of dystrophin. In this work, we examined the case of a missense mutation where lysine at the 18th position in the dystrophin amino acid sequence is replaced by asparagine (Lys18Asn, or K18N). This mutation occurs in the N-terminal actin-binding domain (N-ABD) of dystrophin (Fig. 1A). The K18N mutation is associated with the hallmarks of XLDCM (LVEDD 70 mm, shortening fraction 8%, ejection fraction 19%, and large ventricular thrombus). Among 141 control patients examined, none harbored the K18N mutation indicating its 100% linkage with XLDCM [17]. Creatine kinase levels were 11,300, which were as high as observed in more severe DMD patients; however, no sign of skeletal myopathy were recorded, confirming that the K18N mutation is a XLDCM case. Here, we analyzed the effect of mutation on dystrophin structure and function using various biophysical and structural methods. Our results indicate that the K18N mutation decreases thermodynamic stability, accelerates unfolding, perturbs protein structure, but does not affect protein function.

Materials and Methods

Expression and purification

Wild type (WT) dystrophin N-ABD (residues 1–246) was cloned into pET28a vector as described previously [18]. The K18N mutant was created using WT as the template using quick mutagenesis protocol (Qiagen), and was confirmed by DNA sequencing. Both WT and mutant proteins were transformed into BL21(DE3) by heat shock, and glycerol stocks were made for further expression and purification. WT and mutant were expressed in 2 L lysogeny broth (LB) media and pellets were used for purification. Cells were lysed with sonication and proteins were purified from the supernatant by nickel affinity chromatography. Pure proteins were dialyzed against PBS (0.1 M sodium phosphate, 0.15 M sodium chloride, pH 7) and used for experiments described below.

Fluorescence and circular dichroism spectroscopy

Fluorescence spectra of native and unfolded states (in 8 M urea) of WT and mutant proteins (1 µM) in PBS buffer were recorded by exciting the samples at 280 nm (QuantaMaster, PTI). CD spectra (5 µM protein in PBS) were recorded on an Applied Photophysics ChirascanPlus spectrometer. Mean residue ellipticity (MRE) of the proteins were calculated from the CD values in millidegrees using the equation [19], $[\theta] = $ CD in millidegrees/ (path length in millimeters x molar concentration of protein x number of residue).

Thermal melts

WT and mutant proteins (1 µM each in PBS) were used for calculating the T_m (midpoint of thermal denaturation transition) by using CD signal at 222 nm with increasing temperature in a continuous ramp mode (1°C/min). Normalized signal was plotted against temperature, and the data were fitted to a two-state equilibrium unfolding model in SigmaPlot as described earlier [18].

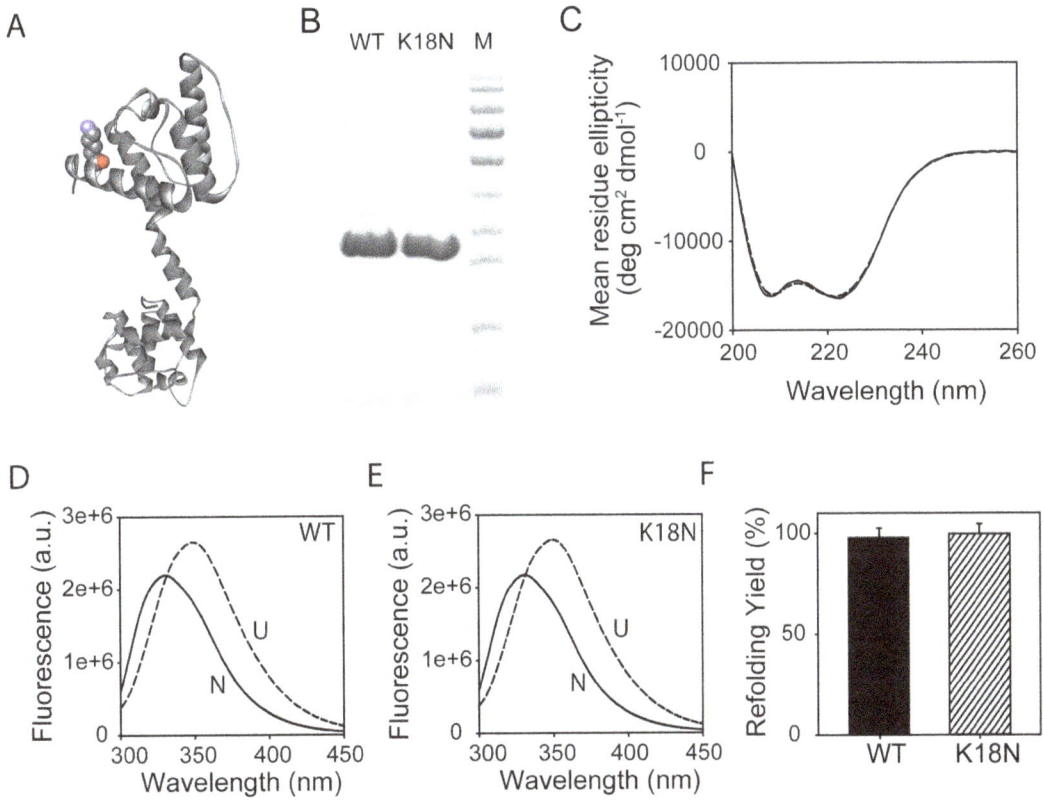

Figure 1. Expression, purification, and biophysical characterization of the K18N mutant and the WT. (A) Ribbon diagram of the X-ray crystal structure of dystrophin N-ABD (1DXX.pdb) with the mutated lysine shown in a space-filling or cpk (Corey, Pauling, and Koltun) model. (B) SDS-PAGE of purified proteins. Lane M corresponds to molecular weight markers (bottom to top: 11, 17, 26, 34, 43, 56, 72, 96, 130, and 170 kDa respectively). (C) CD spectra of the WT (solid line) and the mutant (dashed line). (D) Fluorescence spectra of the native (solid line) and denatured (dashed line) WT. (E) Fluorescence spectra of the native (solid line) and denatured (dashed line) mutant. (F) Refolding yields of the two proteins.

Chemical denaturation melts

WT and mutant proteins (1 μM each in PBS) were subjected to equilibrium denaturation melts. Urea was used as the denaturant. Protein solutions at varying urea concentrations were prepared and equilibrated for 1 hr. CD signal at 222 nm was measured (ChirascanPlus, Applied Photophysics) for all samples. Normalized signal was plotted against increasing urea concentration. Free energy of unfolding, ΔG, was determined by fitting the curve to a two-state unfolding model [20,21].

For fluorescence urea denaturation melts, samples were prepared using the same method as described above. Samples were excited at 280 nm and the change in emission intensity at 350 nm was monitored (QuantaMaster, PTI). Normalized fluorescence signal was fitted to a two-state unfolding model using SigmaPlot.

Structural analysis by NMR

For labeling WT and mutant proteins with ^{15}N, 2 L LB culture (optical density ~2) pellet was resuspended into 1 L sterile M9 media (48 mM Na_2HPO_4, 22 mM KH_2PO_4, 8.5 mM NaCl, 18.6 mM $^{15}NH_4Cl$, 2 mM $MgSO_4$, 0.1 mM $CaCl_2$, 0.4% glucose), and was grown for 2 hrs at 310 K. Protein expression was induced with 0.5 mM isopropyl β-D-1-thiogalactopyranoside (IPTG) overnight, and the labeled proteins were purified to high homogeneity using protocols described above for unlabeled proteins. Proteins were concentrated using Centricon (Amicon Ultra, Millipore) to 150 μM and standard HSQC-TROSY NMR spectra were recorded on a 900 MHz Varian NMR instrument. Spectra were processed using NMRPipe [22]. A peak table was generated with arbitrary index number for WT protein as described earlier [18]. Same peak index number was used for the peaks of mutant protein to calculate the chemical shift differences. Pictures showing overlay of the spectra were generated with NMRView software.

Refolding yield

Mutant and WT proteins were solubilized in PBS buffer containing 8 M urea to a final concentration of 10 μM. Molar extinction coefficient (46075 $M^{-1}cm^{-1}$ @ 280 nm) used for quantification was calculated using ExPASy program (http://www.expasy.org/). Solubilized proteins in 8 M urea were refolded by 10 times dilution into PBS and incubated for one hr to reach equilibrium. Refolded solutions were centrifuged at 30,000 g for 10 min and proteins in the supernatant were measured using absorbance at 280 nm.

Stopped flow kinetics

Unfolding kinetics of the mutant and WT proteins were monitored using an Applied Photophysics stopped flow assembly attached to a ChirascanPlus spectrometer. Proteins (10 μM each) were diluted 10 times into PBS buffer containing high denaturant, and changes in the CD signal at 222 nm were recorded. An average of 20 traces was fit to an exponential function using SigmaPlot to determine the rate constant.

Actin binding affinity

Skeletal muscle G-Actin (Cytoskeleton, Denver, USA) was polymerized (7 μM, final concentration) and incubated at room temperature with varying concentrations of either the WT or the mutant protein. Above mix (final volume 100 μl) was centrifuged at 100,000 g for 30 min (sw55Ti rotor, Beckman Optima LE80K) and pellets were solubilized in 30 μl SDS-PAGE loading buffer. Half of this was boiled and subjected to SDS-PAGE electropho-

resis and was stained with coomassie brilliant blue. Intensities of the individual bands were determined using the QuantityOne software on Biorad Gel Doc XR. These intensity values were corrected for the differential staining of proteins with Coomassie Blue as described before [23,24]. Ratios of the corrected intensities were used to determine the fraction bound of actin and free protein (either the WT or mutant). Fraction bound of F-actin was plotted against free protein to obtain the binding curve.

Actin depolymerization Assay

G-actin labeled with pyrene (40 μM) was polymerized in polymerization buffer (10 mM Tris, 2 mM $MgCl_2$, 50 mM KCl, 1 mM ATP, pH 7.5) and diluted 20 times into G-Actin buffer (5 mM Tris, 0.2 M $CaCl_2$, 0.2 mM ATP, pH 8.0) containing either the WT or the mutant protein (16 μM). Change in pyrene excimer fluorescence signal was monitored against time.

Results

Protein expression, purification, and biophysical characterization

Both proteins (WT & K18N) were expressed and purified to high homogeneity (Fig. 1B). Circular dichroism (CD) in the far-UV region, typically in the wavelength range 200–260 nm where the peptide group absorbs, is generally used to characterize the secondary structure of proteins [19]. Each secondary structure has distinct maxima and minima in this wavelength range. For the K18N mutant, its far-UV CD spectrum is similar to that of the WT (Fig. 1C), indicating that the mutation did not affect the protein's secondary structure. The minima at 222 nm and 208 nm are characteristic of an α-helical structure [19], which are consistent with the crystal structure of the WT protein (Fig. 1A). In addition, the mutant shows similar fluorescence spectra as that of the WT in both native and unfolded states (Figs. 1D & E). Intrinsic protein fluorescence originates mainly from tryptophan residues [25]. WT protein contains eight tryptophans distributed all across the protein. Tryptophan fluorescence is in general sensitive to the local environment around the aromatic sidechain. Identical fluorescence spectra for the mutant and the WT indicate that there is no effect of mutation on the local environment of the tryptophan residues. The native fluorescence of both the WT and mutant are blue-shifted with respect to their unfolded states. For both proteins, fluorescence emission maxima of the native and unfolded states occur at 330 nm and 349 nm respectively. The blue shift in the tryptophan emission maximum indicates that these residues are located in the native protein interior, not accessible to solvent water [25]. For tryptophans exposed to water, emission maximum occurs at around 355 nm, whereas for tryptophans buried in the hydrophobic protein interior, the emission maximum occurs at around 325 nm. The observed blue shift of 19 nm indicates that at least some of the eight tryptophan residues are significantly buried in the folded protein interior, implying that both proteins are well-folded in solution and the mutation has no effect on the local environment around the tryptophan residues. For both proteins, the fluorescence is increased upon unfolding, which indicates that the tryptophan fluorescence is quenched in the native state by the neighboring amino acids, similar to that observed in other proteins [26,27].

The K18N mutant is less stable than the WT

Thermal and chemical denaturation are commonly used to measure protein stability [18,28]. In thermal denaturation, solution temperature is increased to destabilize the protein structure, whereas in chemical denaturation, a denaturant such

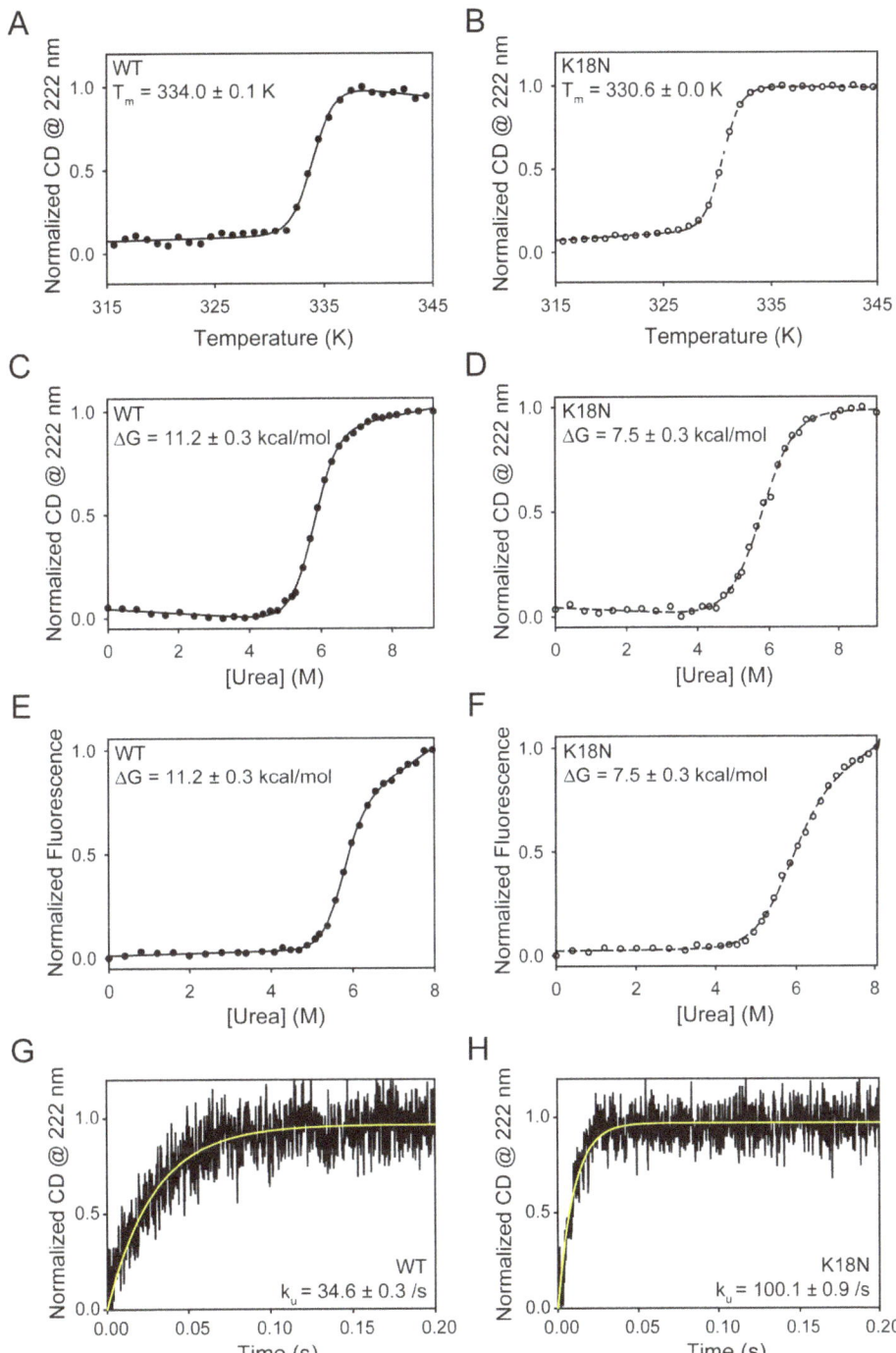

Figure 2. Effect of K18N mutation on protein stability and unfolding. (A) and (B) show the thermal melts of the WT and the mutant respectively. Circular dichroism (CD) @ 222 nm was used to monitor protein unfolding. (C) and (D) show the denaturant melts of the WT and the mutant measured using CD @ 222 nm as the signal. (E) and (F) show the corresponding denaturant melts with fluorescence (λ_{ex} = 280 nm, λ_{em} = 350 nm) as the signal. (G) and (H) show the stopped-flow unfolding kinetics of the WT and the mutant starting from their native states, measured using CD @ 222 nm as the signal.

as urea is added to the solution to destabilize the protein structure. Thermal denaturation melts of the WT and the mutant show that both proteins unfold in a sigmoidal fashion (Figs. 2A & B), indicating their cooperative unfolding. Further, these melting curves indicate that the mutant requires lesser temperatures to unfold compared to the WT, indicating its decreased stability.

Fitting these curves to a standard sigmoidal function [18] resulted in midpoint transition temperature (T_m) values of 334.0±0.1 K (or, 61°C) and 330.6±0.0 K (or, 57.6°C) for the WT and the mutant respectively. However, thermal denaturation is irreversible, and denatured proteins could not be renatured back as they resulted in serious aggregation. Hence, these thermal melts can

A

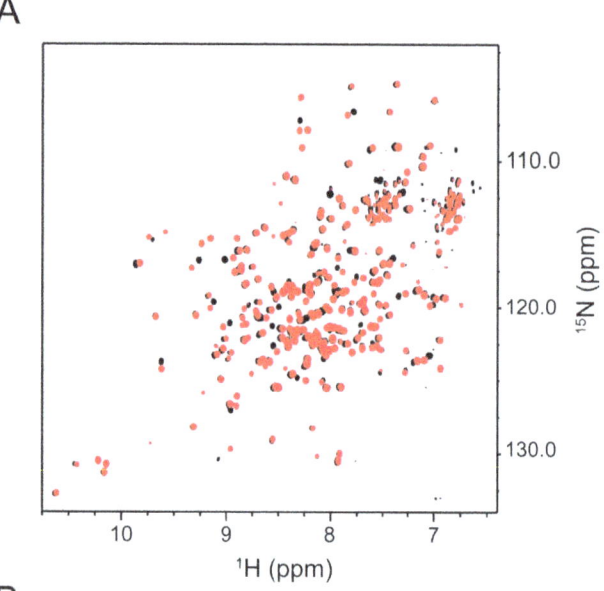

B

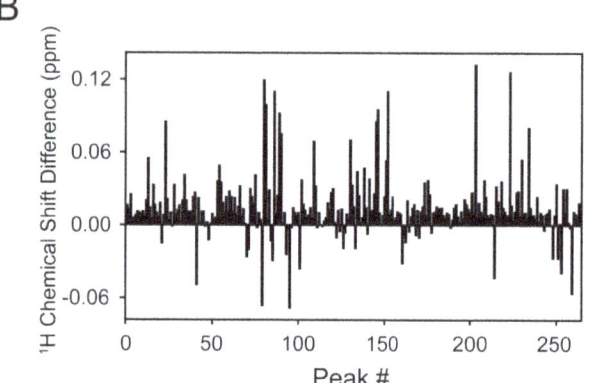

C

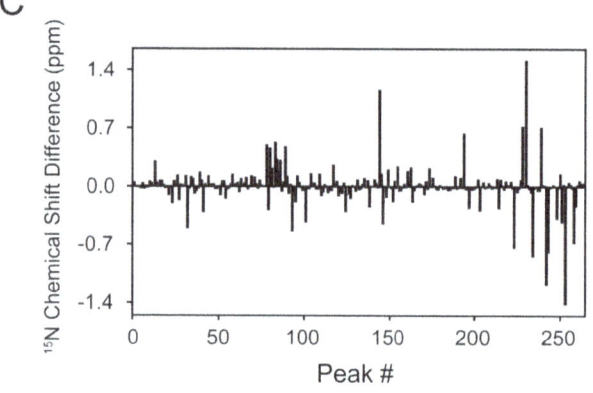

Figure 3. Effect of K18N mutation on protein structure. (A) Overlay of 2D ^{15}N-^{1}H TROSY-HSQC spectra of the WT (black) and the mutant (red). Each cross-peak represents a unique residue in the protein. (B) and (C) show the change in the ^{1}H and ^{15}N chemical shifts of the mutant with respect to the WT.

only be used to qualitatively assess the decreased stability of the mutant.

To obtain the true thermodynamic stability difference, we used equilibrium chemical denaturation [29]. We first showed that the folding of both WT and the mutant are completely reversible. Starting from their unfolded states in high denaturant, both

proteins refold by 100% (Fig. 1F). No aggregation was observed for both proteins during refolding. Therefore, chemical denaturation can be used to measure thermodynamic protein stability. When the protein denaturation was monitored as a function of increasing urea concentration using either CD or fluorescence as the signal, both the WT and the mutant unfolded in a cooperative, sigmoidal transition. Fitting these denaturant melts to a standard two-state equation [20,21] resulted in Gibbs free energy of unfolding, ΔG values of 11.2 ± 0.3 kcal/mol and 7.5 ± 0.3 kcal/mol for the WT and the mutant respectively. These values indicate that the K18N mutant is thermodynamically less stable than the WT by 3.7 ± 0.4 kcal/mol. This stability difference can be understood as follows. The non-functional unfolded state population is given by [Unfolded] = [Native] exp $(-\Delta G/RT)$, where R is the gas constant, and T is absolute temperature in kelvin. That means, the decrease in ΔG by 3.7 kcal/mol increases the unfolded state population by 529 times in the case of mutant when compared to the WT.

The K18N mutant unfolds faster than the WT

Millisecond stopped-flow methods were used to monitor unfolding of the two proteins [30]. Starting from their native states with no denaturant, unfolding was initiated by mixing with high denaturant. The mutant unfolds three times faster than the WT (Figs. 2G & H). Unfolding rate constants for the mutant and the WT were 100.1 ± 0.9/s and 34.6 ± 0.3/s, respectively. These experiments along with the above chemical denaturation melts indicate that the mutant has lower kinetic and thermodynamic stability compared to the WT protein.

The K18N mutant shows structural differences at numerous amino acids

Two-dimensional NMR experiments were used to monitor the amino acid residue-resolved changes in protein structure [18]. Fig. 3A shows the overlay of heteronuclear ^{1}H-^{15}N HSQC spectra of the mutant and the WT. Each cross-peak corresponds to one amide in the protein, which is mainly the mainchain amide coming from an individual amino acid. Any change in the cross-peak position represents a local structural change at that particular amino acid, which in general may not be detectable by global probes such as circular dichroism. The overall cross-peak pattern of the mutant is similar to that of the WT (Fig. 3A). However, when individual cross-peak positions were compared, we do see clear differences (Figs. 3B & C). Both ^{1}H and ^{15}N chemical shifts of many residues show significant changes. The fact that NMR chemical shift differences can be seen at several residues implies that the effect of mutation is propagated over long distances in the protein structure. However, the overall secondary and tertiary structures of the protein were not affected by the mutation, as evident from no changes in the CD and fluorescence spectra (Figs. 1D & E).

The K18N mutant does not affect the actin-binding function

Traditional actin co-sedimentation assays were used to measure the actin binding affinity of the WT and the mutant [23,31,32]. Varying concentrations of the protein (0–60 μM) were added to a solution containing a fixed concentration of F-actin (final concentration of 7 μM) and were centrifuged at high speed (100,000g for 30 min) to separate the free protein from the protein bound to F-actin. The unbound protein remained in the supernatant, whereas F-actin and the bound protein pelleted at the bottom of the tube. The concentration of the bound protein

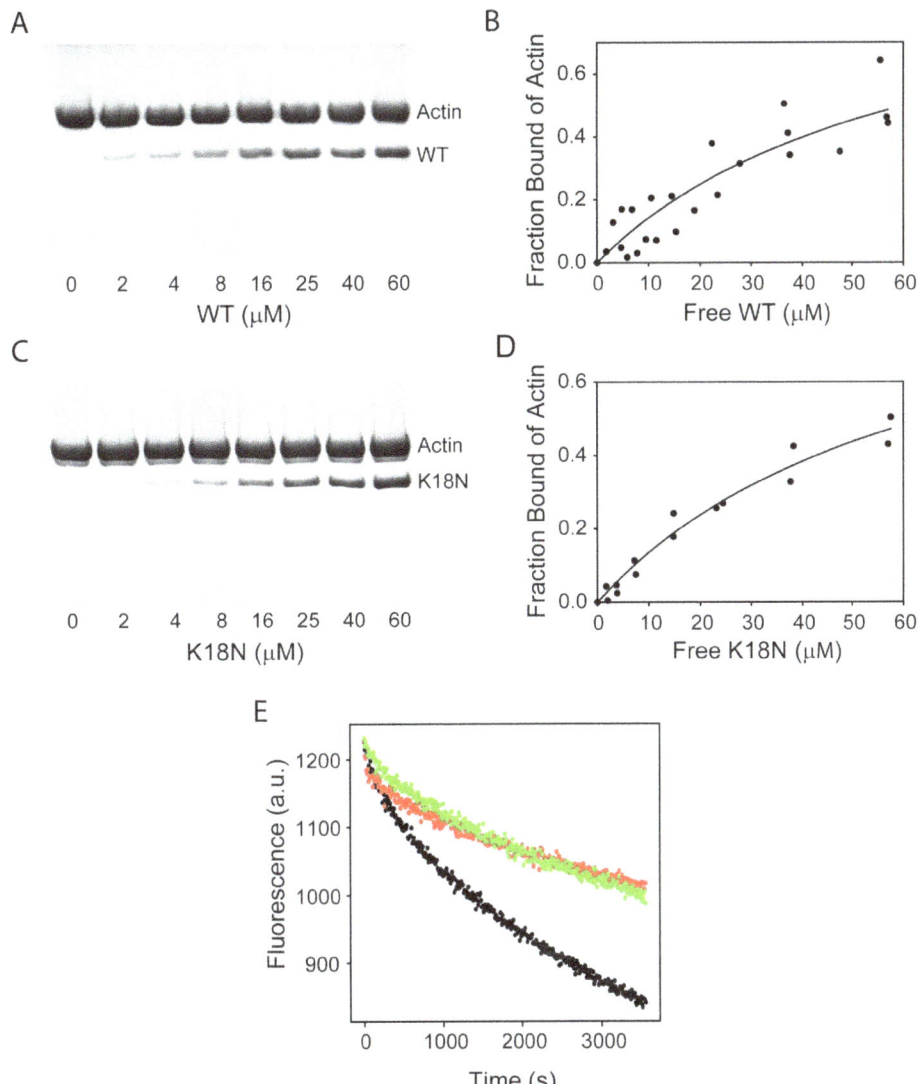

Figure 4. Effect of K18N mutation on protein function. (A) – (D) show the results from actin co-sedimentation assay. (A) & (C) show SDS-PAGE of the pellets after co-sedimentation of varying concentration of the WT or the mutant with a fixed concentration of F-actin. The upper and lower bands in SDS-PAGE indicate actin and bound WT (or mutant) in the pellet. (B) and (D) show the fraction of actin bound as a function of free protein, calculated from the protein band intensities from SDS-PAGE after correcting for the differential staining of the dye. (E) Actin depolymerization assay. Green and red represent the depolymerization kinetics of F-actin in the presence of the WT or the mutant, respectively. Black trace corresponds to the kinetics in the absence of the WT or the mutant.

was determined using SDS−PAGE densitometry (Figs. 4A & C), and was corrected using bovine serum albumin (BSA) as a standard to account for the differential staining of coomassie blue to proteins [23,24]. These corrected values were used to calculate the fraction of actin bound as a function of the free protein concentration (Figs. 4B & D). Actin-binding curves of the mutant and the WT were similar, indicating that the mutation does not affect the actin-binding function. These binding curves were fit to a standard binding equation, and the obtained K_d value of ~45 μM matches the K_d value predicted from earlier cryo-EM studies [33].

Dystrophin binding to F-actin offers partial protection against its depolymerization [14]. We examined the effect of K18N mutation on such depolymerization (Fig. 4E). For this experiment, F-actin with individual actin molecules labeled with pyrene fluorophores was diluted into G-actin buffer, and the decrease in

pyrene excimer fluorescence was followed. The main principle behind pyrene excimer fluorescence is that when an excited pyrene molecule is close in three-dimensional space to a neighboring, ground-state pyrene, their aromatic rings stack against each other to form a dimer (excited-state dimer or excimer) [34,35]. This results in a new fluorescence emission band, which is not present in the fluorescence spectrum of monomeric pyrene. When actin molecules labeled with pyrene are close in F-actin, excimer fluorescence will be higher. With depolymerization, excimer fluorescence will decrease, as the actin molecules in F-actin are falling apart into monomers. Pyrene excimer fluorescence decreases at the same rate for both WT and the mutant implying that both proteins protect F-actin to the same extent (Fig. 4E). This assay along with the above co-sedimentation assay indicates that the mutation did not affect the actin-binding function of the protein.

A

```
Human          MLWWEEVEDCYEREDVQKKTFTKWVNAQFSKFGKQHIENLFSDLQDGRRLLDLLEGLTGQ 60
Dog            MLWWEEVEDCYEREDVQKKTFTKWVNAQFSKFGKQHIENLFSDLQDGRRLLDLLEGLTGQ 60
Mouse          MLWWEEVEDCYEREDVQKKTFTKWINAQFSKFGKQHIDNLFSDLQDGKRLLDLLEGLTGQ 60
Pig            ----MSEVSSDEREDVQKKTFTKWINAQFSKFGKQHIENLFNDLQDGRRLLDLLEGLTGQ 56
Naked Mole Rat -----------EREDVQKKTFTKWINAQFSKFGKPHIENLFSDLNDGRCLLDLLEGLTGQ 49
```

B

```
Dystrophin     YEREDVQKKTFTKWVNAQFSKFGKQHIENLFSDLQDGRRLLDLLEGLTGQKLPK-EKGSTRV 71
Utrophin       DEHNDVQKKTFTKWINARFSKSGKPPINDMFTDLKDGRKLLDLLEGLTGTSLPK-ERGSTRV 87
Plectin        DERDRVQKKTFTKWVNKHLIKAQ-RHISDLYEDLRDGHNLISLLEVLSGDSLPR-EKGRMRF 60
Spectrin-β     DEREAVQKKTFTKWVNSHLARVS-CRITDLYTDLRDGRMLIKLLEVLSGERLPKPTKGRMRT 110
α-Actinin      PAWEKQQRKTFTAWCNSHLRKAG-TQIENIEEDFRDGLKLMLLLEVISGERLAKPERGKMRV 87
```

C

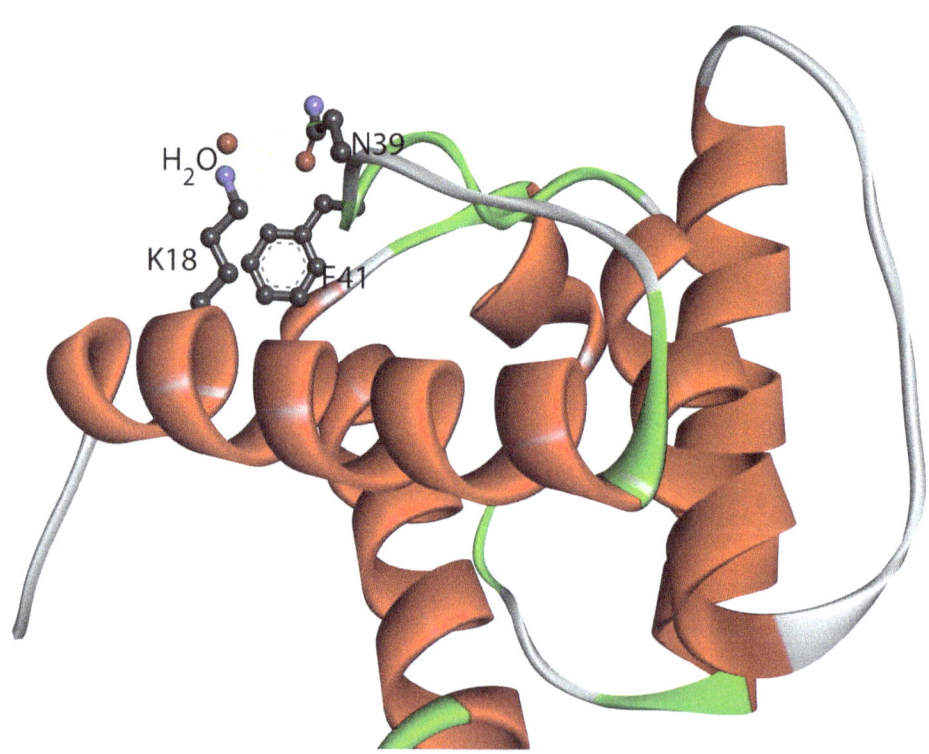

Figure 5. Sequence and structural analysis of the K18N mutation. (A) Sequence alignment of dystrophin from various mammals. Lysine at the 18th position, asparagine at the 39th position, and phenyl alanine at the 41st position are shown in red, green, and blue colors respectively. (B) Sequence alignment of similar actin binding domains from other human proteins. Residues at the 18th, 39th, and 41st positions are shown in red, green, and blue colors respectively. (C) Structural view of stabilizing interactions formed by the sidechain of K18 in the WT dystrophin structure. All atoms that come close to the amide nitrogen of the lysine sidechain are shown by connecting green lines.

Discussion

Our results show that the K18N mutation decreases protein stability and perturbs protein structure, but does not affect function. This is in contrary to common presumption that disease-causing mutations always affect protein function. There-fore, K18N mutation can be classified as a protein stability defect, and not as a protein function defect. Decrease in protein stability increases the non-functional unfolded protein population, given by [Unfolded] = [Native] exp $(-\Delta G/RT)$, where ΔG is the Gibbs free energy of unfolding, R is the gas constant, and T is the absolute temperature in kelvin. Unfolded or improperly folded protein molecules are in general broken into small peptides by the proteasome machinery, and the resultant peptides are used for further protein synthesis [36]. That means, any mutation that decreases the protein stability will result in a depletion of the net functional protein molecules. This might be the disease triggering factor for XLDCM. Consistently, XLDCM patients were reported to have decreased levels of dystrophin in cardiac muscle [11,12,17].

Sequence and structural analysis suggest why K18N mutation can be destabilizing (Fig. 5). The lysine residue at the 18th position is highly conserved in mammals (Fig. 5A) and in similar functional domains from other human proteins (Fig. 5B), which implies that this residue is essential for protein structure and/or function. This residue is part of the first helix in dystrophin (Fig. 1A). It stabilizes the turn connecting the first and second helices by forming a salt bridge with the sidechain of asparagine at position 39 and by forming a cation-pi interaction with the aromatic sidechain of phenylalanine at position 41 (Fig. 5C). Both residues 39 and 41 are also highly conserved in mammals (Fig. 5A) and across similar functional domains (Fig. 5B). Whenever there exists a lysine at the 18th position, the residues at positions 39 and 41 are always asparagine and an amino acid with a phenyl ring (either phenylalanine or tyrosine) respectively. In addition, the K18 sidechain also forms a stabilizing hydrogen bond with a bound water molecule (Fig. 5C). Therefore, mutating this residue is expected to break these various stabilizing interactions, which will destabilize the dystrophin structure.

A similar situation occurs in other diseases where decrease in protein stability appears to be the major triggering factor. In cystic fibrosis, the $\Delta F508$ mutation in cystic fibrosis transmembrane regulator (CFTR) protein, which occurs in 70% of cystic fibrosis patients, has a decreased stability compared to WT, leading to a decrease in the functional protein concentration *in vivo* and hence net decreased function [37,38]. Carcinogenic mutations in the tumor suppressor protein p53 reduce the protein stability. Folded p53 levels *in vivo* were correlated with changes in protein stability [39,40]. Disease-causing mutations in rhodopsin decrease its thermodynamic stability [41]. Although missense mutations can trigger disease by multiple routes, decrease in stability seems to be the major factor that is responsible for ~80% of monogenic missense mutation-triggered diseases in proteins [42].

It is interesting to compare the K18N mutation with other dystrophin mutations that trigger Duchenne/Becker muscular dystrophy (DMD/BMD). Similar to K18N, DMD/BMD mutations also decrease protein stability, which may account for decreased dystrophin levels observed in DMD/BMD patients [18,43,44]. However, the major difference between K18N and DMD/BMD mutations is protein aggregation. DMD/BMD mutants undergo severe protein aggregation, whereas K18N does not aggregate. Whether aggregation plays a major role in DMD/BMD and not in XLDCM is not clear. In addition, why K18N has no effect on skeletal muscles is unclear. These aspects need to be further examined.

One caveat in this study is that we are studying isolated N-ABD instead of the full-length human dystrophin. High-yield expression and purification of full-length human dystrophin to acceptable purity is not yet possible. Further, full-length dystrophin is very large in size (~427 kDa), and is not amenable to many high-resolution structural and biophysical techniques. Therefore, we follow a reductionist approach of studying individual domains, similar to studies on other dystrophin domains that include N-ABDs [18,28,33,34,45], spectrin repeats [46–52], and C-terminal domains [53]. Because the full-length human protein is not yet possible to express and purify, effects of disease-causing mutations on the structure and function of dystrophin are in general probed at the level of individual domains [18,28,50,51,54]. In addition, determining the mutation effects on full-length dystrophin solution structure on the scale of individual amino acids (Fig. 3) is not possible with the available structural methods. Typical protein size for which solution NMR methods can be currently applied needs to be less than 300 amino acids.

Our results on human dystrophin N-ABD agree to some extent with an earlier study on full-length mouse dystrophin [44]. In this earlier work, which involved the expression and purification of full-length mouse dystrophin, the K18N mutant showed an α-helical circular dichroism as that of the WT dystrophin [44], similar to that observed here for human dystrophin N-ABD (Fig. 1C). However, the mutation resulted in the loss of sigmoidal transition for mouse dystrophin during its temperature melt [44], in contrary to the sigmoidal melt observed here for the K18N mutant of human dystrophin N-ABD (Fig. 2B). Further, the K18N mutation resulted in a significant aggregation of mouse dystrophin [44], whereas the human dystrophin N-ABD mutant did not aggregate (Fig. 1F). These differences can be because of two possible reasons. The first reason is that we are studying the mutation effects on isolated N-ABD rather than the full-length protein. The second possible reason is that human dystrophin may differ from mouse dystrophin in terms of protein structure and function. These aspects need to be further probed.

Acknowledgments

We acknowledge the help from the Rocky Mountain 900 MHz NMR Facility (funded by the NIH instrumentation Grant P41GM068928) and the Biophysics Core, University of Colorado Anschutz Medical Campus in carrying out this work.

Author Contributions

Conceived and designed the experiments: SMS SB GA KMGM. Performed the experiments: SMS SB GA. Analyzed the data: SMS SB DS GA KMGM. Contributed reagents/materials/analysis tools: SMS SB GA KMGM. Wrote the paper: SMS KMGM.

References

1. Berko BA, Swift M (1987) X-linked dilated cardiomyopathy. N Engl J Med 316: 1186–1191.

2. Ortiz-Lopez R, Li H, Su J, Goytia V, Towbin JA (1997) Evidence for a dystrophin missense mutation as a cause of X-linked dilated cardiomyopathy. Circulation 95: 2434–2440.

3. Towbin JA, Hejtmancik JF, Brink P, Gelb B, Zhu XM, et al. (1993) X-linked dilated cardiomyopathy. Molecular genetic evidence of linkage to the Duchenne muscular dystrophy (dystrophin) gene at the Xp21 locus. Circulation 87: 1854–1865.

4. Franz WM, Cremer M, Herrmann R, Grunig E, Fogel W, et al. (1995) X-linked dilated cardiomyopathy. Novel mutation of the dystrophin gene. Ann N Y Acad Sci 752: 470–491.

5. Muntoni F, Cau M, Ganau A, Congiu R, Arvedi G, et al. (1993) Deletion of the dystrophin muscle-promoter region associated with X-linked dilated cardiomyopathy. N Engl J Med 329: 921–925.

6. Milasin J, Muntoni F, Severini GM, Bartoloni L, Vatta M, et al. (1996) A point mutation in the 5′ splice site of the dystrophin gene first intron responsible for X-linked dilated cardiomyopathy. Hum Mol Genet 5: 73–79.

7. Bies RD, Maeda M, Roberds SL, Holder E, Bohlmeyer T, et al. (1997) A 5′ dystrophin duplication mutation causes membrane deficiency of alpha-dystroglycan in a family with X-linked cardiomyopathy. J Mol Cell Cardiol 29: 3175–3188.

8. Yoshida K, Nakamura A, Yazaki M, Ikeda S, Takeda S (1998) Insertional mutation by transposable element, L1, in the DMD gene results in X-linked dilated cardiomyopathy. Hum Mol Genet 7: 1129–1132.

9. Muntoni F, Di Lenarda A, Porcu M, Sinagra G, Mateddu A, et al. (1997) Dystrophin gene abnormalities in two patients with idiopathic dilated cardiomyopathy. Heart 78: 608–612.

10. Ferlini A, Galie N, Merlini L, Sewry C, Branzi A, et al. (1998) A novel Alu-like element rearranged in the dystrophin gene causes a splicing mutation in a family with X-linked dilated cardiomyopathy. Am J Hum Genet 63: 436–446.

11. Muntoni F, Wilson L, Marrosu G, Marrosu MG, Cianchetti C, et al. (1995) A mutation in the dystrophin gene selectively affecting dystrophin expression in the heart. J Clin Invest 96: 693–699.

12. Muntoni F, Melis MA, Ganau A, Dubowitz V (1995) Transcription of the dystrophin gene in normal tissues and in skeletal muscle of a family with X-linked dilated cardiomyopathy. Am J Hum Genet 56: 151–157.

13. Aaartsma-Rus A, van Deutekom JCT, Fokkema IF, van Ommen G-JB, Den Dennen JT (2006) Entries in the Leiden Duchenne muscular dystrophy mutation database: An overview of mutation types and paradoxical cases that confirm the reading-frame rule. Muscle Nerve 34: 135–144.

14. Ervasti JM (2007) Dystrophin, its interactions with other proteins, and implications. Biochim Biophys Acta 1772: 108–117.

15. Winder SJ (1997) The membrane-cytoskeleton interface: The role of dystrophin and utrophin. J Muscle Res Cell Motility 18: 617–629.

16. Sweeney HL, Barton ER (2000) The dystrophin-associated glycoprotein complex: what parts can you do without? Proc Natl Acad Sci USA 97: 13464–13466.

17. Feng J, Yan J, Buzin CH, Towbin JA, Sommer SS (2002) Mutations in the dystrophin gene are associated with sporadic dilated cardiomyopathy. Mol Genet Metabol 77: 119–126.

18. Singh SM, Kongari N, Cabello-Villegas J, Mallela KMG (2010) Missense mutations in dystrophin that trigger muscular dystrophy decrease protein stability and lead to cross-β aggregates. Proc Natl Acad Sci USA 107: 15069–15074.

19. Greenfield NJ (2006) Using circular dichroism spectra to estimate protein secondary structure. Nature Protocols 6: 2876–2890.

20. Santoro MM, Bolen DW (1988) Unfolding free energy changes determined by the linear extrapolation method. 1. Unfolding of phenylmethanesulfonyl alpha-chymotrypsin using different denaturants. Biochemistry 27: 8063–8068.

21. Santoro MM, Bolen DW (1992) A test of the linear extrapolation of unfolding free energy changes over an extended denaturant concentration range. Biochemistry 31: 4901–4907.

22. Delaglio F, Grzesiek S, Vuister GW, Zhu G, Pfeifer J, et al. (1995) NMRPipe: a multidimensional spectral processing system based on UNIX pipes. J Biomol NMR 6: 277–293.

23. Singh SM, Bandi S, Winder SJ, Mallela KMG (2014) The actin binding affinity of the utrophin tandem calponin-homology domain is primarily determined by its N-terminal domain. Biochemistry 53: 1801–1809.

24. Tal M, Silberstein A, Nusser E (1985) Why does Coomassie Brilliant Blue R interact differently with different proteins? A partial answer. J Biol Chem 260: 9976–9980.

25. Lakowicz JR (2006) Principles of Fluorescence Spectroscopy. New York: Springer Science.

26. Chen Y, Barkley MD (1998) Toward understanding tryptophan fluorescence in proteins. Biochemistry 37: 9976–9982.

27. Royer CA (2006) Probing protein folding and conformational transitions with fluorescence. Chem Rev 106: 1769–1784.

28. Singh SM, Molas JF, Kongari N, Bandi S, Armstrong GS, et al. (2012) Thermodynamic stability, unfolding kinetics, and aggregation of the N-terminal actin binding domains of utrophin and dystrophin. Proteins: Struct Func & Bioinform 80: 1377–1392.

29. Pace CN (1986) Determination and analysis of urea and guanidine hydrochloride denaturation curves. Methods Enzymol 131: 266–280.

30. Bandi S, Singh SM, Mallela KMG (2014) The C-terminal domain of the utrophin tandem calponin-homology domain appears to be thermodynamically

31. Rybakova IN, Humston JL, Sonnemann KJ, Ervasti JM (2006) Dystrophin and utrophin bind actin through distinct modes of contact. J Biol Chem 281: 9996–10001.

32. Way M, Pope B, Weeds AG (1992) Evidence for functional homology in the F-actin binding domains of gelsolin and α-actinin: Implications for the requirements of severing and capping. J Cell Biol 119: 835–842.

33. Sutherland-Smith AJ, Moores CA, Norwood FLM, Hatch V, Craig R, et al. (2003) An atomic model for actin binding by the CH domains and spectrin-repeat modules of utrophin and dystrophin. J Mol Biol 329: 15–33.

34. Singh SM, Mallela KMG (2012) The N-terminal actin-binding tandem calponin-homology (CH) domain of dystrophin is in a closed conformation in solution and when bound to F-actin. Biophys J 103: 1970–1978.

35. Lehrer SS (1997) Intramolecular pyrene excimer fluorescence: A probe of proximity and protein conformational change. Methods Enzymol 278: 286–295.

36. Glickman MH, Ciechanover A (2002) The ubiquitin-proteasome proteolytic pathway: destruction for the sake of construction. Physiol Rev 82: 373–428.

37. Thibodeau PH, Brautigam CA, Machius M, Thomas PJ (2005) Side chain and backbone contributions of Phe508 to CFTR folding. Nature Struct Mol Biol 12: 10–16.

38. Du K, Sharma M, Lukacs GL (2005) The ΔF508 cystic fibrosis mutation impairs domain-domain interactions and arrests post-translational folding of CFTR. Nature Struct Mol Biol 12: 17–25.

39. Mayer S, Rüdiger S, Ang HC, Joerger AC, Fersht AR (2007) Correlation of levels of folded recombinant p53 in *Escherichia coli* with thermodynamic stability in vitro. J Mol Biol 372: 268–276.

40. Joerger AC, Fersht AR (2008) Structural biology of the tumor suppressor p53. Annu Rev Biochem 77: 557–582.

41. Toledo D, Ramon E, Aguila M, Cordomi A, Perez JJ, et al. (2011) Molecular mechanisms of disease for mutations at Gly-90 in rhodopsin. J Biol Chem 286: 39993–40001.

42. Yue P, Li Z, Moult J (2005) Loss of protein structure stability as a major causative factor in monogenic disease. J Mol Biol 353: 459–473.

43. Henderson DM, Belanto JJ, Li B, Heun-Johnson H, Ervasti JM (2011) Internal deletion compromises the stability of dystrophin. Hum Mol Genet 20: 2955–2963.

44. Henderson DM, Lee A, Ervasti JM (2010) Disease-causing missense mutations in actin binding domain 1 of dystrophin induce thermodynamic instability and protein aggregation. Proc Natl Acad Sci USA 107: 9632–9637.

45. Norwood FL, Sutherland-Smith AJ, Keep NH, Kendrick-Jones J (2000) The structure of the N-terminal actin-binding domain of human dystrophin and how mutations in this domain may cause Duchenne or Becker muscular dystrophy. Structure 8: 481–491.

46. Mirza A, Menhart N (2008) Stability of dystrophin STR fragments in relation to junction helicity. Biochim Biophys Acta 1784: 1301–1309.

47. Mirza A, Sagathevan M, Sahni N, Choi L, Menhart N (2010) A biophysical map of the dystrophin rod. Biochim Biophys Acta 1804: 1796–1809.

48. Ruszczak C, Mirza A, Menhart N (2009) Differential stabilities of alternative exon-skipped rod motifs of dystrophin. Biochim Biophys Acta 1794: 921–928.

49. Sahni N, Mangat K, Rumeur EL, Menhart N (2012) Exon edited dystrophin rods in the hinge 3 region. Biochim Biophys Acta 1824: 1080–1089.

50. Legardinier S, Legrand B, Raguénès-Nicol C, Bondon A, Hardy S, et al. (2009) A two-amino acid mutation encountered in Duchenne muscular dystrophy decreases stability of the rod domain 23 (R23) spectrin-like repeat of dystrophin. J Biol Chem 284: 8822–8832.

51. Acsadi G, Moore SA, Chéron A, Delalande O, Bennett L, et al. (2012) Novel mutation in spectrin-like repeat 1 of dystrophin central domain causes protein misfolding and mild Becker muscular dystrophy. J Biol Chem 287: 18153–18162.

52. Muthu M, Kylie A. Richardson, Sutherland-Smith AJ (2012) The crystal structures of dystrophin and utrophin spectrin repeats: Implications for domain boundaries. PLoS One 7: e40066.

53. Huang X, Poy F (2000) Structure of a WW domain containing fragment of dystrophin in complex with β-dystroglycan. Nature Struct Biol 7: 634–638.

54. Kahana E, Flood G, Gratzer WB (1997) Physical properties of dystrophin rod domain. Cell Motility and the Cytoskeleton 36: 246–252.

and kinetically more stable than the full-length protein. Biochemistry 53: 2209–2211.

17

Unusual Ratio between Free Thyroxine and Free Triiodothyronine in a Long-Lived Mole-Rat Species with Bimodal Ageing

Yoshiyuki Henning[1]*, Christiane Vole[1], Sabine Begall[1], Martin Bens[2], Martina Broecker-Preuss[3], Arne Sahm[2], Karol Szafranski[2], Hynek Burda[1], Philip Dammann[1,4]

1 Department of General Zoology, Faculty of Biology, University of Duisburg-Essen, Essen, Germany, **2** Genome Analysis, Leibniz Institute for Age Research - Fritz Lipmann Institute, Jena, Germany, **3** Department of Endocrinology and Metabolism and Division of Laboratory Research, University Hospital, University of Duisburg-Essen, Essen, Germany, **4** Central Animal Laboratory, University Hospital, University of Duisburg-Essen, Essen, Germany

Abstract

Ansell's mole-rats (*Fukomys anselli*) are subterranean, long-lived rodents, which live in eusocial families, where the maximum lifespan of breeders is twice as long as that of non-breeders. Their metabolic rate is significantly lower than expected based on allometry, and their retinae show a high density of S-cone opsins. Both features may indicate naturally low thyroid hormone levels. In the present study, we sequenced several major components of the thyroid hormone pathways and analyzed free and total thyroxine and triiodothyronine in serum samples of breeding and non-breeding *F. anselli* to examine whether *a*) their thyroid hormone system shows any peculiarities on the genetic level, *b*) these animals have lower hormone levels compared to euthyroid rodents (rats and guinea pigs), and *c*) reproductive status, lifespan and free hormone levels are correlated. Genetic analyses confirmed that Ansell's mole-rats have a conserved thyroid hormone system as known from other mammalian species. Interspecific comparisons revealed that free thyroxine levels of *F. anselli* were about ten times lower than of guinea pigs and rats, whereas the free triiodothyronine levels, the main biologically active form, did not differ significantly amongst species. The resulting fT4:fT3 ratio is unusual for a mammal and potentially represents a case of natural hypothyroxinemia. Comparisons with total thyroxine levels suggest that mole-rats seem to possess two distinct mechanisms that work hand in hand to downregulate fT4 levels reliably. We could not find any correlation between free hormone levels and reproductive status, gender or weight. Free thyroxine may slightly increase with age, based on sub-significant evidence. Hence, thyroid hormones do not seem to explain the different ageing rates of breeders and non-breeders. Further research is required to investigate the regulatory mechanisms responsible for the unusual proportion of free thyroxine and free triiodothyronine.

Editor: Eliseo A. Eugenin, Rutgers University, United States of America

Funding: This work was supported by a grant of the German Research Foundation to PD (DFG-grant DA 992/3-1) (http://www.dfg.de/). The publication fee was funded by the "Publish Open Access"-Program of the University of Duisburg-Essen and the German Research Foundation.The funders had no role in study design, data collection and analysis, decision to publish, or preparation of the manuscript.

Competing Interests: The authors have declared that no competing interests exist.

* Email: yoshiyuki.henning@uni-due.de

Introduction

Most ageing theories assume a link between metabolism and ageing because of several inevitable side effects of metabolic processes that potentially impair somatic integrity in the long term. Examples of such side effects are the production of reactive oxygen species [1,2], formation of advanced glycation end products [3,4,5], and telomere shortening with every cell proliferation cycle [6].

Thyroid hormones (THs) play a major role in development, differentiation and metabolism in vertebrates and are therefore assumed to affect ageing, too [7,8]. Experimental as well as comparative studies on various mammal models support this assumption. For example, experimentally induced hypothyroidism increases lifespan in rats [9], whereas experimentally induced hyperthyroidism decreased lifespan in young and middle-aged rats [10]. Additionally, *Ames dwarf mice* and *Snell dwarf mice*, which have extraordinary low levels of THs, and other hormones related to growth and development (e.g., somatropin, insulin-like growth factor 1), live significantly longer than wild type mice [11]. Whereas the treatment with somatropin does not have any effect on the lifespan of *Snell dwarf mice*, the administration of THs via food throughout adult life diminishes their lifespan, although it is still longer than in non-treated wild type mice [7,12]. Furthermore, longevity in vertebrate species is often associated with low metabolic rates, low TH levels, or both. For example, naked mole-rats (*Heterocephalus glaber*) are the longest-living rodent species (lifespan of >30 years) [13], showing only 79% of the allometrically expected resting metabolic rate of non-subterranean rodents [14] and very low levels of certain THs have been reported [15]. Also some long-lived bat species feature low metabolic rates [16,17]. In humans, there is a significant correlation between low TH metabolism and longevity [7,18].

In all vertebrates, the main THs are thyroxine (T4) and triiodothyronine (T3). Both THs are derivatives of the amino acid tyrosine and are synthesised in the thyroid gland. Synthesis of T4 and T3 is stimulated by the thyroid-stimulating hormone (TSH), which is released from the pituitary gland. The structures of T4 and T3 are strongly conserved in all mammalian species studied thus far, whereas TSH is species-specific. TSH consists of an unspecific alpha-subunit (TSHA), and a beta-subunit (TSHB), which is responsible for biological specificity [19]. TSH is stimulated by the thyrotropin-releasing hormone (TRH), which is secreted by the hypothalamus [20]. This hypothalamic-pituitary-thyroid (HPT) axis is regulated by THs exerting a negative feedback control over the secretion of TRH and TSH [21]. In peripheral tissues, THs are actively transported through the plasma membrane mainly by the monocarboxylate transporters 8 and 10 (MCT8 and MCT10) and, at least in mice and rats, the organic anion-transporting polypeptide 1C1 [22]. In the cytoplasm, specific deiodinases type 1 and 2 (D1, D2) convert T4 into T3 by deiodination of the outer ring of the T4 molecule [23,24].

The main biologically active TH, namely T3, regulates gene expression in the nucleus at various loci by binding to two types of thyroid hormone receptors (THRA and THRB) [25]. In addition to these classical TH functions, some non-nuclear TH actions have been described recently [26]. T4 and T3 are typically secreted into the blood stream in a ratio of about 6:1 (T4:T3) in rats and 14:1 in humans [27,28]. Thus, circulating T4 levels are manifold higher than T3 levels in healthy organisms. After secretion, less than 1% of the THs are circulating as biologically active free hormones (fT4 and fT3), while the major amount of T4 and T3 is bound to transport proteins [29,30].

Ansell's mole-rats (*Fukomys anselli*) are subterranean rodents endemic to Zambia. They show some promising features for ageing and TH studies. Similar to naked mole-rats, which belong to the same family of African mole-rats (Bathyergidae), *F. anselli* live in eusocial families, in which reproduction is usually monopolized by a single breeding pair [31]. The species has an extraordinary maximum lifespan of more than 20 years, which is far more than expected based on their body weight (~60–150 g). Remarkably, reproductive individuals live about twice as long as non-reproductive animals, regarding both their average and maximum lifespan (breeders: mean ca. 10 years, max. >20 years; non-breeders: mean lifespan ca. 4.8 years, max. 11.1 years ([32], own unpublished data). This bimodal ageing pattern of Ansell's mole-rats and other species of the same genus [33,34] contradicts the classic model that assumes a trade-off between reproduction and somatic maintenance [35,36]. Until now, conflicts to this trade-off model have only been reported about eusocial insect species like ants or termites [37,38]. The mechanisms underlying the unusual ageing pattern of *Fukomys* mole-rats are largely unknown.

Several indications suggest that Ansell's mole-rats may be naturally hypothyroid. First, oxygen consumption of *F. anselli* is significantly lower than expected from allometric equations, suggesting a low metabolic rate [39]. Low resting metabolic rates are typical for bathyergid rodents and are probably a physiological adaptation to their subterranean, low-oxygen environment [40]. Second, the retinae of Ansell's mole-rats show a high density of short-wave sensitive S-cone opsin. This is untypical for rodents, as they usually show S-cone opsins, as well as middle-to-long-wave sensitive L-cone opsins in diverse arrangements [41]. Although the adaptive function of colour perception in mole-rats is not yet understood, the high S-cone opsin density in Ansell's mole-rats

would be in line with the expected low TH levels, as fT3 is essential for the expression of L-cone opsins during the prenatal development [42]. Moreover, in athyroid mice, opsin expression can be restored with postnatal T4 treatment [43].

In order to characterize the TH system of Ansell's mole-rats qualitatively and quantitatively, we first sequenced mRNA of several major components of TH related pathways in order to find out whether the TH system of Ansell's mole-rats is evolutionary conserved, or whether it contains qualitative peculiarities compared to other mammalian species. Then, we determined serum TH levels in individuals of different age, sex, and breeding status. Here, we focused mainly on the following two questions: *i*) Do Ansell's mole-rats have lower circulating TH levels than unrelated, euthyroid rodent species (rats, guinea pigs)? *ii*) Are there differences in free TH levels between the slowly ageing reproductive and the faster ageing non-reproductive individuals? We hypothesized that *i*) Ansell's mole-rats have naturally low TH levels in comparison to euthyroid rodents, and *ii*) lower free TH levels in breeding animals (slow ageing) compared to non-breeders (faster ageing) as a possible molecular trigger for the lifespan differences between these two cohorts.

Materials and Methods

Animals

All Ansell's mole-rats used in this study were born, raised and maintained at the animal facilities of the Department of General Zoology, University of Duisburg-Essen, Germany. The age of the animals ranged from 1.2–10.2 years in non-breeders and 5.4–13.5 years in breeders at the day of serum sampling. They were housed as family groups in glass terraria on horticultural peat and fed *ad libitum* with carrots and potatoes every day, apples every second day, and grain and lettuce once a week. Room temperature and humidity were kept constant at 24±1°C and 40±3%, respectively.

Wistar-Unilever rats aged 6–9 months (Central Animal Laboratory of the University Hospital Essen, Germany) and *Dunkin Hartley* guinea pigs aged 12–24 months (Charles River, Wilmington, MA, USA) served as healthy, fully grown euthyroid controls. We did not introduce major age variation within these groups because the intraspecific variation of THs in these species has been summarized and studied elsewhere [44,45] and was not the focus of our study. Both species were housed at 21±1°C and 55±5% humidity in standard macrolon cages and were fed commercial, species-specific food pellets (ssniff).

Ethics Statement

Maintenance and all treatments of the animals were approved by the North Rhine-Westphalia State Environment Agency (Permit number: 87–51.04.2010.A359). Blood sampling was the only invasive treatment and was performed under deep anesthesia (ketamine and xylazine or isoflurane), except for guinea pigs, since anaesthesia is not necessary for blood sampling via the *vena saphena* if the procedure is sufficiently quick and appropriate restraining is possible [46]. All efforts were made to minimize suffering.

Sequencing and sequence analysis

The *F. anselli* transcriptome was characterized by high-throughput sequencing, as will be reported elsewhere. From the resulting data, we extracted transcripts of twelve thyroid-relevant genes. In detail, one male *F. anselli* was deeply anesthetized with isoflurane and killed by cervical translocation. Tissues from thyroid gland, ventral skin, adrenal gland, pancreas, testis, and brain stem were homogenized in a Tissue Lyser (Qiagen), and total mRNA was isolated using RNeasy (Qiagen, Valencia, CA, USA). Thereof, mRNA-Seq libraries were generated using platform-specific chemistry, according to the supplier's instructions (Illumina). Sequencing was performed using an Illumina Genome Analyzer IIx, resulting in a total of 88.7 million (9492 Mbp) single-end reads. Adapter clipping and trimming of low-quality 3′ ends (error probability of 0.5%) was performed with the programs cutadapt [47] and sickle [48], respectively. Reads shorter than 35 nt were removed, resulting in a total of 82.6 million (8,202 Mbp) reads. The transcriptome reads were pooled for a *de-novo* assembly with the Trinity software [49]. Resulting transcript contigs were labelled with gene symbols and were annotated for coding sequences (CDS) based on best bidirectional BLAST mapping against human protein coding genes (National Center for Biotechnology Information [NCBI], *H. sapiens* Annotation Release 104) using in-house scripts. The mRNA sequences were deposited in NCBI GenBank under accession numbers KJ958510-KJ958520 and KM676335. For species comparisons, *F. anselli* mRNA sequences were translated into proteins. These were aligned with orthologous protein sequences from 11 to 17 other mammalian species (RefSeq-database, NCBI) using CLUSTAL W. For genome-wide analysis of evolutionary selection trends we created five-species alignments of the CDS (human, dog, rat, mouse and naked mole-rat) using CLUSTAL W. For the thyroid target genes we additionally created the five-species multiple alignment with the Ansell's mole-rat sequence as the bathyergid representative. We used a parametric model of evolution implemented in the CodeML program of the PAML package [50,51] in performing the branch test on the mole-rat branch against all other branches as background (options CodonFreq = 2 and Kappa = 2). CodeML estimates the ratio of non-synonymous to synonymous mutations (Ka/Ks ratio), among several other parameters, separately for the mole-rat and all other branches. Additionally we used the "M0" model of CodeML to calculate the average Ka/Ks across the whole tree. We estimated the Ka/Ks ratios for the genome-wide orthologous gene set and used these to determine empirical probabilities ("percentiles") for particular Ka/Ks values, as well as Ka/Ks differences between the mole-rat branch and outer branches. This allowed to relate the Ansell's mole-rat Ka/Ks values of specific genes to the genome-wide Ka/Ks spectrum.

Sample sizes and sampling protocols

We sampled a total of 32 Ansell's mole-rats (12 breeders and 20 non-breeders, sex-balanced), 4 male rats and 4 male guinea pigs for the free TH measurements.

Additionally, we sampled a subgroup of 12 mole-rats (sex-balanced; both reproductive groups) plus further 7 rats (4 males, 3 females) and 4 guinea pigs (sex-balanced) to determine total TH levels (free and protein-bound fractions in total; tT4 and tT3). Again, free T4 and free T3 were determined in these serum samples in order to calculate the ratios of the free: total fractions for each hormone.

For blood sampling, mole-rats were anesthetized according to a standard protocol for mole-rats with an intramuscular injection of a 6 mg/kg dose of ketamine (10%, Ceva GmbH) and 2.5 mg/kg xylazine (2%, Ceva GmbH) [52]. Rats were isoflurane anesthetized, while guinea pigs did not receive any anaesthesia. To avoid hypothermia, animals were kept under a heat lamp before and after the treatment.

The mole-rat and guinea pig blood samples were taken from the *vena saphen*a of the hind paw with a capillary (Servoprax, 100 μl) and transferred into a serum test tube (Multivette 600, Sarstedt). The rat blood samples were taken by orbita puncture. All samples

were taken at approximately the same daytime, in order to avoid any bias due to the circadian variation in TRH secretion [21,53]. After approximately 20 minutes, the blood samples were centrifuged (Biofuge Pico, Heraeus Instruments) at $900 \times g$ for 5 minutes [15] and the serum (the clear top layer) was stored at $-80°C$ until use.

Quantification of thyroid hormones

Free and total thyroxine and triiodothyronine levels were quantified by means of a solid phase competitive enzyme immunoassay (EIA) for human serum (DRG Instruments GmbH). The use of a human serum EIA is justifiable since the molecular structures of T3 and T4 are not species specific ([29], results of the present study].

The accuracy of the fT3 and fT4 microplate EIA test system was confirmed by analyzing known hormone values, and by comparing the results with those of a reference method (radioimmunoassay). The correlation coefficient between the concentrations measured by the two methods was 0.95 (fT3) and 0.96 (fT4), which indicates a high accuracy of the test systems. The intra- and inter-assay variances are shown in Table S1. According to the manufacturer the cross-reactivities of the antibodies were as follows: *Triiodothyronine* − triiodothyronine: 100%; thyroxine: 0.02–0.37%; iodothyrosine, diiodothyrosine, phenylbutzone, sodium salicylate: 0.01–0.2%. *Thyroxine* − thyroxine 100%, triiodothyronine 3%, diiodothyronine, diiodotyrosine and iodotyrosine 0.01%. The assay sensitivities (i.e., detection limits) were 0.05 ng/dl (fT4), 8 nmol/l (tT4), 0.05 pg/ml (fT3), and 0.1 ng/ml (tT3).

Statistical analyses

The statistical analyses of the thyroid hormone levels were conducted with the software SPSS Statistics v.20.0.0 (IBM Corp.). Normal distribution was tested with the Kolmogorov-Smirnov-test with Lilliefors correction. For interspecies comparisons, one-way ANOVA with Bonferroni post hoc tests were applied. For intraspecific comparisons, a generalized linear model (GLM) was run with sex, reproductive status, age, and weight as independent factors, and fT4 and fT3 values as dependent variables. We calculated *i*) the main effects of the independent factors alone and *ii*) a two factor model using the interaction of status × age as explaining variable.

In the present study, mole-rats proved to have very low fT4 levels in general: 8 out of 32 mole-rats showed fT4 levels below the detection limit (<0.05 ng/dl), and the majority of fT4 values fell relatively close to the detection limit of the assay. It was impossible to decide whether the 8 missing values were failures or represented fT4 levels lower than 0.05 ng/dl. Therefore, analyses of fT4 were run under two different scenarios: 1) treating missing values as failures (effective n = 24; "first scenario" henceforth), and 2) replacing the missing values with the value of the detection limit (0.05 ng/dl) (effective n = 32; "second scenario" henceforth).

Results

Molecular constituents of TH system are conserved in *F. anselli*

We first characterized the TH system of Ansell's mole-rats on the genetic level and compared protein sequences with those of several other species representing different mammalian subgroups. Starting with RNA-seq of five different *F. anselli* tissues, we obtained full protein sequences for the two TH receptors (THR alpha [THRA, Figure S1]; THR beta [THRB, Figure S2]), one member of the regulation cascade (TSH beta subunit [TSHB, Figure S3]), two metabolizing deiodinases (D1, Figure S4; D2,

Figure S5), two members of the synthesis pathway (thyroglobulin [TG, Figure S6]; thyroperoxidase [TPO, Figure S7]), two transporter proteins (transthyretin [TTR, Figure S8]; thyroxine-binding globulin [TBG, coded by *Serpina7*, Figure S9]), one TH-regulated protein (hypoxia-inducible factor 1 alpha [HIF1A, Figure S10]), one TH transporter (monocarboxylate transporter 8 [MCT8, coded by *Slc16a2*, Figure S11]) and the sodium/iodid symporter (NIS, coded by *SLC5a5*, Figure S12). In order to analyze the conservation status of these molecular markers for the *F. anselli* TH system in comparison to other mammals, we analyzed the ratio of non-synonymous versus synonymous nucleotide changes in the CDS (Ka/Ks; Table 1). In all genes, the Ka/Ks ratio across the selected species was within the 95% percentile of genome-wide Ka/Ks values (≤ 0.443), indicating average levels of purifying selection. For only one gene (*Serpina7*), the Ka/Ks ratio was outside the 90% percentile (≤ 0.340), which is well within bounds accounting for multiple testing on eleven genes. Furthermore, in nearly all cases, the branch-specific Ka/Ks ratio of *F. anselli* was close to the ratio of the background branches as well as the average across the tree (Table 1). Exceptions were *Slc16a2*, which showed a slightly higher purifying selection pressure, and *Ttr*, which showed a much weaker purifying selection in the *F. anselli* branch. Nevertheless, for *Ttr*, the absolute Ka/Ks difference between foreground and background branch is within the 80% percentile ($p = 0.201$, one-sided) that is seen genome-wide.

TSHB, an important upstream molecular target for the study of TH dynamics, shows three species-specific changes in *F. anselli* at positions that are conserved in all 16 other mammalian species (E46D, V112I and F157I; Figure S3). This is comparable to the rate of species-specific changes found in TSHB of other mammals, ranging from 0 to 9. In pairwise comparisons to the sequence of *F. anselli*, the number of amino acid differences varies from 10 (to thirteen-lined ground squirrel, *Ictidomys tridecemlineatus*) to 24 (guinea pig), which underlines that even closely related species (Ansell's mole-rats and guinea pigs) show high variations in the TSHB sequence. Finally, we analyzed the *F. anselli* TH receptors, particularly their ligand-binding domains, as specific markers for conservation of the TH molecules. The ligand-binding domain of *F. anselli* THRA (protein position 190–370) shows only one conservative change from glutamic acid to aspartic acid (E270D) when compared to the consensus of 14 other mammalian species (Figure 1a, Figure S1). The ligand-binding domain of THRB (position 222–464) does not show a single amino acid difference between *F. anselli* and the sequence consensus of 17 other mammalian species (Figure 1b, Figure S2).

Free TH levels: Ansell's mole-rats have low fT4, but normal fT3

Under both scenarios, Ansell's mole-rat fT4 levels (0.18 ± 0.08 ng/dl [n = 24] and 0.15 ± 0.09 ng/dl [n = 32]; Table 2) were about 10 times lower than fT4 measured in rats (2.11 ± 0.67 ng/dl) and guinea pigs (2.25 ± 0.25 ng/dl; one-way ANOVA: F [first/second scenario] = 206.38/273.96, $p < 0.0001$; Bonferroni post hoc comparisons in both scenarios: mole-rat vs. rat and mole-rat vs. guinea-pig: $p < 0.0001$; rat vs. guinea-pig: $p > 0.99$); Figure 2). By contrast, fT3 levels did not differ significantly among the three species (mole-rat: 2.24 ± 0.96 pg/ml; rat: 2.85 ± 0.34 pg/ml; guinea pig: 2.36 ± 0.35 pg/ml; one-way ANOVA: $F = 0.85$, $p = 0.44$; Figure 2).

Total TH levels: Ansell's mole-rats have low tT4 and low tT3

Ansell's mole-rat tT4 levels (20.07 ± 6.32 ng/ml) were significantly lower than those of rats (47.05 ± 10.39 ng/ml) and guinea pigs (42.01 ± 14.77 ng/ml; one-way ANOVA, $F = 19.51$, $p < 0.0001$; Bonferroni post hoc comparison: mole-rat vs. rat: $p < 0.0001$; mole-rat vs. guinea pig: $p = 0.003$; rat vs. guinea pig: $p > 0.99$; Figure 3). Levels of tT3 were significantly lower between mole-rats (1.13 ± 0.25 ng/ml) and rats (3.14 ± 1.34 ng/ml), but not guinea pigs (1.80 ± 1.03 ng/ml; one-way ANOVA: $F = 11.13$, $p = 0.001$; Bonferroni post hoc comparison: mole-rat vs. rat: $p < 0.0001$; mole-rat vs. guinea pig: $p = 0.629$; rat vs. guinea pig: $p = 0.076$; Figure 3).

Hormone ratios

We estimated the percentage of unbound hormone fractions by determining the proportion of fTH in relation to tTH per serum sample. The proportion of fT4 to tT4 was significantly lower in mole-rats ($0.02 \pm 0.01\%$) compared to rats ($0.04 \pm 0.01\%$) and guinea pigs ($0.05 \pm 0.02\%$; one-way ANOVA: $F = 9.60$, $p = 0.001$; Bonferroni post hoc comparison: mole-rats vs. rat: $p = 0.02$; mole-rats vs. guinea pig: $p = 0.002$; rat vs. guinea pig: $p = 0.496$; Figure 4). In contrast, the proportion of fT3 to tT3 was highest in mole-rats ($0.38 \pm 0.14\%$; rats: $0.17 \pm 0.09\%$; guinea pigs $0.23 \pm 0.08\%$), with differences being statistically significant in the comparison with rats and close to significance threshold in the

Table 1. K_a/K_s ratios indicate persisting purifying selection on proteins of the TH system in *F. anselli*.

Protein (Gene)	K_a/K_s ratio across tree	K_a/K_s ratio *F. anselli* branch	K_a/K_s ratio background branches
THRA	0.011	0.017	0.008
THRB	0.036	0.054	0.033
TG	0.312	0.336	0.305
TPO	0.127	0.088	0.108
D1 (Dio1)	0.229	0.227	0.198
D2 (Dio2)	0.144	0.158	0.149
TSHB	0.209	0.176	0.189
TBG (Serpina7)	0.342	0.358	0.333
TTR	0.226	0.531	0.202
MCT8 (Slc16a2)	0.076	0.007	0.098
NIS (*Slc5a5*)	0.074	0.051	0.068

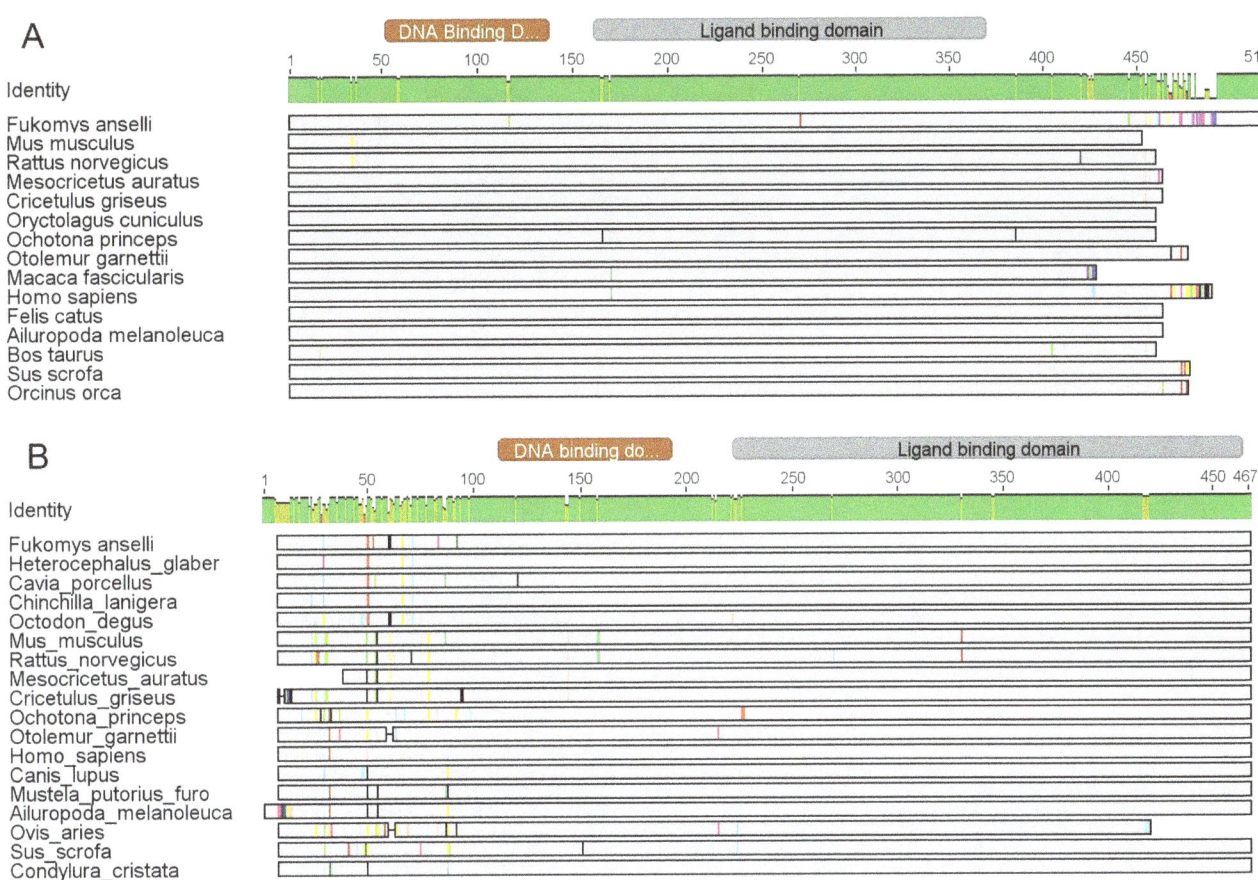

Figure 1. Protein sequence alignment for (A) TH receptor α (THRA) and (B) TH receptor β (THRB) of different mammalian species. *F. anselli* mRNA sequences were derived from RNA-seq and subsequently translated, the other sequences were retrieved from NCBI databases (accession numbers are given in Figures S1 and S2). Sequence differences are marked in gray. "Identity" shows the percentage of amino acid conformity for each position; the protein domain regions correspond to the human sequence entry. A fully resolved representation of the alignment is given as a supplement (Figures S1 and S2).

comparison with guinea pigs (one-way ANOVA: $F = 7.724$, $p = 0.004$; Bonferroni post hoc comparison: mole-rats vs. rats: $p = 0.004$; mole-rats vs. guinea pig: $p = 0.075$; rat vs. guinea pig: $p > 0.99$; Figure 4).

We also calculated the ratios of free T4:T3, and total T4:T3 in all three species (Table 3). Free T4:T3 ratios in mole-rats (0.70 ± 0.45–0.85 ± 0.43, depending on the scenario) were significantly lower than in both other species (rats: 7.28 ± 1.87, guinea pigs: 9.63 ± 1.50; one-way ANOVA [first/second scenario]: $F = 240.53/309.71$, $p < 0.0001$ each; Bonferroni post hoc comparison: mole-rat vs. rat: $p < 0.0001$ [both scenarios]; mole-rat vs. guinea pig: $p < 0.0001$; rat vs. guinea pig: $p = 0.002$). In contrast, tT4:tT3 ratios did not differ between mole-rats and the other species (mole-rats: 18.60 ± 6.98, rats: 16.24 ± 4.28, guinea pigs: 26.43 ± 9.51; one-way ANOVA: $F = 3.00$, $p = 0.07$; Bonferroni post hoc comparison: mole-rat vs. rat: $p > 0.99$; mole-rat vs. guinea pig: $p = 0.17$; rat vs. guinea-pig: $p = 0.08$).

Free thyroid hormone levels do not explain intraspecific ageing differences

Intraspecific comparisons revealed, that the fT4 levels of Ansell's mole-rats were not affected by any of the tested factors sex, reproductive status, age and weight alone in the first scenario ($n = 24$), and neither so by the interaction of reproductive status ×

age (Table 4). In the second scenario ($n = 32$), the fT4 levels were again not affected by sex, reproductive status and weight alone, nor by the interaction of status × age (Table 4). Levels fT4 increased significantly with age under this scenario ($p = 0.041$; Table 4).

Free T3 levels were not influenced by any of the tested factors or interactions in the GLM (Table 4).

Discussion

The aim of our study was to characterize the TH system of Ansell's mole-rats and to determine T4 and T3 levels in this species to investigate if these animals are hypothyroid, and if their hormone concentrations correlate with their extraordinary lifespan and the bimodal ageing pattern of reproductive and non-reproductive animals.

Molecular constituents of TH system are conserved in *F. anselli*

In *F. anselli*, the major molecular constituents of the mammalian TH system could be identified via their mRNAs, namely TG, TPO, TSHB, D1, D2, TTR, MCT8, TBG and NIS. The mRNAs show full protein-coding capacity, and sequence substitution patterns suggest that purifying selection acts on these molecules to a similar extent as found in other mammal species

Table 2. Mean (±SD) fT4 and fT3 values in Ansell's mole-rats.

	fT4 (ng/dl; n=24)			fT4 (ng/dl; n=32)			fT3 (pg/ml; n=32)		
	female	male	all	female	male	all	female	male	all
R[1]	0.20±0.11 (n=5)	0.17±0.08 (n=6)	0.18±0.09 (n=11)	0.18±0.12 (n=6)	0.17±0.08 (n=6)	0.17±0.09 (n=12)	3.01±2.02 (n=6)	2.24±0.49 (n=6)	2.63±1.45 (n=12)
NR[2]	0.18±0.07 (n=9)	0.16±0.06 (n=4)	0.17±0.06 (n=13)	0.17±0.08 (n=10)	0.09±0.07 (n=10)	0.13±0.08 (n=20)	1.97±0.23 (n=10)	2.04±0.43 (n=10)	2.01±0.34 (n=20)

[1]NR = non-reproductive; R = reproductive.

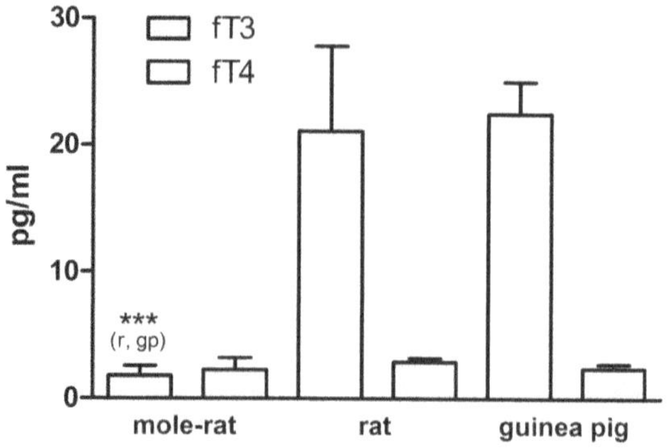

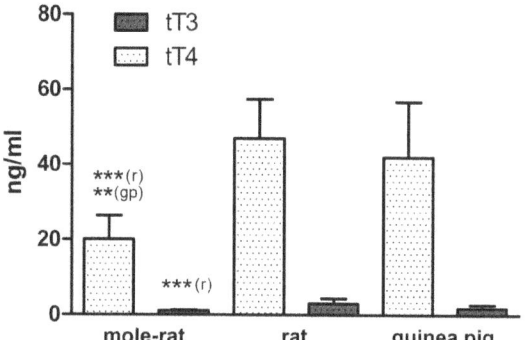

Figure 2. Free T4 and free T3 levels of Ansell's mole-rats (n = 24), rats (n = 4) and guinea pigs (n = 4). Mean ± SD; all data expressed in pg/ml. One-way ANOVA, fT4: $F = 206.38$, $p < 0.0001$; fT3: $F = 0.85$, $p = 0.44$. Significant differences in the Bonferroni post hoc comparisons are indicated by asterisks coupled with the comparison species referred to (mr = mole-rat, r = rat, gp = guinea pig) in parenthesis. See "Results" section for statistical details and Table 3 for TH ratios obtained from these data. Mole-rat fT4 data refer to scenario 1; applying the second scenario (not depicted here) created essentially the same result because mean fT4 values of mole-rats were slightly lower.

(Table 1). However, the mutational drift for some proteins is such high that immunochemical detection methods, e.g. for TSHB, will require development of species-specific antibodies. In addition, preliminary work showed no results for *F. anselli* samples with commercially available TSH assays, both for human and guinea pig, which confirms the species-specificity of TSH as well.

Most notably, the two TH receptor isoforms, THRA and THRB, show a high level of sequence conservation in *F. anselli* compared to 15 mammalian species, especially within their hormone-binding domains (Figure 1, Figures S1 and S2). This finding supports the expectation that just as in other mammals, the structures of T4 and T3 are conserved in *F. anselli*, although these were not directly determined.

Figure 3. Total T4 and total T3 levels of Ansell's mole-rats (n = 12), rats (n = 7) and guinea pigs (n = 4). Mean ± SD; all data expressed in ng/ml. One-way ANOVA, tT4: $F = 19.51$, $p < 0.0001$; fT3: $F = 11.13$, $p = 0.001$. Significant differences in the Bonferroni post hoc comparisons are indicated by asterisks coupled with the comparison species referred to (mr = mole-rat, r = rat, gp = guinea pig) in parenthesis. See "Results" section for statistical details and Table 3 for TH ratios obtained from these data.

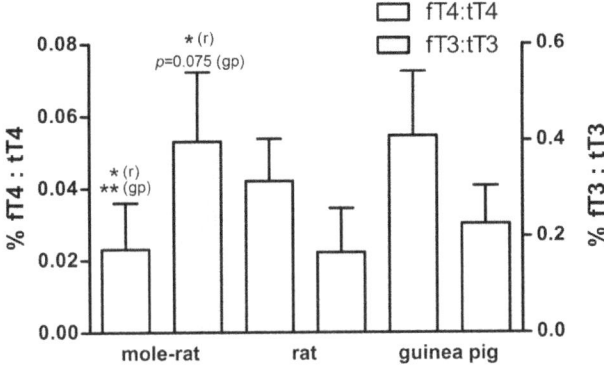

Figure 4. fT4:tT4 and fT3:tT3 ratios in Ansell's mole-rats (n = 12), rats (n = 7) and guinea pigs (n = 4). One-way ANOVA, fT4:tT4: $F = 9.60$, $p = 0.001$; fT3:tT3: $F = 7.724$, $p = 0.004$. Significant differences in the Bonferroni post hoc comparisons are indicated by asterisks coupled with the comparison species referred to (mr = mole-rat, r = rat, gp = guinea pig) in parenthesis. See "Results" section for statistical details.

Interspecies comparison of TH hormone levels

The first striking result of our study is that serum fT4 levels were about 10 times lower in mole-rats than in guinea pigs and rats, regardless of the scenario applied. Low circulating T4 levels are often caused by iodine deficiency in the diet [54], but this explanations appears unlikely in Ansell's mole-rats, because carrots, which they receive *ad libitum*, contain more iodine than required [54]. Moreover, the sequence analysis of the natrium/iodide symporter (NIS; Figure S12) appears to be under strong purifying selection in Ansell's mole-rats as in other mammals (Table 1), thus reducing the possibility of an iodide deficiency in the thyroid gland. Note that typical symptoms of a lifelong iodide deficiency like e.g. goiters [55] have not been observed in *Fukomys* (own unpublished data) or *Heterocephalus* mole-rats [15] so far. Thus it is plausible to assume that the low fT4 and tT4 levels reported here reflect the natural status of these animals, which provides principal support for our first hypothesis that Ansell's mole-rats have naturally low TH levels in comparison to euthyroid rodents. As shown in various vertebrate species, long lifespan is often correlated with low T4 levels, low metabolic rates, or both [7,16–18]. Many long-lived bathyergid species, including *F. anselli*, have very low metabolic rates [39,40,56], and the only member of the family *Bathyergidae* in which TH levels have been quantified so far (the naked mole-rat) has also shown remarkably low fT4 [15]. The T4 levels observed in the present study are in good agreement with these findings.

Considering that the fT4 levels in mole-rats are by an order of magnitude lower than in rats and guinea pigs, it is interesting to note that after 30 years of maintaining and breeding mole-rats, we have no indications for developmental or cognitive impairments of the progeny. This is noteworthy, because in other animal models (chicken, rats and mice), induction of even mild fT4 deficits in the mother during prenatal development affects brain development, potentially leading to significant cognitive and/or motoric impairments in the progeny [57–59]. In humans, maternal fT4 deficits during pregnancy are associated with an elevated risk of cognitive impairments in the child, including severe disorders like e.g. autism [60,61]. Preliminary own data suggests that female mole-rats do not elevate fT4 levels during pregnancy. Should this assumption be verified, it will be worthwhile to investigate the mechanisms that enable Ansell's mole-rats to deal with such low maternal fT4 levels during prenatal development without ontogenetic impairments.

The low fT4 values may, however, help to explain the unexpectedly high S-cone opsin concentration in the retina of Ansell's mole-rats [41], because THs are essentially involved in the expression of L-opsins in the mammalian retina by binding to a THRB isoform in the cones [42,43]. The adaptive value of colour perception for a strictly subterranean rodent is probably residual; studies by Kott et al. [62] suggest that while rods play an important role in the subterranean habitat, cones, especially S-cones, have no specific adaptive function. Our results provide the alternative explanation that the expression of these S-opsins could be a side effect of a natural state of low T4, which has evolved for other reasons (in this case potentially metabolism).

Total thyroxine levels (tT4) were also significantly lower in mole-rats than in the controls, but the differences were less pronounced than in the free hormone fractions; on average, mole-rat tT4 levels reached about 50% of those measured in rats and guinea pigs. Of note, we found that also the fT4/tT4 ratio is significantly lower in mole-rats than in the two control species (Figure 4). It hence appears that mole-rats do not only produce less T4 in their thyroid glands, but also recruit lesser proportions of their total T4 resources into the active form. Taken together, mole-rats seem to possess two distinct mechanisms that work hand in hand to downregulate fT4 levels reliably, which indicates an adaptive function of low T4 levels in these animals. We will discuss potential proximate mechanisms for their maintenance later in this manuscript.

Interestingly, and in sharp contrast to the low T4, fT3 levels were undistinguishable between *F. anselli* and the euthyroid controls (Figure 2). Total T3 levels of mole-rats were also not statistically different from those measured in guinea pigs, but significantly lower than in rats (Figure 3). The ratios between free and total T3 (Figure 4) suggest that mole-rats recruit significantly higher portions of the available T3 into the active unbound form than the other two species, counteracting the much lower T4 levels. Although these results should be treated with some caution because our rat tT3 levels appear atypically high (see e.g.

Table 3. Ratios ($\pm$ SD) of free and total T4:T3 in interspecies comparison.

Species	Ratio fT4:fT3	Ratio tT4:tT3
Mole-rat	0.70±0.45* 0.85±0.43[†]	18.60±6.98
Rat	7.28±1.87	16.24±4.28
Guinea pig	9.63±1.50	26.43±9.51

fT4:fT3 ratios were obtained from the data shown in Figure 2 and tT4:tT3 ratios were obtained from data shown in Figure 3. See "Results" section for statistical details.
*: scenario 2 (n = 32).
[†]: scenario 1 (n = 24).

Table 4. Intraspecific fT4 (both scenarios) and fT3 differences in Ansell's mole-rats.

Factor	fT4 first scenario (n = 24)		fT4 second scenario (n = 32)		fT3 (n = 32)	
	F	p	F	p	F	p
Sex	2.47	0.13	4.00	0.056	0.90	0.35
Reproductive status	0.54	0.47	0.14	0.71	0.62	0.44
Age	2.48	0.13	4.58	0.041*	0.63	0.44
Weight	2.04	0.17	0.59	0.45	0.32	0.58
Reproductive status × age	0.71	0.41	0.04	0.84	1.29	0.28

GLM main effects for all four factors alone and a GLM two factor model with reproductive status × age as independent variable. The p-values and the correlation coefficients F are shown.
* = significant ($p<0.05$).

[29,63,64] where rat tT3 levels between 0.8 ng/ml–1.62 ng/ml have been reported), there is little doubt that the overall T3 pattern differs quite clearly compared to T4.

The combination of low fT4 and "normal" fT3 resulted in a very uncommon fT4:fT3 ratio of only 0.70–0.85 (depending on the scenario applied) in mole-rats, compared to 7.28 and 9.63 in rats and guinea pigs, respectively (Table 3), the latter being in good agreement with published data [28,29]. This phenotype resembles that of hypothyroxinemia, a condition characterized by low levels of fT4 while TSH and often also fT3 are in a normal range or slightly elevated [57]. However, whether mole-rats are naturally hypothyroxinemic cannot be answered until TSH can be quantified reliably also in mole-rats.

Regardless of the terminology, our findings raise interesting questions about the proximate and ultimate mechanisms being responsible for this unusual and hitherto unreported hormone distribution. We have already discussed that mole-rats seem to recruit less T4 and more T3 from their respective resources than other rodents. Both may be linked to higher expression rates and/or binding affinities of the mole-rat TH binding proteins. In rodents, the main known binding proteins are albumin, TTR and TBG [65], and the combination of their expression rates and binding affinities have major influence on the half-life of circulating THs. TBG, for instance, has a high T4 binding affinity, but is expressed at very different levels across the lifetime of rats [66]. Specific expression rates and/or functional mutations affecting binding affinities of TBG and other involved proteins could potentially provide an explanation for the altered ratios between free and total TH fractions observed in Ansell's mole-rats and should therefore be focussed in future investigations. For instance, amino acid changes in TTR, at position 109 or 119, were shown to increase thyroxine affinity and decrease fT4/fT3 ratio in humans [67]. However, in the present study no such changes were observed in Ansell's mole-rats (Figure S8).

The observed TH pattern could be linked to alterations in deiodination rates in and/or efflux rates out of target cells. Deiodination of T4 to T3 takes place in the cytoplasm of target cells [24,68]. A higher D1 and D2 activity, both responsible for converting T4 to T3 [24], and/or a high efflux of T3 out of the cells could lead to a relatively high T3 concentration in the blood stream [69] and help compensate for low levels of T4. Therefore, expression rates of the regulatory components of the TH system as well as D1, D2 and D3 activities in the brain, the thyroid and peripheral organs should be determined in further studies.

The "normal" fT3 concentration is rather unexpected on the basis of the low metabolic rate of Ansell's mole-rats [39,40] and

the low L-opsin density in the retina [41]. Thus, alternative functions of T3 could help explain these contradictions: For instance, novel signalling pathways of T3, which imply indirect activation of transcription as a non-nuclear activity, are discussed. One such pathway initiated by THs is the activation of the transcription of the alpha subunit of hypoxia-inducible factor 1 (HIF1A). It is a transcription factor found in all mammalian cells and responsible for a wide range of cellular responses to hypoxia [70,71]. In human fibroblasts, HIF1A mRNA and protein concentration are upregulated by a pathway which is activated by T3 binding to THRB in the cytoplasm [72,73] without being transported into the nucleus.

Therefore, the maintenance of normal T3 levels despite low T4 levels may be an adaptive cellular mechanism of animals living in hypoxic environments to assure a more specific and continuous availability of HIF1. Of course, this is speculative at the moment. However, the importance of HIF1 in subterranean environments is supported by findings from another strictly subterranean mammal, *Spalax ehrenbergi*. In the skeletal muscles of these animals, the concentration of HIF1A mRNA is significantly higher than in rats [74].

Not surprisingly, a remarkably high concentration of HIF1A was also detected in the brain of old naked mole-rats [75,76]. This suggests that this kind of adaptation to a hypoxic environment is not restricted to *S. ehrenbergi* and may also be found in bathyergid species. Further research has to confirm whether these adaptations also occur in *F. anselli*.

Intraspecies comparison of TH hormone levels

Intraspecific fT4 and fT3 comparisons suggest that THs are not the major determinants of the caste-specific ageing rates found in Ansell's mole-rats. In neither scenario was there a significant difference in hormone levels (fT4 or fT3) between non-breeders and breeders. Likewise, sex and weight of the animals did not have an influence on hormone levels. On the other hand fT4 levels did seem to increase with age, when applying the second scenario (Table 4). Age effects on TH levels are well-known, which is not surprising, because THs play a major role in development and metabolism. However, in other mammalian species, THs usually decline with age. In human for instance, fT3 levels usually decline with age, while fT4 levels remain more or less unchanged [77,78]. Guinea pigs do not show an alteration in serum fT4 levels as well [45].

In summary, our results indicate that in *F. anselli*, euthyroid fT3 levels are coupled with lower circulating levels of T4, which, in combination with their low metabolic rate, may represent a novel mechanism to cope with the hypoxic subterranean environment these animals have adapted to. However, THs do not seem to have a major influence on the intraspecific ageing rates in these mole-rats.

Supporting Information

Figure S1   Protein alignment of thyroid hormone receptor α (THRA) from different mammal species. The mRNA sequence of *F. anselli* was obtained from RNA-seq and subsequently translated, other sequences were retrieved from NCBI databases with the following accession numbers: Mus musculus (CAA30576), Rattus norvergicus (NP_112396), Mesocricetus auratus (XP_005076008), Cricetulus_griseus (XP_003510526), Oryctolagus cuniculus (XP_002719397), Ochotona princeps(XP_004591220), Otolemur garnettii (XP_003786450), Macaca fascicularis (NP_001270601),Homo sapiens (NP_003241),Felis catus (XP_003996845), Ailuropoda melanoleuca (XP_002924975), Bos Taurus (NP_001039794), Sus scrofa (O97716), Orcinus orca (XP_004282800).

Figure S2   Protein alignment of thyroid hormone receptor β (THRB) from different mammal species. The mRNA sequence of *F. anselli* was obtained from RNA-seq and subsequently translated, other sequences were retrieved from NCBI databases with the following accession numbers: Heterocephalus glaber (XP_004892721), Cavia porcellus (XP_005008354), Chinchilla lanigera (XP_005387539), Octodon degus (XP_004634245), Mus musculus (P37242), Rattus norvegicus (P18113), Mesocricetus auratus (XP_005081037), Cricetulus griseus (ERE87100), Ochotona princeps (XP_004588346), Otolemur garnettii (XP_003781795), Homo sapiens (P10828), Canis lupus (XP_862690), Mustela putorius furo (XP_004786805), Ailuropoda melanoleuca (XP_002928077), Ovis aries (Q28571), Sus scrofa (XP_001928500), Condylura cristata (XP_004692336).

Figure S3   Protein alignment of thyreotropin β subunit (TSHB) from different mammal species. The mRNA sequence of *F. anselli* was obtained from RNA-seq and subsequently translated, other sequences were retrieved from NCBI databases with the following accession numbers: Cavia porcellus (XP_003479306), Octodon degus (XP_004641725), Mus musculus (NP_001159412), Rattus norvegicus (NP_037248), Ictidomys tridecemlineatus (XP_005334966), Otolemur garnettii (XP_003793878), Macaca fascicularis (XP_001111873), Nomascus leucogenys (XP_003268073), Gorilla gorilla (XP_004026450), Pan paniscus (XP_003805682), Pan troglodytes (XP_001160337), Homo sapiens (AAB30828), Bos taurus (XP_005204060), Orcinus orca (XP_004263279), Echinops telfairi (XP_004714911).

Figure S4   Protein alignment of Type I iodothyronine deiodinase (D1) from different mammal species. The mRNA sequence of *F. anselli* was obtained from RNA-seq and subsequently translated, other sequences were retrieved from NCBI databases with the following accession numbers: Heterocephalus glaber (XP_004908861), Cavia porcellus (NP_001244903), Octodon degus (XP_004642620), Mus musculus (Q61153), Rattus norvegicus (CAA41063), Cricetulus griseus (NP_001243688), Ochotona princeps (XP_004588749), Otolemur garnettii (XP_003793192), Macaca mulatta (NP_001116124), Pan troglodytes (NP_001116123), Homo sapiens (NP_000783), Canis lupus (NP_001007127), Felis catus (NP_001009267), Bos taurus

(NP_001116065), Sus scrofa (NP_001001627), Equus caballus (NP_001159924), Orcinus orca (XP_004273874).

Figure S5   Protein alignment of Type II iodothyronine deiodinase (D2) from different mammal species. The mRNA sequence of *F. anselli* was obtained from RNA-seq and subsequently translated, other sequences were retrieved from NCBI databases with the following accession numbers: Heterocephalus glaber (XP_004900438), Chinchilla lanigera (XP_005390287), Octodon degus (XP_004624767), Mus musculus (NP_034180), Rattus norvegicus (NP_113908), Ochotona princeps (XP_004584413), Homo sapiens (AAC95470), Canis lupus (NP_001116117), Ovis aries (XP_004011138), Sus scrofa (NP_001001626), Equus caballus (NP_001159927), Orcinus orca (XP_004262346), Condylura cristata (XP_004681708), Echinops telfairi (XP_004698804).

Figure S6   Protein alignment of thyroglobulin (TG) from different mammal species. The mRNA sequence of *F. anselli* was obtained from RNA-seq and subsequently translated, other sequences were retrieved from NCBI databases with the following accession numbers: Cavia porcellus (XP_003467392), Chinchilla lanigera (XP_005398080), Octodon degus (XP_004642544), Mus musculus (AAB53204), Rattus norvegicus (BAL14775), Ochotona princeps (XP_004580794), Otolemur garnettii (XP_003792914), Macaca mulatta (EHH28780), Pan troglodytes (XP_003311969), Homo sapiens (AAC51924), Canis lupus (XP_005627864), Felis catus (XP_004000173), Sus scrofa (NP_001161890), Equus caballus (XP_001916622), Orcinus orca (XP_004265356), Echinops telfairi (XP_004697442).

Figure S7   Protein alignment of thyroperoxidase (TPO) from different mammal species. The mRNA sequence of *F. anselli* was obtained from RNA-seq and subsequently translated, other sequences were retrieved from NCBI databases with the following accession numbers: Cavia porcellus (XP_003464975; patched), Octodon degus (XP_004644658), Mus musculus (EDL36934), Rattus norvegicus (EDM03234), Cricetulus griseus (XP_003501455), Ochotona princeps (XP_004582879), Otolemur garnettii (XP_003798602), Macaca mulatta (XP_001117795), Homo sapiens (XP_005264756), Canis lupus (Q8HYB7), Felis catus (XP_003984594), Bos taurus (XP_603356), Sus scrofa (P09933), Equus caballus (XP_001918216), Orcinus orca (XP_004274968), Echinops telfairi (XP_004709888).

Figure S8   Protein alignment of transthyretin (TTR) from different mammal species. The mRNA sequence of *F. anselli* was obtained from RNA-seq and subsequently translated, other sequences were retrieved from NCBI databases with the following accession numbers: Heterocephalus glaber (XP_004905241), Chinchilla lanigera (XP_005372800), Octodon degus (XP_004623610), Rattus norvegicus (AAA41801), Mesocricetus auratus (XP_005065406), Cricetulus griseus (XP_003510202), Ictidomys tridecemlineatus (XP_005337518), Oryctolagus cuniculus (XP_002713532), Chlorocebus aethiops (BAL44398), Homo sapiens (CAG33189), Equus caballus (XP_001495232), Echinops telfairi (XP_004702987).

Figure S9   Protein alignment of thyroxine-binding globin (TBG) from different mammal species. The mRNA sequence of *F. anselli* was obtained from RNA-seq and subsequently translated, other sequences were retrieved from

NCBI databases with the following accession numbers: Hetero-cephalus glaber (EHB09876), Octodon degus (XP_004646260), Mus musculus (P61939), Rattus norvegicus (AAA42205), Cricetulus griseus (ERE65740), Otolemur garnettii (XP_003801681), Gorilla gorilla (XP_004064693), Pan troglodytes (NP_001009109), Homo sapiens (NP_783866), Canis lupus (XP_538128), Bos taurus (AAI03464), Ovis aries (NP_001094390), Sus scrofa (Q9TT35), Equus caballus (XP_001493492), Orcinus orca (XP_004285286), Echinops telfairi (XP_004710081).

Figure S10 Protein alignment of hypoxia-induced factor (HIF1A) from different mammal species. The mRNA sequence of *F. anselli* was obtained from RNA-seq and subsequently translated, other sequences were retrieved from NCBI databases with the following accession numbers: Hetero-cephalus glaber (XP_004837489), Octodon degus (XP_00462 4861), Mus musculus (CAA70305), Rattus norvegicus (O35800), Ochotona princeps (XP_004597684), Otolemur garnettii (XP_003 794480), Pan troglodytes (XP_001168972), Homo sapiens (NP_00 1521), Canis lupus (XP_003639249), Felis catus (XP_003987765), Bos taurus (NP_776764), Sus scrofa (NP_001116596), Orcinus orca (XP_004262152).

Figure S11 Protein alignment of Monocarboxylate transporter 8 (MCT8) from different mammal species. The mRNA sequence of *F. anselli* was obtained from RNA-seq and subsequently translated, other sequences were retrieved from NCBI databases with the following accession numbers: Hetero-cephalus glaber (XP_004905154), Cavia porcellus (XP_0050042 29), Octodon degus (XP_004648070), Mus musculus (AAC40078), Rattus norvegicus (EDM07172), Ochotona princeps (XP_0045 92817), Otolemur garnettii (XP_003802275), Macaca mulatta (XP_001096017), Homo sapiens (NP_006508), Bos taurus (NP_0 01193868), Orcinus orca (XP_004283857). Suggestions of NCBI for translation starts of Octodon degus (XP_004648070), Ocho-

tona princeps (XP_004592817), Macaca mulatta (XP_001096017) and Bos taurus (NP_001193868) were changed to the position of the orthologous sequences.

Figure S12 Protein alignment of Natrium-Iodid-Symporter (NIS) from different mammal species. The mRNA sequence of *F. anselli* was obtained from RNA-seq and subsequently translated, other sequences were retrieved from NCBI databases with the following accession numbers: Hetero-cephalus glaber (XP_004873534), Cavia porcellus (XP_003465 226), Octodon degus (XP_004646953), Mus musculus (NP_444 478), Rattus norvegicus (Q63008), Otolemur garnettii (XP_00 3796602), Macaca mulatta (EHH29802), Pan troglodytes (XP_524 154), Homo sapiens (NP_000444), Canis lupus (XP_541946), Sus scrofa (NP_999575), Orcinus orca (XP_004277608).

Table S1 Intra- and inter-assay variances of the Enzyme Immunoassays for fT3, fT4, tT3 and tT4. Shown are the coefficients of variances (%CV) of the assays used in the present study, according to the manufacturer (DRG Instruments GmbH).

Acknowledgments

The authors thank Christine Krueger, Gero Hilken and Andreas Wissmann for their help at blood sampling, and Lilia Kufeld and Andrea Jaeger for conducting the hormone measurements.

Author Contributions

Conceived and designed the experiments: YH CV SB MB MBP AS KS HB PD. Performed the experiments: YH CV AS KS PD. Analyzed the data: YH CV SB MB AS KS PS. Contributed reagents/materials/analysis tools: SB HB MBP KS PD. Wrote the paper: YH CV SB AS KS HB PS.

References

1. Harman D (1956) Aging - A theory based on free-radical and radiation-chemistry. Journals of Gerontology 11: 298–300.
2. Balaban RS, Nemoto S, Finkel T (2005) Mitochondria, oxidants, and aging. Cell 120: 483–495.
3. Monnier VM (1989) Toward a Maillard reaction theory of aging. Progress in clinical and biological research 304: 1–22.
4. Baynes JW (2001) The role of AGEs in aging: causation or correlation. Experimental Gerontology 36: 1527–1537.
5. Dammann P, Sell DR, Begall S, Strauch C, Monnier VM (2011) Advanced Glycation End-Products as Markers of Aging and Longevity in the Long-Lived Ansell's Mole-Rat (Fukomys anselli). Journals of Gerontology Series a-Biological Sciences and Medical Sciences 67: 573–583.
6. Harley CB, Vaziri H, Counter CM, Allsopp RC (1992) The telomere hypothesis of cellular aging. Experimental Gerontology 27: 375–382.
7. Bowers J, Terrien J, Clerget-Froidevaux MS, Gothié JD, Rozing MP, et al. (2013) Thyroid Hormone Signaling and Homeostasis During Aging. Endocrine Reviews 34: 556–589.
8. Zhang J, Lazar MA (2000) The mechanism of action of thyroid hormones. Annu Rev Physiol 62: 439–466.
9. Ooka H, Fujita S, Yoshimoto E (1983) Pituitary-thyroid activity and longevity in neonatally thyroxine-treated rats. Mechanisms of Ageing and Development 22: 113–120.
10. Ooka H, Shinkai T (1986) Effects of chronic hyperthyroidism on the lifespan of the rat. Mechanisms of Ageing and Development 33: 275–282.
11. Brown-Borg HM, Borg KE, Meliska CJ, Bartke A (1996) Dwarf mice and the ageing process. Nature 384: 33–33.
12. Vergara M, Smith-Wheelock M, Harper JM, Sigler R, Miller RA (2004) Hormone-treated snell dwarf mice regain fertility but remain long lived and disease resistant. Journals of Gerontology Series a-Biological Sciences and Medical Sciences 59: 1244–1250.
13. Liang S, Mele J, Wu Y, Buffenstein R, Hornsby PJ (2010) Resistance to experimental tumorigenesis in cells of a long-lived mammal, the naked mole-rat (Heterocephalus glaber). Aging Cell 9: 626–635.
14. Lovegrove BG (1986) The metabolism of social subterranean rodents: adaptation to aridity. Oecologia 69: 551–555.
15. Buffenstein R, Woodley R, Thomadakis C, Daly TJM, Gray DA (2001) Cold-induced changes in thyroid function in a poikilothermic mammal, the naked mole-rat. American Journal of Physiology-Regulatory Integrative and Comparative Physiology 280: R149–R155.
16. Willis CKR, Brigham RM, Geiser F (2006) Deep, prolonged torpor by pregnant, free-ranging bats. Naturwissenschaften 93: 80–83.
17. Becker NI, Encarnacao JA, Tschapka M, Kalko EKV (2013) Energetics and life-history of bats in comparison to small mammals. Ecological Research 28: 249–258.
18. Rozing MP, Houwing-Duistermaat JJ, Slagboom PE, Beekman M, Frolich M, et al. (2010) Familial Longevity Is Associated with Decreased Thyroid Function. Journal of Clinical Endocrinology & Metabolism 95: 4979–4984.
19. Shupnik MA, Chin WW, Habener JF, Ridgway EC (1985) Transcriptional regulation of the thyrotropin subunit genes by thyroid hormone. J Biol Chem 260: 2900–2903.
20. Kelly GS (2000) Peripheral metabolism of thyroid hormones: a review. Alternative medicine review: a journal of clinical therapeutic 5: 306–333.
21. Costa-e-Sousa RH, Hollenberg AN (2012) Minireview: The Neural Regulation of the Hypothalamic-Pituitary-Thyroid Axis. Endocrinology 153: 4128–4135.
22. Visser WE, Friesema EC, Visser TJ (2011) Minireview: thyroid hormone transporters: the knowns and the unknowns. Mol Endocrinol 25: 1–14.
23. Crantz FR, Larsen PR (1980) Rapid thyroxine to 3,5,3′-triiodothyronine conversion and nuclear 3,5,3′-triiodothyronine binding in rat cerebral-cortex and cerebellum. Journal of Clinical Investigation 65: 935–938.
24. Bianco AC, Larsen PR (2005) Cellular and structural biology of the deiodinases. Thyroid 15: 777–786.
25. Harvey CB, Williams GR (2002) Mechanism of thyroid hormone action. Thyroid 12: 441–446.
26. Moeller LC, Broecker-Preuss M (2011) Transcriptional regulation by nonclassical action of thyroid hormone. Thyroid research 4 Suppl 1: S6.

27. Gereben B, Zavacki AM, Ribich S, Kim BW, Huang SA, et al. (2008) Cellular and Molecular Basis of Deiodinase-Regulated Thyroid Hormone Signaling. Endocrine Reviews 29: 898–938.

28. Ma C, Xie J, Huang X, Wang G, Wang Y, et al. (2009) Thyroxine alone or thyroxine plus triiodothyronine replacement therapy for hypothyroidism. Nucl Med Commun 30: 586–593.

29. Davies DT (1993) Assessment of rodent thyroid endocrinology - Advantages and pit-falls. Comparative Haematology International 3: 142–152.

30. Schussler GC (2000) The thyroxine-binding proteins. Thyroid 10: 141–149.

31. Burda H, Honeycutt RL, Begall S, Locker-Grutjen O, Scharff A (2000) Are naked and common mole-rats eusocial and if so, why? Behavioral Ecology and Sociobiology 47: 293–303.

32. Dammann P, Burda H (2006) Sexual activity and reproduction delay ageing in a mammal. Current Biology 16: R117–R118.

33. Dammann P, Sumbera R, Massmann C, Scherag A, Burda H (2011) Extended Longevity of Reproductives Appears to be Common in Fukomys Mole-Rats (Rodentia, Bathyergidae). Plos One 6.

34. Schmidt CM, Jarvis J. U. M.; Bennett NC (2013) The Long-Lived Queen: Reproduction and Longevity in Female Eusocial Damaraland Mole-Rats (Fukomys damarensis). African Zoology 48: 193–196.

35. Kirkwood TB (1977) Evolution of ageing. Nature 270: 301–304.

36. Edward DAaC, T. (2011) Mechanisms underlying reproductive trade-offs: Costs of reproduction. In: Flatt THA, editor. Mechanisms of Life History Evolution - The Genetics and Physiology of Life History Traits and Trade-Offs New York Oxford University Press Inc. pp. 137–152.

37. Keller L, Genoud M (1997) Extraordinary lifespans in ants: a test of evolutionary theories of ageing. Nature 389: 958–960.

38. Corona M, Velarde RA, Remolina S, Moran-Lauter A, Wang Y, et al. (2007) Vitellogenin, juvenile hormone, insulin signaling, and queen honey bee longevity. Proceedings of the National Academy of Sciences of the United States of America 104: 7128–7133.

39. Marhold S, Nagel A (1995) The energetics of the common mole-rat Cryptomys, a subterranean eusocial rodent from Zambia. Journal of Comparative Physiology B-Biochemical Systemic and Environmental Physiology 164: 636–645.

40. Zelová J, Sumbera R, Sedlácek F, Burda H (2007) Energetics in a solitary subterranean rodent, the silvery mole-rat, Heliophobius argenteocinereus, and allometry of RMR in African mole-rats (Bathyergidae). Comparative biochemistry and physiology Part A, Molecular & integrative physiology 147: 412–419.

41. Peichl L, Němec P, Burda H (2004) Unusual cone and rod properties in subterranean African mole-rats (Rodentia, Bathyergidae). European Journal of Neuroscience 19: 1545–1558.

42. Glaschke A, Gloesmann M, Peichl L (2010) Developmental Changes of Cone Opsin Expression but Not Retinal Morphology in the Hypothyroid Pax8 Knockout Mouse. Investigative Ophthalmology & Visual Science 51: 1719–1727.

43. Glaschke A, Weiland J, Del Turco D, Steiner M, Peichl L, et al. (2011) Thyroid Hormone Controls Cone Opsin Expression in the Retina of Adult Rodents. The Journal of Neuroscience 31: 4844–4851.

44. Choksi NY, Jahnke GD, St Hilaire C, Shelby M (2003) Role of thyroid hormones in human and laboratory animal reproductive health. Birth Defects Research Part B-Developmental and Reproductive Toxicology 68: 479–491.

45. Mueller K, Mueller E, Klein R, Brunnberg L (2009) Serum thyroxine concentrations in clinically healthy pet guinea pigs (Cavia porcellus). Veterinary Clinical Pathology 38: 507–510.

46. Hoff J RL (2000) Methods of blood collection in the mouse. Lab animals 29: 47–53.

47. Martin M (2011) Cutadapt removes adapter sequences from high-throughput sequencing reads. 2011 17.

48. Joshi NA FJ (2011) Sickle: A sliding-window, adaptive, quality-based trimming tool for FastQ files (Version 1.21).

49. Grabherr MG, Haas BJ, Yassour M, Levin JZ, Thompson DA, et al. (2011) Full-length transcriptome assembly from RNA-Seq data without a reference genome. Nat Biotechnol 29: 644–652.

50. Yang Z (1997) PAML: A program package for phylogenetic analysis by maximum likelihood. Comput Appl Biosci 13: 555–556.

51. Yang Z (2007) PAML 4: Phylogenetic Analysis by Maximum Likelihood. Molecular Biology and Evolution 24: 1586–1591.

52. Garcia Montero A, Burda H, Begall S (2014) Chemical restraint of African mole-rats (Fukomys sp.) with a combination of ketamine and xylazine. Veterinary Anaesthesia and Analgesia: doi:10.1111/vaa.12180.

53. Campos-Barros A, Musa A, Flechner A, Hessenius C, Gaio U, et al. (1997) Evidence for circadian variations of thyroid hormone concentrations and type II 5′-iodothyronine deiodinase activity in the rat central nervous system. Journal of Neurochemistry 68: 795–803.

54. Negro R, Soldin O, Obregon M-J, Stagnaro-Green A (2011) Hypothyroxinemia and Pregnancy. Endocrine Practice 17: 422–429.

55. Bauch K, Meng W, Ulrich FE, Grosse E, Kempe R, et al. (1986) Thyroid status during pregnancy and post partum in regions of iodine deficiency and endemic goiter. Endocrinol Exp 20: 67–77.

56. Buffenstein R (2005) The naked mole-rat: a new long-living model for human aging research. J Gerontol A Biol Sci Med Sci 60: 1369–1377.

57. Opazo MC, Gianini A, Pancetti F, Azkcona G, Alarcón L, et al. (2008) Maternal Hypothyroxinemia Impairs Spatial Learning and Synaptic Nature and Function in the Offspring. Endocrinology 149: 5097–5106.

58. Dong H, You S-H, Williams A, Wade MG, Yauk CL, et al. (2014) Transient Maternal Hypothyroxinemia Potentiates the Transcriptional Response to Exogenous Thyroid Hormone in the Fetal Cerebral Cortex Before the Onset of Fetal Thyroid Function: A Messenger and MicroRNA Profiling Study. Cerebral Cortex.

59. Darras VM, Van Herck SLJ, Geysens S, Reyns GE (2009) Involvement of thyroid hormones in chicken embryonic brain development. General and Comparative Endocrinology 163: 58–62.

60. Haddow JE, Palomaki GE, Allan WC, Williams JR, Knight GJ, et al. (1999) Maternal Thyroid Deficiency during Pregnancy and Subsequent Neuropsychological Development of the Child. New England Journal of Medicine 341: 549–555.

61. Román GC, Ghassabian A, Bongers-Schokking JJ, Jaddoe VWV, Hofman A, et al. (2013) Association of gestational maternal hypothyroxinemia and increased autism risk. Annals of Neurology 74: 733–742.

62. Kott O, Moritz RE, Šumbera R, Burda H, Němec P (2014) Light propagation in burrows of subterranean rodents: tunnel system architecture but not photoreceptor sensitivity limits light sensation range. Journal of Zoology 294: 67–75.

63. Moreno M LA, Lombardi A GF (1997) How the thyroid controls metabolism in the rat: different roles for triiodothyronine and diiodothyronines. J Physiol 505(Pt 2): 529–538.

64. Hood A, Liu YP, Gattone VH 2nd, Klaassen CD (1999) Sensitivity of thyroid gland growth to thyroid stimulating hormone (TSH) in rats treated with antithyroid drugs. Toxicol Sci 49: 263–271.

65. Kaneko JJ (2008) Thyroid Function. In: Kaneko JJH, John W.; Bruss, Michael L., editor. Clinical Biochemistry of Domestic Animals. 6 ed. Amsterdam; Boston: Elsevier Academic Press. pp. 623–634.

66. Savu L, Vranckx R, Rouaze-Romet M, Nunez EA (1992) The pituitary control of rat thyroxine binding globulin. Acta Med Austriaca 19 Suppl 1: 88–90.

67. Refetoff S, Marinov VS, Tunca H, Byrne MM, Sunthornthepvarakul T, et al. (1996) A new family with hyperthyroxinemia caused by transthyretin Val109 misdiagnosed as thyrotoxicosis and resistance to thyroid hormone–a clinical research center study. J Clin Endocrinol Metab 81: 3335–3340.

68. Pohlenz J, Maqueem A, Cua K, Weiss RE, Van Sande J, et al. (1999) Improved radioimmunoassay for measurement of mouse thyrotropin in serum: Strain differences in thyrotropin concentration and thyrotroph sensitivity to thyroid hormone. Thyroid 9: 1265–1271.

69. Visser WE, Friesema EC, Jansen J, Visser TJ (2008) Thyroid hormone transport in and out of cells. Trends Endocrinol Metab 19: 50–56.

70. Weidemann A, Johnson RS (2008) Biology of HIF-1 alpha. Cell Death and Differentiation 15: 621–627.

71. Otto T, Fandrey J (2008) Thyroid hormone induces hypoxia-inducible factor 1 alpha gene expression through thyroid hormone receptor beta/retinoid X receptor alpha-dependent activation of hepatic leukemia factor. Endocrinology 149: 2241–2250.

72. Moeller LC, Dumitrescu AM, Refetoff S (2005) Cytosolic action of thyroid hormone leads to induction of hypoxia-inducible factor-1 alpha and glycolytic genes. Molecular Endocrinology 19: 2955–2963.

73. Storey NM, Gentile S, Ullah H, Russo A, Muessel M, et al. (2006) Rapid signaling at the plasma membrane by a nuclear receptor for thyroid hormone. Proceedings of the National Academy of Sciences of the United States of America 103: 5197–5201.

74. Avivi A, Shams I, Joel A, Lache O, Levy AP, et al. (2005) Increased blood vessel density provides the mole rat physiological tolerance to its hypoxic subterranean habitat. The FASEB Journal 19: 1314–1316.

75. Edrey YH, Park TJ, Kang H, Biney A, Buffenstein R (2011) Endocrine function and neurobiology of the longest-living rodent, the naked mole-rat. Experimental Gerontology 46: 116–123.

76. Kim EB, Fang X, Fushan AA, Huang Z, Lobanov AV, et al. (2011) Genome sequencing reveals insights into physiology and longevity of the naked mole rat. Nature 479: 223–227.

77. Peeters RP (2008) Thyroid hormones and aging. Hormones-International Journal of Endocrinology and Metabolism 7: 28–35.

78. Gesing A, Lewinski A, Karbownik-Lewinska M (2012) The thyroid gland and the process of aging; what is new? Thyroid Research 5: 1–5.

18

Inferring Meaningful Communities from Topology-Constrained Correlation Networks

Jose Sergio Hleap[1]*, Christian Blouin[1,2]

1 Department of Biochemistry and Molecular Biology, Dalhouise University, Halifax, Nova Scotia, Canada, **2** Department of Computer Science, Dalhouise University, Halifax, Nova Scotia, Canada

Abstract

Community structure detection is an important tool in graph analysis. This can be done, among other ways, by solving for the partition set which optimizes the modularity scores Q. Here it is shown that topological constraints in correlation graphs induce over-fragmentation of community structures. A refinement step to this optimization based on Linear Discriminant Analysis (LDA) and a statistical test for significance is proposed. In structured simulation constrained by topology, this novel approach performs better than the optimization of modularity alone. This method was also tested with two empirical datasets: the Roll-Call voting in the 110th US Senate constrained by geographic adjacency, and a biological dataset of 135 protein structures constrained by inter-residue contacts. The former dataset showed sub-structures in the communities that revealed a regional bias in the votes which transcend party affiliations. This is an interesting pattern given that the 110th Legislature was assumed to be a highly polarized government. The α-amylase catalytic domain dataset (biological dataset) was analyzed with and without topological constraints (inter-residue contacts). The results without topological constraints showed differences with the topology constrained one, but the LDA filtering did not change the outcome of the latter. This suggests that the LDA filtering is a robust way to solve the possible over-fragmentation when present, and that this method will not affect the results where there is no evidence of over-fragmentation.

Editor: Kay Hamacher, Technical University Darmstadt, Germany

Funding: This study was funded by NSERC through the grant No. 120504858. This work was partially supported by The Departamento Administrativo de Ciencia y Tecnología - Colciencias (Colombia) through the CALDAS scholarship. The funders had no role in study design, data collection and analysis, decision to publish, or preparation of the manuscript.

Competing Interests: The authors have declared that no competing interests exist.

* Email: jshleap@dal.ca

Introduction

Many problems in science can be abstracted as networks. For example, in biological sciences, protein structures can be abstracted as graphs of connected residues [1], metabolic networks can be created by connecting enzymes by their interactions in a given pathway [2], or food webs can be created by joining species with their trophic interactions [3]. Networks are common models for the Internet [4] and social networks [5]. Any kind of data that can be summarized into vertices (nodes) and connections (edges), can be abstracted as a graph. An special case of graphs can be constructed when one is interested in the correlation among variables. In this case, a correlation network can be constructed by assigning each variable to a vertex (or node), and the connections between are defined by the correlation. Since correlation is a measure of strength of relationship, the actual correlation value can be use as a weight in the edge, therefore representing such relationship. This graph abstraction is useful since allow us to analyze the relationships using the graph invariants. There are many such properties, but one of special interest here is the community structure which represents how the vertices are arranged in groups densely connected internally and sparsely connected externally [6].

Many networks have heterogeneous edge densities, which may imply a community structure. Communities are groups of nodes whose associations imply new insights in the understanding of a system [7]. A community can be loosely defined as groups of nodes that share more among themselves than to the rest of the graph. The most commonly used algorithm (and the one of focus in this paper) to detect communities in graphs is the modularity optimization proposed by Newman and Girvan [8]. In this algorithm, the modularity score Q is optimized to obtain a partition scheme. Intuitively, Q evaluates the excess of the number of edges inside a group against the expected connectivity of a randomly connected graph with similar properties. It can be calculated with:

$$Q = \frac{1}{2m} \sum_{vw} \left[A_{vw} - \frac{\sum_w A_{vw} \sum_v A_{vw}}{2m} \right] \delta(C_v, C_w) \qquad (1)$$

where m is the number of edges in the graph, A_{vw} represents the weight of the edge between vertices v, and w, $\sum_w A_{vw}$ and $\sum_v A_{vw}$ are the weighted degree of a vertex (v or w), defined to be the sum of the edge weights of the adjacent edges for each vertex. C_v and C_w are communities to which the vectors v and w belong to, and the δ is a binary function where $\delta(C_v, C_w)$ is 1 if $C_v = C_w$ and 0 otherwise.

This approach has been applied to numerous problems [7,9,10]. Despite its wide use, exact algorithms for modularity optimization are computationally expensive. Some caveats also exist [7]: One example is the fact that high Q can be found in random graphs

[11]. This issue might create either an over-fragmentation of the graph into smaller communities, or a failure to detect a small community which size is below a preset resolution limit [12]. Despite these caveats, modularity optimization (and in general community structure detection) is still an important tool in science if the confidence in the robustness of the solution can be assessed. Other methods to re-construct graphs and assess their structure exist, particularly dealing with high-dimensional data. Methods such as sparse graphical models [13] and LASSO-type problems [14] can be applied in graph reconstruction, and sometimes in community structure detection [15]. However, most of these methods rely on the assumption of independence of the variables [14] (or at least that the covariates are not highly correlated [16]), on the *a priori* determination of the number and size of the communities [15], and a full sparcity of the covariation among traits in the data. These kind of limitations makes these particular methods of limited in use in correlation networks, where the covariates are normally correlated, non-independent, and not completely sparse.

There is no guarantee that a community based on correlation is actually meaningful. It is posited here that asserting the statistical significance of a community enhances the odds that such structure provides insight. An application in protein structures exploring this with a Cholesky decomposition-based simulation have previously been shown [1]. After the membership vector is created by the optimization of Q, a pairwise permutation test is used to evaluate the statistical significance of each bipartition between modules. If the test fails, the two modules are merged and the membership vector is iteratively refined. In this work [1], the performance of community inference was shown to be high for simulated data.

Let us consider the case of correlation networks, where the edges are defined as the correlation between two nodes. These networks are important in biological sciences [1,17–19] and economics [20,21] since they constitute an intermediate between topology and the dynamics of the system [22]. Analyzing the community structures of these networks can help identify clusters of co-expressed genes causing a disease, or groups of stocks that are co-varying in the market. It is important to know whether such clustering partition has any significance. In some cases it is also appropriate to constraint a graph to a meaningful topology. For example: let's define a correlation network as a graph where two vertices are connected by an edge with a weight determined by the correlation of a pair of properties. It is also possible to further define a topologically-constrained correlation graph as a graph where an edge would exist only if the two incident vertices are connected by another meaningful property. The extra constraint in topology will create a sparser graph. Sparser graphs show an intrinsic level of modularity due to their topologies [23]. This is a problem if the modularity is inferred on the assumption that the community structure is dictated by correlation. It has also been shown that sparser graphs tend to cluster into more modules than predicted before [24]. Let's define this effect as over-fragmentation. In some cases the sparsity caused by the constraint is not complete; that is, not the majority of entries in the adjacency matrix are zero. Given this and coupled with the fact that in correlation networks covariates are correlated and most of them are not zero, methods that can be more robust against over-fragmentation (such as LASSO-based and sparse graphical methods) are not easily applicable.

Here the effect of the topology-constraint in the community structure detection by modularity (Q) optimization is analyzed, and a strategy to mitigate the over-fragmentation is proposed. Such an effect will be evaluated in a simulation, a protein dataset, and in the 110th US Legislature roll-call votes. In the first two cases, the additional property or constraint property, will be the contact between points in the simulation or residues in the protein. For the roll-call votes, the constraining property is the geographic adjacency of the state of origin of each senator.

Results and Discussion

To compare with the topology-unconstrained simulations in [1] a shape-structured simulation using Cholesky decomposition (See Methods) is developed. The simulation uses two contiguous letters "H" (Figure 1) to create a heterogeneous shape. The topology constraint is based on contacts since the points in simulation lay on a unit grid. The shape was chosen since it creates a point of contact between the two clusters as well as bottlenecks of contacts which make it a more difficult clustering problem for the topology constraint.

Table 1 shows the results of the performance (mean F-score $\pm$ standard deviation; refer to Methods for details) of the methods in [1] in a topology (contacts) constrained simulation. As can be seen, the results here differ from that in [1] simulations, which has no contact constraints. It appears that the reduced number of edges, given the constraint, creates an over-fragmentation by the modularity optimization that cannot be corrected by the 95% confidence permutational t-test reported by [1].

Addressing the over-fragmentation problem: Linear discriminant filtering

Linear discriminants are a standard multivariate statistical tool to reduce the dimensionality by finding a suitable linear subspace in which the the groups or classes are optimally separated by maximizing the variance between groups while minimizing the intraclass variance. It has been commonly used as a preprocessing step in pattern recognition systems [25] and is commonly used in other sciences to explore the variate space to find shared properties of samples and variables [26]. It is based on a linear model where a given dependent variable can be explained by a linear combination of factors given by the independent variables. Such factors can be a clustering scheme itself. By providing a membership vector derived from the optimization of the modularity score, the linear discriminant analysis (LDA) will provide a set of linear discriminants that better fit the data. Such linear discriminants can be analyzed for the differences between groups. When the differences between groups are not large enough given a particular clustering

Figure 1. Starting shape for simulation. Letters and colors represent the true clustering. The chokepoints (gray arrows) create weakly linked sub-clusters that should not be fragmented.

Table 1. Performance of the structured simulation without LDA pre-filtering.

Cluster A

Corr.	0.15	0.2	0.25	0.3	0.35	0.4	0.45	0.5	0.55
0.15	0.80±0.15	0.83±0.12	0.83±0.09	0.79±0.14	0.82±0.13	**0.86±0.09**	0.80±0.11	0.81±0.12	0.83±0.12
0.20	**0.86±0.11**	**0.85±0.12**	**0.87±0.13**	0.83±0.11	**0.85±0.12**	0.81±0.09	0.77±0.14	0.78±0.12	0.82±0.11
0.25	**0.86±0.13**	**0.86±0.12**	0.83±0.13	0.82±0.14	0.81±0.12	0.74±0.14	0.82±0.13	0.81±0.11	0.82±0.07
0.30	**0.85±0.12**	**0.88±0.10**	**0.86±0.10**	0.81±0.13	0.83±0.10	**0.85±0.13**	0.83±0.11	0.77±0.11	0.81±0.11
0.35	0.80±0.13	**0.85±0.11**	0.84±0.11	0.82±0.14	0.83±0.12	0.81±0.11	0.76±0.12	0.79±0.11	0.77±0.11
0.40	0.80±0.12	**0.86±0.11**	**0.89±0.11**	**0.86±0.13**	0.81±0.13	0.84±0.09	0.80±0.13	0.76±0.10	0.74±0.10
0.45	0.83±0.11	**0.85±0.08**	**0.88±0.10**	0.82±0.13	0.80±0.15	0.84±0.09	0.81±0.14	0.80±0.11	0.73±0.12
0.50	0.78±0.11	0.81±0.11	0.81±0.14	0.79±0.11	0.82±0.13	0.80±0.11	0.81±0.08	0.80±0.08	0.79±0.09
0.55	0.79±0.12	0.83±0.12	0.78±0.11	0.84±0.09	0.81±0.13	0.80±0.13	0.79±0.12	0.77±0.13	0.75±0.12
0.60	0.76±0.12	0.82±0.08	0.80±0.10	**0.86±0.07**	0.79±0.10	0.80±0.10	0.78±0.15	0.72±0.13	0.75±0.12
0.65	0.83±0.08	0.82±0.12	0.79±0.13	0.79±0.11	0.83±0.12	0.79±0.12	0.83±0.10	0.81±0.14	0.78±0.10
0.70	0.80±0.10	0.79±0.11	0.79±0.11	0.81±0.15	0.79±0.12	0.78±0.13	0.78±0.13	0.83±0.09	0.81±0.13
0.75	0.79±0.12	0.82±0.10	0.82±0.13	0.76±0.12	0.80±0.11	0.80±0.13	0.77±0.11	0.77±0.11	0.83±0.08
0.80	0.76±0.14	0.80±0.12	0.80±0.12	0.75±0.11	0.78±0.11	0.73±0.13	0.82±0.13	0.76±0.13	0.77±0.12
0.85	0.81±0.11	0.83±0.08	0.83±0.12	0.76±0.09	0.79±0.11	0.77±0.13	0.78±0.12	0.74±0.12	0.76±0.12
0.90	0.77±0.11	0.81±0.12	0.78±0.11	0.83±0.15	0.78±0.11	0.76±0.09	0.76±0.15	0.76±0.10	0.79±0.09
0.95	0.78±0.09	0.79±0.13	**0.85±0.11**	0.80±0.13	0.79±0.13	0.73±0.11	0.70±0.11	0.73±0.12	0.77±0.10

Continuation Cluster A

Cluster B

Corr.	0.6	0.65	0.7	0.75	0.8	0.85	0.9	0.95
0.15	0.80±0.12	0.81±0.16	0.81±0.12	**0.88±0.11**	0.83±0.08	0.83±0.10	0.80±0.09	0.84±0.09
0.20	0.80±0.09	0.81±0.10	0.84±0.09	0.82±0.13	0.82±0.09	**0.85±0.09**	0.79±0.11	0.77±0.08
0.25	0.81±0.12	0.79±0.13	0.83±0.12	0.76±0.10	0.82±0.09	0.79±0.12	0.77±0.12	0.81±0.10
0.30	0.81±0.08	0.76±0.12	0.77±0.11	0.82±0.11	0.79±0.12	0.79±0.11	0.80±0.15	0.80±0.10
0.35	0.79±0.12	0.76±0.10	0.77±0.12	0.83±0.14	0.76±0.14	0.80±0.11	0.78±0.14	0.82±0.10
0.40	0.79±0.13	0.74±0.12	0.79±0.10	0.75±0.12	0.71±0.11	0.75±0.11	0.75±0.11	0.81±0.09
0.45	0.79±0.13	0.78±0.13	0.76±0.12	0.72±0.14	0.84±0.11	0.74±0.13	0.79±0.12	0.76±0.11
0.50	0.79±0.13	0.80±0.12	0.78±0.10	0.82±0.12	0.72±0.13	0.78±0.12	0.78±0.11	0.77±0.14
0.55	0.79±0.13	0.79±0.11	0.78±0.12	0.73±0.13	0.76±0.11	0.77±0.09	0.78±0.11	0.79±0.14
0.60	0.77±0.12	0.79±0.10	0.77±0.09	0.74±0.11	0.74±0.10	0.76±0.12	0.73±0.10	0.74±0.13
0.65	0.79±0.11	0.78±0.14	0.81±0.11	0.77±0.14	0.74±0.12	0.72±0.11	0.76±0.10	0.69±0.13
0.70	0.78±0.10	0.81±0.11	0.81±0.12	0.80±0.11	0.77±0.13	0.78±0.11	0.66±0.11	0.75±0.10
0.75	0.81±0.10	0.79±0.11	0.77±0.13	0.80±0.14	0.80±0.14	0.78±0.10	0.69±0.11	0.75±0.14
0.80	0.79±0.11	0.77±0.14	0.83±0.08	0.77±0.14	0.79±0.12	0.79±0.11	0.77±0.12	0.80±0.10
0.85	0.78±0.12	0.78±0.13	0.82±0.09	0.79±0.13	0.83±0.10	0.81±0.11	0.80±0.10	0.80±0.12
0.90	0.83±0.10	0.77±0.15	0.76±0.11	0.81±0.10	0.79±0.09	0.83±0.12	0.78±0.13	0.81±0.08
0.95	**0.85±0.12**	0.77±0.13	0.75±0.14	0.78±0.10	0.80±0.10	0.84±0.09	0.80±0.10	0.81±0.12

Mean F-score and standard deviation of 20 replicates of a Cholesky-based structured simulation.
Each entry corresponds to the mean F-Score ± the standard deviation for 20 replicates in each pair of intracorrelations. Corr. = Intracorrelation. Bolded numbers correspond to F-Scores higher than 0.85.

Table 2. Performance of the structured simulation using LDA pre-filtering.

Cluster A

Corr.	0.15	0.2	0.25	0.3	0.35	0.4	0.45	0.5	0.55
0.15	0.96±0.06	0.96±0.07	0.92±0.09	0.93±0.08	0.93±0.09	0.93±0.09	0.93±0.09	0.93±0.09	0.93±0.07
0.20	0.96±0.06	0.97±0.07	0.97±0.06	0.98±0.05	0.96±0.06	0.94±0.11	0.95±0.09	0.96±0.06	0.97±0.06
0.25	0.92±0.09	0.99±0.04	0.98±0.05	0.99±0.03	0.96±0.07	0.93±0.10	0.92±0.12	0.96±0.04	0.95±0.06
0.30	0.94±0.07	0.97±0.05	0.93±0.13	0.95±0.10	0.93±0.07	0.91±0.11	0.96±0.07	0.93±0.10	0.96±0.08
0.35	0.95±0.07	0.97±0.05	0.94±0.10	0.93±0.10	0.94±0.12	0.94±0.07	0.94±0.10	0.96±0.07	0.90±0.10
0.40	0.93±0.07	0.95±0.10	0.96±0.07	0.92±0.10	0.96±0.06	0.96±0.07	0.94±0.07	0.95±0.07	0.96±0.09
0.45	0.94±0.09	0.96±0.08	0.93±0.09	0.95±0.09	0.93±0.09	0.92±0.14	0.93±0.10	0.96±0.11	0.92±0.10
0.50	0.93±0.07	0.96±0.09	0.92±0.08	0.92±0.11	0.93±0.10	0.92±0.10	0.93±0.10	0.94±0.10	0.95±0.07
0.55	0.95±0.08	0.96±0.06	0.96±0.06	0.96±0.09	0.93±0.11	0.93±0.10	0.91±0.12	0.87±0.09	0.92±0.12
0.60	0.93±0.08	0.93±0.09	0.91±0.10	0.96±0.06	0.95±0.06	0.94±0.09	0.96±0.07	0.96±0.07	0.93±0.12
0.65	0.96±0.06	0.95±0.05	0.94±0.08	0.96±0.07	0.91±0.10	0.95±0.07	0.93±0.11	0.90±0.12	0.92±0.09
0.70	0.94±0.06	0.94±0.09	0.96±0.05	0.96±0.07	0.91±0.12	0.92±0.09	0.94±0.09	0.91±0.11	0.92±0.06
0.75	0.91±0.08	0.97±0.05	0.96±0.07	0.94±0.07	0.94±0.06	0.92±0.08	0.91±0.08	0.94±0.08	0.95±0.06
0.80	0.94±0.08	0.93±0.09	0.88±0.10	0.89±0.10	0.92±0.10	0.98±0.04	0.94±0.10	0.91±0.10	0.91±0.10
0.85	0.95±0.08	0.89±0.12	0.93±0.08	0.93±0.08	0.92±0.10	0.90±0.08	0.87±0.11	0.95±0.07	0.93±0.08
0.90	0.93±0.09	0.94±0.07	0.94±0.06	0.96±0.06	0.93±0.09	0.94±0.07	0.96±0.08	0.90±0.10	0.93±0.09
0.95	0.95±0.07	0.92±0.09	0.95±0.07	0.90±0.07	0.93±0.08	0.92±0.08	0.93±0.07	0.94±0.08	0.94±0.07

Cluster B / Continuation Cluster A

Corr.	0.6	0.65	0.7	0.75	0.8	0.85	0.9	0.95
0.15	0.94±0.08	0.92±0.09	0.92±0.09	0.91±0.10	0.91±0.08	0.94±0.09	0.96±0.06	0.96±0.06
0.20	0.95±0.09	0.92±0.08	0.91±0.08	0.94±0.06	0.91±0.09	0.97±0.05	0.95±0.08	0.92±0.09
0.25	0.93±0.07	0.94±0.08	0.87±0.11	0.92±0.10	0.95±0.06	0.96±0.08	0.95±0.07	0.94±0.08
0.30	0.92±0.11	0.94±0.07	0.94±0.09	0.91±0.12	0.93±0.09	0.92±0.09	0.94±0.08	0.92±0.06
0.35	0.96±0.06	0.93±0.07	0.93±0.08	0.93±0.06	0.95±0.06	0.93±0.09	0.94±0.07	0.94±0.07
0.40	0.96±0.05	0.91±0.12	0.89±0.10	0.91±0.11	0.97±0.06	0.90±0.11	0.93±0.05	0.89±0.09
0.45	0.94±0.11	0.93±0.07	0.92±0.08	0.95±0.08	0.92±0.11	0.92±0.12	0.94±0.07	0.94±0.07
0.50	0.97±0.05	0.93±0.09	0.94±0.09	0.91±0.11	0.90±0.12	0.91±0.07	0.92±0.09	0.94±0.09
0.55	0.94±0.09	0.93±0.11	0.93±0.11	0.91±0.11	0.89±0.14	0.90±0.09	0.93±0.07	0.93±0.07
0.60	0.92±0.10	0.96±0.10	0.92±0.09	0.88±0.13	0.94±0.10	0.91±0.10	0.94±0.08	0.91±0.09
0.65	0.95±0.09	0.94±0.08	0.90±0.12	0.93±0.11	0.91±0.08	0.93±0.12	0.95±0.07	0.93±0.10
0.70	0.95±0.06	0.92±0.11	0.91±0.09	0.91±0.12	0.92±0.08	0.93±0.08	0.96±0.07	0.94±0.11
0.75	0.94±0.07	0.90±0.14	0.95±0.08	0.91±0.12	0.92±0.08	0.94±0.08	0.95±0.06	0.94±0.10
0.80	0.88±0.11	0.95±0.07	0.88±0.10	0.91±0.11	0.96±0.09	0.93±0.10	0.90±0.12	0.97±0.05
0.85	0.94±0.08	0.91±0.09	0.93±0.09	0.93±0.11	0.88±0.14	0.91±0.11	0.92±0.11	0.95±0.08
0.90	0.96±0.06	0.93±0.10	0.92±0.11	0.90±0.12	0.93±0.09	0.95±0.07	0.95±0.08	0.97±0.08
0.95	0.91±0.09	0.95±0.06	0.94±0.07	0.98±0.04	0.97±0.06	0.94±0.06	0.99±0.04	0.99±0.04

Mean F-score and standard deviation of 20 replicates of a Cholesky-based structured simulation. Each entry corresponds to the mean F-Score ± the standard deviation for 20 replicates in each pair of intracorrelations. Corr. = Intracorrelation. Bolded numbers correspond to a significant (Mann-Whitney pvalue ≤0.05) improvement of the F-Score with respect to the one obtained without LDA-prefiltering.

scheme, some collision between classes may occur in which case it can be hypothesized that there is not enough information in the data to support their separation.

After obtaining the membership vector for a topology-constrained dataset, and before performing significance testing as explored in [1], a filtering step is introduced using LDA:

1. Given the membership vector of the modularity optimization, fit the data to the grouping using LDA.
2. Using the first two linear discriminants find the 95% confidence ellipses of each group.
3. Determine if there is a collision between all pairs of ellipses.
4. Merge groups if a collision is found.

The Methods section contains the details for each of these steps. Table 2, shows the results of 20 replicates of the simulation of topology-constrained correlation networks with the implementation of LDA filtering. As can be seen, the improvement is significant ($Mann-Whitney\ U : 5287314.000; p-value : <0.0001$) obtaining the true answer in most cases (even in intra-community correlations as low as 0.15). Is important to keep in mind that our simulations also include correlation between clusters (inter-community correlation) drawn from a random uniform distribution with minimum of 0 and maximum of 0.1. This means that the discrimination with the LDA filtering is robust even with correlation noise.

Despite the usefulness of the LDA in topology-constrained correlation network analysis, it is important to state that in a fully or nearly-fully connected graph, LDA tends to cluster everything in a single group. This is particularly true when the variance is small (data not shown). However, as shown in Table 2, LDA dramatically increases the performance when there is a topology constraint in the graph.

Case studies

Now some case studies that have been analyzed previously [1,27] will be considered. In this section it will shown how there are some real cases in which a topology-constrained correlation network community structure is over-fragmented. It is also shown how LDA can address fragmentation without systematically merging every partition scheme.

Voting in the United States 110th Senate. A great effort has been placed into analyzing the political partisanship in the US congress, particularly on how polarized Legislatures can influence the voting on non-particular issues [28]. In the 110th Legislature of the United States, in the second government of G.W. Bush, the polarization was evident. It has been suggested that in highly polarized Legislatures the representatives tend to vote more strongly with their party. Figure 2 shows that not only the polarization played an important roll. In Figure 2a, it is evident that the vote of individual representatives fell along party lines. Each color represents the cluster and the party, with the exception of the independent representatives whose votes are indistinguishable from the Democrats, and Senator Snowe, that despite being a Republican voted more similarly to Democrats. Figure 2a). If this correlation graph is constrained to geographical adjacency (i.e. neighboring states), the clustering is modified. In Figure 2b six clusters are found. The singleton (black node) corresponds to senator Nelson, a Democrat representative the Republican dominated region of Florida (South USA; Figure 3). Nodes in cyan and magenta correspond to Alaska and Hawaii, which have no neighbors. The yellow cluster includes Maine and New Hampshire senators who (as can be seen in Figure 3) are Republicans in a Democrat/independent neighborhood. When

the LDA prefiltering is used (Figure 2c), the clusters corresponding to Hawaii and Alaska are merged with the blue cluster which mainly contains Democrats, while Alaska had a Republican representation. However in Figure 2a, the Alaskan representatives had a voting profile closer to the Democrat along with Senator Collins (Maine), Senator Specter (Pennsylvania) and Senator Smith (Oregon), who were also Republicans with an intermediate voting profile between Democrats and their party. In figure 2c, Senator Collins (Maine) and Senator Specter (Pennsylvania) actually cluster with a few other Republicans and Democrats following a neighborhood voting profile. Despite polarization, there is still a neighborhood signal driving some of the votings. However, most Republicans and Democrats have a clear partisan profile of voting, and the differences rely on particular bills and motions that might have a regional scope.

In this example it can be seen that after the LDA filtering, the number of clusters obtained is reduced. Given the results of the simulation show that the heuristic to optimize Q does over-fragment the graph, the observed reduction is likely a more accurate description of the community structure giving a regional focus. The LDA filtering proposed here have no information of the topology constraint, therefore the results shown in this section demonstrate that there is a geographic signal in the US votes, and that does not follow a party-strict pattern. In this particular case, the correlation graph in Figure 2a shows that the polarization plays the major role, splitting most Democrats and Republicans in different groups. However, Figures 2b and 2c show that a regional bias remains in some of the motions voted.

α-Amylase homologs sub-domain architecture. In Hleap et al. [1], a dataset of 85 protein structures was analyzed to find a sub-domain architecture. They found four significant clusters, one of which comprises the minimum functional TIM-barrel [1]. In this manuscript that search has been broaden gathering 135 structures. To show a biological application of the LDA prefiltering, the algorithm described in [1] without contacts restrains was performed, with inter-residue contacts constraint, and the latter with LDA pre-filtering. Figure 4 shows the results for this case, where each color represents a cluster of residues within the protein. In the absence of contact restrains (Figure 4a) bigger clusters are found. Some clusters are made of disconnected components (orange cluster). There are significant smaller clusters than in the other cases (Figures 4b and 4c), and the biological meaning for the lack of contiguity is obscure. It can be ascribed that disjoint components in a cluster reflect a higher level community, which is not interesting from a protein modularity perspective. Figure 4b, shows the result for the same algorithm, when considering topology constraint based on the inter-residue contacts. Here, more sensible results are gathered returning the minimal functional TIM barrel topology obtained in [1] (yellow cluster). Figure 4c corresponds to the same topology-constrained network in Figure 4b, but with LDA pre-filtering, however the result is identical. This suggests that the LDA-filtered community structure at the protein level is strong and significant enough to avoid merging. This observation makes sense since Hleap et al. [1] were testing for correlation among residues and this information can be correlated with the contact between them. It is also important to state that when no over-fragmentation occurs (like in this particular dataset) LDA will not affect the result.

Conclusions

Here, by means of structured simulations, it is shown that topological constraints in a correlation network can lead to over-fragmentation, which supports the claims in [24].

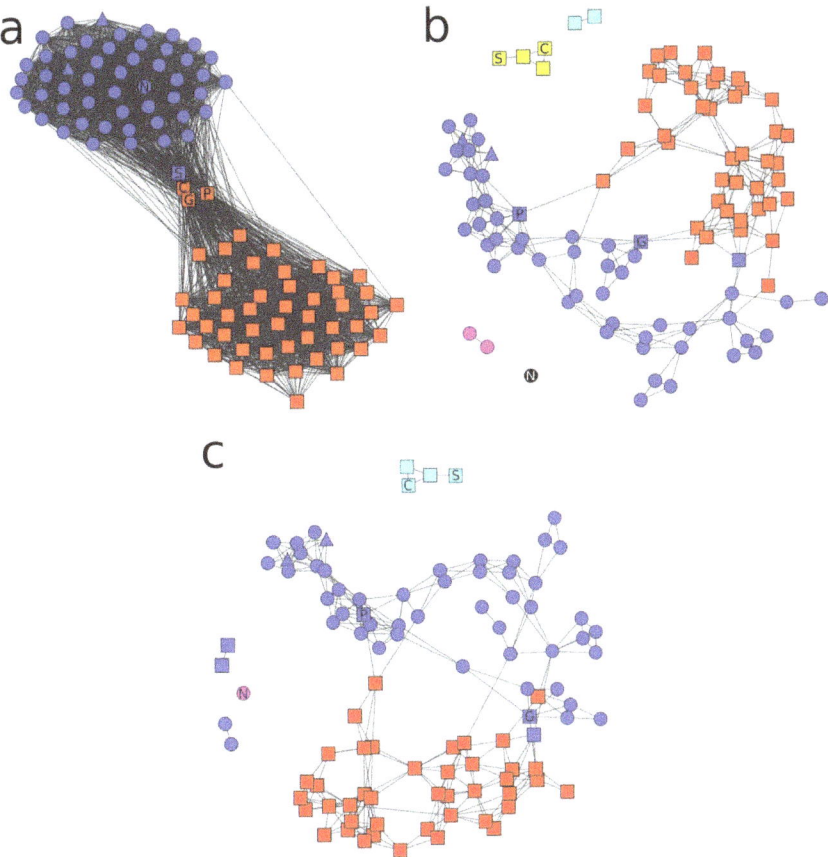

Figure 2. Networks of correlations of roll-call votings in the 110th US senate. 2a: Correlation network without state neighborhood constrain and without the use of LDA pre-filtering; 2b: Correlation network with state neighborhood constrain but without the use of LDA pre-filtering; 2c: Correlation network with state neighborhood constraint and using of LDA pre-filtering. The nodes are colored by cluster and each party is denoted with a given shape. Triangle: Independent; Square: Republican; Circle: Democrat. Letters inside nodes represents some senators names mentioned in text. S: Snowe; N: Nelson (FL); G: Smith (OR); Collins (ME); P: Specter (PA).

It also has been shown that topological constraints can be used to mine correlation graphs to obtain particular insights. The Roll-Call voting results demonstrate that there is a more complex structure than partisan politics alone, and in the LDA-filtered graph there is less fragmentation than in the non-filtered one. The inter-residue correlation network in protein structures needs to be considered with contacts to obtain biologically meaningful results. This can be a problem if artificial fragmentation is being created. However, it has been shown that LDA filtering does not merge

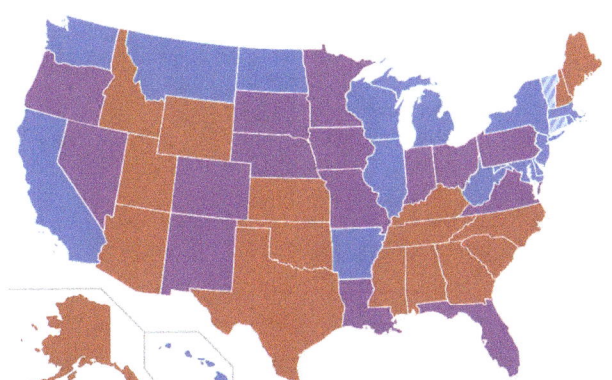

Figure 3. 110th US Congress Senate. USA map colored by the party who holds the seats in the 110th Senate (between January 3, 2007, and January 3, 2009). Blue: fully Democratic state; Red: Fully Republican state; Purple: Half Republican, half Democratic; Striped blue: Independent senator. Image taken from http://commons.wikimedia.org/wiki/File:110th_US_Congress_Senate.svg.

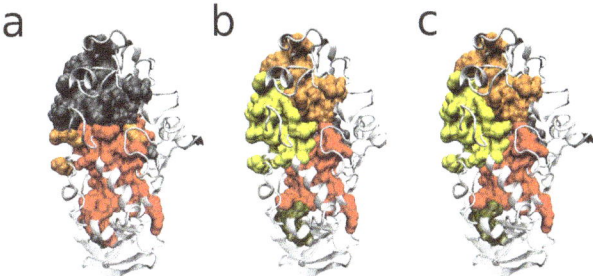

Figure 4. α-amylase homologs. Clusters (modules) found in an extension of the modularity inference performed in [1], including 135 homologs of the catalytic domain of the α-amylase. a) Modules inferred without constraining the topology with inter-residue contacts. b) Modules inferred constraining the topology in A with inter-residue contacts. c) Modules inferred by prefiltering the results in B, before significance testing.

clusters that were found to be meaningful in the first place.

It can be argued that other methods, such as sparse graphical models and LASSO-based methods [15,29], exist to cope with the over-fragmentation in sparser graphs. However, correlation graphs normally do not fulfill the assumptions of such methods like independence of the variables, *a priori* knowledge of some community properties, and a high degree of sparseness of the covariation among variables. Furthermore, optimization of Q has been an important tool for community detection in graph theory. Solving the problem of over-fragmentation by LDA and statistical testing is an important contribution to the study of correlation graphs in a data-driven way, without the need of a model, and where the distributional properties of the variables are not the main driving force of inference.

Methods

Multivariate normal structured simulations

To create the true clustering shown in Figure 1, the same approach done in [1] non-structured or topology-unconstrained simulations will be applied. However, to retain the shape (topology), the following procedure will be done:

1. Create a $1 \times k$ vector with original shape coordinates (k).
2. Create a $n \times k$ shape matrix, where each row is a repetition of the vector in the previous step. n is the number of desired samples.
3. Obtain a $n \times k$ multivariate normal ($MVN(0, U(0,1))$) matrix as performed in [1].
4. Create a $k \times k$ correlation matrix following the structure of each true module.
5. Perform the Cholesky decomposition on the random matrix (multivariate normal matrix) as explained in [1].
6. Sum the factorized random (and therefore now correlated) and shape matrices.

For the Cholesky decomposition, the intracorrelation in both clusters was controlled, starting in 0.15 to 0.95, in 0.05 increments. The intercorrelations in between clusters were drawn from a uniform distribution ($U(0,0.1)$). Given that [1] showed that 500 samples were enough to resolve most of the correlations, only as many samples were used.

This simulation was repeated 20 times for each intracorrelation pairs.

Performance measure

To quantify the performance of the simulation, an F-Score was calculated as:

$$F - score = 2 \frac{Sn \times Sp}{Sn + Sp} \qquad (2)$$

where Sn stands for sensitivity which can be expressed as $\frac{TP}{TP+FN}$; and Sp stands for specificity which can be estimated as $\frac{TP}{TP+FP}$.

In all cases, TP are the true positives, FN are the false negatives, and FP are the false positives.

The results of the 20 simulations are summarized as the mean F-score $\pm$ the F-score standard deviation for each intracorrelation pair.

Contact definition

In structured (shape-defined) datasets, a contact matrix can be inferred. Each point in a given configuration is said to be in contact with any other point in the dataset if the distance between a given pair is not greater than one unit plus the standard deviation of the simulation. This holds true only if the shape being constructed lays on a grid of one unit per square cell (like ours does). In the Roll-Call voting dataset, the contact was defined as touching (neighbors) states. In the case of the protein dataset, the contact matrix was inferred as in Hleap et al. [1].

Filtering the Q optimization output

The output of the modularity (Q) optimization developed by [8] is a membership vector. Here as in [1], the optimization is performed using a fast-greedy algorithm, which has been shown to be a good and fast heuristic for the optimization of Q [30]. After such a membership vector is obtained, the refinement proposed by [1] can be performed. However, some over-fragmentation may occur when a topology-constrained graph is used. To deal with this issue, here it is proposed a Linear Discriminants (LD) pre-filtering of the modularity membership vector.

Linear Discriminant Analysis (LDA). The LDA for the present paper was performed using the lda function available in the package MASS [31] in R [32]. Here the fit will be done between the correlation magnitude matrix (as performed in [1]), where each entry row/column corresponds to each variable, and each entry is the magnitude of the correlation vector as the square root of the sum of squared correlations in each dimension (X, Y for 2D, and X, Y, Z for 3D). The latter two cases are generalizations of the simpler case of one dimension in which case the data is the $n \times n$ correlation matrix, n being the variables in the dataset. In any of the cases, a fisher transformation and a significant test of the correlation is performed, as suggested in [1]. This data matrix is the same matrix that represents the graph, where the non-zero entries correspond to an edge and the actual value represents the weight of that edge.

Collision test and membership refinement. After the first two LD are obtained, a 95% confidence ellipse is computed. Here, the package ellipse [33] implemented in R [32] is used to compute the ellipses. After the ellipse have been estimated, a collision test is made. A point will be inside or at the edge of any given ellipse if the following inequality [34] is satisfied:

$$\frac{(x-h)^2}{r_x^2} + \frac{(y-k)^2}{r_y^2} <= 1 \qquad (3)$$

where x and y are the coordinates of a given point, h and k are the coordinates of the center of the ellipse, and rx and ry are the semi-minor and semi-major axes of the ellipse.

If the inequality in equation 3 is satisfied, the two ellipses are colliding and therefore the groups/classes they represent should be merged, otherwise the groups are not touched.

With this approach some of the over-fragmentation created by the lost of edges in a topology-constrained network might be dealt with.

Case studies datasets

Voting in the United States 110th Senate. The Roll-Call voting of 110th United States Senate (available online at [35] or in Supporting Information File S1) was used to construct the network. First a data matrix is created where each row represents each senator and each column represents a vote for a given motion or amendment. With that data matrix a correlation matrix Ξ is created, where each entry have been tested for significance using a Z test of a fisher transformation of the correlation. If the significance test failed, the corresponding entry is set to zero, otherwise the correlation value is recorded. Let $S = (N, f)$ be an undirected graph, where N is a list of nodes (senator) and f is a function $f : N \times N \to \Bbbk$ that assigns an edge weight to each senator pair. An edge E_{ij} is assigned only if $\Xi_{ij} > 0$. To create a topology-constrained graph a fixed topology accounting for neighboring states is applied to the edge assignment as an extra condition. In the topology-constrained weighted network, an edge will be drawn only if $\Xi_{ij} > 0$, *and if* the senators represent neighboring states. This constraint will allow to test the hypothesis if there is any subdivision that is determined by the geography more than by only party affiliation.

α-Amylase structures homologs. The α-Amylase-like family catalyzes the hydrolysis of α-(1,4) glycosidic bonds of polysaccharides, therefore being classified as glycoside hydrolases [36] in the family 13 [37]. It is a multi-reaction catalytic family since its members can catalyze different reactions (hydrolysis, transglycosylation, condensation and cyclization) [38]. All members of this family share a symmetrical TIM-barrel $((\beta/\alpha)_8)$ catalytic domain [39], including those without any catalytic activity [40]. This fold is highly versatile and widespread among the structurally characterized enzymes, being present in almost 10% of them [41–44]. There has been a debate about the type of evolution that this fold has been through: convergent, divergent or a mixture of both mechanisms [41]. However, there is some evidence suggesting the divergent evolution hypothesis is the most likely [42]. The catalytic activity and substrate binding residues occurs at the C-termini of β-strands and in loops that extend from these strands [39]. The catalytic site includes aspartate as a catalytic nucleophile, glutamate as an acid/base, and a second aspartate for stabilization of the transition state [45]. The catalytic triad plus an arginine residue are totally conserved in this family across all catalysis-active members [37].

In [1], the protein structures belonging to the α-Amylase catalytic domain were gathered from the Homstrad database [46] and these seeded a Blast search restricted to the protein data bank. Here, the search is broaden by seeding a PSI-BLAST [47] search with a PFAM [48] seed alignment of α-Amylase structures (PFAM code PF00128). The PSI-BLAST search was restricted to structures available at the protein data bank (http://www.rcsb.org/pdb/). There were in total 135 structures gathered which homology and membership to the α-amylase family (the Glycoside Hydrolase Family 13, GH13) was guaranteed (Available in File S1).

Those 135 structures were aligned using the algorithm proposed by [49] that modifies the pairwise MATT flexible structure aligner [50] to complete the multiple structure alignment.

After the alignment, the procedure explained in [1] was used, where the coordinates of the centroid of homologous residues are recorded in a data matrix. The graph construction is performed as before, but one correlation matrix is created per dimension, and then the matrix of magnitudes of the correlation vectors (Ξ) is computed as the euclidean distance between the three matrices. Edges will be assigned, as before, if two residues correlate *and if* they are in contact in the structure (topology constraint).

Supporting Information

File S1 Data File. The data is available as supporting information as a compressed TAR file named File S1.tar.gz containing the files Amy135.gm and sen110kh.2008.USA.roll.call.txt. **File sen110kh.2008.USA.roll.call.txt.** It contains the information of the Roll-Call votings in 2008 for the US Senate. This information is available also in VoteView [35]. The file is space-delimited text file where each line represents a Senator. The first field corresponds to the Senator's code, followed by the state they represent. After the state, a number indicating party affiliation, followed by the lastname of the Senator. The last field correspond to the Roll-Call votes. **File Amy135.gm.** It contains the centroid coordinates in a semicolon-delimited format. In this format the first field correspond to the name of the structure and the X, Y, and Z coordinates for the centroid of each homologous aminoacids are stored sequentially. There is one line per structure (135 in this dataset), and 3 times the number of homologous residues coordinates entries.

Acknowledgments

The authors thank Professor N. Zeh, Professor R. Beiko, Conor Mehan, and the members of Dr. Beiko's Lab in Dalhousie University for some helpful suggestions. The members of the Blouin Lab for helpful comments and critical review of this manuscript. We also thank Liz Mackay for the editorial revision of the manuscript.

Author Contributions

Conceived and designed the experiments: JSH CB. Performed the experiments: JSH. Analyzed the data: JSH CB. Contributed reagents/materials/analysis tools: CB. Wrote the paper: JSH.

References

1. Hleap JS, Susko E, Blouin C (2013) Defining structural and evolutionary modules in proteins: a community detection approach to explore sub-domain architecture. BMC structural biology 13: 20.
2. Stanford NJ, Lubitz T, Smallbone K, Klipp E, Mendes P, et al. (2013) Systematic construction of kinetic models from genome-scale metabolic networks. PLOS ONE 8: -79195.
3. Navia AF, Cortés E, Mejía-Falla PA (2010) Topological analysis of the ecological importance of elasmobranch fishes: A food web study on the gulf of tortugas, colombia. Ecological modelling 221: 2918–2926.
4. Gorman SP, Malecki EJ (2000) The networks of the internet: an analysis of provider networks in the usa. Telecommunications Policy 24: 113–134.
5. Burt RS, Kilduff M, Tasselli S (2013) Social network analysis: Foundations and frontiers on advantage. Annual review of psychology 64: 527–547.
6. Diestel R (2012) Graph Theory, volume 173 of *Graduate Texts in Mathematics*. Heidelberg: Springer-Verlag, 4rd. edition.
7. Fortunato S, Castellano C (2012) Community structure in graphs. In: Meyers RA, editor, Computational Complexity, New York: Springer. pp. 490–512.
8. Newman M, Girvan M (2004) Finding and evaluating community structure in networks. Phys Rev E 69: 026113.
9. Newman ME (2004) Detecting community structure in networks. The European Physical Journal B-Condensed Matter and Complex Systems 38: 321–330.

10. Danon L, Diaz-Guilera A, Duch J, Arenas A (2005) Comparing community structure identification. Journal of Statistical Mechanics: Theory and Experiment 2005: -09008.
11. Guimera R, Sales-Pardo M, Amaral LAN (2004) Modularity from fluctuations in random graphs and complex networks. Physical Review E 70: 025101.
12. Fortunato S, Barthelemy M (2007) Resolution limit in community detection. Proceedings of the National Academy of Sciences 104: 36–41.
13. Dobra A, Hans C, Jones B, Nevins JR, Yao G, et al. (2004) Sparse graphical models for exploring gene expression data. Journal of Multivariate Analysis 90: 196–212.
14. Tibshirani R (1996) Regression shrinkage and selection via the lasso. Journal of the Royal Statistical Society Series B (Methodological): 267–288.
15. Mukherjee S, Hill SM (2011) Network clustering: probing biological heterogeneity by sparse graphical models. Bioinformatics 27: 994–1000.
16. Zhao P, Yu B (2006) On model selection consistency of lasso. The Journal of Machine Learning Research 7: 2541–2563.
17. Fidelak J, Ferrer S, Oberlin M, Moras D, Dejaegere A, et al. (2010) Dynamic correlation networks in human peroxisome proliferator-activated receptor-γ nuclear receptor protein. European Biophysics Journal 39: 1503–1512.
18. Bernhardt BC, Chen Z, He Y, Evans AC, Bernasconi N (2011) Graph-theoretical analysis reveals disrupted small-world organization of cortical thickness correlation networks in temporal lobe epilepsy. Cerebral Cortex 21: 2147–2157.

19. Friedman J, Alm EJ (2012) Inferring correlation networks from genomic survey data. PLoS Computational Biology 8: e1002687.

20. Kenett DY, Tumminello M, Madi A, Gur-Gershgoren G, Mantegna RN, et al. (2010) Dominating clasp of the financial sector revealed by partial correlation analysis of the stock market. PLoS one 5: e15032.

21. Keskin M, Deviren B, Kocakaplan Y (2011) Topology of the correlation networks among major currencies using hierarchical structure methods. Physica A: Statistical Mechanics and its Applications 390: 719–730.

22. Müller-Linow M, Weckwerth W, Hütt MT (2007) Consistency analysis of metabolic correlation networks. BMC Systems Biology 1: 44.

23. Reichardt J, Bornholdt S (2009) Innovation Networks: New Approaches in Modelling and Analyzing, Springer, chapter Tools from Statistical Physics for the Analysis of Social Networks. pp. 149–187.

24. Reichardt J, Bornholdt S (2007) Partitioning and modularity of graphs with arbitrary degree distribution. Physical Review E 76: 015102.

25. Jain AK, Duin RPW, Mao J (2000) Statistical pattern recognition: A review. Pattern Analysis and Machine Intelligence, IEEE Transactions on 22: 4–37.

26. Rao CR (2009) Linear statistical inference and its applications, volume 22. John Wiley and sons.

27. Onnela JP, Fenn DJ, Reid S, Porter MA, Mucha PJ, et al. (2012) Taxonomies of networks from community structure. Physical Review E 86: 036104.

28. Cho WKT, Fowler JH (2010) Legislative success in a small world: Social network analysis and the dynamics of congressional legislation. The Journal of Politics 72: 124–135.

29. Jacob L, Obozinski G, Vert JP (2009) Group lasso with overlap and graph lasso. In: Proceedings of the 26th Annual International Conference on Machine Learning. pp. 433–440.

30. Clauset A, Newman MEJ, Moore C (2004) Finding community structure in very large networks. PHYSREVE 70: 066111.

31. Venables WN, Ripley BD (2002) Modern Applied Statistics with S. New York: Springer, fourth edition. URL http://www.stats.ox.ac.uk/pub/MASS4. ISBN 0-387-95457-0.

32. R DCT (2011) R: A Language and Environment for Statistical Computing. Vienna, Austria. URL http://www.R-project.org/. ISBN 3-900051-07-0.

33. Murdoch D, Chow ED (2013) ellipse: Functions for drawing ellipses and ellipse-like confidence regions. URL http://CRAN.R-project.org/package=ellipse. R package version 0.3-8.

34. Berger M, Pansu P, Berry JP, Saint-Raymond X (1984) Euclidean conics. In: Problems in Geometry, New York: Springer. pp. 102–105.

35. Poole KT (2013) data available at http://voteview.com.

36. Davies G, Henrissat B (1995) Structures and mechanisms of glycosyl hydrolases. Structure 3: 853–859.

37. Svensson B, Janecek S (2013) Glycoside hydrolase family 13. available at URL http://www.cazypedia.org/.

38. Ben Ali M, Khemakhem B, Robert X, Haser R, Bejar S (2006) Thermostability enhancement and change in starch hydrolysis profile of the maltohexaose-forming amylase of bacillus stearothermophilus us100 strain. Biochem J 394: 51–6.

39. Svensson B (1994) Protein engineering in the α-amylase family: catalytic mechanism, substrate specificity, and stability. Plant molecular biology 25: 141–157.

40. Fort J, Laura R, Burghardt HE, Ferrer-Costa C, Turnay J, et al. (2007) The structure of human 4f2hc ectodomain provides a model for homodimerization and electrostatic interaction with plasma membrane. Journal of Biological Chemistry 282: 31444–31452.

41. Farber GK (1993) An α/β-barrel full of evolutionary trouble. Current opinion in structural biology 3: 409–412.

42. Höcker B, Jürgens C, Wilmanns M, Sterner R (2001) Stability, catalytic versatility and evolution of the $(\beta/\alpha)_8$-barrel fold. Current opinion in biotechnology 12: 376–381.

43. Wierenga RK (2001) The tim-barrel fold: a versatile framework for efficient enzymes. FEBS letters 492: 193–198.

44. Gerlt JA, Raushel FM (2003) Evolution of function in $(\beta/\alpha)_8$-barrel enzymes. Current opinion in chemical biology 7: 252–264.

45. Uitdehaag JC, Mosi R, Kalk KH, van der Veen BA, Dijkhuizen L, et al. (1999) X-ray structures along the reaction pathway of cyclodextrin glycosyltransferase elucidate catalysis in the α-amylase family. Nature Structural & Molecular Biology 6: 432–436.

46. Mizuguchi K, Deane CM, Blundell TL, Overington JP (1998) Homstrad: a database of protein structure alignments for homologous families. Protein Sci 7: 2469–71.

47. Altschul SF, Madden TL, Schäffer AA, Zhang J, Zhang Z, et al. (1997) Gapped blast and psi-blast: a new generation of protein database search programs. Nucleic acids research 25: 3389–3402.

48. Finn RD, Mistry J, Tate J, Coggill P, Heger A, et al. (2010) The pfam protein families database. Nucleic acids research 38: D211–D222.

49. Hleap JS, Nguyen KN, Safatli A, Blouin C (2013) Reference matters: An efficient and scalable algorithm for large multiple structure alignment. In: Saeed F, DasGupta B, editors, Proceedings of the 5th International Conference on Bioinformatics and Computational Biology (BICOB–2013). Winona, MN, USA, pp. 153–158.

50. Menke M, Berger B, Cowen L (2008) Matt: local flexibility aids protein multiple structure alignment. PLoS Comput Biol 4: e10.

19

Cupulin Is a Zona Pellucida-Like Domain Protein and Major Component of the Cupula from the Inner Ear

Jens Dernedde[1]*, Christoph Weise[2], Eva-Christina Müller[3], Akira Hagiwara[4], Sebastian Bachmann[5], Mamoru Suzuki[4], Werner Reutter[1], Rudolf Tauber[1], Hans Scherer[6]

1 Institut für Laboratoriumsmedizin, Klinische Chemie und Pathobiochemie, Charité -Universitätsmedizin Berlin, Berlin, Germany, **2** Institut für Chemie und Biochemie, Freie Universität Berlin, Berlin, Germany, **3** Max Delbrück Center for Molecular Medicine, Berlin-Buch, Germany, **4** Department of Otolaryngology, Tokyo Medical University, Shinjuku-ku, Tokyo, Japan, **5** Institut für Vegetative Anatomie, Charité - Universitätsmedizin Berlin, Berlin, Germany, **6** Klinik für Oto-Rhino-Laryngologie, Charité - Universitätsmedizin Berlin, Berlin, Germany

Abstract

The extracellular membranes of the inner ear are essential constituents to maintain sensory functions, the cupula for sensing torsional movements of the head, the otoconial membrane for sensing linear movements and accelerations like gravity, and the tectorial membrane in the cochlea for hearing. So far a number of structural proteins have been described, but for the gelatinous cupula precise data are missing. Here, we describe for the first time a major proteinogenic component of the cupula structure with an apparent molecular mass of 45 kDa from salmon. Analyses of respective peptides revealed highly conserved amino-acid sequences with identity to zona pellucida-like domain proteins. Immunohistochemistry studies localized the protein in the ampulla of the inner ear from salmon and according to its anatomical appearance we identified this glycoprotein as Cupulin. Future research on structure and function of zona pellucida-like domain proteins will enhance our knowledge of inner ear diseases, like sudden loss of vestibular function and other disturbances.

Editor: Silvio C. E. Tosatto, Universita' di Padova, Italy

Funding: The work was supported by internal funding of Charité. The funders had no role in study design, data collection and analysis, decision to publish, or preparation of the manuscript.

Competing Interests: The authors have declared that no competing interests exist.

* Email: jens.dernedde@charite.de

Introduction

The vestibular organ of vertebrates has five mechanical sensors that convert acceleration into electrical signals. They are located in the labyrinth organ of the inner ear. Three of them function as membranes (cupulae) in a liquid-filled cavity (Fig. 1A). The sensors are located in a widened part (ampulla) of a fluid filled ring system, the semicircular canals. The cupulae are fixed at the roof of the ampulla and ride on a barrel-like structure, the crista ampullaris. Kino- and stereocilia growing out from the top of hair cells connect the gelatinous cupula with the underlying neuroepithelium. A torsional acceleration of the head leads to a counter rotation of the fluid resulting in a deflection of the cupula and thereby in a stimulation of the hair cells [1–3]. Shearing of tip-links between the hairs opens mechanosensitive ion channels. The result is a potassium influx into the cells which causes a generator potential and in the afferent bipolar nerve an alteration of the action potential rate. A detachment of the cupula from the roof or a leak in the membrane impedes stimulation [4,5].

In humans suffering from a sudden loss of vestibular function, a malfunction in the ampulla is considered to be a possible explanation. Experiments in pigeons [3,4] have demonstrated that the mechanical detachment of the cupula from the roof of the ampulla results in the clinical picture of a vestibular loss of function. Furthermore a membrane leak could develop if the structural integrity of the cupula is compromised, e.g. by a lack of structural material production, or by an elevation of the ampulla roof caused by increased pressure of incoming liquid [6].

To gain further knowledge on the origin of sudden loss of vestibular function, we started to analyze the cupula material from salmon and chicken. Goodyear and Richardson [7] have compared the protein composition of acellular matrices of the inner ear, the tectorial and otoconial membranes and the cupula. In the mouse inner ear, α- and β-Tectorin are the major components of the tectorial and otoconial membranes, but are missing in the cupula [7,8]. Although Otogelin, a 313 kDa protein related to mucins, was found in all acellular structures of the inner ear in mice [7], in otogelin-null mutant the cupula is still present, but detached from the crista ampullaris [9]. Therefore, a so far unknown structural component must be responsible to build the macromolecular cupula structure. Goodyear and Richardson [7] postulated the existence of this structural protein and suggested the name "Cupulin".

In order to identify this missing component of the cupula, we investigated the inner ear from salmon and chicken. These animals were selected because their vestibular organs are relatively easily accessible. In birds the bony layer of the vestibular organ is very thin and can be removed carefully giving access to the membranous structures. In addition, in fish the vestibular organ is not embedded in bone at all.

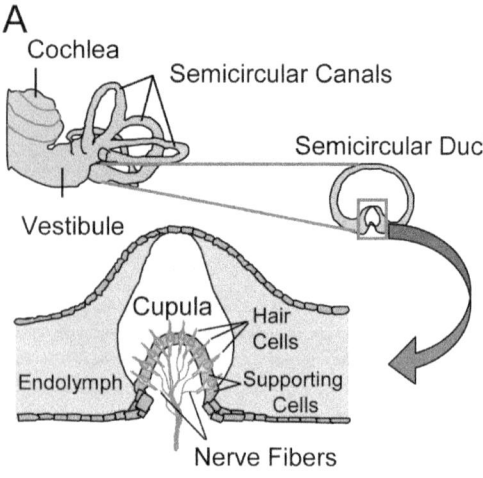

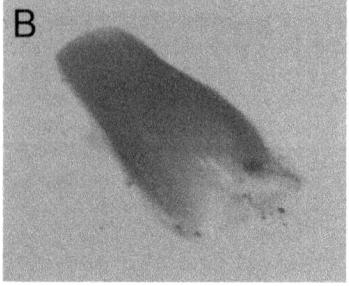

Figure 1. The cupula. A, localization of the cupula in the inner ear. **B**, dissected cupula from salmon stained with Evans blue.

Methods and Chemicals

All chemicals were from Sigma-Aldrich (München, Germany) if not otherwise stated.

Preparation of cupulae from salmon and chicken

The heads of commercially slaughtered salmon were opened. After suction of the brain, the free endocranial part of the vertical canal was cut and the labyrinth was removed carefully without touching the ampullas. The labyrinth was immediately immerged in artificial endolymph (126 mM KCl, 1 mM NaCl, 25 mM KHCO$_3$, 0.025 mM MgCl$_2$, 0.025 mM CaCl$_2$, 1.4 mM K$_2$HPO$_4$, 25 mM mannitol, pH 7.4), as described previously by Marcus et al. [10]. For asservation of the salmon cupulae the semicircular canal was cut 3 mm away from the ampulla. With a micropipette the canal was filled on the side of the ampulla with Evans-blue to stain the cupulae for better visualization. Subsequently the canal was cut again just at the site where it enters the ampulla. By an additional longitudinal short cut on the roof, the ampulla was opened. Small movements of the specimen with micro-forceps on both sides detached the cupula. The cupula has almost the same specific weight as the endolymph and does therefore not sink. This effect and its shape allowed distinguishing the cupula from various parts of the specimen, formed during preparation. As long as the acellular cupula was kept in endolymph it did not change either its form or size.

For preparation of cupulae from commercially slaughtered chicken the heads were fixed in an upright position. After removal of the bone of the posterior lateral portion of the head, the semicircular canals and the ampullas were identified. The very

thin bony layer was removed with needles and micro-forceps. The labyrinth was removed and stored in artificial endolymph solution. Further preparation steps were identical to those described for the salmon preparation.

Trypsin digestion and mass-spectrometric analyses

Crude cupula material from salmon and chicken was dissolved in denaturing 2 x SDS sample buffer and boiled for 5 min. Remained debris was removed by centrifugation (14,000×g, 10 min). Soluble extract was separated by SDS-PAGE under reducing conditions and the gel was stained with Coomassie Brilliant Blue. The dominant 45 kDa band was excised from the gel with a scalpel and cut into small 1 mm gel cubes.

Peptides were obtained by trypsin in-gel digestion as described previously [11] and peptide masses were analysed by matrix-assisted laser desorption ionization-time of flight mass spectrometry (MALDI-TOF-MS) using an Ultraflex-II TOF/TOF instrument (Bruker Daltonics, Bremen, Germany) equipped with a 200 Hz solid-state Smart beam laser. The mass spectrometer was operated in the positive reflector mode. Mass spectra were acquired over an m/z range of 600–4,000.

α-cyano-4-hydroxycinnamic acid (CHCA) was used as the matrix and protein digest samples were spotted using the dried-droplet technique. MS/MS spectra of selected peptides were acquired in the LIFT mode [12].

Database searches were performed using Mascot (Matrix Science Ltd., http://www.matrixscience.com). Mass tolerance was typically set at ±75 ppm and we allowed for one missed cleavage. Annotation of the MS/MS spectra was done manually.

Data analysis

By using the Basic Local Aligment Search Tool BLASTP [13] peptide sequences were screened for similarity in the protein database to assign the protein. For comparison of protein sequence data we used the ClustalW2 program from the European Bioinformatics Institute, EBI [14]. For the detection of the signal peptide we applied the SignalP algorithm [15], the transmembrane domain was predicted by the TMHMM 2.0 software [16]. The protein sequence was further screened for potential N-glycosylation sites with the program NetNGlyc 1.0 (Center for Biological Sequence Analysis, Technical University of Denmark).

Zona pellucida-like domain protein specific peptide antibodies and Western blotting

Peptide-specific antibodies were obtained by standard immunization of guinea pigs with a mixture of two synthetic peptides linked to the KLH antigen (Pineda, Berlin, Germany). The peptide sequences were: P1: NH2-**C**DANFHSRFPAERDI, and P2: NH$_2$-VKHKNQKMS TVFLH**C** respectively. Cysteine residues (**C**) were added to the sequence to achieve further peptide coupling. Serum was prepared and the total IgG fraction was first isolated by affinity chromatography on protein A sepharose (GE-Healthcare, München, Germany). The peptide specific antibodies were then purified by peptide affinity chromatography. Therefore 1 mg of both peptides was coupled to 2 ml of thiol sepharose, according to the manufacturers' description (GE-Healthcare, München, Germany). Reactivity to crude cupula preparations was analysed by Western blotting at a 1:10,000 dilution of purified peptide-antibodies (0.7 mg/ml). The secondary peroxidase-labelled anti-guinea pig antibody was from Dianova (Hamburg, Germany). For blot development the Amersham ECL Western Blotting System Kit form GE Healthcare (München, Germany) was applied.

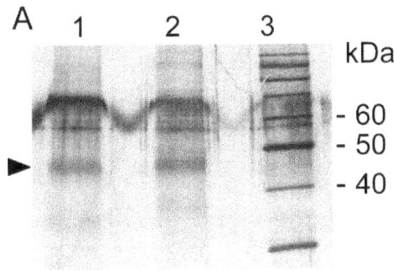

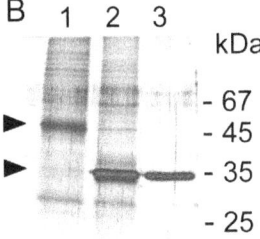

Figure 2. Visualization of cupula proteins. A, crude extracts from isolated cupulae from salmon, (lane 1) and chicken (lane 2) were separated on a 12% SDS-PAGE under reducing conditions and silver stained. The arrowhead highlights a dominant protein (~45 kDa) chosen for further analyses. Lane 3, marker proteins. In the 60 kDa range additional yet unidentified protein components are visible. **B**, deglycosylation of salmon cupula protein extract. Lane 1, cupula extract untreated; lane 2, cupula extract+PNGase F (100 NEB units), lane 3, PNGase F control (500 NEB units). Arrowheads indicate molecular weight shift of the 45 kDa protein due to the N-deglycosylation.

N-Deglycosylation Analysis

Crude cupula material was first boiled for 5 min in a 1% SDS, 1% β-mercaptoethanol solution, next diluted to a final concentration of 0.1% SDS in 20 mM sodium phosphate, pH 7.4, 1% Nonidet P-40, and digested with peptide:N-glycosidase F (PNGase F, New England Biolabs GmbH, Frankfurt, Germany) for 2 hours at 37°C. A typical analytical sample contained 2–3 cupulae and

was digested with 0.2 μl enzyme (100 NEB units) in a final reaction volume of 20 μl. Deglycosylation of samples was demonstrated by SDS-PAGE and subsequent protein staining.

Histochemistry

The vestibular organ of salmon heads was removed from the cerebral cavity and kept in a fixation solution (4% formaldehyde) for 30 min. The ampullas were separated from the stony otoliths and the specimens were embedded in paraffin. Cross-sections of the ampullas from salmon were alternating prepared for either HE-staining or immunohistology with anti-zona pellucida-like domain protein antibodies (1:100). The secondary antibody was a commercial PE-labeled anti-guinea pig antibody (Dianova, Hamburg, Germany)

Results

The cupula is a jelly-like extracellular matrix of the inner ear and part of the sensor system that measures torsional accelerations (Fig. 1). When we started to analyze the cupula protein composition from salmon and chicken by gel electrophoresis, a comparable protein pattern with ~10–15 bands, depending on the quality of sample preparation was detected for both organisms (Fig. 2). The existence of a dominant protein that constitutes the cupula structure was predicted by Goodyear and Richardson [7]. Here we identified a prominent fuzzy band in the range of approximately 45 kDa after sample separation from salmon and chicken under denaturing and reducing conditions (Fig. 2A). Although the protein was always visible, the distinctness varied between different preparations. As extracellular matrices usually consist of glycosylated proteins we treated the cupula sample from salmon with the N-glycosidase PNGase F. Indeed the size of the dominant 45 kDa cupula protein was reduced by ~11 kDa to a size of 34 kDa (Fig. 2 A,B). The PNGase F control (lane 3) migrated at the same position, but here we loaded the fivefold quantity of enzyme compared to the amount in lane 2 to visualize the protein. From the disappearance of the 45 kDa band and the intense staining at 34 kDa we concluded that the deglycosylated salmon protein and PNGase F run at the same position. Faintly stained protein bands in lanes 1 and 2 that migrate approximately

Table 1. Peptide sequences obtained from the 45 kDa gel band from salmon.

Peptide No.	Mass observed (g/mol)	Mass calculated (g/mol)	Peptide sequences
1	1713.71	1712.71	pyroQF**D**GYNCDANFHSR (+cam)
1a	1729.72	1728.71	pyroQF**D**GYNCDAN**Y**HSR (+cam)
2	619.26	618.31	FPAER
3	1596.87	1595.80	DISVYCGVQTITLK (+cam)
4	852.35	851.32	HGDAHCR (+cam)
5	2519.13	2518.09	WNVLMDYCYTTASGNPNDELR (+cam)
5a	2559.16	2558.12	WN**I**LMDYCYTT**P**SGNPNDELR (+cam)
6	915.47	914.47	FAFEVFR
6*	972.50	971.47	FAFEVFR (+cam)
7	1322.70	1321.65	MSTVFLHCVTK (+cam)
7*	1338.70	1337.65	MSTVFLHCVTK (+cam+ox)

Data base searches were performed using Mascot and annotation of the MS/MS spectra was done manually. Amino-acid residues that differ from the published salmon sequence (C0H9B6) are bold. Amino-acid modifications: pyroQ, pyroglutamate, (delta mass: −17); ox, oxidized methionine (delta mass: +16); cam, carbamidomethyl, (delta mass: +57).
apeptide with additional amino-acid exchange.
*peptide with additional modification.

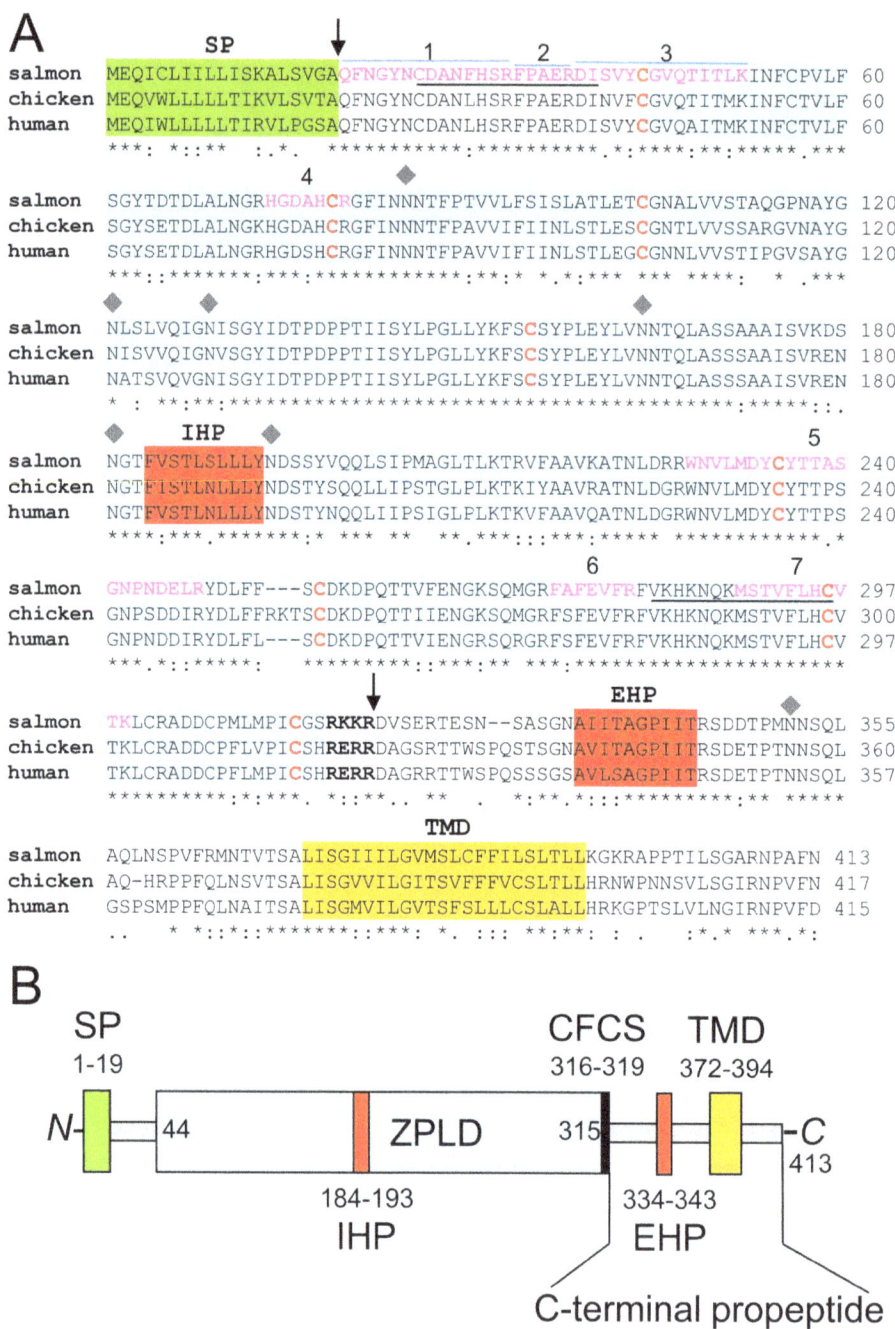

Figure 3. Zona pellucida-like domain protein homology and structure. A, protein sequences from UniProt database: salmon: C0H9B6; chicken: E1C8E6; human: Q8TCW7 were aligned by applying the ClustalW2 program. Asteriks (★) marked below the sequence highlight conserved amino-acids (~70%) between the three organisms. Conserved cysteine residues 1–8 that constitute the zona pellucida-like domain (blue box) are shown in red letters. Arrows mark the mature protein after cleavage of the *N*-terminal signal sequence (SP, green box) predicted by the SignalP algorithm and the *C*-terminal furin cleavage site (CFCS, black and bold letters). IHP and EHP (red boxes) show the potential internal and external hydrophobic patches of the zona pellucida-like-domain (ZLPD). The transmembrane domain (TMD, orange box) was predicted by the TMHMM 2.0 software. In pink and bold are highlighted the peptides 1–7, identified by mass spectrometry. Individual peptides 1, 2 and 3 are highlighted with blue lines. Underlined peptide sequences were used for immunization. Grey diamonds indicate the asparagine residue of potential *N*-glycosylation sites (NXS/T) determined with the program NetNGlyc 1.0. **B**, scheme of zona pellucida-like domain protein structure.

20 kDa below the glycosylated or deglycosylated protein, may reflect immunoreactive degradation products (Fig. 2B).

To identify specific peptide sequences from the salmon protein, gel electrophoresis was performed, the 45 kDa band was cut-out, and the protein was trypsinized and further analyzed by mass spectrometry. The peptide mass fingerprint analysis with annotated peptide sequences is shown in Fig. S1 and exemplarily a detailed MS/MS spectrum for one peptide is presented in Fig. S2. Database searches revealed identity to several predicted open reading frames of zona pellucida-like domain proteins. Here, to

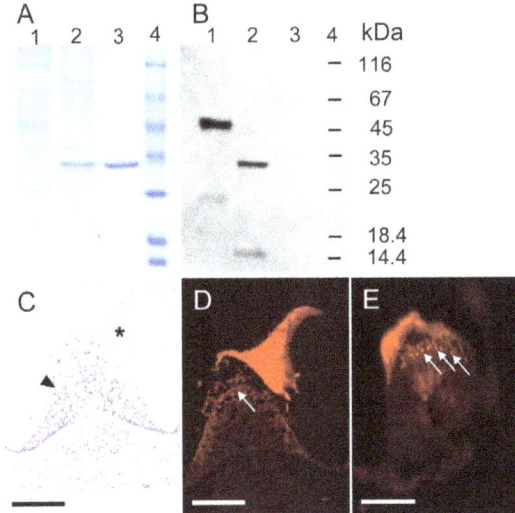

Figure 4. Immunodetection of zona pellucida-like domain protein from salmon samples. A, Coomassie stained SDS-polyacrylamide gel and B, corresponding Western blot of separated crude cupula extracts. Lane 1, untreated sample, lane 2, PNGase F treated sample with faster migration of zona pellucida-like domain protein. **C,** HE stained inner ear cross section, asterisk marks the cupula and the arrowhead the subcupulary region with sensory and supporting cells. **D,** immunostaining of inner ear cross section, the arrow probably indicates staining of supporting cells which produce the zona pellucida-like domain protein. This is even more pronounced in **E**.

the best of our knowledge we identified for the first time corresponding peptide sequences from a zona pellucida-like domain protein (Table 1). The overall match of seven identified peptides with the predicted sequences from salmon, chicken, and human origin is convincing (Fig. 3A) and covers about 26% of the mature extracellular protein ranging from amino-acids 21 to 319 (Fig. 3) With peptide 1 we most probably identified the N-terminus of the secreted zona pellucida-like domain protein. Cleavage of the signal peptide between amino-acids A_{19} and Q_{20} is predicted by the SignalP algorithm (data not shown). Furthermore the conversion of glutamine to pyroglumate (Table 1) argues for the N-terminal position, where spontaneous intramolecular cyclization can occur. In addition it is interesting to note, that we observed one difference to the published sequence (C0H9B6) from the salmon zona pellucida-like domain protein. Instead of the uncharged asparagine (N_{22}) we identified the acidic aspartic acid residue (D_{22}) in peptides 1 and 1^a. Further amino-acid changes might be explained by the interindividual variations due to our randomly pooled material from wild and farm-raised salmon from diverse origin. In detail, in addition to F_{30} present in peptide 1 a substitution to Y_{30} was identified in peptide 1^a. Further in addition to the correctly matching sequence from peptide 5, two variations were identified compared to the database entry within peptide 5^a, where V_{229} was changed to I_{229} and A_{239} to P_{239}. In each case the amino-acid substitution was conservative and hydrophobic apolar amino-acids were used.

The alignment of protein sequences from salmon (C0H9B6), chicken (E1C8E6) and human (Q8TCW7) was performed with the ClustaW algorithm. Overall, the zona pellucida-like domain proteins are highly conserved from fish to human, with a sequence identity of 72%. According to the nomenclature by Bork and Sander [17] which is based on the positioning of conserved amino-acids, i.e. structuring cysteine residues, the protein contains a zona

pellucida-like domain (Figure 3), but exhibits only minor amino-acid sequence identity to the zona pellucida (ZP) domain, when compared to the well-studied murine sperm receptor mZP3 (data not shown). On the other hand, the conserved regions of zona pellucida-like domain protein and the ZP of mZP3 imply a similar structure and therefore hint at comparable function of both proteins as described [18–22].

When we compared the salmon zona pellucida-like protein to the zona pellucida protein mZP3 following similarities were obvious: i) an N-terminal signal peptide directing the protein to the endoplasmatic reticulum (ER) and Golgi apparatus for posttranslational modification, ii) for the zona pellucida-like domain we also predict an N-terminal internal hydrophobic patch (IHP) and C-terminal external hydrophobic patch (EHP) as demonstrated for mZP3 [19], iii) a consensus furin cleavage site (CFCS) that separates IHP from EHP and enables the extracellular delivery of the mature protein and its polymerization, iv) a transmembrane domain necessary for initial anchoring at the cell membrane.

The zona pellucida-like domain protein from salmon is probably modified after translation. Initially the protein consists of 413 amino-acids (aa) and has a calculated molecular mass of 45.2 kDa. Cleavage of the signal peptide and further processing at the C-terminal furin cleavage site (CFCS), could deliver the mature secreted protein consisting of 299 aa with a molecular mass of 33.4 kDa. As described above, separation of extracted cupula material by SDS-PAGE displayed a dominant fuzzy protein band at 45 kDa (Fig. 2, lane 1). After deglycosylation with PNGase F, the 45 kDa band disappeared and a new band appeared at about 34 kDa (Fig. 2, lane 2), which is consistent with the calculated molecular mass of the mature protein. The difference of 11 kDa can therefore be attributed to posttranslational modification by 3 to 4 N-glycan chains depending on their individual structure. In Fig. 3 A potential N-glycosylation sites are depicted.

To further characterize the protein, antibodies were generated from two different peptide sequences. Antibodies were peptide affinity purified from serum of guinea pig and rabbit. As expected the antibodies recognized the 45 kDa and the 33 kDa deglycosylated protein band (Fig. 4B). So it is obvious that the zona pellucida-like domain protein is one major structural protein of the complex cupula structure. Further, we analyzed the protein expression on histological sections of the inner ear in the region of the ampulla. In Fig. 4C the hematoxylin-stained tissue shows the detached and shrunken cupula which sits on top of the neuroepithelium *in vivo* connected with the hair cells. In Fig. 4D the cupula is intensely stained by the zona pellucida-like domain protein-specific antibodies and in the neuroepithelium faint red-stained dots are visible, which are more pronounced in Fig. 4E. We assume that the red dots represent protein-loaded vesicles that derive from the zona pellucida-like domain protein-producing supporting cells (Fig. S3).

Discussion

Cupulin is the missing link of cupula structural material predicted by Goodyear and Richardson [7]. Here we show for the first time that this zona pellucida-like domain protein from salmon is a major structural component of the ampullary cupula which senses torsional accelerations of the head. A similar protein profile (Fig. 2A) of salmon and chicken samples indicates a common cupula architecture, but this has to be proven in the future in more detail.

Zona pellucida-like domain proteins are highly conserved. The comparison of the deduced amino-acid sequences from salmon, chicken and human reveals a high overall identity (Fig. 3A). A minor identity to ZP proteins, e.g. murine ZP3, is predominantly based on the conserved arrangement of cysteine residues. Nevertheless, both proteins are synthesized as precursor polypeptides with an N-terminal signal sequence and a C-terminal propeptide that contains the consensus furin cleavage site (CSFS), the external hydrophobic patch (EHP), a transmembrane region and a short cytoplasmic tail (Fig. 3B). To avoid ZP polymerisation, it is assumed for ZP3 that the EHP binds to the internal hydrophobic patch (IHP) during intracellular vesicular transport [21,22]. After fusion of secreted ZP3-containing vesicles with the cellular membrane the propeptide is thereafter released by cleavage of the CSFS enabling the mature protein to start homopolymerisation [19,23]. We propose a similar mechanism for zona pellucida-like domain protein assembly. It is likely that the release of the mature secreted protein to the ECM triggers polymerization and is a prerequisite for assembly of a macromolecular structure in the inner ear, the cupula. Here we have described a main building material of the salmon cupula, the zona pellucida-like domain protein at the molecular level. Protein expression takes place in the supporting cells of the crista ampullaris which surround the sensory hair cells (Fig. 1 and Fig. S3). The synthesis must be strictly controlled otherwise sensing of torsional accelerations could not be accurately measured. A sudden loss of vestibular function may originate from a failure of zona pellucida-like protein synthesis. In agreement with this assumption, treatment with antibiotics that block protein synthesis leads to a loss of vestibular function and to an atrophy of the cupula structure [24,25].

References

1. Dohlman GF (1969) The shape and function of the cupula. J Laryngol Otol 83: 43–53.
2. Takumida M (2001) Functional morphology of the crista ampullaris: with special interests in sensory hairs and cupula: a review. Biol Sci Space 15: 356–358.
3. Helling K, Clarke AH, Watanabe N, Scherer H (2000) [Morphological studies of the form of the cupula in the semicircular canal ampulla]. Hno 48: 822–827.
4. Scherer H, Watanabe S (2001) Introductory remarks on this issue. On the role of the ampulla in disturbances of vestibular function. Biol Sci Space 15: 350–352.
5. Helling K, Watanabe N, Jijiwa H, Mizuno Y, Watanabe S, et al. (2002) Altered cupular mechanics: a cause of peripheral vestibular disorders? Acta Otolaryngol 122: 386–391.
6. Iimura Y, Suzuki M, Otsuka K, Inagaki T, Konomi U, et al. (2010) Effect of cupula shrinkage on the semicircular canal activity. Acta Otolaryngol 130: 1092–1096.
7. Goodyear RJ, Richardson GP (2002) Extracellular matrices associated with the apical surfaces of sensory epithelia in the inner ear: molecular and structural diversity. J Neurobiol 53: 212–227.
8. Killick R, Legan PK, Malenczak C, Richardson GP (1995) Molecular cloning of chick beta-tectorin, an extracellular matrix molecule of the inner ear. J Cell Biol 129: 535–547.
9. Simmler MC, Cohen-Salmon M, El-Amraoui A, Guillaud L, Benichou JC, et al. (2000) Targeted disruption of otog results in deafness and severe imbalance. Nat Genet 24: 139–143.
10. Marcus DC, Rokugo M, Ge XX, Thalmann R (1983) Response of cochlear potentials to presumed alterations of ionic conductance: endolymphatic perfusion of barium, valinomycin and nystatin. Hear Res 12: 17–30.
11. Shevchenko A, Wilm M, Vorm O, Mann M (1996) Mass spectrometric sequencing of proteins silver-stained polyacrylamide gels. Anal Chem 68: 850–858.
12. Suckau D, Resemann A, Schuerenberg M, Hufnagel P, Franzen J, et al. (2003) A novel MALDI LIFT-TOF/TOF mass spectrometer for proteomics. Anal Bioanal Chem 376: 952–965.
13. Altschul SF, Gish W, Miller W, Myers EW, Lipman DJ (1990) Basic local alignment search tool. J Mol Biol 215: 403–410.

Supporting Information

Figure S1   MS/MS spectrum of trypsin digested 45 kDa salmon protein. Peptide mass fingerprint with annotated peptide sequences. The protein was identified as zona pellucida-like protein (C0H9B6). Peptide numbers correspond to numbers given in Figure 3A.

Figure S2   Detailed MS/MS spectrum of peptide: FAFEVFR (peptide 6). b and y ion series with inserted fragment ion table.

Figure S3   Electron microscope image of sensory tissue below the cupula. Sensory cells (white arrows) with hairbundles (black arrows) are shown, adjacent to supporting cells (asterix).

Acknowledgments

We are grateful to Prof. HJ Merker (Anatomisches Institut der Freien Universität Berlin) for providing Fig. S3. For mass spectrometry (performed by CW), we would like to acknowledge the assistance of the Core Facility BioSupraMol supported by the Deutsche Forschungsgemeinschaft (DFG).

Author Contributions

Conceived and designed the experiments: JD MS RT SB WR HS CW. Performed the experiments: JD AH SB ECM HS CW. Analyzed the data: JD ECM SB HS CW. Contributed reagents/materials/analysis tools: JD AH SB ECM MS HS CW. Wrote the paper: JD RT WR HS CW.

14. Larkin MA, Blackshields G, Brown NP, Chenna R, McGettigan PA, et al. (2007) Clustal W and Clustal X version 2.0. Bioinformatics 23: 2947–2948.
15. Petersen TN, Brunak S, von Heijne G, Nielsen H (2011) SignalP 4.0: discriminating signal peptides from transmembrane regions. Nat Methods 8: 785–786.
16. Krogh A, Larsson B, von Heijne G, Sonnhammer EL (2001) Predicting transmembrane protein topology with a hidden Markov model: application to complete genomes. J Mol Biol 305: 567–580.
17. Bork P, Sander C (1992) A large domain common to sperm receptors (Zp2 and Zp3) and TGF-beta type III receptor. FEBS Lett 300: 237–240.
18. Jovine L, Darie CC, Litscher ES, Wassarman PM (2005) Zona pellucida domain proteins. Annu Rev Biochem 74: 83–114.
19. Jovine L, Qi H, Williams Z, Litscher ES, Wassarman PM (2004) A duplicated motif controls assembly of zona pellucida domain proteins. Proc Natl Acad Sci U S A 101: 5922–5927.
20. Legan PK, Rau A, Keen JN, Richardson GP (1997) The mouse tectorins. Modular matrix proteins of the inner ear homologous to components of the sperm-egg adhesion system. J Biol Chem 272: 8791–8801.
21. Jovine L, Qi H, Williams Z, Litscher E, Wassarman PM (2002) The ZP domain is a conserved module for polymerization of extracellular proteins. Nat Cell Biol 4: 457–461.
22. Llorca O, Trujillo A, Blanco FJ, Bernabeu C (2007) Structural model of human endoglin, a transmembrane receptor responsible for hereditary hemorrhagic telangiectasia. J Mol Biol 365: 694–705.
23. Monne M, Han L, Schwend T, Burendahl S, Jovine L (2008) Crystal structure of the ZP-N domain of ZP3 reveals the core fold of animal egg coats. Nature 456: 653–657.
24. Konomi U, Suzuki M, Otsuka K, Shimizu A, Inagaki T, et al. (2010) Morphological change of the cupula due to an ototoxic agent: a comparison with semicircular canal pathology. Acta Otolaryngol 130: 652–658.
25. Quint E, Furness DN, Hackney CM (1998) The effect of explantation and neomycin on hair cells and supporting cells in organotypic cultures of the adult guinea-pig utricle. Hear Res 118: 157–167.

20

Existence of Molten Globule State in Homocysteine-Induced Protein Covalent Modifications

Tarun Kumar, Gurumayum Suraj Sharma, Laishram Rajendrakumar Singh*

Dr. B. R. Ambedkar Center for Biomedical Research, University of Delhi, Delhi, India

Abstract

Homocysteine thiolactone is a toxic metabolite produced from homocysteine by amino-acyl t-RNA synthetase in error editing reaction. The basic cause of toxicity of homocysteine thiolactone is believed to be due to the adduct formation with lysine residues (known as protein N-homocysteinylation) leading to protein aggregation and loss of enzyme function. There was no data available until now that showed the effect of homocysteine thiolactone on the native state structural changes that led to aggregate formation. In the present study we have investigated the time dependent structural changes due to homocysteine thiolactone induced modifications on three different proteins having different physico-chemical properties (cytochrome-c, lysozyme and alpha lactalbumin). We discovered that N-homocysteinylation leads to the formation of molten globule state—an important protein folding intermediate in the protein folding pathway. We also found that the formation of the molten globule state might be responsible for the appearance of aggregate formation. The study indicates the importance of protein folding intermediate state in eliciting the homocysteine thiolactone toxicity.

Editor: Rizwan H. Khan, Aligarh Muslim University, India

Funding: This work is supported by the grant provided by University of Delhi (DRCH/R&D). LRS, TK and GSS acknowledge Council of Scientific and Industrial Research (File No.: 09/045(1131)/2011-EMR-1) and Department of Science and Technology (Ref No.: SR/SO/BB-0003/2011) for the financial assistance provided in the form of research fellowship. The funders had no role in study design, data collection and analysis, decision to publish, or preparation of the manuscript.

Competing Interests: The authors have declared that no competing interests exist.

* Email: lairksingh@gmail.com

Introduction

Homocysteine is a sulfur containing toxic metabolite produced as a byproduct in methionine metabolism pathway. This toxic homocysteine is known to metabolize to methionine by remethylation or to cystiene by trans-sulfurylation [1]. However, mutations in the homocysteine metabolizing enzymes, cystathionine β-synthase (CBS) or methylene tetrahydrofolate reductase (MTHFR) cause an impaired ability to metabolize the toxic homocysteine resulting in an increased levels of cellular and plasma homocysteine [2,3]. The increased accumulation of plasma homocysteine results in homocystinuria and the symptoms include arteriosclerosis, osteoporosis, mental retardation, thrombosis, dislocated eye lenses and neurodegenerative pathologies such as dementia, Parkinson's and Alzheimer's diseases [4,5,6,7]. The serum level of homocysteine in healthy adults is $5–10$ μM, it may range from $15–20$ μM in mild form but in case of hyperhomocysteinimia it may rise up to 500 μM [8,9]. The toxic effect of homocysteine has been believed due to the formation of homocysteine thiolactone (HTL), which is synthesized from homocysteine by amino acyl t-RNA synthetase in proofreading or error editing reaction mechanism whenever homocysteine gets incorporated in place of methionine by mistake during translation process [10,11,12,13]. It has been demonstrated that HTL preferentially forms amide bonds with ε-amino group of lysine residues of protein in a non-enzymatic mechanism; a process referred to as "protein N-homocysteinylation" [13] which results in various kinds of toxic effects on macromolecules including loss of enzymatic activity,

generation of oxidative stress within the cell and autoimmune response generation against N-homocysteinylated self proteins [14].

It has been known that protein N-homocysteinylation results in the formation of toxic multimers, aggregate or amyloid and therefore, this has been considered to be a risk factor for neurodegeneration [15,16,17,18,19,20,21]. Various studies have also shown that the protein aggregation can also be induced *in vitro* by partially folded; molten globule (MG) states [22,23]. Molten globule states are indeed compact denatured protein folding intermediates that exist between the native and denatured states [24]. MG state has attracted much attention in recent years because it is believed to be involved in many biological processes [25,26,27,28,29,30]. It has the following common structural characteristics: (i) presence of intact secondary structure, (ii) loss of tertiary interactions, (iii) presence of loosely packed hydrophobic core that increases the hydrophobic surface accessibility to solvent environment and (iv) $10–30\%$ increase in the radius of gyration [29,31,32,33]. Studies on covalent modification of native proteins by glycating agents that target lysine residues (similar to HTL induced covalent modifications) have also been shown to result in the formation of molten globule states ultimately leading to aggregate formation [23]. Since covalent modification by HTL has been known to induce aggregate/amyloid formation, it is important to investigate if HTL induced covalent modification of proteins also results in formation of MG states for which the detailed structural consequences that HTL modification has on the native state of proteins have to be understood.

Most studies reported earlier on the structural consequences of homocysteine on proteins were however made at time intervals ranging from 12–72 hours [15,17,19,34,35,36] where proteins have already formed aggregate. To the best of our knowledge there are no reports of the effect of protein N-homocysteinylation on the structural characteristics of native proteins at different time intervals. Therefore, what effects HTL exposure results in the native state conformation of the protein has not been properly understood. Here, we investigated the effect of protein N-homocysteinylation on the conformation of three different proteins having different physico-chemical properties and different lysine contents (cytochrome-c from bovine heart, bovine alpha lactalbumin and lysozyme from chicken egg white) at different time intervals. The present work clearly demonstrates that protein N-homocysteinylation results in the formation of molten globule state- an important protein folding intermediate in case of cytochrome-c (cyt-c) and alpha lactalbumin (α-LA) while there exists no structural changes on lysozyme. We also found that molten globule state might be responsible for the formation of aggregate due to protein N-homocysteinylation. Our results indicate the importance of the protein folding intermediate in eliciting the toxic effect of protein N-homocysteinylation.

Materials and Methods

Materials

Cytochrome c (from bovine heart), lysozyme (from chicken egg white), alpha lactalbumin (from bovine milk), DL-homocysteine thiolactone hydrochloride (HTL) and 8-Anilinonaphthalene-1-sulphonic acid were purchased from Sigma-Aldrich chemical Co. Potassium chloride, di-potassium hydrogen phosphate and potassium di-hydrogen phosphate were purchased from Merck, India. Dithiobis(2-nitrobenzoic acid), the Ellman's reagent was also purchased from Sigma-Aldrich chemical Co. Guanidium hydrochloride (Gdmcl) was purchased from M.P. Biomedicals. Unless otherwise stated, all chemicals were used without further purification and double distilled water was used as aqueous phase.

Analytical Procedures

Lysozyme, cyt-c and α-LA solutions were dialyzed extensively against 0.1 M KCl at pH 7.0 at ~4°C. Protein stock solutions were filtered using 0.22-μm Millipore syringe driven filter. Cytochrome c was oxidized using 0.01% potassium ferrocyanide before dialysis. All the proteins gave a single band during polyacrylamide gel electrophoresis (see Figure S1). Concentration of the protein solutions was determined experimentally using the molar absorption coefficient (ε) values 3.9×10^4 M^{-1} cm^{-1} at 280 nm for Lysozyme [37], 1.06×10^5 M^{-1} cm^{-1} at 409 nm for cyt c [38] and 29210 M^{-1} cm^{-1} at 280 nm for α-LA [39]. All solutions for optical measurements were prepared in the desired degassed buffer (0.05 M phosphate buffer, pH 7.0). Since pH of the protein solution may change on the addition of HTL, pH of each solution was also measured after addition of HTL. We observed no significant change in the pH of the protein solutions after addition of HTL.

Modification of proteins (cyt c, α-LA and lysozyme) by HTL and sulfhydryl estimation using Ellman's reagent

Proteins were treated with different concentrations of HTL (1 mM, 2 mM, 5 mM and 10 mM) and then incubated (18 hours for cyt-c, 4 hours for α-LA and 24 hours for lysozyme) at 25°C in 0.05 M phosphate buffer of pH 7.0. Aliquotes of proteins

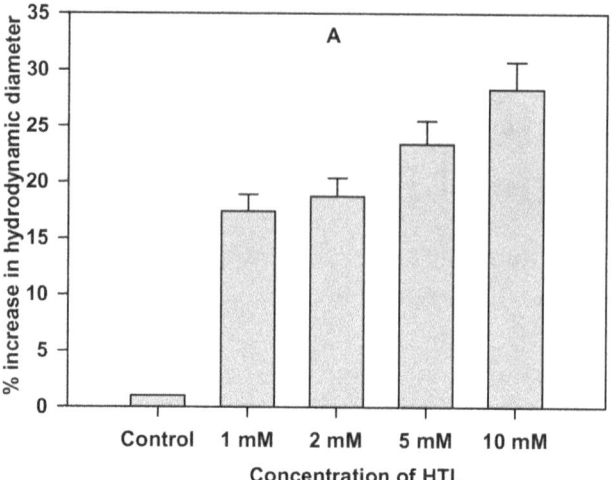

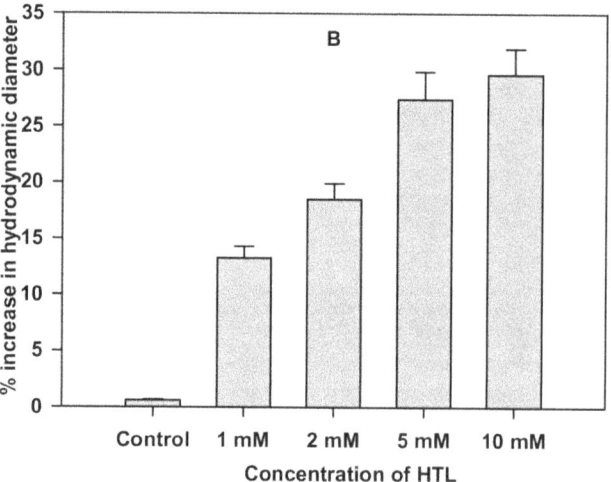

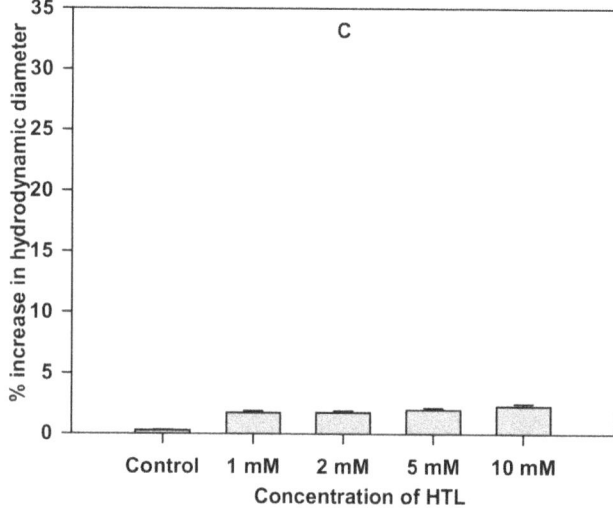

Figure 1. Effect of homocysteine thiolactone (HTL) on the hydrodynamic diameter of the native state of proteins. Percent increase in hydrodynamic diameter of (A) cyt-c (at 18 hours) and (B) α-LA (at 4 hours) and (C) lysozyme (at 24 hours) after modification with different concentrations of HTL.

Table 1. Hydrodynamic diameter of cyt-c, α-LA and lysozyme*.

| Time in hours | Hydrodynamic diameter of cyt-c (nm) | | | | |
	Native	1 mM HTL	2 mM HTL	5 mM HTL	10 mM HTL
0	2.84	2.81	2.82	2.86	2.79
1	2.81	2.83	2.81	2.84	2.78
2	2.82	2.88	2.85	2.8	2.83
3	2.75	2.8	2.85	2.85	2.84
4	2.88	2.89	2.82	2.83	2.87
6	2.81	2.79	2.72	2.85	2.98
8	2.83	2.71	2.82	2.93	3.07
10	2.78	2.78	2.88	2.98	3.28
12	2.85	2.93	2.91	3.05	3.34
14	2.83	2.95	3.03	3.24	3.41
16	2.8	3.13	3.16	3.48	3.54
18	2.87	3.3	3.35	3.53	3.58, 90.12
20	2.83	3.56	3.4	3.49, 25.26	140.8, 365.13, 831.54
22	2.91	3.45, 18.13	3.58, 20.15	116.7, 260.8, 741.8	160.56, 370.24, 840.34
24	2.88	3.52, 20.12	3.54, 24.65	120.5, 278.9, 801.4	165.45, 369.59, 845.21

| Time in hours | Hydrodynamic diameter of α-LA (nm) | | | | |
	Native	1 mM HTL	2 mM HTL	5 mM HTL	10 mM HTL
0	3.42	3.55	3.46	3.54	3.58
1	3.53	3.49	3.49	3.67	3.69
2	3.52	3.56	3.58	3.69	4.14
3	3.51	3.67	3.63	4.02	4.66
4	3.44	3.98	4.1	4.51	4.64
6	3.57	4.02	4.52	6.51, 20.58	6.98, 26.47
8	3.68	6.91, 15.32	6.93, 20.13	6.88, 23.46	6.83, 30.15

| Time in hours | Hydrodynamic diameter of lysozyme (nm) | | | | |
	Native	1 mM HTL	2 mM HTL	5 mM HTL	10 mM HTL
0	3.38	3.45	3.44	3.55	3.49
1	3.41	3.55	3.33	3.52	3.52
2	3.38	3.33	3.38	3.52	3.49
3	3.39	3.43	3.32	3.58	3.53
4	3.39	3.52	3.37	3.42	3.53
6	3.38	3.58	3.49	3.42	3.64
8	3.39	3.49	3.55	3.53	3.54
10	3.32	3.51	3.49	3.53	3.54
12	3.47	3.44	3.46	3.55	3.62
14	3.52	3.43	3.49	3.46	3.57
16	3.52	3.55	3.52	3.34	3.61
18	3.44	3.56	3.49	3.46	3.51
20	3.46	3.44	3.45	3.49	3.56
22	3.39	3.53	3.57	3.55	3.47
24	3.39	3.51	3.5	3.62	3.57

*Experimental error in hydrodynamic diameter measurement is in the range of 7–9%.

modified with HTL were first precipitated down with 10% TCA to remove unbound HTL. Protein pellets were collected and then resolubilized in phosphate buffer, pH 7.0. The levels of thiol groups in control and homocysteinylated protein samples were assayed using Dithiobis(2-nitrobenzoic acid), the Ellman's reagent. The absorbance of the samples were measured at 412 nm,

Table 2. Sulfhydryl content measurement of HTL treated cyt-c, α-LA and lysozyme*.

Concentration of HTL	Sulfhydryl content (μM/mg)		
	cyt-c (At 18 hours)	α- LA (At 4 hours)	Lysozyme (At 24 hours)
0 mM	19	36	40
1 mM	753	305	391
2 mM	972	865	819
5 mM	2052	1958	1736
10 mM	2982	3098	2013

*Error in -SH content measurement is in the range of 5–8%.

using a 1 cm path-length cuvette. The amount of 5′-nitrothio-benzoate released was calculated from the molar extinction coefficient of 13,700 M^{-1} cm^{-1} [40,41,42].

Dynamic light scattering (DLS) measurements

Size distribution of the particles present in the protein sample were obtained using a Zetasizer Micro V/ZMV 2000 (Malvern, UK). Measurements were made at a fixed angle of 90° using an incident laser beam of 689 nm. Fifteen measurements were made with an acquisition time of 30 seconds for each sample at sensitivity of 10%. The data was analysed using Zetasizer software provided by the manufacturer to get hydrodynamic diameters, and polydispersity which is a measure of the standard deviation of the size of the particles. The protein concentration was 2.0 mg/ml. All measurements were performed at 25°C.

Circular Dichroism (CD) Measurements

CD measurements were made in a Jasco J-810 spectropolarimeter equipped with peltier controller at 25°C with six accumulations. Protein concentration used for the CD measurements was 0.5 mg/ml. Cells of 0.1 and 1.0 cm path length were used for the measurements of the far- and near-UV CD spectra respectively. Necessary blanks were subtracted. The CD instrument was routinely calibrated with D-10-camphorsulfonic acid. The molar ellipticity in units of Deg cm^2 $dmol^{-1}$ was determined for average residue molecular weight for the proteins used in the study.

Fluorescence Measurements

Fluorescence spectra of the protein samples were measured in a Perkin Elmer LS 55 Spectrofluorimeter in a 3 mm quartz cell, with both excitation and emission slits set at 10 nm. Protein concentration for all the experiments was in the range of 0.06–0.07 mg/ml. For intrinsic fluorescence measurements, cyt-c, α-LA and lysozyme were excited at 295 nm, while the emission spectra were recorded from 300–500 nm. For ANS-protein binding experiments the excitation wavelength was 360 nm, and emission spectra were recorded from 400–600 nm. ANS concentration was kept 16 fold than that of protein concentration.

Absorption spectroscopy measurements

The light scattering intensity of the protein samples were measured by monitoring the absorbance at 500 nm for cyt-c [43], 400 nm for α-LA and 450 nm for lysozyme [44] in a JASCO V-660 spectrophotometer equipped with temperature controlled cuvette holder at 25°C. The protein concentration was 1 mg/ml.

Results

We have investigated the effect of protein N-homocysteinylation on three different proteins (cyt-c, α-LA and lysozyme) having different physico-chemical properties. The hydrophobicity indices, pI values and lysine content of these proteins are 1110, 10.5, 19 for cyt-c, 1050, 4.2, 12 for α-LA and 890, 10.7, 6 for lysozyme respectively. Concentrations of HTL for the treatment used for all the studies were 1 mM, 2 mM, 5 mM and 10 mM.

Hydrodynamic diameter of proteins increases upon treatment with HTL

In the present study, we have initially studied the conformational changes due to N-homocysteinylation at different concentrations of HTL using DLS as a tool at different time intervals by monitoring the hydrodynamic diameter of the proteins (Table 1). We observed significant changes in the hydrodynamic diameter of cyt-c in the time interval 4–18 hours depending on HTL concentrations and α-LA in the time interval 1–4 hours after treatment with HTL beyond which both the proteins tend to get aggregated. However, lysozyme shows no differences in the hydrodynamic diameter due to N-homocysteinylation up to the maximum time of incubation (Table 1) at all HTL concentrations used in this study. Figure 1 also shows that the changes in the maximum increase in hydrodynamic diameter of cyt-c and α-LA is around 13–30% at all HTL concentrations while in the case of lysozyme there is negligible effect on hydrodynamic diameter at all concentrations of HTL. The result suggests that hydrodynamic diameter of different proteins increases at different time intervals depending on the HTL concentration.

Estimation of -SH groups in all the three proteins at specified time periods confirmed protein N-homocysteinylation

To verify if the changes in the conformation of the native states is due to N-homocysteinylation, we have further measured -SH content of the protein samples treated with HTL. Increase in the -SH content has been previously reported to be a signature of the protein N-homocysteinylation [5]. Table 2 shows increase in the sulfhydryl content of all the three proteins due to treatment with HTL at 4 hours for α-LA, 18 hours for cyt-c and 24 hours for lysozyme. It is seen in this table that there is an increase in the -SH groups of the proteins upon treatment with HTL at the specified time periods mentioned in the previous section. The results confirmed that each of the native protein has been N-homocysteinylated at the specified time periods.

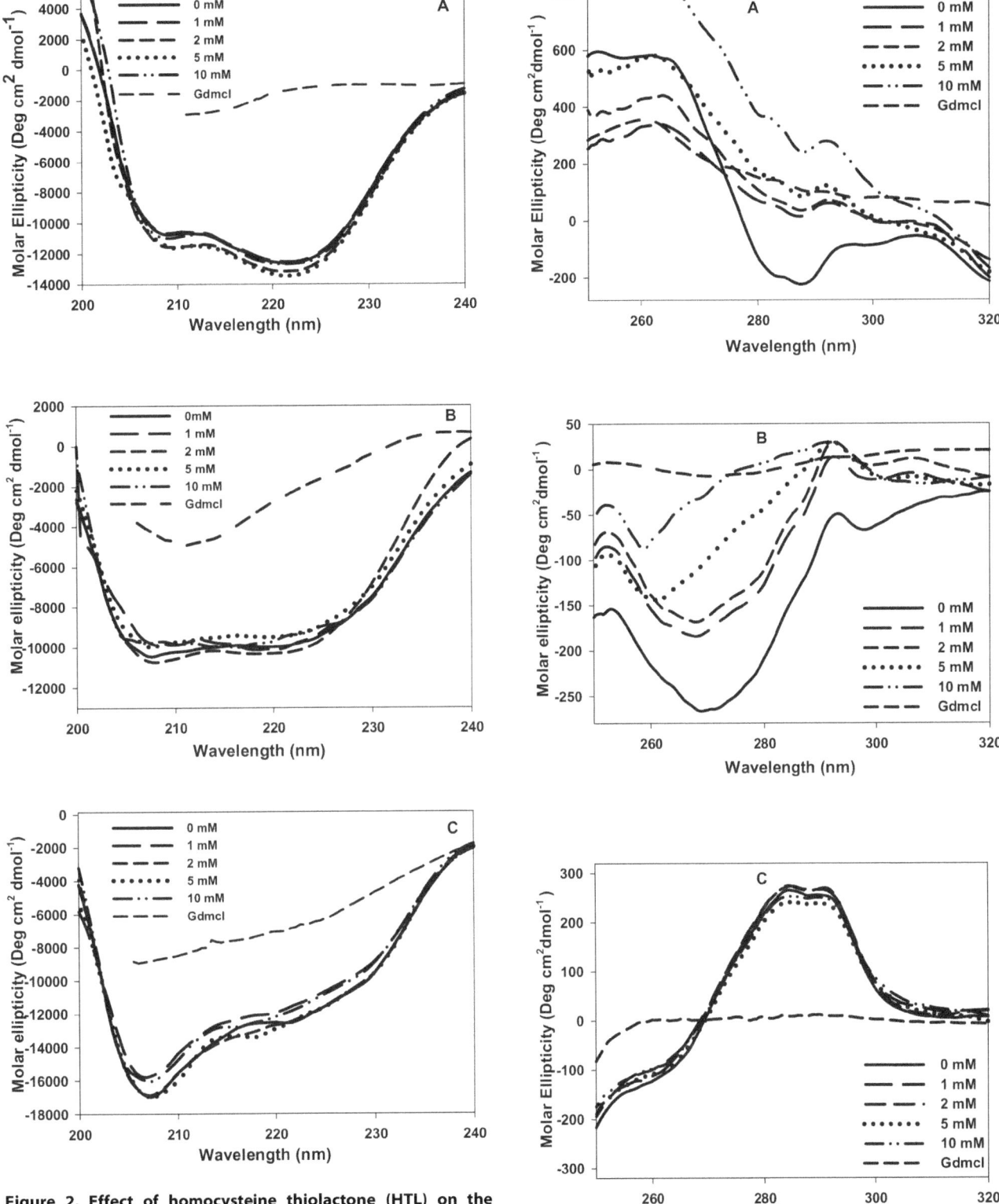

Figure 2. Effect of homocysteine thiolactone (HTL) on the secondary structure of the native state of proteins. Far UV CD spectra of (A) cyt-c (at 18 hours), (B) α-LA (at 4 hours) and (C) lysozyme at (24 hours) modified with different concentrations of HTL. Unfolded state control (7.0 M GdmCl) have also been shown for each protein.

Figure 3. Effect of homocysteine thiolactone (HTL) on the tertiary structure of the native state of proteins. Near UV CD spectra of (A) cyt-c (at 18 hours), (B) α-LA (at 4 hours) and (C) Lysozyme (at 24 hours) modified with different concentrations of HTL. Unfolded state control (7.0 M GdmCl) have also been shown for each protein.

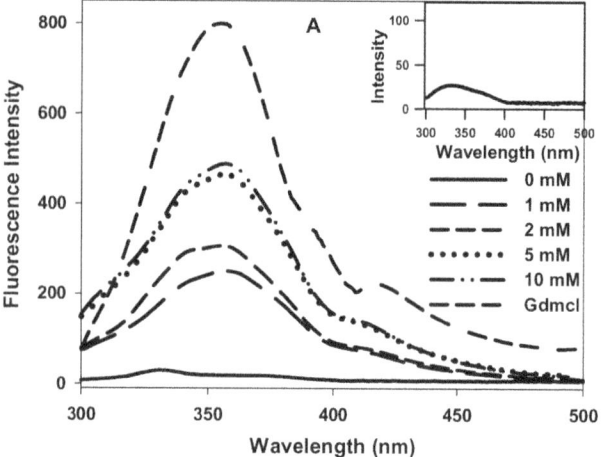

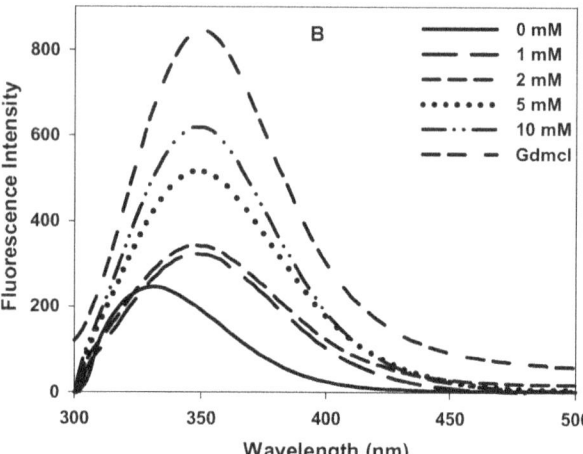

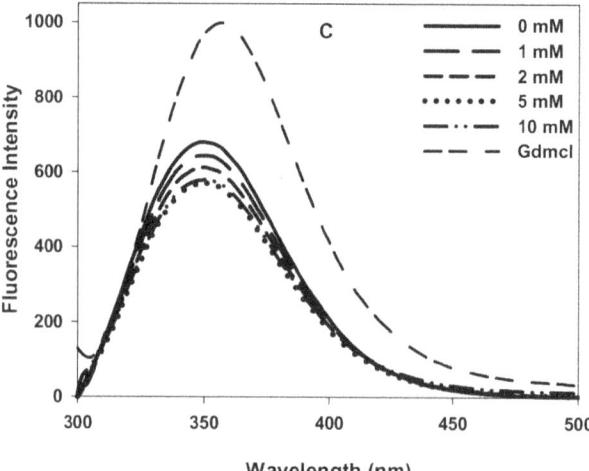

Figure 4. Effect of homocysteine thiolactone (HTL) on the intrinsic fluorescence of the native state of proteins. Intrinsic fluorescence spectra of (A) cyt-c (at 18 hours), (B) α-LA (at 4 hours) and (C) lysozyme (at 24 hours) modified with different concentrations of HTL. Unfolded state control (7.0 M GdmCl) have also been shown for

each protein. Inset in Figure 4 (A) shows the enlarged view of the native state emission spectra of cyt-c. Excitation wavelength of 295 nm was used for all the three proteins and emissions were recorded at the wavelength range of 300–500 nm.

N-homocysteinylation results in the formation of molten globule state in cyt-c and α-LA but not in lysozyme

To characterize the structural changes in the proteins due to covalent modification by HTL, we measured the changes in secondary and tertiary structures of the proteins at 4 hours for α-LA, 18 hours for cyt-c and 24 hours for lysozyme by measuring the far-, near-UV CD and intrinsic fluorescence spectra of all the three HTL treated proteins. It may, however, be noted that in the case of cyt-c there appears a non-native species in the presence of 10 mM HTL at 18 hours (Table 1). We have intentionally taken 18 hours incubation time for cyt-c for structural characterization although it has one additional non-native species in our DLS study, as this time interval represents the maximum change in the native state hydrodynamic diameters of protein at all HTL concentrations and the volume fraction of the non-native species observed is negligible (volume distribution less than 1%). Figure 2 shows the secondary structural changes of the proteins treated and incubated for the respective time periods (as shown in figure legend) with different concentrations of HTL. It is seen in this figure that there is (i) neither significant alterations in the spectral properties in terms of the respective peaks of each of the protein (ii) nor the intensity (molar ellipticity) of the homocysteinylated and non-homocysteiylated proteins are changed at different concentrations of HTL. The result leads us to believe that there are no changes in the secondary structure of all the three proteins due to N-homocysteinylation. However, the near-UV CD spectra shown in Figure 3 suggests that the tertiary structures of both cyt-c and α-LA have been unfolded due to N-homocysteinylation at the specified time intervals, while the tertiary structure of lysozyme is not affected due to N-homocysteinylation at different concentrations of HTL. Figure 4 also shows an increase in the tryptophan fluorescence of cyt-c and α-LA along with very prominent red shift in emission maxima (also see Table S1) but in case of lysozyme there was no alteration in the fluorescence properties indicating that N-homocysteinylation (at the respective time periods) disrupts the tertiary interactions of the native states of both cyt-c and α-LA but does not have any perturbing effect on lysozyme. Taken together, the results indicate that N-homocysteinylation leads to different consequences on the native state structure of different proteins. Absence of tertiary structure with intact secondary structure in case of cyt-c and α-LA might indicate a molten globule state. To further verify for this possibility we have intentionally performed ANS binding experiment of these N-homocysteinylated proteins at the respective time periods (Figure 5). It is seen in Figure 5, that there was increase in the intensity of ANS fluorescence and a blue shift in case of both cyt-c and α-LA upon HTL treatment after the respective time of incubations confirming that ANS binds to both the proteins indicating the existence of molten globule state in case of cyt-c and α-LA at all HTL concentrations used in this study. No such ANS binding was observed in case of lysozyme as there was neither an apparent increase in ANS fluorescence intensity nor a blue shift. The results indicate that N-homocysteinylation at the specified time periods results in molten globule state formation in case of cyt-c and α-LA but not in lysozyme.

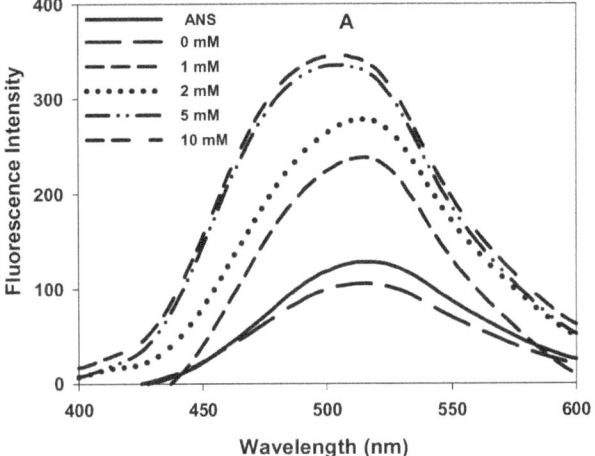

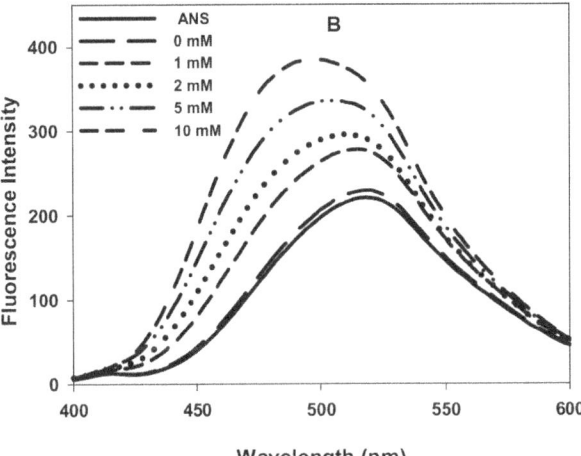

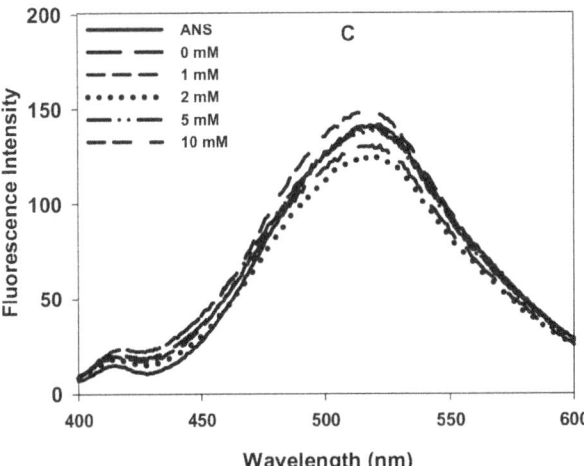

Figure 5. Effect of homocysteine thiolactone (HTL) on the extrinsic fluorescence of the native state of proteins. ANS binding study (A) cyt-c (at 18 hours), (B) α-LA (at 4 hours) and (C)

lysozyme (at 24 hours) modified with different concentrations of HTL. Excitation wavelength of 360 nm was used and emissions were recorded at the wavelength range of 400–600 nm.

N-homocysteinylation induces aggregate formation in cyt-c and α-LA but not in lysozyme at 24 hours

In order to study the aggregation propensities of the proteins due to N-homocysteinylation, the turbidity of the modified protein samples was measured using UV-visible spectroscopy. Figure 6 shows the light scattering intensity of the proteins at the maximum time of incubation with HTL (24 hours). It is seen in this figure that both cyt-c and α-LA have been aggregated due to N-homocysteinylation. However, lysozyme was protected against HTL-induced protein aggregation (figure not shown). Since the proteins chosen in the study have different physico-chemical properties, the results indicate that protein N-homocysteinylation has different consequences on the aggregation propensities of the proteins having different physico-chemical properties.

Discussion

To investigate for the possible effect of protein N-homocysteinylation on the native state structure of different proteins, we have first of all measured the change in the hydrodynamic diameter of the proteins at different time intervals. Results shown in Table 1 and Table 2 suggest that HTL has been incorporated to the proteins resulting in the adduct formation that affects hydrodynamic diameter of the native proteins in a time dependent manner. There is increase in the hydrodynamic diameter of the cyt-c (up to 18 hours) and α-LA (up to 4 hours). Beyond these time intervals aggregate formation starts as evidenced by large increase in the hydrodynamic diameter of the HTL modified proteins. There was no significant change in the hydrodynamic diameter of lysozyme till 24 hours. The results indicate that homocysteinylation has different consequences on the native states of different proteins. To investigate for the difference in the native state structure of the proteins due to N-homocysteinylation, we have further measured the secondary and tertiary structures of the native proteins after incubation with HTL for 4 hours in case of α-LA and 18 hours in case of cyt-c. Our results indicate that the secondary structural content (see Figure 2) for the proteins are not at all changed due to N-homocysteinylation while the tertiary structures (Figure 3 and Figure 4) have been modified differently by HTL on the different proteins. In case of lysozyme the tertiary interactions are not significantly changed while in case of cyt-c and α-LA, the tertiary interactions have been disrupted completely at the highest concentrations of the HTL used in the study. Loss of tertiary structure while having no effect on the secondary structure suggests that the native state of both the cyt-c at 18 hours and α-LA at 4 hours of incubation might be a molten globule (MG) state. Indeed MG states are characterized by the (i) presence of intact secondary structure, (ii) the loss of tertiary interactions and (iii) 10–30% increase in the radius of gyration [29,31,32,33]. Interestingly, increase in the hydrodynamic diameter shown in Figure 1 (17–28% for cyt-c and 12–29% for α-LA) further evidenced that the modified proteins might exist in the MG state at the specified time periods. There is no increase in the hydrodynamic diameter of lysozyme suggesting that lysozyme does not lose its native conformation by HTL modification. If the HTL modified native state of cyt-c (at 18 hours) and α-LA (at 4 hours) is really a molten globule state then hydrophobic clusters must be exposed to the solvent due to disruption of tertiary structure. To probe for this, we have further measured the ANS binding to the proteins at the

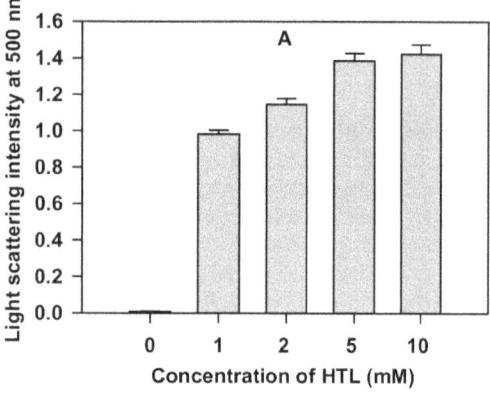

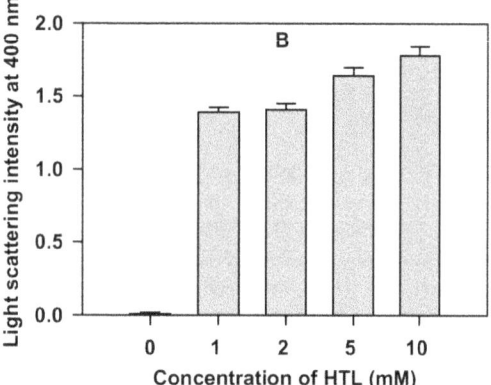

Figure 6. Homocysteine thiolactone (HTL) induced aggregation study of cyt-c and α-LA. Turbidometric study of (A) cyt-c (at 500 nm) and (B) α-LA (at 400 nm) modified with different concentrations of HTL at 24 hours.

given experimental conditions as ANS specifically binds to the hydrophobic clusters exposed to the solvent [45]. Results in Figure 5 suggest that ANS has been bound to both cyt-c and α - LA but not in lysozyme. Taken together the results lead us to believe that N-homocysteinylation results in the formation of MG state in cyt-c and α -LA. It is however seen in case of lysozyme that there is no formation of such MG state since neither the tertiary structure gets unfolded nor there is increase in the hydrodynamic diameter which are prerequisite characteristic features reported for the proteins to be in molten globule state [46]. We therefore conclude that the formation of MG state due to HTL modification is dependent on the amino acid sequence of the proteins. In accordance with our results, glycating agents that target and modify proteins in the same manner by forming covalent adduct with lysine residues, also have been reported to induce MG state formation and hence aggregation [23]. At present we do not have a concrete explanation for having different effects on the native state of different proteins due to HTL induced modification. It might be possible that disruption of the tertiary structure due to N-homocysteinylation might be the limiting step to the conversion of the N-state to the MG state as the tertiary structure of lysozyme could not be disrupted due to the modification. Perhaps the incorporation of HTL at a specific lysine in each of the proteins might be responsible for opening of the tertiary structure which in the case of α-LA and cyt-c is easily accessible while is difficult to target in case of lysozyme. In support to our argument it has been shown earlier in cyt-c that incorporation of HTL in certain lysine

residues does not result in significant change in the native structure while there are four lysine residues (Lys 8 or 13, Lys86 or 87, Lys 99, and Lys 100) which are susceptible to N-homocysteinylation resulting in subtle structural changes in the native structure of protein [34]. Interestingly, modification of only one single lysine (Lys 29) residue in bovine pancreatic insulin results in aggregation [15]. It has also been reported that some of the lysine residues in hemoglobin which are directly accessible to solvent environment do not get modified by HTL [35]. Based on these studies we can conclude that the modification of only selective lysine residues results in the perturbation in the structure of proteins.

It has been reported earlier that most of the end products of protein N-homocysteinylation is the formation of aggregates or amyloids [15,16,17,18,19]. We speculate that since MG states are very unstable and have exposed hydrophobic clusters, they might play an important role in the formation of aggregate or amyloid. If our speculation is true, then it is expected that both cyt-c and α - LA should form aggregate while lysozyme should not aggregate (as there are no existence of MG states in lysozyme due to N-homocysteinylation). To pursue for this possibility we have measured the aggregation propensity of the HTL adducted proteins after 24 hours by using light scattering as a tool. It is seen in Figure 6 that there is increase in the light scattering intensity in case of both cyt-c and α-LA while in case of lysozyme the scattering intensity is not at all increased (not shown in Figure). This result is also in agreement with the DLS measurement that there is existence of 2–3 aggregated species at the maximum time of incubation. Thus we conclude that the formation of MG state is responsible for different appearance of aggregates in two different proteins. One of the basic causes of homocysteine toxicity in the cells is the formation of protein aggregates or amyloids and loss of enzyme activity. The results therefore indicate the role of protein folding intermediates in eliciting the HTL toxicity. Indeed such MG state induced by HTL might be more toxic than the aggregates induced by HTL modifications. Further research should focus on these issues.

Conclusions

In summary, we are sure of at least two things (i) N-homocysteinylation results in different consequences on the native state of different proteins: N-homocysteinylation of cyt-c and α-LA induces formation of molten globule state while there is no significant alteration in the native state structure of lysozyme. (ii) The formation of the protein aggregates due to N-homocysteinylation is protein dependent and appears to be related with the formation of molten globule state. Protein N-homocysteinylation has earlier been reported to result in different consequences on the aggregation propensities of different proteins (no aggregation to multimer and aggregate/amyloid formation) and formation of aggregates/amyloids is considered to be the basic cause of neurodegeneration [15,16,17,18,19,20,21]. The study therefore indicates the importance of the MG state as one of the mechanisms in eliciting homocysteine toxicity and neurodegeneration. Studies aimed at preventing the MG state formation induced by HTL modification might yield clinical implications to prevent or delay the onset of homocysteine induced proteopathies and associated disorders.

Author Contributions

Conceived and designed the experiments: LRS. Performed the experiments: TK GSS. Analyzed the data: TK GSS LRS. Contributed reagents/ materials/analysis tools: LRS. Wrote the paper: LRS.

References

1. Brosnan JT, Brosnan ME (2006) The sulfur-containing amino acids: an overview. J Nutr 136: 1636S–1640S.
2. Jakubowski H (1991) Proofreading in vivo: editing of homocysteine by methionyl-tRNA synthetase in the yeast *Saccharomyces cerevisiae*. EMBO J 10: 593–598.
3. Jakubowski H (2002) The determination of homocysteine-thiolactone in biological samples. Anal Biochem 308: 112–119.
4. Seshadri S (2006) Elevated plasma homocysteine levels: risk factor or risk marker for the development of dementia and Alzheimer's disease? J Alzheimers Dis 9: 393–398.
5. Jakubowski H (1999) Protein homocysteinylation: possible mechanism underlying pathological consequences of elevated homocysteine levels. FASEB J 13: 2277–2283.
6. White AR, Huang X, Jobling MF, Barrow CJ, Beyreuther K, et al. (2001) Homocysteine potentiates copper- and amyloid beta peptide-mediated toxicity in primary neuronal cultures: possible risk factors in the Alzheimer's-type neurodegenerative pathways. J Neurochem 76: 1509–1520.
7. Jacobsen DW (1998) Homocysteine and vitamins in cardiovascular disease. Clin Chem 44: 1833–1843.
8. Gellekink H, den Heijer M, Heil SG, Blom HJ (2005) Genetic determinants of plasma total homocysteine. Semin Vasc Med 5: 98–109.
9. Refsum H, Ueland PM, Nygard O, Vollset SE (1998) Homocysteine and cardiovascular disease. Annu Rev Med 49: 31–62.
10. Jakubowski H (1997) Metabolism of homocysteine thiolactone in human cell cultures. Possible mechanism for pathological consequences of elevated homocysteine levels. J Biol Chem 272: 1935–1942.
11. Jakubowski H, Zhang L, Bardeguez A, Aviv A (2000) Homocysteine thiolactone and protein homocysteinylation in human endothelial cells: implications for atherosclerosis. Circ Res 87: 45–51.
12. Jakubowski H, Goldman E (1993) Synthesis of homocysteine thiolactone by methionyl-tRNA synthetase in cultured mammalian cells. FEBS Lett 317: 237–240.
13. Jakubowski H (2000) Calcium-dependent human serum homocysteine thiolactone hydrolase. A protective mechanism against protein N-homocysteinylation. J Biol Chem 275: 3957–3962.
14. Undas A, Perla J, Lacinski M, Trzeciak W, Kazmierski R, et al. (2004) Autoantibodies against N-homocysteinylated proteins in humans: implications for atherosclerosis. Stroke 35: 1299–1304.
15. Jalili S, Yousefi R, Papari MM, Moosavi-Movahedi AA (2011) Effect of homocysteine thiolactone on structure and aggregation propensity of bovine pancreatic insulin. Protein J 30: 299–307.
16. Khazaei S, Yousefi R, Alavian-Mehr MM (2012) Aggregation and fibrillation of eye lens crystallins by homocysteinylation; implication in the eye pathological disorders. Protein J 31: 717–727.
17. Paoli P, Sbrana F, Tiribilli B, Caselli A, Pantera B, et al. (2010) Protein N-homocysteinylation induces the formation of toxic amyloid-like protofibrils. J Mol Biol 400: 889–907.
18. Stroylova YY, Chobert JM, Muronetz VI, Jakubowski H, Haertle T (2012) N-homocysteinylation of ovine prion protein induces amyloid-like transformation. Arch Biochem Biophys 526: 29–37.
19. Stroylova YY, Zimny J, Yousefi R, Chobert JM, Jakubowski H, et al. (2011) Aggregation and structural changes of alpha(S1)-, beta- and kappa-caseins induced by homocysteinylation. Biochim Biophys Acta 1814: 1234–1245.
20. Obeid R, Herrmann W (2006) Mechanisms of homocysteine neurotoxicity in neurodegenerative diseases with special reference to dementia. FEBS Lett 580: 2994–3005.
21. Mattson MP, Shea TB (2003) Folate and homocysteine metabolism in neural plasticity and neurodegenerative disorders. Trends Neurosci 26: 137–146.
22. Rochet JC, Lansbury PT, Jr. (2000) Amyloid fibrillogenesis: themes and variations. Curr Opin Struct Biol 10: 60–68.
23. Iram A, Alam T, Khan JM, Khan TA, Khan RH, et al. (2013) Molten Globule of Hemoglobin Proceeds into Aggregates and Advanced Glycated End Products. PloS one 8: e72075.
24. Rabbani G, Ahmad E, Zaidi N, Fatima S, Khan RH (2012) pH-Induced molten globule state of Rhizopus niveus lipase is more resistant against thermal and chemical denaturation than its native state. Cell Biochem Biophys 62: 487–499.
25. Jennings PA, Wright PE (1993) Formation of a molten globule intermediate early in the kinetic folding pathway of apomyoglobin. Science 262: 892–896.
26. Arai M, Kuwajima K (2000) Role of the molten globule state in protein folding. Advances in Protein Chemistry = Advances in Protein Chemistry 53: 209–282.
27. Hameed M, Ahmad B, Fazili KM, Andrabi K, Khan RH (2007) Different molten globule-like folding intermediates of hen egg white lysozyme induced by high pH and tertiary butanol. J Biochem 141: 573–583.
28. Arai M, Kuwajima K (2000) Role of the molten globule state in protein folding. Adv Protein Chem 53: 209–282.
29. Ptitsyn OB (1995) Molten globule and protein folding. Adv Protein Chem 47: 83–229.
30. Naeem A, Khan RH (2004) Characterization of molten globule state of cytochrome c at alkaline, native and acidic pH induced by butanol and SDS. Int J Biochem Cell Biol 36: 2281–2292.
31. Ansari MA, Zubair S, Atif SM, Kashif M, Khan N, et al. (2010) Identification and characterization of molten globule-like state of hen egg-white lysozyme in presence of salts under alkaline conditions. Protein Pept Lett 17: 11–17.
32. Kuwajima K (2002) The role of the molten globule state in protein folding: the search for a universal view of folding. PROCEEDINGS-INDIAN NATIONAL SCIENCE ACADEMY PART A 68: 333–340.
33. Kuwajima K (1989) The molten globule state as a clue for understanding the folding and cooperativity of globular' protein structure. Proteins: Structure, Function, and Bioinformatics 6: 87–103.
34. Perla-Kajan J, Marczak L, Kajan L, Skowronek P, Twardowski T, et al. (2007) Modification by homocysteine thiolactone affects redox status of cytochrome C. Biochemistry 46: 6225–6231.
35. Zang T, Dai S, Chen D, Lee BW, Liu S, et al. (2009) Chemical methods for the detection of protein N-homocysteinylation via selective reactions with aldehydes. Anal Chem 81: 9065–9071.
36. Yousefi R, Jalili S, Alavi P, Moosavi-Movahedi AA (2012) The enhancing effect of homocysteine thiolactone on insulin fibrillation and cytotoxicity of insulin fibril. Int J Biol Macromol 51: 291–298.
37. Hamaguchi K, Kurono A (1968) Structure of muramidase (lysozyme). I. The effect of guanidine hydrochloride on muramidase. J Biochem 54: 111–122.
38. Margoliash E, Frohwirt N (1959) Spectrum of horse-heart cytochrome c. Biochem J 71: 570–572.
39. Sugai S, Yashiro H, Nitta K (1973) Equilibrium and kinetics of the unfolding of alpha-lactalbumin by guanidine hydrochloride. Biochim Biophys Acta 328: 35–41.
40. Riener CK, Kada G, Gruber HJ (2002) Quick measurement of protein sulfhydryls with Ellman's reagent and with 4,4'-dithiodipyridine. Anal Bioanal Chem 373: 266–276.
41. Butterworth PH, Baum H, Porter JW (1967) A modification of the Ellman procedure for the estimation of protein sulfhydryl groups. Arch Biochem Biophys 118: 716–723.
42. Ellman GL (1959) Tissue sulfhydryl groups. Arch Biochem Biophys 82: 70–77.
43. Fedunová D, Antalík Mn (2006) Prevention of thermal induced aggregation of cytochrome c at isoelectric pH values by polyanions. Biotechnology and bioengineering 93: 485–493.
44. Raman B, Ramakrishna T, Rao CM (1996) Refolding of denatured and denatured/reduced lysozyme at high concentrations. Journal of Biological Chemistry 271: 17067–17072.
45. Rabbani G, Ahmad E, Zaidi N, Khan RH (2011) pH-dependent conformational transitions in conalbumin (ovotransferrin), a metalloproteinase from hen egg white. Cell Biochem Biophys 61: 551–560.
46. Rabbani G, Kaur J, Ahmad E, Khan RH, Jain SK (2014) Structural characteristics of thermostable immunogenic outer membrane protein from Salmonella enterica serovar Typhi. Appl Microbiol Biotechnol 98: 2533–2543.

Permissions

List of Contributors

Xuejun Yao, Ulrich H. N. Dürr, Zrinka Gattin, Yvonne Laukat, Rhagavendran L. Narayanan, Ann-Kathrin Brückner, Adam Lange and Stefan Becker
Max Planck Institute for Biophysical Chemistry, Göttingen, Germany

Chris Meisinger
Institut für Biochemie und Molekularbiologie, ZBMZ and BIOSS Centre for Biological Signalling Studies, Universität Freiburg, Freiburg, Germany

Markus Zweckstetter
Max Planck Institute for Biophysical Chemistry, Göttingen, Germany
German Center for Neurodegenerative Diseases (DZNE), Goöttingen, Germany
Center for Nanoscale Microscopy and Molecular Physiology of the Brain (CNMPB), University Medical Center, Göttingen, Germany

Reem R. Al Olaby and Hassan M. Azzazy
Department of Chemistry, The American University in Cairo, New Cairo, Egypt

Laurence Cocquerel, Laure Saas and Jean Dubuisson
Center for Infection and Immunity of Lille, CNRS-UMR8204/Inserm-U1019, Pasteur
Institute of Lille, University of Lille North of France, Lille, France

Adam Zemla
Pathogen Bioinformatics, Lawrence Livermore National Laboratory, Livermore, CA, United States of America

Jost Vielmetter
Protein Expression Center, Beckman Institute, California Institute of Technology, Pasadena, CA, United States of America

Joseph Marcotrigiano and Abdul Ghafoor Khan
Department of Chemistry and Chemical Biology, Rutgers University, Piscataway, NJ, United States of America

Felipe Vences Catalan and Shoshana Levy
Department of Medicine, Stanford University Medical Center, Stanford, CA, United States of America

Alexander L. Perryman
Department of Medicine, Division of Infectious Diseases, Center for Emerging & Re-emerging Pathogens, Rutgers University-New Jersey Medical School, Newark, NJ, United States of America

Joel S. Freundlich
Department of Medicine, Division of Infectious Diseases, Center for Emerging & Re-emerging Pathogens, Rutgers University-New Jersey Medical School, Newark, NJ, United States of America
Department of Pharmacology and Physiology, Rutgers University-New Jersey Medical School, Newark, NJ, United States of America

Stefano Forli
Department of Integrative Structural and Computational Biology, The Scripps Research Institute, La Jolla, CA, United States of America

Rod Balhorn
Department of Applied Science, University of California Davis, Davis, CA, United States of America

Jeremy Weaver, Tylan Watts, Pingwei Li and Hays S. Rye
Department of Biochemistry and Biophysics, Texas A&M University, College Station, Texas, United States of America

Jinpu Yu
Centre for Cancer Molecular Diagnosis, Tianjin Medical University Cancer Institute and Hospital, National Clinical Research Center for Cancer, Tianjin, China

Ziding Zhang
State Key Laboratory of Agrobiotechnology, College of Biological Sciences, China Agricultural University, Beijing, China

Lei Han
Centre for Cancer Molecular Diagnosis, Tianjin Medical University Cancer Institute and Hospital, National Clinical Research Center for Cancer, Tianjin, China
State Key Laboratory of Agrobiotechnology, College of Biological Sciences, China Agricultural University, Beijing, China

Yong-Jun Zhang
State Key Laboratory for Biology of Plant Diseases and Insect Pests, Institute of Plant Protection, Chinese Academy of Agricultural Sciences, Beijing, China

Long Zhang
Key Lab for Biological Control of the Ministry of Agriculture, China Agricultural University, Beijing, China

Xu Cui
Beijing Computing Center, Beijing, China

Ming S. Liu
CSIRO - Computational Informatics & Digital Productivity Flagship, Private Bag 10, Clayton South, Australia

Ricksen S. Winardhi
NUS Graduate school for Integrative Sciences and Engineering, Singapore, Singapore
Mechanobiology Institute, National University of Singapore, Singapore, Singapore
Centre for Bioimaging Sciences, National University of Singapore, Singapore, Singapore

Sandra Castang and Simon L. Dove
Division of Infectious Diseases, Boston Children's Hospital, Harvard
Medical School, Boston, Massachusetts, United States of America

Jie Yan
Mechanobiology Institute, National University of Singapore, Singapore, Singapore
Centre for Bioimaging Sciences, National University of Singapore, Singapore, Singapore
Department of Physics, National University of Singapore, Singapore, Singapore

Stefania Correale
Kedrion S.p.A., Sant 'Antimo (Na), Italy

Ivan de Paola Laura Zaccaro and Emilia Pedone
Istituto di Biostrutture e Bioimmagini, Consiglio Nazionale delle Ricerche, Napoli, Italy

Carmine Marco Morgillo, Aldo Galeone and Bruno Catalanotti
Dipartimento di Farmacia, Universita` degli Studi di Napoli Federico II, Napoli, Italy

Antonella Federico, Pierlorenzo Pallante and Alfredo Fusco
Istituto di Endocrinologia ed Oncologia Sperimentale Consiglio Nazionale delle Ricerche, Napoli, Italy

F. Javier Luque
Departament de Fisicoquímica and Institut de Biomedicina (IBUB), Facultat de Farmàcia, Universitat de Barcelona, Santa Coloma de Gramenet, Spain

Manuel Correia
Department of Physics and Nanotechnology, Aalborg University, Aalborg, Denmark

Viruthachalam Thiagarajan
BioPhotonics Group, Department of Nanomedicine, International Iberian
Nanotechnology Laboratory (INL), Braga, Portugal
School of Chemistry, Bharathidasan University, Tiruchirappalli, India

Isabel Coutinho, Gnana Prakash Gajula and Maria Teresa Neves-Petersen
BioPhotonics Group, Department of Nanomedicine, International Iberian
Nanotechnology Laboratory (INL), Braga, Portugal

Steffen B. Petersen
Department of Health Science and Technology, Aalborg University, Aalborg, Denmark
The Institute for Lasers, Photonics and Biophotonics, University at Buffalo, The State University of New York, New York, United States of America

Marharyta Petukh, Li Wang, Emil Alexov, Shannon Stefl and Nick Smith
Computational Biophysics and Bioinformatics, Physics Department, Clemson University, Clemson, South Carolina, United States of America

Bohua Wu
School of Nursing, Clemson University, Clemson, South Carolina, United States of America

David Hyde-Volpe
Department of Chemistry, Clemson University, Clemson, South Carolina, United States of America

Chi-Wen Lee and Jenn-Kang Hwang
Institute of Bioinformatics and Systems Biology, College of Biological Science and Technology, National Chiao Tung University, Hsinchu, Taiwan, Republic of China

Hsiu-Jung Wang and Ching-Ping Tseng
Department of Biological Science and Technology, College of Biological Science and Technology, National Chiao Tung University, Hsinchu, Taiwan, Republic of China

Amalia Muńoz-Gómez
Grupo Interdisciplinario de Estudios Moleculares (GIEM), Instituto de Química, Universidad de Antioquia, Medellín, Antioquia, Colombia
Genetic and Biochemistry of Microorganisms group (GEBIOMIC), Instituto de Biología, Universidad de Antioquia, Medellín, Antioquia, Colombia
Bioinformatic Analysis Group (GABi), Centro de Investigación y Desarrollo en Biotecnología, CIDBIO, Bogotá, Distrito Capital, Colombia

Mauricio Corredor
Genetic and Biochemistry of Microorganisms group (GEBIOMIC), Instituto de Biología, Universidad de Antioquia, Medellín, Antioquia, Colombia

Alfonso Benítez-Páez
Bioinformatic Analysis Group (GABi), Centro de Investigación y Desarrollo en Biotecnología, CIDBIO, Bogot, Distrito Capital, Colombia

Carlos Peláez
Grupo Interdisciplinario de Estudios Moleculares (GIEM), Instituto de Química, Universidad de Antioquia, Medellín, Antioquia, Colombia

Michał A. Surma, Andrzej Szczepaniak and Jarosław Króliczewski
Faculty of Biotechnology, University of Wroclaw, Wroclaw, Poland

Craig R. Miller
Department of Biological Sciences, University of Idaho, Moscow, Idaho
Department of Mathematics, University of Idaho, Moscow, Idaho

Institute for Bioinformatics and Evolutionary Studies, University of Idaho, Moscow, Idaho

Kuo Hao Lee
Institute for Bioinformatics and Evolutionary Studies, University of Idaho, Moscow, Idaho, Department of Biochemistry and Molecular Biophysics, Kansas State University, Manhattan, Kansas

Holly A. Wichman
Department of Biological Sciences, University of Idaho, Moscow, Idaho
Institute for Bioinformatics and Evolutionary Studies, University of Idaho, Moscow, Idaho

F. Marty Ytreberg
Institute for Bioinformatics and Evolutionary Studies, University of Idaho, Moscow, Idaho
Department of Physics, University of Idaho, Moscow, Idaho

Philip Lössl, Knut Kölbel, Dirk Tänzler, Andrea Sinz and Christian H. Ihling
Department of Pharmaceutical Chemistry and Bioanalytics, Institute of Pharmacy, Martin Luther University Halle-Wittenberg, Halle (Saale), Germany

David Nannemann and Jens Meiler
Department of Chemistry and Center for Structural Biology, Vanderbilt University, Nashville, TN, United States of America

Manuel V. Keller and Frank Zaucke
Center for Biochemistry, Medical Faculty, University of Cologne, Cologne, Germany

Marian Schneider
Research Group Artificial Binding Proteins, Institute of Biochemistry and Biotechnology, Martin Luther University Halle-Wittenberg, Halle (Saale), Germany

Chao He and Damien M. O'Halloran
Department of Biological Sciences, The George Washington University, Washington, D.C., United States of America
Institute for Neuroscience, The George Washington University, Washington, D.C., United States of America

Vendula Tvrdonova
Department of Cellular and Molecular Neuroendocrinology, Institute of Physiology Academy of Sciences of the Czech Republic, Prague, Czech Republic
Department of Physiology of Animals, Faculty of Science, Charles University, Prague, Czech Republic

Milos B. Rokic
Department of Cellular and Molecular Neuroendocrinology, Institute of Physiology Academy of Sciences of the Czech Republic, Prague, Czech Republic
Section on Cellular Signaling, Program in Developmental Neuroscience, National Institute of Child Health and Human Development, National Institutes of Health, Bethesda, Maryland, United States of America

Stanko S Stojilkovic
Section on Cellular Signaling, Program in Developmental Neuroscience, National Institute of Child Health and Human Development, National Institutes of Health, Bethesda, Maryland, United States of America

Hana Zemkova
Department of Cellular and Molecular Neuroendocrinology, Institute of Physiology Academy of Sciences of the Czech Republic, Prague, Czech Republic

Surinder M. Singh, Swati Bandi and Dinen D. Shah
Department of Pharmaceutical Sciences, Skaggs School of Pharmacy and Pharmaceutical Sciences, University of Colorado Anschutz Medical Campus, Aurora, Colorado, United States of America

Geoffrey Armstrong
Department of Chemistry and Biochemistry, University of Colorado Boulder, Boulder, Colorado, United States of America

Krishna M. G. Mallela
Department of Pharmaceutical Sciences, Skaggs School of Pharmacy and Pharmaceutical Sciences, University of Colorado Anschutz Medical Campus, Aurora, Colorado,
United States of America

Program in Structural Biology and Biochemistry, University of Colorado Anschutz Medical Campus, Aurora, Colorado, United States of America

Yoshiyuki Henning, Christiane Vole, Sabine Begall and Hynek Burda
Department of General Zoology, Faculty of Biology, University of Duisburg-Essen, Essen, Germany

Martin Bens, Arne Sahm and Karol Szafranski
Genome Analysis, Leibniz Institute for Age Research - Fritz Lipmann Institute, Jena, Germany

Martina Broecker-Preuss
Department of Endocrinology and Metabolism and Division of Laboratory Research, University Hospital, University of Duisburg-Essen, Essen, Germany

Philip Dammann
Department of General Zoology, Faculty of Biology, University of Duisburg-Essen, Essen, Germany
Central Animal Laboratory, University Hospital, University of Duisburg-Essen, Essen, Germany

Jose Sergio Hleap
Department of Biochemistry and Molecular Biology, Dalhouise University, Halifax, Nova Scotia, Canada

Christian Blouin
Department of Biochemistry and Molecular Biology, Dalhouise University, Halifax, Nova Scotia, Canada
Department of Computer Science, Dalhouise University, Halifax, Nova Scotia, Canada

Jens Dernedde, Werner Reutter and Rudolf Tauber
Institut für Laboratoriumsmedizin, Klinische Chemie und Pathobiochemie, Charité -Universitätsmedizin Berlin, Berlin, Germany

Christoph Weise
Institut für Chemie und Biochemie, Freie Universität Berlin, Berlin, Germany

Eva-Christina Müller
Max Delbrück Center for Molecular Medicine, Berlin-Buch, Germany

Akira Hagiwara and Mamoru Suzuki
Department of Otolaryngology, Tokyo Medical University, Shinjuku-ku, Tokyo, Japan

Sebastian Bachmann
Institut für Vegetative Anatomie, Charité - Universitätsmedizin Berlin, Berlin, Germany

Hans Scherer
Klinik für Oto-Rhino-Laryngologie, Charité-Universitätsmedizin Berlin, Berlin, Germany

Tarun Kumar, Gurumayum Suraj Sharma and Laishram Rajendrakumar Singh
Dr. B. R. Ambedkar Center for Biomedical Research, University of Delhi, Delhi, India

Index

www.ingramcontent.com/pod-product-compliance
Lightning Source LLC
Chambersburg PA
CBHW080408190526
45161CB00003B/168